Smart Stimuli-Responsive Polymers, Films, and Gels

Smart Stimuli-Responsive Polymers, Films, and Gels

Edited by

Liang Hu
Yongfeng Gao
Michael J. Serpe

WILEY-VCH

Editors

Prof. Liang Hu
Soochow University
RAD-X
199 Ren'ai Road
215123 Suzhou
China

Dr. Yongfeng Gao
University of Alberta
Department of Chemistry
11227 Saskatchewan Drive
T6G 2G2 NK
Canada

Prof. Michael J. Serpe
University of Alberta
Department of Chemistry
11227 Saskatchewan Drive
T6G 2G2 NK
Canada

Cover Image: Courtesy of Liang Hu

Library of Congress Card No.: applied for

British Library Cataloguing-in-Publication Data
A catalogue record for this book is available from the British Library.

Bibliographic information published by the Deutsche Nationalbibliothek
The Deutsche Nationalbibliothek lists this publication in the Deutsche Nationalbibliografie; detailed bibliographic data are available on the Internet at <http://dnb.d-nb.de>.

Print ISBN: 978-3-527-34901-2
ePDF ISBN: 978-3-527-83237-8
ePub ISBN: 978-3-527-83239-2
oBook ISBN: 978-3-527-83238-5

Cover Design: ADAM DESIGN, Weinheim, Germany
Typesetting Straive, Chennai, India
Printing and Binding CPI Group (UK) Ltd, Croydon, CR0 4YY

C113101_250722

Contents

Preface *xi*

1 From Mechanochemistry to Mechanoresponsive Materials *1*
Lasith S. Kariyawasam, Connor Filbin, Cameron Locke, and Ying Yang
1.1 Introduction *1*
1.2 Mechanochemistry in Biological Systems *2*
1.2.1 A Stressful Environment During Heart Development *3*
1.2.2 Protein Unfolding by Force *4*
1.2.3 Stress Mitigation by Tissue *6*
1.2.4 Sensing by Ion Channel Opening *6*
1.3 Mechanistic View of Mechanochemistry *7*
1.4 Polymer Covalent Mechanochemistry *15*
1.4.1 Pyran-Based Mechanochromophores *16*
1.4.2 Retro-Cycloadditions *19*
1.4.3 Ladderenes *23*
1.4.4 Stable Radical Systems *24*
1.4.5 Other Types of Mechanophores *27*
1.5 Polymer Noncovalent Mechanochemistry *29*
1.5.1 Mechanoresponses of Metal–Ligand Bonds *30*
1.5.2 Mechanochemistry of Other Noncovalent Interactions and Their Applications in Functional Polymers *37*
1.6 Conclusions *41*
References *42*

2 Photoresponsive Polymers *53*
Hossein Roghani-Mamaqani and Zeinab Tajmoradi
2.1 Introduction *53*
2.2 Photoresponsive Polymers *55*
2.2.1 Photoluminescent Polymers *55*
2.2.1.1 Fluorescent Polymers *56*
2.2.1.2 Phosphorescent Polymers *58*
2.2.2 Photochromic Polymers *62*
2.2.3 Photocleavable Polymers *71*

2.2.4 Photodimerizable Polymers *74*
2.2.5 Photoadaptable Polymers *76*
2.3 Applications of Photoresponsive Polymers *83*
2.3.1 Smart Polymeric Inks *83*
2.3.2 Polymer Sensors *89*
2.3.3 Photolithography *93*
2.3.4 Surface Active Agents *95*
2.3.5 Photorheological Polymers *98*
2.3.6 Self-Healing Polymers *103*
2.3.7 Shape-Changing Polymers *106*
2.3.8 Photoconductive Polymers *109*
2.3.9 Drug Delivery *111*
2.3.10 Membranes, Films, and Textiles *112*
2.4 Summary and the Future *115*
2.4.1 Water Contact Angle Variation *116*
2.4.2 Viscosity Variation *116*
2.4.3 Color Change and Emission *117*
2.4.4 Sol–Gel Transition *117*
Abbreviations *118*
References *119*

3 Polymer Systems for Ionizing Radiation Dosimetry and Radiotherapy *135*
Li Jiang, Chengfang Zhang, Rensheng Wang, and Liang Hu
3.1 Introduction *135*
3.2 Interaction of Radiation with Matter *136*
3.2.1 α-Particles *136*
3.2.2 Electrons *137*
3.2.3 Photons *137*
3.3 Polymer Systems for Ionizing Radiation Dosimetry *139*
3.3.1 Polymer-Based Dosimeters *139*
3.3.2 Polymer/Dye Dosimeters *139*
3.3.3 Fluorescent Polymer Dosimeters *141*
3.3.4 Polymer/Metal Nanomaterials Dosimeters *143*
3.4 Ionizing Radiation-Responsive Polymer Systems for Therapy *146*
3.5 Conclusion *149*
Acknowledgments *151*
References *151*

4 Shrink and Wrinkle – Thermally Responsive Substrates for Thin-Film Structuring *157*
Eduardo González-Martínez and Jose Moran-Mirabal
4.1 Structured Thin Films *157*
4.2 Measuring the Mechanical Properties of Thin Films Using Thermal Wrinkling *159*
4.2.1 Thermally Structured Thin Films for Cell Culture *162*

4.2.2 Wrinkled Conductive Thin Films for Wearable Electronics *167*
4.2.3 Wrinkled Electrochemical Sensors *173*
4.2.4 Current Challenges and Future Perspectives for the Use of Wrinkled Thin Films *175*
References *176*

5 Design of Nanocomposite Microgels Prepared by Seeded Emulsion Polymerization in the Presence of Microgels *181*
Takumi Watanabe and Daisuke Suzuki
5.1 Background on Composite Hydrogels *181*
5.2 Background on Composite Microgels *182*
5.3 Conventional Emulsion Polymerization and SEP *184*
5.4 Nanocomposite Microgels Prepared by SEP in the Presence of Microgels *186*
5.5 Design of the Internal Structure of the Nanocomposite Microgels *188*
5.6 Synthesis of Multi-layered Nanocomposite Microgels *189*
5.7 Characterization of Nanocomposite Microgels *190*
5.8 Applications of Nanocomposite Microgels *193*
5.9 Summary and Perspective *195*
References *196*

6 Compressible Microgels in Concentrated Suspensions: Phase Behavior, Flow Properties, and Scattering Techniques to Probe Their Structure and Dynamics *203*
A. Scotti, U. Gasser, B. Zhou, A. Arenas-Gullo, A. de la Cotte, J. Rojo González, and A. Fernandez-Nieves
6.1 Introduction *203*
6.2 Swelling Thermodynamics *207*
6.2.1 Polymer/Solvent Mixing *208*
6.2.2 Elasticity *208*
6.2.3 Ionic Effects *209*
6.2.4 Equation of State *210*
6.3 Experimental Techniques *210*
6.3.1 Dynamic Light Scattering *211*
6.3.1.1 Auto-correlation Experiments *213*
6.3.1.2 Cross-correlation and 3D-DLS Experiments *214*
6.3.2 Small-angle Neutron-scattering *216*
6.3.2.1 SANS Setup *218*
6.3.2.2 Scattering Theory *218*
6.3.2.3 Form Factor and Structure Factor *220*
6.3.2.4 Contrast Variation *222*
6.4 Suspension Phase Behavior *228*
6.5 Flow Properties *231*
6.6 Final Remarks *235*
References *236*

7 Structure and Properties of Smart Micro- and Nanogels Determined by (Neutron) Scattering Methods *241*
Julian Oberdisse and Thomas Hellweg
7.1 Introduction *241*
7.2 Scattering Techniques Applied to Microgels *242*
7.2.1 Static and Dynamic Light Scattering Applied to Microgels *242*
7.2.1.1 Static Light Scattering (SLS) *243*
7.2.2 Dynamic Light Scattering (DLS/PCS) *245*
7.2.3 Small-Angle Neutron and X-Ray Scattering Applied to Microgels *247*
7.3 Multicompartment and Multi-Stimuli-Responsive Microgels *254*
7.4 Time-Resolved Small-Angle Scattering *263*
7.5 Crowded Microgel Systems *266*
7.6 Conclusion and Outlook *270*
Appendix: Absolute Intensity for Fuzzy Sphere Form Factors *270*
References *271*

8 Stimuli-Responsive Fluorescent Polymeric Hydrogels *281*
Wei Lu, Shuxin Wei, and Tao Chen
8.1 Introduction *281*
8.2 Strategies for Preparing Fluorescent Polymeric Hydrogels (FPHs) *282*
8.2.1 Physically Incorporating Fluorogens into Polymeric Hydrogels *282*
8.2.2 Covalently Bonding Fluorogens into Polymeric Hydrogels *284*
8.2.3 Supramolecular Polymerizing/Crosslinking Monomeric Fluorogens *286*
8.2.4 Comparison of Different Synthetic Strategies *290*
8.3 Promising Applications *290*
8.3.1 Optical Sensing and Bio-imaging *290*
8.3.2 Information Encoding and Encryption *293*
8.3.3 Bioinspired Mechanosensing Systems and Soft Actuators/Robotics *294*
8.4 Conclusions *297*
References *298*

9 The Fabrication and Applications of Bioinspired Hydrogel Actuators *301*
Baoyi Wu, Jiawei Zhang, and Tao Chen
9.1 Introduction *301*
9.2 The Classification of Hydrogel Actuators *302*
9.2.1 Addition of Active Ingredient *302*
9.2.2 Pneumatic/Hydraulic Actuators *305*
9.2.3 Stimuli-Responsive Hydrogel Actuator Derived from Asymmetric Swelling *306*
9.2.3.1 Single-Stimulus-Responsive Hydrogel Actuators *307*
9.2.3.2 Multi-stimuli-Responsive Hydrogel Actuators *308*
9.3 Anisotropic Structures *310*
9.3.1 1D/2D Anisotropic Structures *310*

9.3.1.1 Bilayer *310*
9.3.1.2 Oriented *311*
9.3.1.3 Gradient *312*
9.3.1.4 Patterned *315*
9.3.2 3D Anisotropic Structures *315*
9.4 Methods to Fabricate Anisotropic Structures *318*
9.4.1 Traditional Technology *318*
9.4.1.1 Stepwise Polymerization *318*
9.4.1.2 3D Printing *320*
9.4.1.3 Macromolecular Assembly *322*
9.4.2 Innovative Technology *322*
9.5 Applications *325*
9.5.1 Soft Robots *325*
9.5.2 Artificial Muscles *327*
9.5.3 Biomimetic Devices *329*
9.5.4 Information Storage Materials *329*
9.6 Conclusion *332*
Conflict of Interest *333*
Acknowledgments *333*
References *333*

10 Hydrogels-Based Electronic Devices for Biosensing Applications *339*
Quanduo Liang, Yuyuan Lu, and Qiang Zhang
10.1 Introduction *339*
10.2 Flexible Hydrogel-Based Sensors *342*
10.2.1 Principles of Conductive Hydrogel Sensors *343*
10.2.2 Improved Mechanical Properties of Hydrogel-Based Sensors *347*
10.2.3 Prolonged Longevity of Hydrogel Sensors *351*
10.2.4 Expanded Usage Scenarios of Hydrogel-Based Sensors *353*
10.2.5 Multifunctionalization and Expanding Application of Hydrogel Sensor *354*
10.3 Tissue–Machine Interfaces *356*
10.3.1 Design and Mechanism of the Neural Interfaces *356*
10.3.2 Multifunctional Applications of Biointerfaces *360*
10.4 The Prospects of Hydrogel Bioelectronic Devices *363*
Acknowledgments *364*
References *364*

Index *375*

Preface

In nature, many living systems can react to changes in their external environment. Inspired by this natural responsivity, scientists have endeavored to understand stimuli-responsive polymers, which are able to change their solubility, volume, and/or conformation in response to external stimuli. Over the past few decades, interest in these "smart" responsive polymer-based systems has been increasing, and hence a book on this topic is warranted, to capture recent trends in this burgeoning area.

This book was assembled primarily with the needs of both junior and senior chemists and materials scientists in mind. We first wish to acknowledge the contribution of all authors from all over the world. Without all your support, this book would not have been published. This book is divided into 10 chapters. Chapters 1–3 discuss the mechanoresponsive, photoresponsive, and ionizing-radiation-responsive polymers. Chapter 4 highlights thermal responsive film. Chapters 5–10 discuss the stimuli-responsive gels from fundamental synthesis strategy to scattering techniques characterization and applications.

This book was published by Wiley, whose support and help we gratefully acknowledge, especially the warm welcome and continued aid given to us by our editors, Katherine Wong, Anne Brennführer, and Elke Maase.

14 April 2022

Liang Hu
Soochow University, China

Yongfeng Gao, Michael J. Serpe
University of Alberta, Canada

1

From Mechanochemistry to Mechanoresponsive Materials

Lasith S. Kariyawasam, Connor Filbin, Cameron Locke, and Ying Yang

University of Nevada, Department of Chemistry, Virginia Street, Reno, NV 89557, USA

1.1 Introduction

Our skin can sense the touch by a series of mechanotransduction mechanisms. Kneading bread dough uncoils gluten proteins, creating an elastic macromolecular network that gives the dough toughness. Stretching or scratching a piece of plastic is likely to break covalent bonds. These processes involve reactions that are activated by mechanical energy, which are prevalent in our daily lives. However, they are less commonly discussed compared to thermochemical, photochemical, or electrochemical reactions. Mechanoactivated reactions have been reported dating back to 315 BCE. These early accounts describe grinding native cinnabar in a copper mortar with a copper pestle in the presence of vinegar to yield the reduction product, mercury. However, it was not until the nineteenth century that systematic studies were conducted [1]. In 1860s, Carey Lea showed that grinding mercury and silver halides in a pestle and mortar at room temperature favors decomposition, whereas heating only leads to melting or sublimation without any decomposition [2]. This discovery provided clear evidence that mechanochemical reactions are distinctively different from thermal ones. Mechanochemistry was, therefore, classified as the fourth type of chemical reaction by Ostwald in 1919 [3].

The first widely accepted definition for mechanochemistry was formulated by Heinicke in 1984 [1], that mechanochemistry is a branch of chemistry, which is concerned with chemical and physicochemical transformations of substances in all states of aggregation produced by the effect of mechanical energy. IUPAC defines it as the chemical reaction that is induced by the direct absorption of mechanical energy [4]. In fact, the definition is still under extensive debate. Molecular motors, which convert chemical energy into mechanical work, certainly do not fit into these definitions. However, the motions generated by molecular motors can apply force to the surrounding molecules to induce a cascade of subsequent reactions. Such topic is of great interest to chemists and engineers working in the field of mechanochemistry. The lack of unification and slow progress since the establishment of mechanochemistry in the nineteenth century reflects the complexity and

Smart Stimuli-Responsive Polymers, Films, and Gels, First Edition.
Edited by Liang Hu, Yongfeng Gao, and Michael J. Serpe.

the lack of understanding of the scope and mechanisms for such reactions. Recently, mechanochemical research has intensified through the use of new tools developed for mechanistic studies and the opportunities to create functional materials [5–12]. As mechanochemical reactions occur for both small and macromolecules, ongoing research can be clarified in four sub-areas: developing novel and scalable mechanochemical synthesis to make useful chemicals via environmental-friendly solvent-free processes; understanding biomechanochemistry, such as motor proteins, mechanosensing, and mechanotransduction mechanisms; creating mechanoresponsive polymeric materials for which mechanical forces become constructive for technological advances; and investigating the molecular mechanisms through simulations and single-molecular force experiments.

In this chapter, we will focus on the fundamental aspects of polymer mechanochemistry that are the key elements in designing mechanoresponsive materials. We will start with a brief introduction of the role of mechanochemistry in biological systems as a springboard for inspiration. The mechanistic aspects of mechanochemistry in general terms, from small molecules to polymer mechanochemistry, will then be discussed to show the unique bond-activation mechanisms. The force-responsive molecules, named mechanophores, can depend on the cleavage of either covalent or noncovalent bonds. The activation energy, dynamics, and reversibility can be tuned via various structural properties. Therefore, the chemistry of these two classes of mechanophores will be covered in detail. There are different mechanical sources for generating mechanical energy, such as shearing, stretching, grinding, ball milling, and sonication. These methods differ in the direction of force, frequency, and heat formation, leading to different effects on molecular distortion and kinetics. However, this chapter will focus on the chemistry of mechanoresponsive bonds within polymer materials regardless of the type of applied force.

1.2 Mechanochemistry in Biological Systems

A powerful source of inspiration for improving the design of polymer materials is Nature as mechanochemical systems are ubiquitous in organisms. A wealth of knowledge can be gained because biological systems have evolved elegant mechanoresponsive arrangements that are critical for supporting and maintaining life. They often involve complicated processes via coherently organized biopolymer networks. A cell, for example, is constantly under mechanical stress, including tension, osmosis, compression, and shear forces. Upon mechanical deformation, feedback from proteins in the cell cytoskeleton activates a variety of mechanosensors that work in unison to create a response in the cell nucleus via multiple mechanotransduction events [13]. As shown in Figure 1.1, the mechanism begins by transducing force through the cell membrane to microfilaments and microtubules of the cytoskeleton in the cytoplasm. Subsequently, cytoskeletal changes directly affect nucleoskeletal proteins called lamina. This has an explicit effect on the spatial arrangement of lamin-bound intermediate filaments as well as

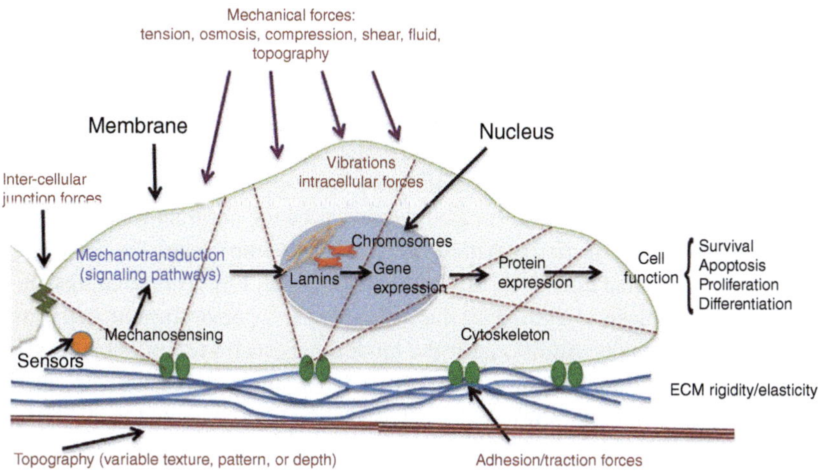

Figure 1.1 Intra- and extracellular forces stimulate a cell in an interconnected system of reactions causing complete change of structure and resulting cell function. Source: Adapted with permission from Tsimbouri [13]. Copyright 2015 MDPI.

chromosomes, as they are anchored to nuclear lamina [14]. Shifts in chromosome packing affect gene expression, which allows key biological functions in response to force, such as survival, motility, reproduction, and differentiation. These events in cells play important roles in maintaining homeostasis and preventing disease in the body. Although understanding of many of these biological pathways is still limited, we will discuss a few chemistries known to be involved in these processes as inspirations for material design.

1.2.1 A Stressful Environment During Heart Development

In a vertebrate embryo, the heart first starts as a tube composed of primarily early cardiomyocytes, the cardiac muscle cells that drive the heart contraction. It quickly differentiates into different parts and morphologies. As the embryo grows, the contractile capacity of the cardiomyocytes increases to provide greater driving forces to pump more blood. Meanwhile, the extracellular matrix (ECM) surrounding the cardiomyocytes must increase in its stiffness parallelly to keep a proper tissue mechanical integrity with the increasing contractile stress. Cells in connective tissue, called fibroblasts, secrete collagen and other matrix proteins to maintain the structural framework. Therefore, during the development of the heart, a balance between cardiac fibroblast and cardiomyocyte cell populations must be established to maintain muscle contraction along with a significant collagenous matrix. There is mounting evidence suggesting that mechanical stress itself plays important role in directing tissue growth with mechanochemical feedback loops for gene and protein expression [15]. In comparison, the brain tissue in a low-stress environment does not show the same development in stiffness, although a recent study found strong mechanical interactions of the synapses [16] which may be a critical mechanism in brain functions, indicating the broad involvements of mechanochemistry and mechanotransduction in numerous bioprocesses.

During embryonic heart development, many mechanosensitive pathways have been linked to the proper functions of the myocyte and fibroblast cells. Majkut et al. surveyed literature evidence and proposed the network model for understanding how contraction against tissue stiffness affords a functional equilibrium between the cell types [15]. On one hand, cardiomyocytes produce contractile stress that promotes the expression of matrix structural proteins by fibroblasts. On the other hand, as contraction must effectively strain the heart tissue, it is postulated that the proliferation of fibroblasts is limited by the stiffness of their environment and thus collagenous matrix density. Additionally, the model suggests that stabilizing matrix collagen and degradation of motor proteins under strained conditions are also important in regulating tissue stiffness. The stabilization may be related to inhibited protease binding to collagen fibers or kinase binding to myosin minifilaments when they are under tension, thus preventing their dissociation and digestion. A model of dynamic cell–matrix interaction is also extended to nuclear mechanics because during development there are variations in lamin levels that appear to correlate with ECM mechanics [15]. As previously discussed, mechanical signals from the extracellular environment can be physically transmitted by the contractile cytoskeleton to the nucleus by connections through the nuclear membrane to the nuclear lamina. Lamina can interact with chromatin and various proteins that regulate transcription. Therefore, lamin expression is also regulated and affects tissue maturation.

1.2.2 Protein Unfolding by Force

Proteins are folded into various three-dimensional structures to perform biological functions. Many of them are highly resistant to unfolding under mechanical stress to avoid denaturation. However, it has been observed that certain proteins containing hidden binding sites (cryptic sites) that rely on force for a conformational change via unfolding to enable protein activity and initiation of signaling pathways [17–19]. One of the most extensively studied mechanically responsive proteins is the von Willebrand factor (vWF), which plays a crucial role in blood clotting. When vascular systems are ruptured, a decrease in pressure triggers hydrodynamic forces from changing velocity in the blood. These hydrodynamic forces induce a shear force on vWF in the blood, causing a conformation change. At a critical shear stress of approximately 50 dyn/cm^2, the globular protein is significantly elongated [20–22]. Figure 1.2 shows the domain structures and the force-induced conformational change. This leads to exposure and activation of the A1 domain to bind to platelets and A3 domain to attach to the collagen protein on the surface of vascular damage to initiate repair of the vessel injury [23]. The protein unfolding occurs at the A2 domain and is suggested to involve mechanoactivated cleavage of disulfide bonds in combination with hydrophobic interactions. The vWF contains an eight-membered ring linked by a disulfide bond from vicinal cysteines (Cys^{1669}–Cys^{1670}), which is the lowest energy barrier to initiate force-induced unfolding of the A2 domain

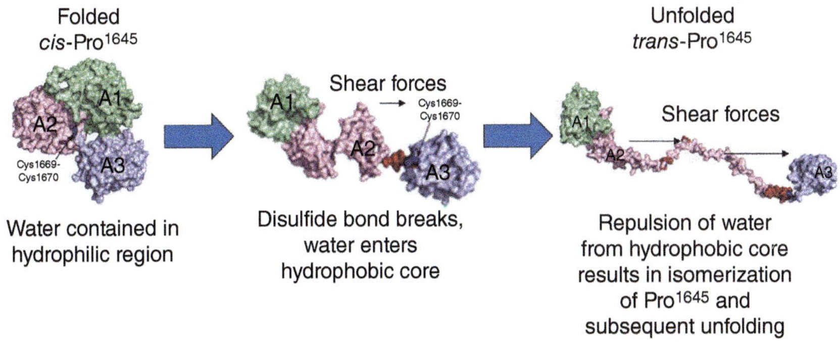

Figure 1.2 Model of vWF protein mechanochemical unfolding from shear forces. When vWF protein is in the unfolded conformation the A1 domain binds platelets (green), A2 domain elongates (pink), and A3 domain binds collagen (blue). Source: Adapted with permission from Crawleyet al. [21]. Copyright 2011 American Society of Hematology.

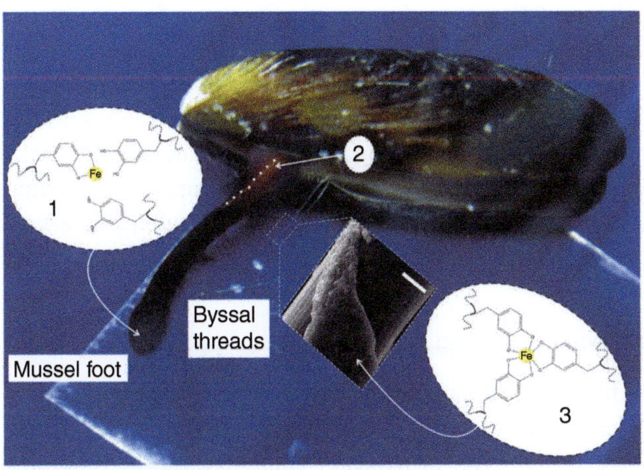

Figure 1.3 Schematic of marine mussel cuticle demonstrating high density of dopa-Fe^{3+} crosslinks in granule compared to matrix. This results in the formation of microcracks following increased strain, preventing complete material fracture. Source: Adapted with permission from Holten-Andersen et al. [24].

(Figure 1.3). When applied force is transduced through the protein, the disulfide bond is mechanically cleaved by shear forces. It is speculated that after cleavage, buried water molecules are admitted to the hydrophobic core of the vWF protein commencing unfolding [25]. Additionally, the unfolded state of the protein is stabilized by a force-induced isomerization of a proline residue (Pro^{1645}). The peptide *cis*-Pro^{1645} in the A2 domain is stabilized in a cis conformation by hydrogen bonding with residue Arg^{1618} when the vWF protein is in the folded conformation. After initial unfolding, the stabilization from hydrogen bonding is lost and the proline residue undergoes isomerization to trans. Conversion to *trans*-Pro^{1645} notably delays A2 domain refolding and allows unfolded vWF to continue the process of blood vessel coagulation and repair [23]. The force-induced unfolding of vWF shows an organized process involving mechanoscissions of covalent and noncovalent bonds, hydrophobic interactions, and subsequent isomerization.

1.2.3 Stress Mitigation by Tissue

Biological tissue has shown to undergo stress mitigation to avoid complete material failure through the use of noncovalent interactions. One specific example is the metal–ligand interactions in the outer coating of byssal threads in marine mussels [24, 26]. These byssal threads are secreted as bundles of 50–100 individual threads and are covered in a proteinaceous outer coating, called the cuticle, which is five times stronger than the internal threads themselves. Mussels use byssal threads to attach to rocks; the cuticle's toughness prevents the byssal threads from disconnecting following applied force. The primary cuticle protein is a modified tyrosine, which includes the molecule 3,4-dihydroxyphenylalanine (dopa). Included among the proteins are naturally existing metal ions, such as iron and calcium. Up to three dopa ligands can crosslink to the iron(III) ions (Fe^{3+}), resulting in an increased hardness (Figure 1.3). Granules are areas of high-concentration dopa-Fe^{3+} crosslinking within the cuticle. Lighter dopa-Fe crosslinking occurs in the matrix surrounding the granules. Following increased strain (>30%), the granules deform slightly, whereas the surrounding matrix crosslinks dissociate, forming microcracks. These microcracks allow the cuticle to absorb mechanical force without complete fracture. Following removal of strain, the granules instantly regain their original shape, while the microcrack crosslinks slowly self-heal over time.

1.2.4 Sensing by Ion Channel Opening

Mechanotransduction is important for signaling in sensory systems. Hair cells in the ear called stereocilia are mechanosensory biological structures that convert mechanical energy from sound waves into electrochemical signals that can be processed by the nervous system as hearing [19]. Stereocilia are linked together at their tips into bundles by proteins called tip links. These tip links are anchored into the cell membrane connecting an ion channel of one stereocilia to the tip of another [27]. Vibrational perturbations from sound deflect stereocilia tips cause opening and closing of ion channels (Figure 1.4). Opening and closing of the channels result in a change of ion flux leading to an electric potential that can stimulate the auditory nerve [28].

Another fascinating example of mechanotransduction in human senses is found in the mechanism of touch. Touch is essential for life as tactile input guides behavior. Discriminative touch allows uniquely human activities where finesse is needed. Delicate touch is allowed by the many sensory receptors imbued in the skin. A diversity of somatosensory neurons in the skin allows for a variety of sensations by taking advantage of varying mechanical thresholds in different proteins. For example, light touch is mediated by Aβ afferents that have a low mechanical threshold compared to the high mechanical threshold of nociceptors that mediate painful touch [29]. Moreover, an assortment of mechanoreceptors under the skin detects stretches,

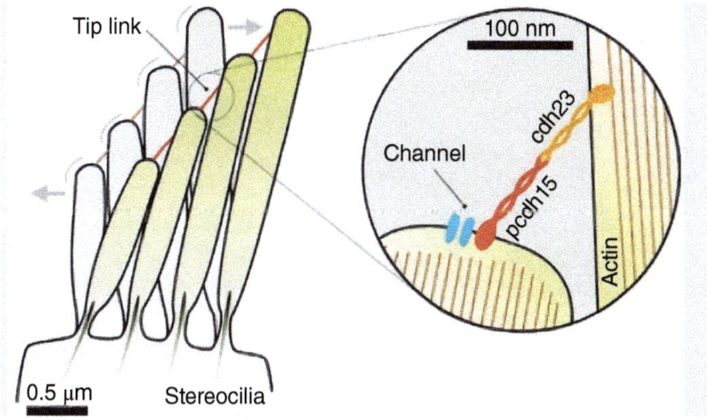

Figure 1.4 Illustration showing sensory cells called stereocilia located in the inner ear of mammals. The stereocilia deflect back and forth upon mechanical stimulation by sound waves. Source: Hoffmann et al. [27]/with permission of Royal Society of Chemistry.

vibrations, slip, and motion that enable us to determine an object's shape and texture. Little is known about the mechanisms for mechanotransduction in touch; however, the recent discovery of piezo proteins may help better understand how proteins are activated by membrane deformation. Piezo proteins are ion channels mechanically activated by touch, suction, and shear stress [30]. High concentrations of piezo proteins around receptors and neurons implicate their responsibility for encoding transduction channels in touch receptors.

As can be seen from these examples, mechanoresponsiveness in biological systems relies on mechanotransduction, which transforms mechanical energy into a biochemical signal that induces specific cellular responses, which frequently takes advantage of noncovalent interactions [19]. Also commonly found in living systems are kinetically trapped states that are structures locked far away from thermodynamic equilibrium in a high-energy conformation, such as folded proteins or prestressed cell membranes [31]. Organisms take advantage of these structures to enable sophisticated dynamic responses to force. Precisely mimicking biological structures and mechanisms is challenging and unnecessary because of the complexity. However, similar mechanoresponsive bonding domains utilizing the chemistry and structure characteristics learned from biomacromolecules can be employed, and living-like functionaries can be created through the design principles of synthetic materials.

1.3 Mechanistic View of Mechanochemistry

One of the most interesting aspects of mechanochemistry is that it yields products that are different from thermal- and photo-reaction pathways. This phenomenon has been observed in both inorganic and organic compounds. Here, we will use an extensively studied mechanochemical reaction, the pericyclic reaction, as an example

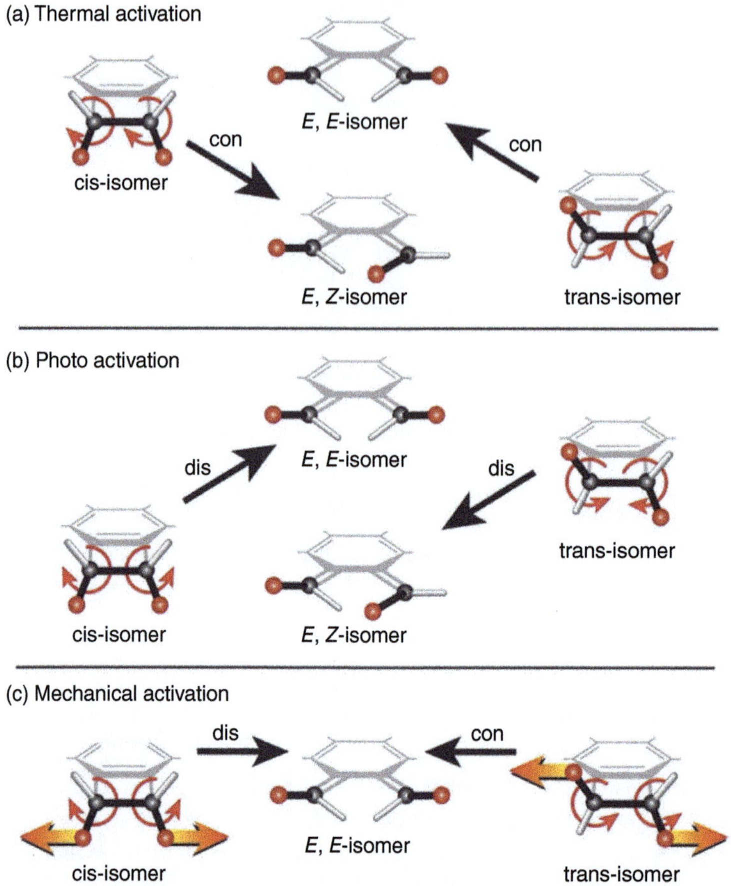

Figure 1.5 Ring opening of 1,2-disubstituted benzocyclobutenes (BCB) under different energy inputs. Source: Hickenboth et al. [32]/with permission of Springer Nature.

to discuss its mechanistic origin. Thermal activation of 1,2-disubstituted benzocyclobutene (BCB) induces conrotatory (con) ring openings, so that the cis and trans isomers give different ring-opening products (Figure 1.5a). When activated by light, disrotatory (dis) ring opening is favored instead of con for both cis and trans isomers (Figure 1.5b). This stereospecificity under thermal and photo-activations is described by the well-known Woodward–Hoffmann (WH) rules. However, when BCB is placed within long polymer chains, the mechanoproducts do not follow WH rules [32]. Mechanical forces induce a formal disrotatory ring opening in the cis isomer and a formal conrotatory ring opening in the trans isomer, yielding the same *E*,*E*-isomer (Figure 1.5c). The lack of selectivity was also observed for mechanoactivation of *gem*-dihalocyclopropanes [33–35]. If the force is viewed as pulling in opposite directions at the two molecular anchoring points, the reaction pathway favors bond breaking to increase the distance between the pulling points.

These stunning mechanochemical phenomena intrigued the study for the underlying mechanisms. Photoirradiation activates thermally forbidden pathways by promoting the reactants to electronically excited states. Force, however, does not directly alter the energy of the electrons. Thus, how does mechanochemical reaction overcome the WH rule? This is discussed in great detail in several studies via quantum chemical methods [36–39]. The results show that applied stresses do not alter the electronic structure. Instead, they lower the activation barrier for WH-forbidden pathways. Figure 1.6 shows the minimum energy pathways for mechanical ring opening of BCB under various forces from quantum chemical calculations [36]. For directional pulling of the cis isomer, beyond a critical force of 1.5 nN, the barrier for WH-forbidden disrotatory pathway drops below the conrotatory pathway, and disrotatory becomes the mechanically favored pathway. For the trans isomer, conrotatory is favored. Therefore, the two isomers yield the same ring-opening products as shown in Figure 1.5c. Since mechanical work alters the potential energy surface (PES), lowering the activation barrier without changing electronic structures, electronic considerations, such as WH rules, is not directly applicable.

The next question is how much force it takes to break a chemical bond. Mechanical strength of macroscopic materials can be characterized by a rupture force. This holds true for single chemical bonds only if they do not undergo any vibrations. The bonds will rupture when dissociation force exceeds the bond strength. However, bonds undergo thermal fluctuations. Combined with a pulling force, they led to a far more complex dependence on force conditions. Developed to describe the strength of cell adhesion, the well-known Bell's model provides a simple explanation for the effects of stress on bond rupture [40]. When a bond is pulled, the mechanical force deforms the PES, reducing the activation energy for bond rupture, which can be overcome by additional energy from thermal fluctuations. As a result, molecular bonds have no single rupture force, which is not so intuitive. Mathematically, Bell's model is described as

$$\tau = \tau_0 \exp[(E_0 - \gamma f)/kT]$$

where τ is the lifetime of a bond, which is the reciprocal of oscillation frequency, and τ_0 is the bond lifetime without load. The change of activation energy is expressed by $E_0 - \gamma f$, where E_0 is the bond energy at zero force, f is the applied force, and γ is a structural parameter that accounts for the distance to the transition state along the reaction coordinate. Thus, the activation barrier changes linearly with force, and the bond lifetime decreases exponentially with increasing force and temperature. Based on this equation, it was proposed that force impulse is the highest when $F = kT/\gamma$ at which force transduction is maximized [41]. A limitation in applying Bell's theory to computational calculations is the selection of the structural parameter γ. Since γ itself is force dependent, the model fails to account for the mechanically induced distortion of transition-state geometry, resulting in an overestimate of the bond rupture rate when there is considerable elongation of the scissile bond. Many extensions have been developed to improve Bell's theory. The extended models along with other quantum chemical treatments to understand the relations of geometries, energies, and transition states in mechanochemistry are summarized in a review article by Stauch and Dreuw [42].

Experimental studies evolved parallelly with the theoretical predictions to measure the rupture force of a single bond. In late 1990s, the use of atomic force microscopy (AFM) made it possible to probe the mechanical responses of covalent and noncovalent bonds in a single macromolecule [43–48]. Force responses are probed on a nanoscale with forces on nano Newton (nN) to sub-nN levels. In a force-probe AFM experiment, the molecule is anchored between a surface and an AFM tip. It is then stretched at a constant force rate until rupture [48]. A force–extension curve for a single chemical bond is thus generated, which reveals conformational changes, supramolecular rearrangements, along with bond rupture steps. Bond rupture of the BCB, which is shown in Figure 1.6, is measured using single-molecule force microscopy [49]. It provides the first experimental evidence

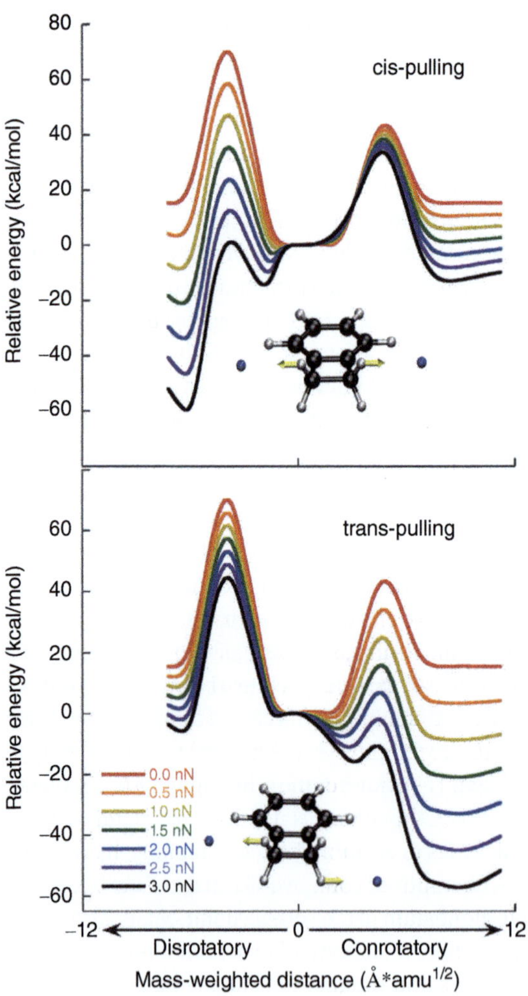

Figure 1.6 Minimum energy pathways for disrotatory and conrotatory ring opening of BCB under various force magnitudes calculated via steered molecular dynamics combined with *ab initio* steered molecular dynamics (AISMD). Source: Ong et al. [36]/with permission of American Chemical Society.

that the symmetry-forbidden conrotatory ring opening can be mechanically accelerated, requiring a 130 pN less force compared to disrotatory pathway at a ~0.1 second experimental timescale.

Force dependency of reaction rate can also be measured using single-molecule force spectroscopy. A constant force is applied, and the deformation is recorded as a function of time. This approach is called force-clamp AFM. The technique has been used to study the kinetics of thiol/disulfide exchange reactions within a protein [50]. The reaction is known to occur during protein unfolding in mechanically stressed proteins and is crucial in regulating protein function. The experiment utilized protein engineered with a precisely positioned disulfide bond. Disulfide reduction events can thus be correlated with a signature protein contour length, which can be identified in the extension experiment. As shown in Figure 1.7a, mechanical force first triggers the unfolding of the protein (unsequestered unfolding), exposing the disulfide bond. In the presence of a reducing agent, 1,4-DL-dithiothreitol (DTT), the disulfide reduction occurs. The rates of the biomolecular reduction were measured by fitting the extension-time curve signaturing the disulfide exchange under varying force magnitudes. The exchange rate showed an exponential increase with the applied force (Figure 1.7b). Fitting the experimental results using Bell-like model gives a γ value, the lengthening of the transition state, of 0.34 Å. An energy landscape is plotted showing an 8.2 kJ/mol reduction of activation energy when a 400 pN force is applied (Figure 1.7c). The reversibility of disulfide exchange was recorded by force-clamp AFM [51]. Mechanical force enabled the thermodynamically unfavorable S_N2 substitution of a disulfide with weak nucleophilic thiols. Upon removal of load, the bonding returned to the original disulfide of lower energy. Reversibility is critical in mechanical-responsive materials so that perturbations can be repeated in a highly dynamic manner.

Going back to Bell's model, as a bond undergoes constant thermal fluctuations, how fast the force is applied relative to the thermal fluctuation affects the breaking force. Therefore, bond-breaking force is dependent on the force-loading rate (df/dt). This dependency can be characterized into three regimes, the spontaneous, force-assisted, and activationless regimes, as shown in Figure 1.8 [52, 53]. When the loading rate is low relative to bond thermal dissociation rate, bonds break spontaneously due to thermal fluctuations. This is called a spontaneous regime. In this region, the breaking force is low and independent of the force rate. Bonds with rapid thermal oscillations and short bond lifetime, such as host–guest interactions, fall within this regime [54]. As force rate increases, it becomes sufficiently fast that the bond is stretched to some extent before breaking. This is called the force-assisted regime, and the breaking force increases rapidly with force rate. This has been observed for avidin–biotin complex [55], H-bonded carboxylic acid groups [56], and gold–octanedithiol–gold links [57]. In the activationless regime under high-loading rate, the energy barrier to breaking is reduced to zero as the bond is maximally stretched, and bond breaks solely by the applied force. In this region, the breaking force reaches the highest and becomes constant again. While single-molecule experiments allow us to validate the theoretical predictions, caution must be used when comparing the rupture force values, which are highly dependent on experimental

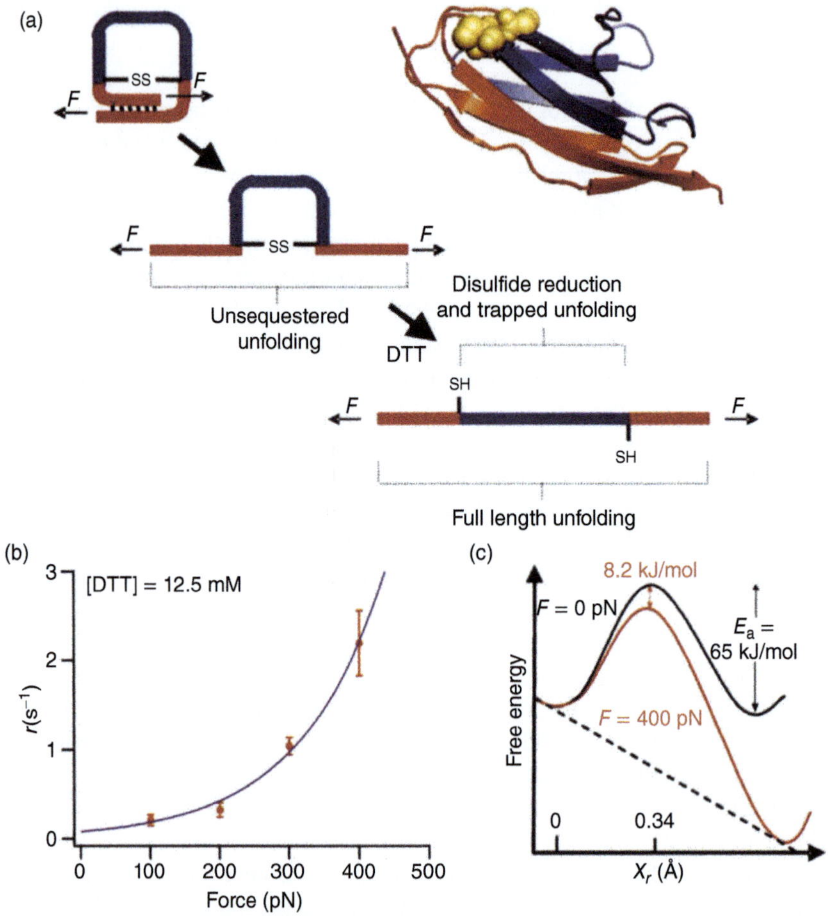

Figure 1.7 (a) Unfolding and disulfide (SS) reduction for an engineered protein with a precisely positioned disulfide bond. (b) Rate of reduction of the disulfide bond as a function of applied force measured by force-clamp AFM. Concentration of the reducing agent, DTT, is kept constant at 12.5 mM. (c) Calculated energy landscape with and without forces. A 400 pN force reduces the activation energy of disulfide reduction by 8.2 kJ/mol. Source: Wiita et al. [50]/with permission of National Academy of Sciences.

conditions. A comprehensive review by Ribas-Arino and Marx summarized the single-molecule spectroscopy and the theoretical treatments in great detail [58].

In the context of bulk polymers, sensitivity of the mechanoresponsive bonds becomes dependent on force transduction and heterogeneities of force distribution. For a single chain, the weaker bonds break first followed by the stronger bonds. The rupture forces for common covalent bonds are within a few nN (Figure 1.9) [59]. For noncovalent interactions, the rupture forces are in a much lower range, around 10–100 pN [58]. When mechanocleavable bonds are placed within a polymer strand, however, the force transduction can be far more complicated. An often-asked question is which bond is most likely to rupture within a polymeric material under mechanical deformation. This is determined by both the bond strength and the

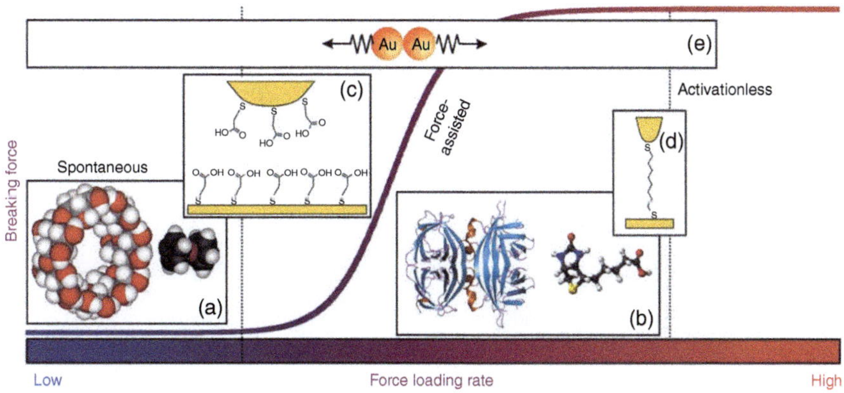

Figure 1.8 Dependence of bond-breaking force on force rate. (a) Host-guest complex representative of spontaneous breaking. (b) Bindings between biomolecules such as avidin and biotin representative of force-assisted breaking. (c) Hydrogen bonding between carboxylic acids falls within the transition regime from spontaneous to force-assisted breaking. (d) Gold-octanedithiol-gold interaction falls within the transition regime from force-assisted to activationless breaking. (e) breaking of the bond between two gold atoms at room temperature spans across the entire spectrum. Source: Pobelov et al. [52]/with permission of Springer Nature.

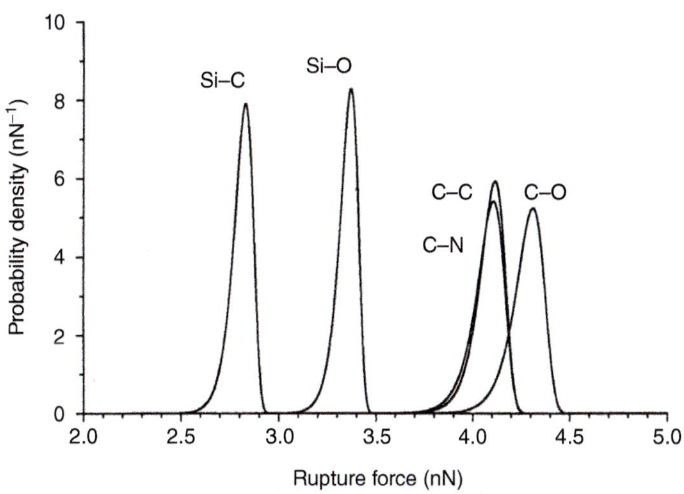

Figure 1.9 Bond rupture probability densities (nN^{-1}) as a function of force F (nN) calculated by density functional theory. Force-loading rate is 10 nN/s. Source: Beyer [59]/with permission of American Institute of Physics.

force distribution. The former can be predicted or measured, the latter is anisotropic within a polymeric material. One study incorporated coumarin dimer (CD) into specific locations of well-defined polymers as a means to determine the role of mechanophore spatial distribution on the efficiency of mechanoactivation [60]. The experiment was carried out in solution under ultrasonication. When CD was incorporated near the center of a chain, selective breaking of the CD dimmer

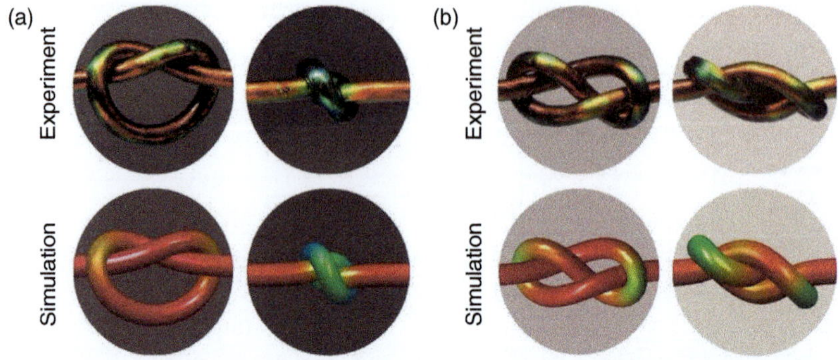

Figure 1.10 Color-changing mechanoresponsive fibers confirm the stress patterns predicted by continuum simulations for the trefoil knot (a) and the figure-of-eight knot (b) during the tightening process. The color is coded in the CIE 1931 XYZ color space. Approximately red end of the spectrum corresponds to low strain and the blue end corresponds to high strain. Source: Reproduced with permission from Patil et al. [63]/American Association for the Advancement of Science – AAAS.

occurred. However, when the CD was located away from the chain center, random cleavages of backbone bonds were dominant. In the solid state, force transduction is further complicated by entanglements, knots, and other defects. Early studies suggested that the highest stress is formed at the entrance position to the knot [61]. However, the chains considered in this simulation are relatively short; thus, it is possible that being close to the chain end also contributes to the high stress at the immediate vicinity of the knot. When polyethylene of 40 carbon atoms is simulated for end-to-end stretching, the stress is concentrated at the torsions around the curved part of the knot [62]. A mechanistic study of tying ropes into different topologies [63] may shed some light on this disagreement. Figure 1.10a, b shows a color-changing photonic fiber with two different knots. The fiber is coated with a periodic cladding. When the knot is tightened, thickness of the cladding changes upon elongation or bending, leading to color variations. Note this is a structural color and does not involve any mechanochemistry. Moving from the red to the blue end of the color spectrum, both strain and stress increase, and the stress is localized at the curved parts of the tightened knot. The stimulation results show a similar stress distribution pattern. Interestingly, the study showed that the stress localization within the knot can be more substantial for certain knots than others under the same pulling force. If we draw an analogy of entangled polymer chains with knotted ropes, one can clearly see the extent of anisotropy in stress distribution. Additionally, cumulative forces of numerous intermolecular interactions with certain configurations can shield a weak bond from mechanical activation. Crystalline domains, covalent crosslinks, physical crosslinks, and topological defects, such as loops, can further contribute to the heterogeneities. The effects of side groups, backbone rigidity, backbone creep, and relaxation cannot be neglected either. Furthermore, an interesting behavior is that the cleavable bonds do not have to directly align with the pulling axis. This is further discussed in Section 1.4 of this chapter. Therefore, the complex and rich behaviors of polymer mechanochemistry are an area to be further explored.

1.4 Polymer Covalent Mechanochemistry

In this section of the chapter, we will consider mechanical activation of numerous mechanophores based on covalent bonds within polymeric materials. Such labile covalent bonds include C—C, C-heteroatom (e.g. C—O, C—S, C—N, C—Cl, and C—Br), and heteroatom–heteroatom (e.g. O—O, S—S, and Se—Se) bonds. The covalent mechanophores described in this section are classified as pyrans, retro-cycloadditions, ladderenes (LDEs), stable radical systems, and other types. The mechanochemical rupture can generally occur via pericyclic (e.g. electrocyclic ring opening of pyrans and retro-cycloadditions), homolytic (e.g. radical formations), or heterolytic (e.g. benzoxazole [Bz's] ester bond cleavage) reactions. The occurrences of mechanochemistry are often associated with distinctive changes in materials' properties which can be utilized to perform certain functions. For example, color change and fluorescent emissions are involved in mechanoactivation of pyran-based compounds. Such mechanophores are termed mechanochromophores and mechanofluorophores, and they enabled the detection and mapping of stresses in polymer materials [64–67]. Mechanoradicals are generated via homolytic cleavage of covalent bonds under force and have been employed in polymerizations to enable self-healing properties [68]. Ultrasonication-induced selective cleavage of disulfide-centered polymers gave thiol-terminated polymer chains, which could then undergo thia-Michael addition to Diels–Alder adducts of furan-functionalized drugs and acetylenedicarboxylates, followed by a retro-Diels–Alder reaction in which the liberation of a small molecular drug occurred [69]. Likewise, another study reported on a release of an alkaloid-type anticancer drug by an intramolecular cyclization of mechanochemically generated thiol-terminated polymers [70].

Constrained-geometry-simulating external force (CoGEF) [59] that is based on density functional theory (DFT) has been carried out to computationally determine the threshold forces required to activate covalent mechanophores. In this CoGEF method, the system begins with an unstrained state of the mechanophore. Two anchoring points from the opposite sides of the mechanophore are then selected, the distance between these two points is increased in small steps, and the geometry with the minimum energy is identified for each elongation step. A plot of relative energy versus equilibrium displacement is subsequently constructed, and the threshold force is extracted from the slope of the plot just before the bond cleavage occurs. The threshold activation forces for numerous mechanophores have been assessed via CoGEF calculations, and the predictive power of this method for polymer mechanochemistry has been demonstrated [71]. For a range of mechanophores, the CoGEF predictions are in good agreement with the experimental values obtained from the single-molecule force spectroscopy measurements. In this section, we will focus on discussing the chemistries involved and the tunability of the bond dynamics.

1.4.1 Pyran-Based Mechanochromophores

Spiropyran (SP) is a well-known photochromic molecule. Under UV radiation, the colorless and nonfluorescent SP undergoes 6π electrocyclic ring opening to form merocyanine (MC) species (Scheme 1.1). The increase in conjugation in MC shifts the absorption to longer wavelengths, giving rise to visible color and fluorescence [72]. Ring opening of SP can also be mechanically activated when incorporated into polymer backbones by tethering its two ends. Under uniaxial stress, the conversion of SP to MC resulted in a color change of the polymer from yellow to purple and to red once failed [72]. SP-functionalized polymeric materials have also shown mechanochromism induced by grinding, ultrasonication [73], and shear stress [74]. Even though SP incorporated polymers are typically colorless or yellow, their MC forms can show different colors, such as blue, purple, or red, under different strains. As shown in Figure 1.11, the polymer turned blue when stretched. Upon release, its color changed to purple via the isomerization of the methine bridge that links the two cyclic components [75]. The colorless form is restored after radiation with visible light.

It is crucial that the tensile stress is applied across the spiro-junction to pull apart the indoline and benzopyran moieties to exclusively cleave the C_{spiro}—O bond. This can be achieved by tethering the polymer chains to the opposing sides of the spiro-junction, such as positions 7 or 8 of the benzopyran and 5′ or 6′ of the indoline [72]. Functionalizing SP at other positions, such as 1′and 5′ of the indoline component, prevents the mechanical force transduction across the reactive

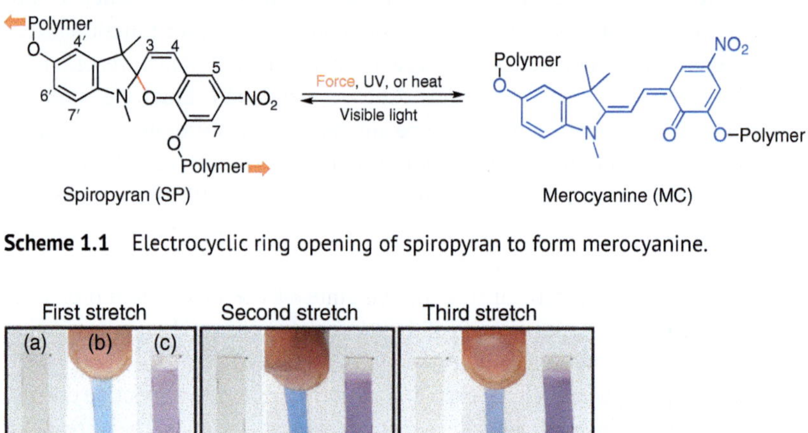

Scheme 1.1 Electrocyclic ring opening of spiropyran to form merocyanine.

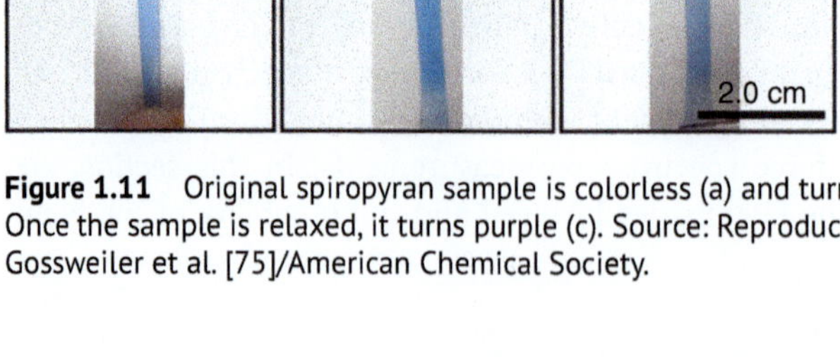

Figure 1.11 Original spiropyran sample is colorless (a) and turns blue when stretched (b). Once the sample is relaxed, it turns purple (c). Source: Reproduced with permission from Gossweiler et al. [75]/American Chemical Society.

C_{spiro}—O bond, although SP to MC transformation can still take place by heating or UV radiation [76]. Regiochemical effects on mechanical activation have been quantified by using two SP regioisomers via single-molecule force spectroscopy measurements [77]. In one isomer the polymer chains were tethered to positions 5′ and 8, while in the second isomer the polymer chains were attached to positions 8 and indole's N atom, and the threshold forces for activation were determined to be 260 and 240 pN, respectively. The rate of mechanochromism is governed by several factors that include the nature of the materials (e.g. elastomeric polymers have demonstrated faster decay of the color change than those of glassy ones) [72], local environment (e.g. temperature and plasticizers) [78], and substituents on the SP (i.e. the higher the electron density on the aromatic ring of benzopyran, the faster the ring closure) [79].

Spirothiopyran (STP), an SP analog, displays both mechanochromism and stress-induced addition reactions rendering it a versatile mechanophore. Ring opening of the thiopyran ring forms the corresponding thiomerocyanine (TMC) in its thiophenolate form (Scheme 1.2), which can take part in thiol-ene click reactions in the presence of olefinic double bonds. Ultrasonication-activated polyester-functionalized STP showed yellow to green color change because of the formation of the TMC form; however, the green color quickly faded in the presence of *N*-methylmaleimide due to the thiol-ene click reaction [80]. When *N*-methylmaleimide is replaced with the bifunctional 1,6-bismaleimidehexnae crosslinker, mechanical-induced crosslinking of the linear polymers was achieved, giving insoluble networks.

Mechanochromism can be affected by both electronic and steric effects. This is shown for naphthopyran (NP) in Scheme 1.3. The activated merocyanines not only exhibit different colors but also their threshold forces vary depending on the nature of the substituents. NP species (Scheme 1.3a) forms a yellow MC, whereas NP modified with electron-donating pyrrolidine (Scheme 1.3b) gives a purple MC [81]. CoGEF calculations indicate that electron-donating and bulky pyrrolidine decreases the activation force for mechanical scission [71]. Once the mechanical force is eliminated, heat-induced ring closure results in the reformation of NP and NP-Pyr.

As is true for SP and NP, oxazine (OX)-derived mechanophores also demonstrate regioisomer-dependent mechanochromism [82]. As shown in Scheme 1.4, the bulk

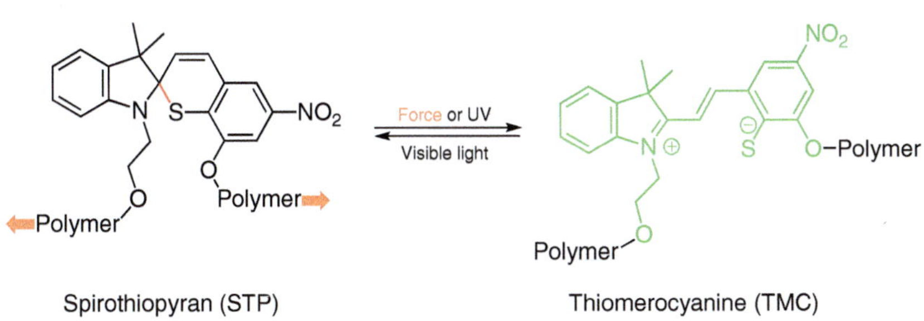

Scheme 1.2 Electrocyclic ring opening of spirothiopyran to form thiomerocyanine.

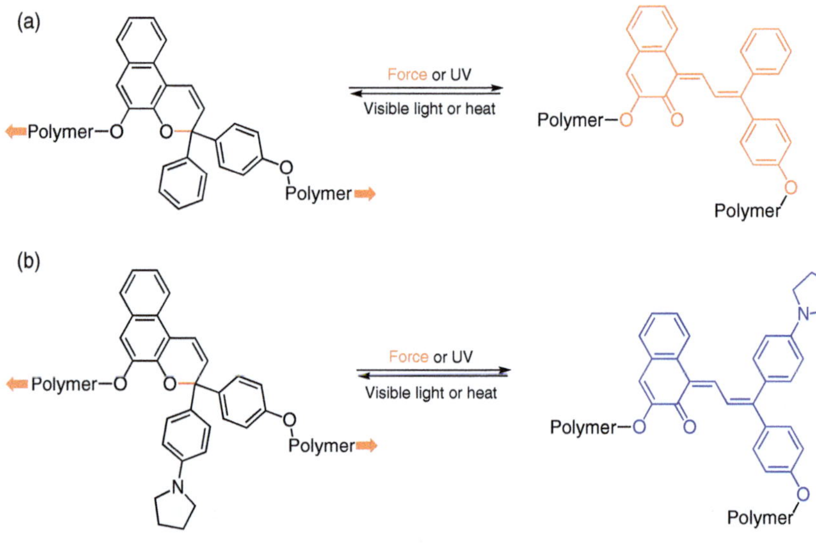

Scheme 1.3 Electrocyclic ring opening of (a) naphthopyrans (NP) and (b) pyrrolidine-appended naphthopyrans (NP-Pyr).

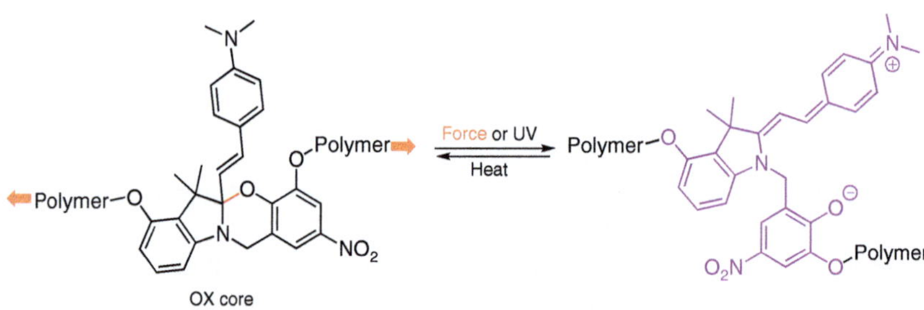

Scheme 1.4 Electrocyclic ring opening of oxazine to give zwitterionic indolium species.

polymer-embedded OX core undergoes mechanical stress-induced pericyclic ring opening to generate a colored, zwitterionic indolium isomer, which reverts to the ring-closed form upon removal of the tensile stress within less than a second. This markedly faster mechanoresponsiveness of OX without any phase lag or fatigue compared to SP or NP is striking. It should be noted that the ring opening of both SP and NP is followed by a double-bond isomerization process to form the corresponding *trans*-MCs, but such an isomerization does not take place with OX, which might explain OX's faster mechanoresponse. In comparison to SP or NP, OX mechanophore may hold better promise as a molecular force probe since OX has a shorter cleavage displacement (e.g. SP and NP frameworks need to be stretched about thrice as much as OX for the bond rupture) as well as its less complicated ring-opening process without the double-bond isomerization.

The above-mentioned mechanophores all contain one pyran ring and give the same mechano- and photo-product. Mechanochromic bis(naphthopyran) (BNP) featuring doubly closed pyran ring (BNP_{C-C}) configuration exhibits a different

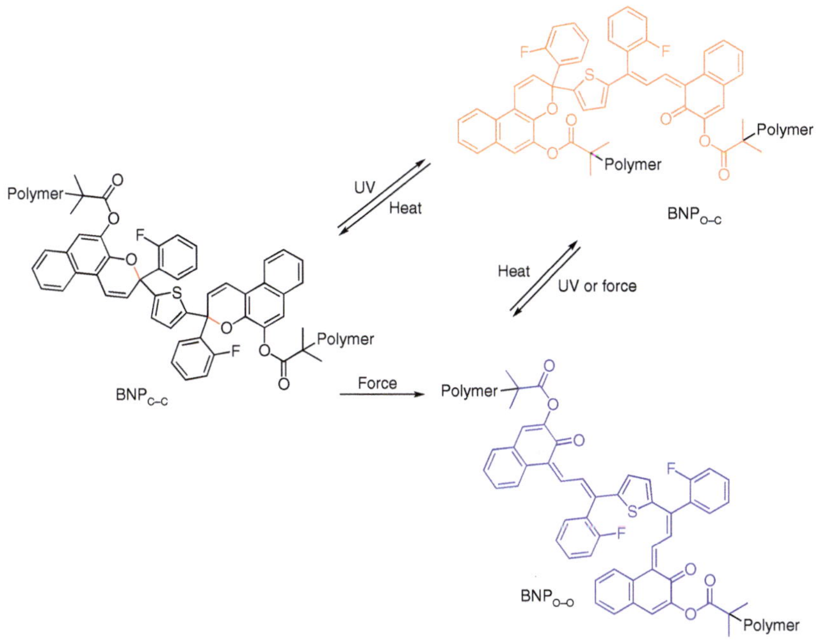

Scheme 1.5 Electrocyclic ring opening of bis(naphthopyran). UV irradiation leads to stepwise ring opening via BNP$_{O-C}$, while mechanical force directly transforms BNP$_{C-C}$ to BNP$_{O-O}$.

mechanochemical pathway [83]. As shown in Scheme 1.5, under continuous UV irradiation, BNP$_{C-C}$ undergoes ring opening of one of the pyran rings to first form an open–closed BNP$_{O-C}$ MC species, followed by a fully open MC form (BNP$_{O-O}$). In contrast to this sequential photochemical ring-opening process, under mechanical activation, an equilibrium of BNP$_{C-C}$, BNP$_{O-C}$, and BNP$_{O-O}$ is reached but through a different mechanism. BNP$_{O-O}$ is formed directly from BNP$_{C-C}$ by mechanical activation; therefore, its concentration is force dependent. BNP$_{O-C}$ is produced predominately from the thermal electrocyclization of BNP$_{O-O}$. Meanwhile, BNP$_{O-C}$ can be mechanically activated to generate BNP$_{O-O}$ again. Therefore, the distribution of BNP$_{O-C}$ and BNP$_{O-O}$ varies with the magnitude of force. Since the two MC products have distinctly different absorptions in the visible range, BNP exhibits gradient multicolor mechanochromism, which can be used as a stress sensor. Systematic studies of these compounds have been conducted to illustrate the substituent effects [43, 71], stereochemistry, and regioselectivity [71]. Computation and experimental studies allow us to understand mechanisms of mechanochemical reactions, as discussed in Section 1.3.

1.4.2 Retro-Cycloadditions

In a cycloaddition reaction, two π reactants react to form a cyclic adduct that contains two new σ bonds. The adducts are susceptible for reverse reactions (i.e.

retro-cycloadditions) under sufficient mechanical force. Here, we will summarize the use of retro-cycloadditions in designing polymeric materials for sensing, release of small molecules, gated ring opening, and degradable materials.

Anthracene-derived Diels–Alder adducts have been widely used as mechanofluorophores because of their high fluorescent quantum yields. Anthracene dimers [84], anthracene–maleimide adducts [85–88], as well as π-extended anthracene–maleimide adducts [89] incorporated into polymers can be mechanically activated. The liberation of fluorescent anthracene derivatives takes place via stress-induced retro Diels–Alder of the anthracene adducts, which in turn leads to the scission of the polymer chains. Using this concept, anthracene–maleimide adducts have been employed as damage sensors at heterointerfaces [90, 91]. An anthracene–maleimide adduct-embedded polymer chain was grafted onto the surface of silica nanoparticles. Here, anthracene species was liberated via ultrasound. This showed the selective rupture or activation of mechanophores under mechanochemical stress at a heterointerface. The mechanochemical activation of π-extended anthracene–maleimide adducts (Scheme 1.6a) is particularly interesting because they give a fluorescence quantum yield of 0.72, which is about 2 orders of magnitude higher than those reported for merocyanines derived from spiropyrans [89]. Besides, the fluorescence is stable in the presence of excess oxygen and the system is not susceptible to thermal reversibility. The appropriate functionalization of this mechanofluorochromophore (via the substituents in anthracene) gives the opportunity to tune excitation and emission wavelengths while maintaining mechanochemical properties of the system. This highly sensitive π-extended anthracene–maleimide adduct was used as a crosslinker in a poly(*N*-isopropylacrylamide) hydrogel network to detect and accurately localize the covalent bond scission induced by a needle-puncturing (hand) [92]. Anthracene–maleimide adducts have been employed as damage sensors in polymers using them at heterointerfaces [90, 91]. For example, an anthracene–maleimide adduct-embedded polymer chain has been linked to a surface of silica nanoparticles, and ultrasound activation of the mechanophore has led to the liberation of the anthracene species thereby ensuring the rupture of the polymer chain.

Mechanochemical activation of the mechanophore-centered polymer chains not only yields optical responses but also can display chemical responses by the release of small molecular cargos, such as furans [93, 94], coumarin [95], and phenyltriazolinedione [75]. A difunctional crosslinker comprising mechanophore oxanorbornadiene, which is a Diels–Alder adduct of furan and acetylenedicarboxylate, was incorporated into a polymethyl acrylate (PMA) matrix in which compression-induced retro-[4+2] Diels–Alder reaction liberated the benzyl furfuryl ether small molecules (Scheme 1.6b) [93]. In this example, the cleaved covalent bonds were not an integral part of the polymer main chain, which is quite counterintuitive. The activation occurred by bond bending instead of stretching. Hence the overall polymer architecture is retained intact, while indeed reinforcing the bonds in the polymer backbone. Force-accelerated dissociation of unloaded bonds also occurred for phosphotriesters [96]. In another study,

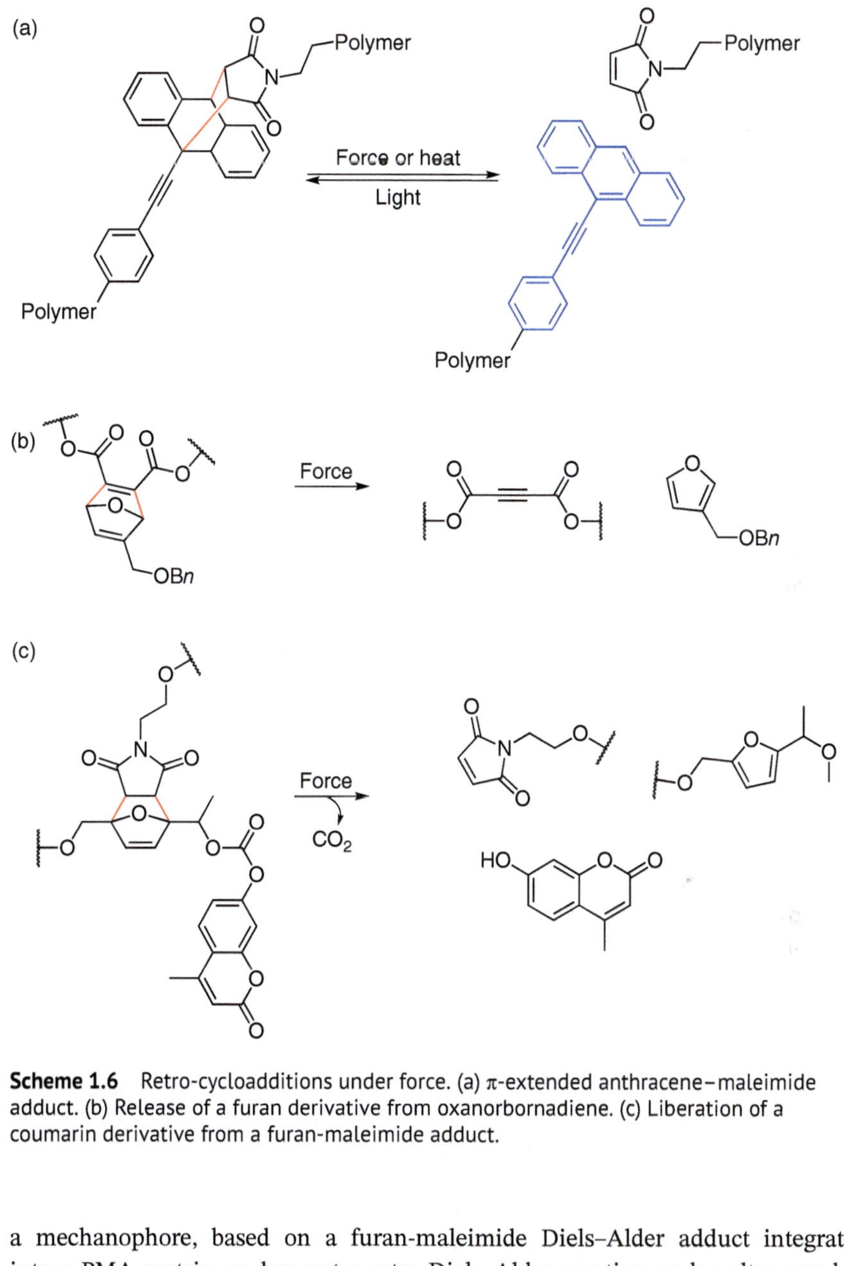

Scheme 1.6 Retro-cycloadditions under force. (a) π-extended anthracene–maleimide adduct. (b) Release of a furan derivative from oxanorbornadiene. (c) Liberation of a coumarin derivative from a furan-maleimide adduct.

a mechanophore, based on a furan-maleimide Diels–Alder adduct integrated into a PMA matrix, underwent a retro-Diels–Alder reaction under ultrasound to first form a polymer-chain-terminated unstable furfuryl carbonate intermediate, which underwent decarboxylation in polar protic solvents at room temperature, eventually liberating a covalently bound cargo molecule hydroxycoumarin, as shown in Scheme 1.6c [95]. Fluorescence spectroscopy can be used here to measure the mechanical activity since the hydroxycoumarin is fluorescent active. As the hydroxycoumarin is disconnected from the polymer chain, it can now diffuse freely into the solution. The systems that enable release of molecular cargos should find applications in areas such as self-healing, depolymerization, and drug delivery [12].

A dual mechanophore – 5,5-dichlorotricyclo(7.2.0.0)undecane (DCTCU) has been employed to demonstrate a mechanochemically gated reaction, where one mechanophore (i.e. cyclobutane) served as a molecular gate to initially prevent the transmission of mechanical force to a second mechanophore, *cis-gem*-dichlorocyclopropane (*cis*-gDCC) [97]. Once the gate was mechanically removed, *cis*-gDCC could experience the force, which in turn led to a cascade of mechanochemical reactions. Thus, the two different mechanophores were designed to be activated in succession under sonication. It has been determined that a higher force is required for a DCTCU polymer (where *cis*-gDCC is gated) than that of a free *cis*-gDCC polymer for the mechanical isomerization of *cis*-gDCC to dichloroalkene to occur at an equal rate. The additional force required to first dissociate the cyclobutane gate before the activation of *cis*-gDCC agrees with the higher force necessary for DCTCU polymer. In a separate study, ultrasonication-induced mechanochemical activation of a thermally and photochemically stable cyclopentadiene-maleimide Diels–Alder adduct embedded polymer has been reported, where the adduct underwent retro-[4 + 2] cycloaddition to furnish dithienylethene photoswitch that could isomerize between colorless (ring-open) and colored (ring-closed) forms with UV and visible light, respectively [98]. The mechanophore (adduct) lacked a 6π electronic arrangement for cyclization, rendering it photochemically latent. Therefore, we can realize that mechanochemistry followed by photochemistry here offers the only way to achieve this transformation from the adduct to the ring-closed form of dithienylethene. Note that the example shown in Scheme 1.6c was yet another scenario where mechanochemical gating has been exploited. According to these examples we discussed, mechanochemical gating shows promise to be applied in various areas, such as stress sensing, release of small molecules, catalysis, and tracing of the mechanical history of bulk polymers [97, 98].

Traditional degradable polymers usually contain reversible dissociable bonds on the backbone of the polymer. However, these cleavable bonds can be susceptible to external stimuli (e.g. light and pH) under ambient conditions, making the polymer material less stable. Therefore, polymers that demonstrate both durability and degradability can be difficult to implement. To tackle this conundrum, degradability can be "locked" during the use of a polymer, while it can be "unlocked" when degradation is required. To achieve this goal, a cyclobutane mechanophore-fused lactone polymer (poly(CBL)) has been employed [99]. Here, cyclobutane kept the polymer backbone intact (i.e. cyclobutane served as a "lock" or "gate") under lactone labile conditions (i.e. strongly basic), while the lactone function could be fully restored from the hydrolyzed product and hence the molecular weight of the polymer was not affected. To degrade the material, poly(CBL) could be mechanochemically "unlocked" that resulted in a copolymer of CBL and a linear polyester, which could be degraded under alkaline conditions. A similar study was simultaneously reported on a degradable polymer functionalized with a cyclobutane-fused cyclic acetal [100]. These two examples demonstrate a promising application of cyclobutane-based mechanophores in mechanoresponsive degradable polymers.

1.4.3 Ladderenes

In response to mechanical forces, cyclic rings, such as cyclopropane [97] and cyclobutenes [36], undergo electrocyclic ring opening and structural rearrangements. The mechanical activation of these mechanophores results in the formation of double bonds in the backbone, which results in significant increases in their mechanical properties or conductivities. For example, 2,3-dihaloalkenes are generated from gDCC. These double bonds can potentially take part in force-induced crosslinking that can reinforce the materials [101]. Also featuring formation of double bonds via mechanoactivated bond rupture, another class of well-studied covalent mechanophores is polyladderene (PLDE), which demonstrates an entirely distinct type of functionality than the above monocyclic species. A molecular zipper, PLDE, is synthesized from ring-opening metathesis polymerization (ROMP) of the LDE that consists of multiple fused cyclobutene rings (see Scheme 1.7a) and can be mechanically unzipped [102]. Formation of PLDE is driven by relief of the ring strain in cyclobutene. The ROMP provided a π-bond containing polymer backbone where the PLDE mechanophore was efficiently incorporated in the center of the polymer. Under sonication, the formation of highly conjugated polyacetylene (PA) block copolymers with a uniform *trans*-olefin configuration from nonconjugated soluble PLDE has been reported (Scheme 1.7b). This particular transformation could not be achieved with heat or light. PA-centered and PLDE termini ABA-type triblock copolymers were expected to form via the mechanical unzipping taking place from the middle of the polymer chain, where the mechanophore was embedded. The colorless PLDE solution initially turned blue after 20 second of sonication, and after sonicating for two hours, it became dark blue with simultaneous formation of black precipitate. A set of PLDEs with varying molecular weights was employed to prove their mechanochemical activation to yield PAs. For example, the formation of PAs took place more slowly as the size of the PLDE decreased though the absorption profiles were the same regardless of the polymer size. The mechanochemical nature of this process was hence validated because the thermal activation did not depend on the size of the polymer. The generation of highly conjugated (i.e. more than 100 conjugated C=C bonds) *trans*-PA was inferred according to Raman and UV–Vis absorption spectroscopy. The resulting PA polymers underwent self-assembly presumably via π–π stacking interactions to generate semiconducting nanowires with a measured conductivity of 2.6×10^{-7} S/cm. Ultrasonication-induced formation of semiconducting fluoropolyacetylene from fluoropolyladderene polymechanophore has also been reported [103].

This family of mechanophores holds high potential due to the ability of the monomer (i.e. LDE) to undergo direct polymerization, easy mechanochemical activation of PLDE, and substantial alteration of polymer properties upon mechanoactivation. Nevertheless, difficulties in synthesis (i.e. synthesis of chloroladderene monomer LDE was hard with a poor overall yield, and the yields were limited to milligram amounts) [102, 104] and derivatization are some of the drawbacks of this class. In addition, since ROMP is followed by an HCl elimination to yield the PLDE it could result in the shortening of some polymer backbones,

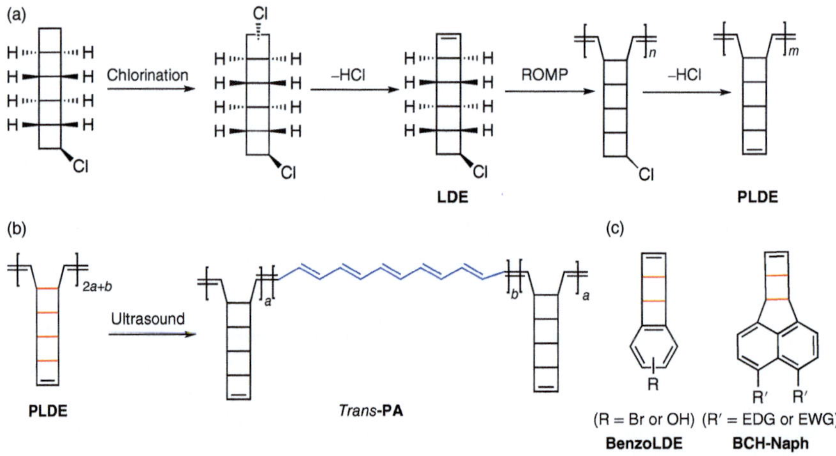

Scheme 1.7 (a) Synthetic scheme for the formation of polyladderene via ROMP (third step) followed by elimination of HCl (last step). (b) Formation of conjugated polyacetylene from polyladderene by mechanical stress. (c) Chemical structures of benzoladderene (left) and bicyclohexene-peri-naphthalene (right).

which in turn results in a broad molecular weight distribution [102]. Therefore, benzoladderene (benzoLDE, Scheme 1.7c, left) was introduced to address the above limitations [105]. Synthesis of the monomer as well as the poly(benzoLDE) was easier (i.e. higher overall yields for benzoLDEs and gram quantities could be synthesized) and could achieve better functionalization with benzoLDE compared to LDE. Furthermore, a more controlled polymerization (i.e. narrower dispersity) was observed with p(benzoLDE) because the ROMP process was not followed by an elimination step for p(benzoLDE) as opposed to PLDE.

An even superior ladder-type mechanophore called bicyclohexene-*peri*-naphthalenes (BCH-Naph, see Scheme 1.7c, right) that can also be effectively polymerized via ROMP to yield polymechanophores (p(BCH-Naph)) with extremely high molecular weights with low dispersity has been introduced [106]. Like PLDE and p(benzoLDE) mechanochromophores, p(BCH-Naph) also transforms into conjugated polymers upon exposure to mechanical stress. The ability to synthesize in multigram scale, broad functionality with both electron-donating and electron-withdrawing groups, enhanced stability, efficient polymerization of BCH-Naph as well as versatile mechanoactivation of p(BCH-Naph) via ultrasonication, stirring, and grinding renders p(BCH-Naph) an excellent polymechanophore reported in the literature.

1.4.4 Stable Radical Systems

A covalent bond can undergo force-activated cleavage to generate two free radicals. These radicals are usually short-lived and are readily quenched. However, when the radicals are stabilized, they can serve as mechanochromophores by the formation of colored radicals. The dramatic color changes accompanied as the radicals are formed provide a way to qualitatively visualize the mechanical stress, while the use of electron paramagnetic resonance (EPR) measurements enables quantitative detection of radical species both in solution and bulk materials. The use of nitrone spin traps also provides a means to quantify the mechanically generated radicals with a colorimetric response [107]. Therefore, they can serve as probes for detecting and evaluating the stress experienced by the polymer materials. Stable radical systems are also utilized in self-healing materials. Mechanical damage causes scissions of sacrificial covalent bonds in the network, resulting in the generation of a large number of free radicals. By controlling their reactivity, self-healing can be achieved by either free radical combination [108–111] or initiation of polymerization when vinyl monomers are present [68].

The radical-based mechanophores can be broadly classified as carbon- and heteroatom-centered radical forming systems. For the formation of radicals from radical-type mechanophores described in this section, it is critical that the weak, central cleavable bond, such as C—C bond, possesses a lower bond-dissociation energy (BDE) than covalent bonds in the polymer backbone to enable exclusive cleavage. Typical BDE of a C—C bond is c. 350 kJ/mol. The reduction of the BDE is achieved by introducing bulk side groups to prestretch the target C—C bond and the resonance stabilization of the generated radicals via the neighboring groups, such as aromatic rings, heteroatoms, and carbonyl groups. Stability of the radicals can be further enhanced by incorporating them into polymer chains with high glass transition temperatures (T_g) to minimize recombination of the radicals. It leads to superior mechanochromism compared to just the mechanophores themselves. Scheme 1.8 lists the radical-based mechanochromophores that have been studied.

Diarylbibenzofuranone (DABBF) is a homodimer of arylbenzofuranone linked by a central head-to-head C—C bond (Scheme 1.8a). The bond length and BDE of the central C—C bond are 1.586 Å and 95.4 kJ/mol, respectively, suggesting it is much weaker than that of a typical C—C bond [112]. When incorporated into polyurethane as a mechanochromophore, it formed blue-colored arylbenzofuranone radicals under tensile stress via homolysis of the central C—C dynamic covalent bond, as shown in Figure 1.12 [113]. A similar mechanoactivated C–C scission was observed for tetraarylsuccinonitrile (TASN) [114] and diarylbibenzothiophenonyl (DABBT) [115]. Their structures are shown in Scheme 1.8b,c. The color of the radicals is pink for TASN and green for DABBT. The central C—C bond in TASN is somewhat longer (i.e. 1.608 Å); nonetheless, its BDE is 110 kJ/mol, which is higher than that of DABBF [112]. This higher BDE is consistent with a higher threshold force of 4.5 nN predicted by CoGEF [71]. The central C–C of DABBT has a BDE of 96 kJ/mol and this value can be modified by the substituents on the phenyl ring, for example, for the para-bromo-functionalized DABBT, the BDE is dropped to 86 kJ/mol [116]. The stable free radicals formed can be quenched by adding a good solvent that helps improve the molecular mobility or heating above the T_g of the polymer.

The poor thermal stability of the radical-based mechanochromophores described above limits their applications. Particularly, these radical-type mechanophores cannot be concurrently employed with free radical polymerization methods because the formed radical species can act as radical scavengers. For this reason,

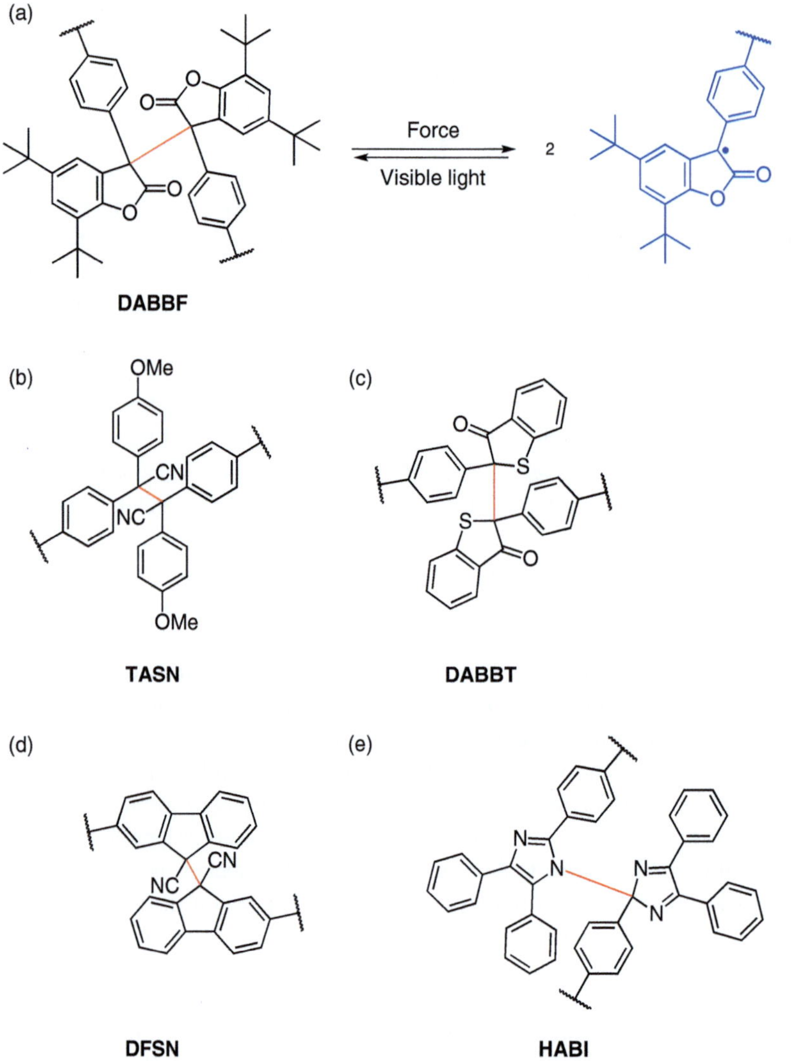

Scheme 1.8 Radical-based mechanochromophores. (a) Mechanical activation of DABBF gives colored radicals. Chemical structures of (b) TASN, (c) DABBT, (d) DFSN, and (e) HABI.

these mechanophores have been introduced into polymer chains mainly by click reactions where heating is not usually required [114, 115]. To tackle this issue, a more promising mechanophore with a higher thermal resistance has been reported. Difluorenylsuccinonitrile (DFSN), a homodimer of fluorenylacetonitrile (Scheme 1.8d), has been exploited as a radical-type mechanochromophore centered in poly(methyl methacrylate) (PMMA) that changed color from white to pink upon grinding [117]. A DFSN dibromide was used as a bifunctional initiator to synthesize DFSN-centered PMMA via atom transfer radical polymerization (ATRP) at 50 °C, where the central C—C bond in the DFSN moiety was retained intact during the

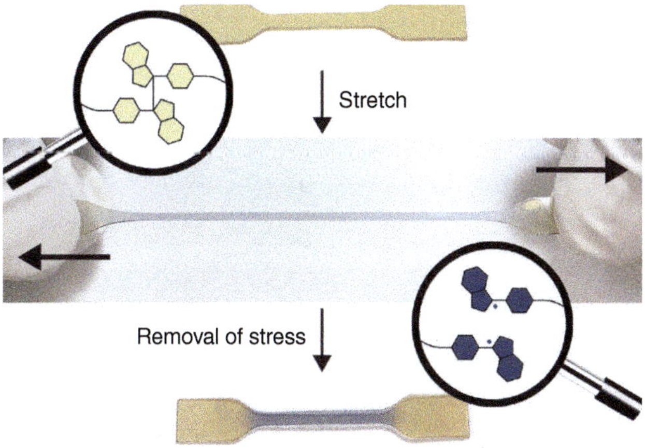

Figure 1.12 Optical images for manually stretched DABBF-linked polymer. Source: Reproduced with permission from Imato et al. [113]/American Chemical Society.

polymerization process. In this study, the DFSN-embedded polymer demonstrated superior thermal stability (e.g. the polymer did not show any color change even after heating to 150 °C).

A heteroatom-centered radical forming molecule is hexaarylbiimidazole (HABI) (Scheme 1.8e). The mechanical activation of mechanochromophore-embedded polymers has been reported to generate triphenylimidazolyl (TPI) radicals via the homolytic cleavage of central C—N bond of HABI moiety with a color change from yellow to green [118]. Here, the HABI units have been incorporated into the backbones of linear poly(methacrylate) and physically crosslinked polyurethane to show the mechanochemical activation of the HABI unit in solution state (via sonication) and solid state (via compression), respectively. Notably, this study demonstrated the use of mechanically generated TPI radicals to initiate radical polymerizations facilitated by sufficiently high amounts of a thiol-based chain transfer agent (e.g. methyl 3-mercaptopropionate).

In contrast to stable radical forming systems discussed above, we can find other heteroatom involved in weak dynamic covalent bonds, such as diselenides. These covalent bonds homolytically cleave to form radicals, which are not that stabilized. For example, ultrasound-induced selective scission of Se—Se bond-centered polystyrene chains has been reported, where the resulting polymeric selenium radicals could subsequently undergo metathesis reactions with an added small molecule comprising a diselenide functionality [119]. Osmotic pressure force-responsive rupturing of vesicles consisting of Se–Se-linked amphiphilic polymers has also been demonstrated, where a diselenide metathesis reaction was observed [120].

1.4.5 Other Types of Mechanophores

Any bond can break when the force is large enough, but polymer mechanochemistry deals with bond breakage that creates functionalities. In this section, we will

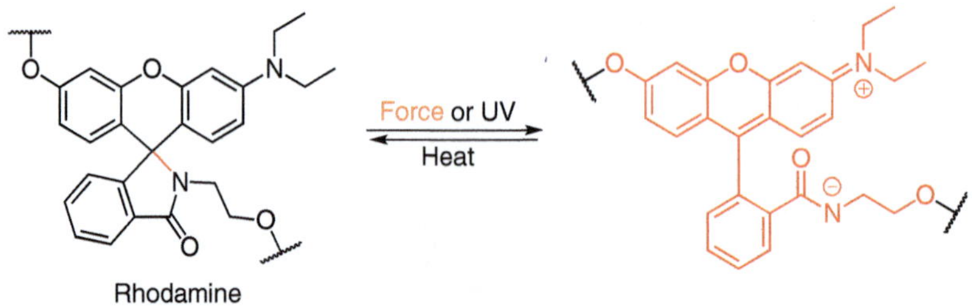

Scheme 1.9 Electrocyclic ring opening of rhodamine.

briefly describe mechanophores, including rhodamine, Bz, and dioxetane, that do not fall within any of the above categories. They offer unique optical properties upon mechanoactivation. Rhodamine is a fluorescent dye that undergoes isomerization from its ring-closed twisted spirolactam form to a ring-open planarized zwitterionic form via the cleavage of C_{spiro}—N bond, as shown in Scheme 1.9. Rh–OH was covalently linked to a polyurethane network. When the polymer was scratched using a pen-like pestle, only the scratched region turned red, demonstrating the mechanochromism of Rh–OH mechanochromophore via its stress-induced isomerization [121]. The color and fluorescence intensity of polymeric Rh–OH were studied as a function of the applied stress. Initially, a fluorescence emission band at 440 nm was observed with a blue color, and later a band at 550 nm emerged with a red color whose intensity increased with increasing applied stress. Importantly, fluorescence intensity remained invariant for the polymer control without Rh–OH. Apart from that, the fluorescence remained the same before and after applying force for samples containing blended crosslinked polymer and Rh–OH, further corroborating the stress-induced isomerization of Rh–OH in the embedded polymer. The color change observed in the polymer material was thermally reversible. It should be noted that this particular system was mechanically inactive for tensile stress.

Ultrasonication-induced Bz-appended aliphatic esters have been reported as a new class of photoluminescent mechanofluorophores [122]. In this study, because of the cleavage of relatively weak ester's C—O bond, the PMA-embedded Bz aliphatic ester (blue) mechanophore formed the free Bz, which has green photoluminescence (Scheme 1.10). The aliphatic ester of Bz emits blue light. This luminescence was a result of intramolecular proton transfer taking place from the electronic excited state. Unlike aromatic esters, the aliphatic ester groups make the polymer embedded mechanophores emissive. Therefore, the mechanoresponsiveness was not simply a turn-on or -off effect, yet a change in photoluminescence color. Here, gel-permeation chromatography (GPC) was used to monitor that mechanical activation indeed took place from the Bz moiety embedded in the PMA material, as shown in Figure 1.13.

Bis(adamantyl)-1,2-dioxetane mechanoluminophore-functionalized polymers exhibit ultrasonication and tensile stress-induced autochemiluminescence in solution and bulk materials, respectively [123]. Under the mechanical force, the ring opening of strained, four-membered 1,2-dioxetane generated two ketones of which one was in the electronic excited state, which emitted blue light (420 nm) as

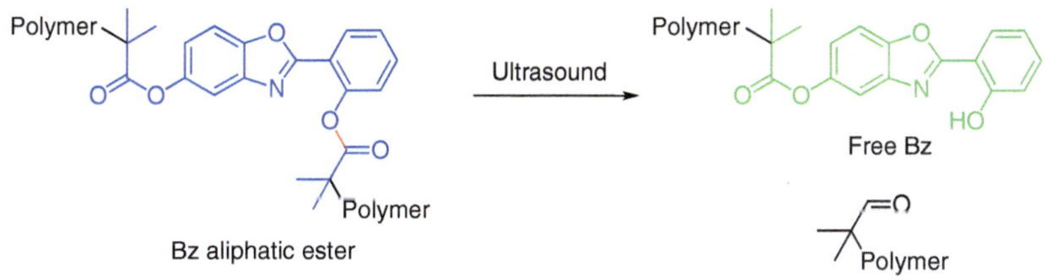

Scheme 1.10 Force-induced cleavage of benzoxazole aliphatic ester.

Figure 1.13 Representative GPC chromatograms of a PMA-Bz-PMA solution after ultrasonication at the indicated times. Source: Karman et al. [122]/with permission of American Chemical Society.

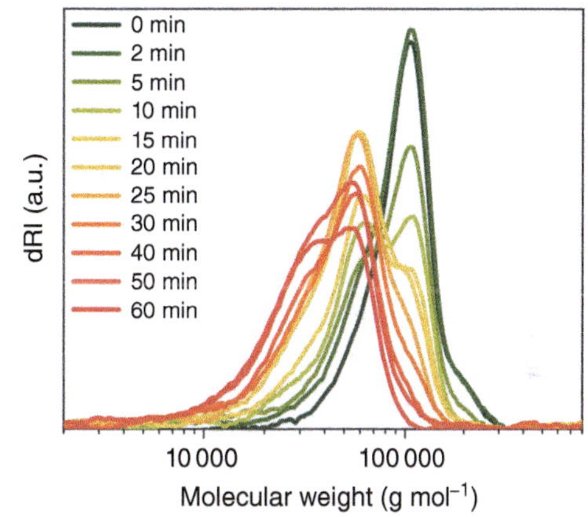

it relaxed to the ground state. It is possible to tune the emission color by transferring the excited-state energy to numerous acceptors [123, 124]. Here, the luminescence occurred in the absence of an excitation light source; meaning, it was exclusively the mechanical force that provided the necessary energy for the photon emission. Therefore, the use of conditions with minimal background signals is viable that in turn leads to improved sensitivity. The activation of 1,2-dioxetane permits for monitoring of bond scissions in real time with high tempo-spatial resolution. These mechanophores have been used as molecular damage indicators to map the damage regions in multiple network elastomers [125].

1.5 Polymer Noncovalent Mechanochemistry

When considering a covalent mechanophore, the bonds affected by mechanochemical stimuli are the weakest in the molecule [126]. Although the weakest in the system, the bond-dissociation energies are consistently hundreds of kilojoules per mole [127, 128]. These high-bonding energies, when reached by mechanochemical stimulation, do not favor spontaneous reformation. Lower energy noncovalent bonding systems, such as metal–ligand bonds, π-stacking, hydrogen bonding, and ionic interactions, feature shorter bond lifetimes, lower activation barriers, and the ability to readily re-associate upon removal of load and therefore should also be considered as mechanophores. Biological systems rely largely on noncovalent interactions to control biomechanics. Synthetic polymer mechanochemistry has focused more on dissociable covalent bonds. However, mechanoresponses of secondary bonds offer unique functionalities due to fast dynamics and reversibility and have found applications in self-healing materials, mechanoenhancement of gels and elastomers, mechanochromism, and force-activated catalysts. Among them, metal–ligand bonds gained the most popularity. In many early studies, the metal–ligand bonds were observed to break following a mechanochemical stimulation [129]. Given the proper proximity of metal and the ligands, bonds can spontaneously form. The range of bond strengths can be tuned widely by both the identity and concentration of the metal and the ligand [130, 131]. Their maximum BDE can be near that of a covalent bond and down to that of a hydrogen bond [5]. Due to the tunability, metal–ligand bonds are the most studied noncovalent mechanophore and will be the main focus of discussion here.

1.5.1 Mechanoresponses of Metal–Ligand Bonds

Ferrocene (FC) has been studied to have high mechanical activity [132]. Initial accounts of ferrocene mechanochemistry in a polymer matrix demonstrated iron(III) ion (Fe^{3+}) liberation following ultrasonication [133]. The liberation of the Fe^{3+} induced a color change of a potassium thiocyanate solution from clear to red. This demonstrates the ability of ferrocene mechanopolymers to be mechanochromic. Sensitivity of ferrocene mechanopolymers is improved through the use of a conformational lock [134]. The "lock" covalently connects the ferrocene's cyclopentadienyl rings together through a hydrocarbon chain. The connection allows the metallopolymer to respond to mechanical force without initial rotation around the central ligand–metal–ligand axis. This lowers the force needed to liberate the Fe^{2+} while increasing the rate of liberation (Figure 1.14a). The conformational lock can vary in the number of atoms as well as cis or trans conformation of the anchoring points. (Figure 1.14b-1). Both cis and trans conformations were locked by a three-atom chain to determine the effects of conformation, while a five-atom chain locked only cis conformation to determine the dependence of chain strain on dissociation energy. As shown in Figure 1.14b-2, the least force for dissociation occurred at ~800 pN in the *cis*-ferrocenophane with a three-carbon conformational lock (*cis*-3FCP). The most force (1140 pN) was required for the same three-atom lock but in the trans position, indicating the critical role of anchoring points in force transduction. Furthermore, the cis conformational lock containing five atoms requires a greater force (960 pN) than its three-atom lock counterpart, indicating that increased strain provided by fewer atom locks is beneficial. Cis conformations were also determined to have the lowest dissociation rates, nearly 7 orders of magnitude faster than ferrocene without conformation locks (Figure 1.14b-3). A heavy object is dropped from a fixed distance onto a polydimethylsiloxane (PDMS) embedded with *cis*-3FCP to liberate Fe^{3+} (Figure 1.14c). The liberated ions can then complex with surrounding phenanthroline in the bulk polymer matrix, resulting in a color change of the bulk material from yellow to red orange. The color change of a covalently embedded

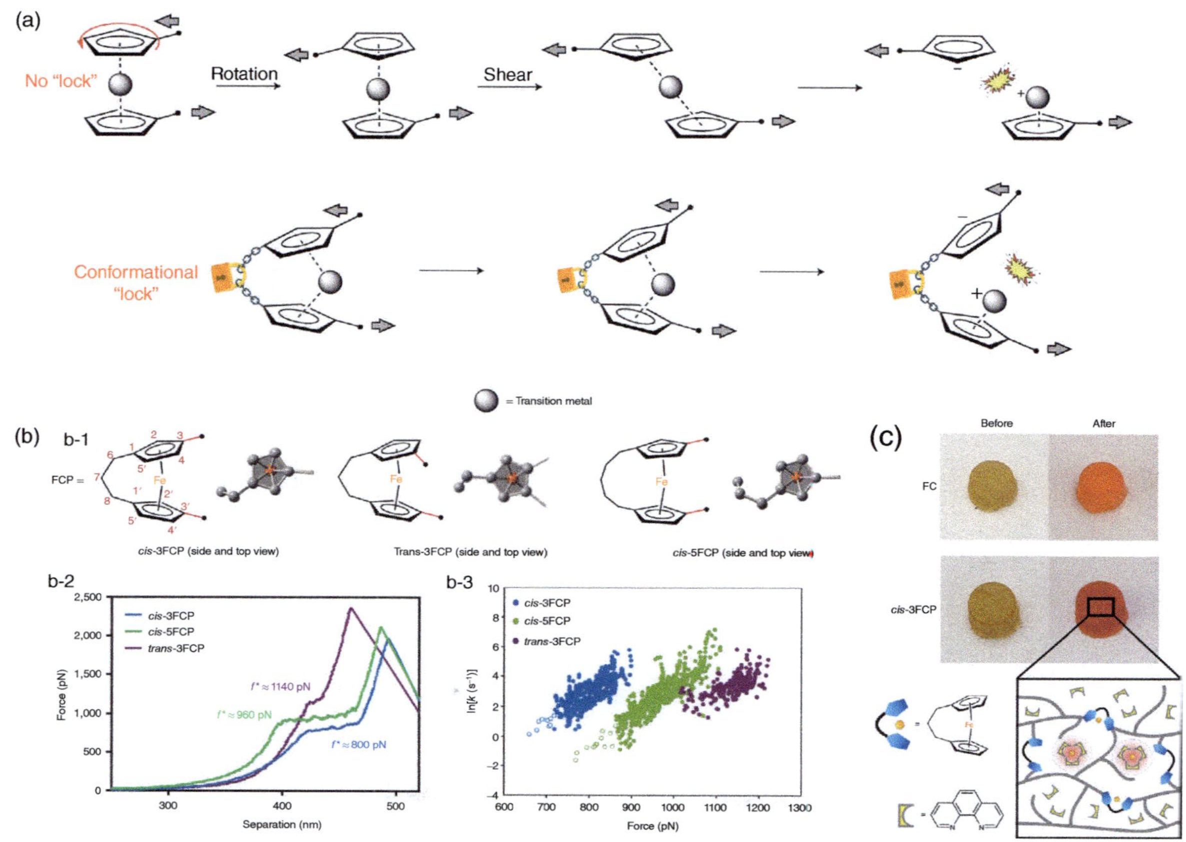

Figure 1.14 (a) Schematic of mechanochemical dissociation resulting in the Fe^{3+} liberation in unconstrained ferrocene and conformationally locked ferrocene. (b) The rate–force plot demonstrating the lowest energy Fe^{3+} liberation occurs in the cis conformation locked with a three-atom "lock" (*cis*-3FCP, blue). (c) Color of PDMS plug functionalized with *cis*-3FCP or unconstrained ferrocene (FC) before and after drop test (top) and schematic of phenanthroline-capturing liberated Fe^{3+} (bottom). Source: Reproduced with permission from Zhang et al. [134]/Springer Nature.

ferrocene (FC) without a conformation lock is less intense, from yellow to light orange. This confirms that unconstrained ferrocene results in fewer Fe^{3+} being liberated from ferrocene under the same force. However, this mechanochromism is irreversible because the cyclopentadienyl rings can dimerize by a Diels–Alder reaction to establish a covalent crosslinking, and the bonds between phenanthroline and Fe^{2+} are relatively stable. But the study demonstrated a promising approach to tune molecular mechanosensitivity.

Metal–ligand mechanopolymers are being widely implemented as catalysts. They can exist within a substrate in the latent state for extended periods of time until activated by stimuli [135]. Through ultrasound sonication, mechanocatalysts experience mechanically induced ligand dissociation, resulting in a switch to a coordinatively unsaturated active catalyst. The activated catalyst can then enter the catalytic cycle to perform the desired reaction. A polymetric ruthenium-based Grubbs-type olefin metathesis catalyst has been reported to induce ROMPs and ring-closing metathesis (RCM) [136]. The mechanically activated catalyst is based on a ruthenium alkylidene complex containing two *N*-heterocyclic carbene (NHC) ligands, which is known as a thermal latent catalyst [137]. It is then modified with polymer chains attached to the NHC ligands to afford a mechanocatalyst. Scheme 1.11 shows the ultrasonic scission of the mechanocatalyst to yield the active catalyst. The metal–ligand bond dissociation is more effective when the attached polymer chains are of higher molecular weight. This is demonstrated through the catalyzed RCM of diethyldiallyl malonate achieving 40% conversion using a high-molecular-weight polymer catalyst compared to 1% conversion using a low-molecular weight, given the same sonication time. Discontinuing ultrasonication is shown to immediately stop the catalyzed RCM, indicating limited lifetime of the active catalyst for that reaction. However, catalyzed ROMP has

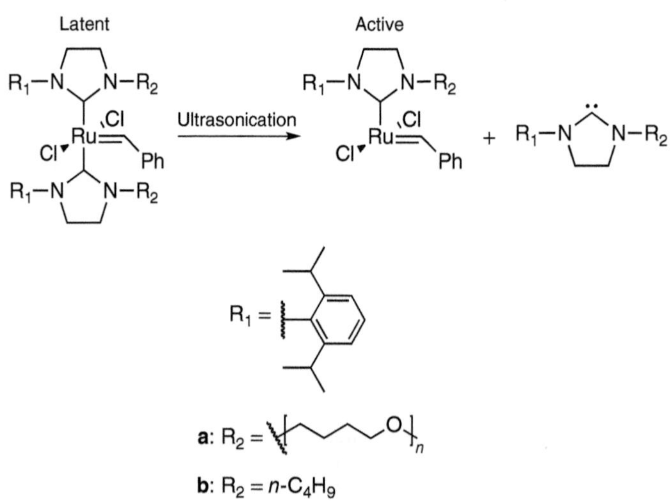

Scheme 1.11 Ultrasonic scission of latent mechanocatalyst to yield the active catalyst. Source: Piermattei et al. [136]/with permission of Springer Nature.

shown to continue hours following ultrasonication has stopped. This difference is consistent with other ROMP and RCM thermal catalysts [138]. Metal–ligand mechanopolymers have also been implemented as drug-delivery devices. Schmidt and coworkers discussed the mechanochemically activated drug delivery from an octahedral Pd cage [139]. All six vertexes contained a polyethylene glycol ($n = 220$)-decorated bipyridine to assemble the octahedral $Pd^{II}{}_6(TPT)_4$ cage around one molecule of ibuprofen. Upon ultrasound stimulation, the polymer chains experienced enough force to dissociate the Pd-2,4,6-tris(4-pyridyl)-1,3,5-triazine (TPT) metal–ligand bond. This resulted in cage fragmentation and subsequent release of pharmaceutically active cargo.

The presence of metal–ligand bonds, along with their dynamic dissociation and reassociation under mechanical forces, can significantly affect the mechanical properties of polymers. A recent example of a highly elastic self-healable material utilizes a dynamic tridentate 2,6-pyridinedicarboxamide ligand (pdca) on a PDMS polymer chain and complexed it with an Fe^{3+} transition metal (Figure 1.15a) [140]. Pdca contains three different atoms that can bind to the Fe^{3+} transition metal center with varying strengths. The order of the strongest to weakest bonds to Fe^{3+} is $N_{pyridine}$, N_{amide}, and O_{amide} verified by calculations using known valence parameters [141]. The transition metal center can bind to two pdca ligands allowing for interchain and intrachain crosslinking. The binding energies were verified

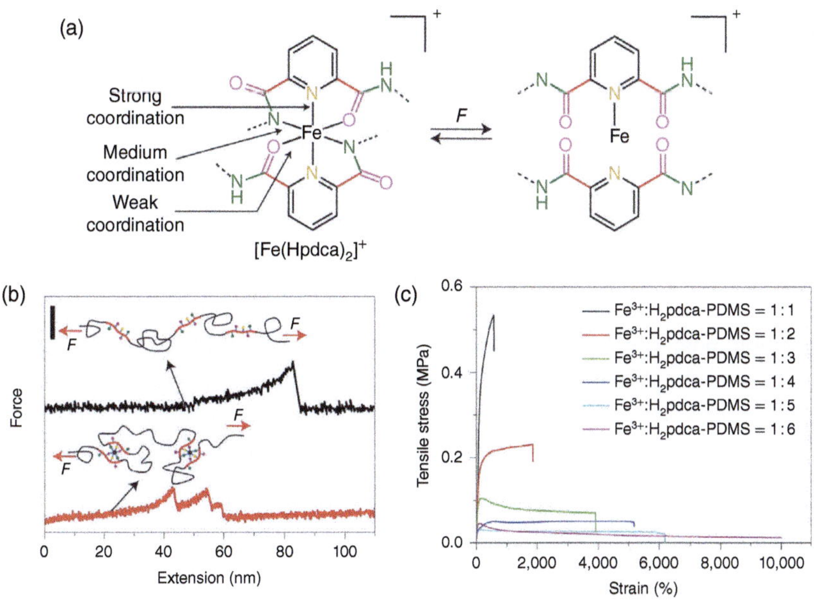

Figure 1.15 (a) Dynamic bonds pdca ligand to Fe^{3+} metal center that individually disconnects following input of mechanical force. (b) Force–extension curve that demonstrates one high-energy bond breakage before addition of Fe^{3+} (black) and three individual bond breakages following addition of Fe^{3+} ligand crosslinks (red). (c) Stress–strain curves demonstrating the dependence of Fe^{3+} concentration to pdca ligand for strength and elasticity. Source: Li et al. [140]/with permission of Springer Nature.

by a force–extension curve measured via single-molecule force spectroscopy (Figure 1.15b). The force–extension curve of a linear pdca–PDMS chain shows a single rupture at 80 nm, signifying a nonspecific covalent bond dissociation. When $FeCl_3$ is added to the linear pdca–PDMS chain, intramolecular crosslinking occurs. This results in a force–extension curve containing three individual bond dissociations (Figure 1.15b). Each of the dissociations is attributed to the three unique bonding energies of the atoms within the tridentate ligand. The covalent rupture no longer appears at 80 nm, suggesting a stabilization of the polymer backbone due to force dissipation metal–ligand bond dissociation. As the material is stretched, the dynamic bond strengths result in controlled bond dissociation, instead of harsh rupture. When all three metal–ligand bonds are broken, the chain slides allowing nonbonded pdca to be in a favorable position to attach to a newly available transition metal center. This strengthens the molecule as it stretches, allowing for further stretching without material failure. Due to these metal–ligand interactions, this material can be stretched beyond 10 000% of its original length with a relatively low Fe^{3+} concentration. The higher concentration of Fe^{3+} increases the crosslink density, resulting in a material with high tensile strength (Figure 1.15c). The dynamic breaking and reforming of metal–ligand bonds during mechanical deformation of a bulk material can be probed using nonlinear rheology [142, 143]. It complements the single-molecule force spectroscopy and provides insights on bond mechanosensitivity in highly anisotropic force environments within highly coiled and entangled systems. For the material shown in Figure 1.15a, rheology measurements confirmed the chain extensions and alignments when the shear rate is much higher than chain relaxation rate, allowing metal–ligand bonds to reform after extension cessation at high strains. Materials that exhibit high stretchability can also exhibit self-healing capabilities [144–146]. When the material was cut and placed back together for 48 hours at room temperature, it could stretch to 90% of what the uncut material could achieve. When allowed to self-heal at −20 °C for 72 hours, it could stretch 68% of an uncut material. This is reported as the first autonomous self-healing material at such low temperatures without additional stimuli.

Transition metals can vary in coordination number, ionic radius, and oxidation state. Thus, mechanical properties differ between two metallopolymers composed of the same ligands but crosslinked with different metals. Monodentate imidazole containing polymers (ICPs) have shown variation in mechanical properties, such as Young's modulus and ultimate extensibility, when crosslinked to different metal centers (Figure 1.16a,b) [147]. In this case, the metals analyzed were Co^{2+} and Zn^{2+}, which are similar in ionic radius and oxidation but differ in coordination number. The coordination numbers of Co^{2+} and Zn^{2+} are six and four, respectively. The crosslink density can be easily tuned by changing the ligand-to-metal ratio (L/M). As expected, increase in metal concentration resulted in higher Young's modulus (E) due to a higher crosslink density. However, the ultimate extensibility (ε_{ult}) is not exclusively related to crosslink density. Rather it is based on metal–ligand-exchange kinetics. The ε_{ult} for zinc does not significantly change when the L/M concentration is reduced from 8 to 4.5. This indicates fast ligand-exchange kinetics. However,

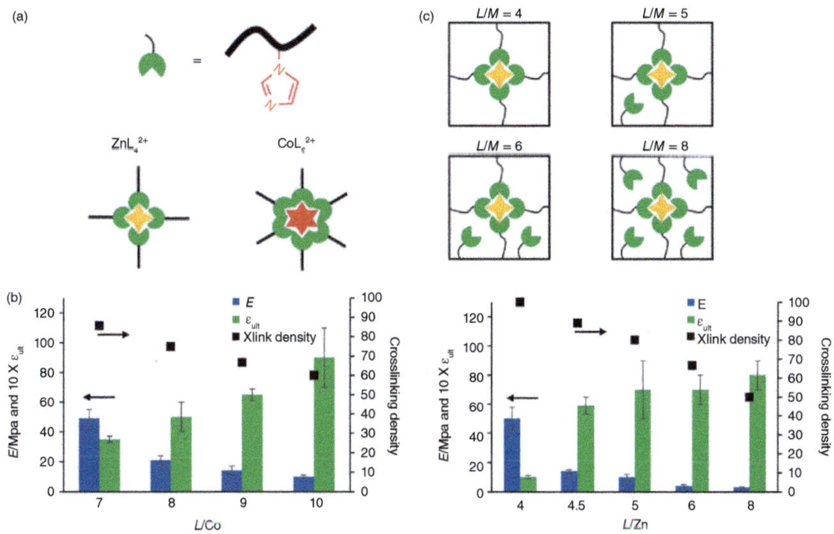

Figure 1.16 (a) Schematic of imidazole ligand binding to zinc (yellow) and cobalt (red) metal centers. (b) Comparison of Young's modulus (E) and ultimate extensibility (ε_{ult}) for various concentrations of ICP crosslinked to cobalt (left) and zinc (right). (c) Schematic of imidazole-zinc ligand exchange in the presence of increasing concentration of ligand. Source: Mozhdehi et al. [147]/with permission of Springer Nature.

when L/Zn decreases to 4, Zn's coordination number, ε_{ult} decreases significantly as there are not any available ligands to exchange, causing the polymer to stiffen (Figure 1.16c). Considering the ε_{ult} of Co, there is a consistent decrease with decreasing L/Co. This suggests concentration-dependent, slow ligand-exchange kinetics. The L/M crosslinks behave more like covalent bonds. If the L/Co reduces to Co's coordination number, one would expect another substantial decrease in ε_{ult}, which is not observed experimentally. The difference in ligand-exchange rates originates from their coordination number and transition states. In the presence of excess ligands, Co^{2+} prefers to form octahedral complexes. Because of the steric hindrance of the octahedral complex, it undergoes dissociative ligand exchange. In comparison, Zn^{2+} prefers complexes with a lower coordination number; therefore, ligand exchange proceeds through associative pathways. Ligand-exchange kinetics directly controls the molecule's ability to self-heal. Zn ICP experiences complete recovery of Young's modulus after 3 hours of room-temperature healing, while Co ICP only experiences ~50% recovery after 24 hours of healing.

Hydrogels have broad applications in biomedical fields, but they often suffer from poor mechanical strengths. The presence of noncovalent interactions significantly enhances the strength, stretchability, and toughness of hydrogels as in bulk polymers due to their dynamics under stress or strain. This has been demonstrated in a double-network gel composed of polyacrylamide (PAA) and alginate crosslinked by three different types of bonds (Figure 1.17a) [148]. The alginate chain comprises mannuronic acid (M blocks) and guluronic acid blocks (G blocks). The G blocks can form the ionic crosslinks in the presence of a divalent cation, such as Ca^{2+}. The M

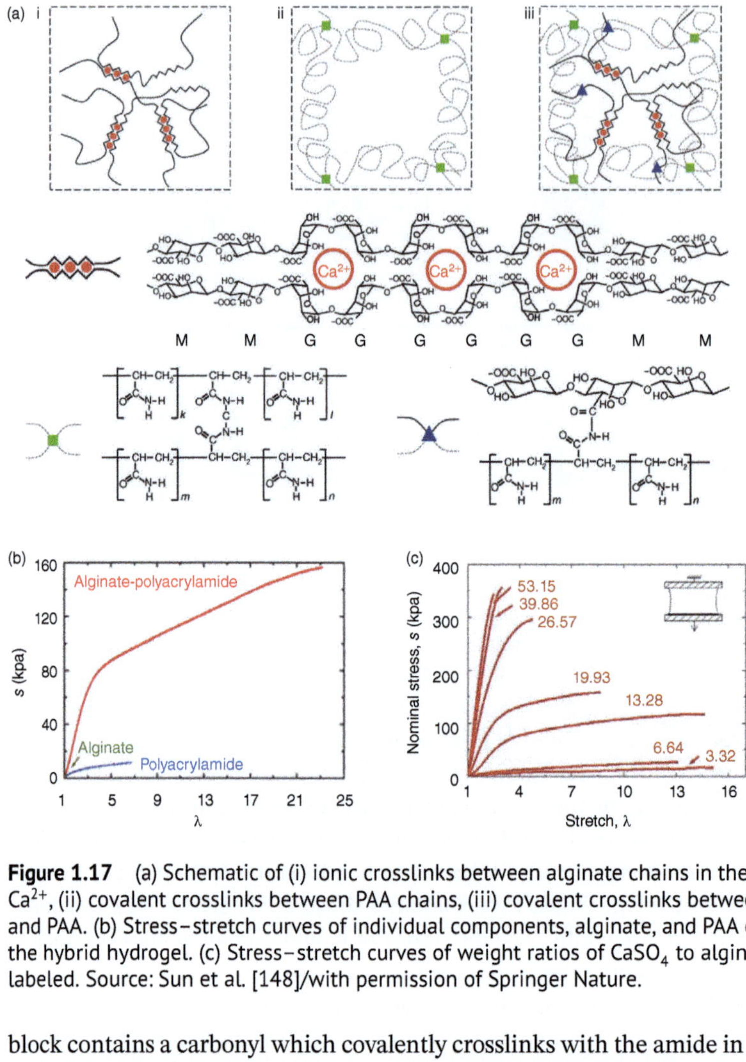

Figure 1.17 (a) Schematic of (i) ionic crosslinks between alginate chains in the presence of Ca^{2+}, (ii) covalent crosslinks between PAA chains, (iii) covalent crosslinks between alginate and PAA. (b) Stress–stretch curves of individual components, alginate, and PAA compared to the hybrid hydrogel. (c) Stress–stretch curves of weight ratios of $CaSO_4$ to alginate, as labeled. Source: Sun et al. [148]/with permission of Springer Nature.

block contains a carbonyl which covalently crosslinks with the amide in PAA in the presence of a *N*,*N*-methylenebisacrylamide crosslinker. Meanwhile, the PAA is also covalently crosslinked. The hybrid gel of these two components can stretch much further and withstand much more stress than as individuals (Figure 1.17b). The enhanced elasticity is attributed to the load sharing of both alginate and PAA chains. As mechanical force increases, the bonds between the Ca^{2+} ions and the G blocks unzip progressively, while the covalent crosslinks remain unbroken. Selective breaking of ionic crosslinks allows for force dissipation without complete material failure. When Ca^{2+} is present around free G blocks, they can rebind to stabilize the gel as it is stretched. The tensile strain (λ) of the hydrogel decreases with increasing Ca^{2+} concentration, but the tensile stress increases (Figure 1.17c). Given water's role in a hydrogel, many diminish in stretchability and toughness below their freezing point at 0 °C; however, the increased concentration of calcium salts induces freezing-point depression which allows for low-temperature mechanoactivity [149]. When soaked

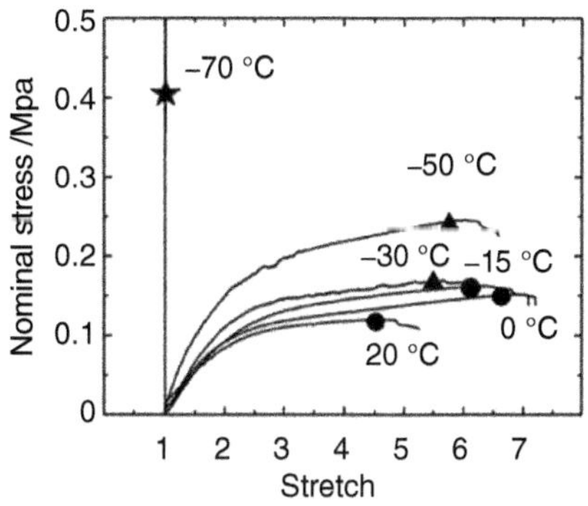

Figure 1.18 Stress–stretch curves of PAA-alginate hydrogel in 30 wt% $CaCl_2$ solution under reduced temperature. Circles are regular hydrogels, triangles are slurry gels, and stars are frozen gels. Source: Morelle et al. [149]/with permission of John Wiley & Sons.

in a 30 wt% $CaCl_2$ solution and temperature is reduced below −30 °C, the hydrogel transforms into a "slurry gel" (Figure 1.18). A slurry gel's increase in toughness is attributed to the formation of ice crystals. The ice crystals increase the volume fraction of the gel, allowing deformation to occur by a stronger shear yielding between them. More remarkably, stretchability also increases due to the formation of microcavities. The microcavities nucleate at the interface between ice crystals and the liquid saltwater solution, which provides a new force dissipation mechanism, resulting in enhanced stretchability.

1.5.2 Mechanochemistry of Other Noncovalent Interactions and Their Applications in Functional Polymers

The development of polymer mechanochemistry is largely driven by the opportunity to create new functional materials. One of the most extensively studied is self-healing polymers. Dynamic breaking and reformation of noncovalent interactions under mechanical force can effectively enhance material strength and lower the activation energy for self-healing to take place. Most noncovalent interactions, including metal–ligand bonds, hydrogen bonding, ionic interactions, π–π stacking, host–guest interactions, and van der Waals forces, have been utilized. The inspiration comes from the healing ability of natural tissue. If materials are capable of sensing and repairing damages on their own, their performance and lifetime will be greatly enhanced.

Noncovalent interactions enable autonomous healing by playing dual roles. One is that they enable healing by breaking and reforming bonds. When the material is mechanically damaged, the noncovalent bonds break, creating nonbonded motifs at the freshly cut surface. When the damaged parts are in contact, chains diffuse and randomize while noncovalent bonds reform, thus erasing the interface and restoring the original material properties. In this case, the healing efficiency is dependent on the resting time from the time the cut is made till the contact is generated. If the bonds undergo fast thermal fluctuations (spontaneous regime, Figure 1.8), self-healing is less dependent on mechanical scission and resting time. The fast dynamics can allow self-healing at temperatures as low as −40 °C [140, 150]. The other role that noncovalent interactions play is providing viscoelastic

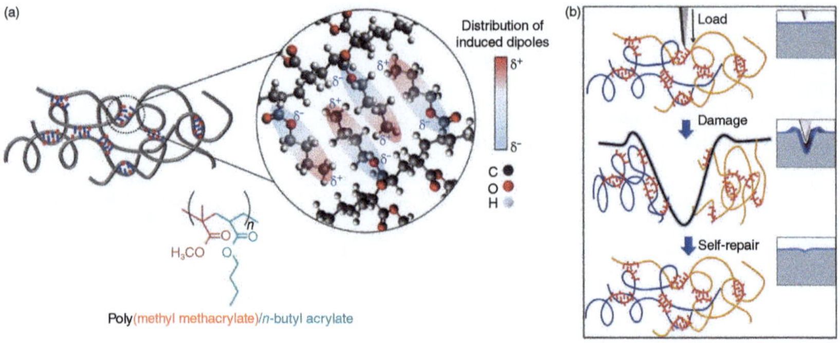

Figure 1.19 Schematic illustration of (a) key-and-lock interaction of poly(methyl methacrylate-*co*-*n*-butyl acrylate) and (b) the self-healing of a scratch driven by both network elasticity and reformation of key-and-lock interactions. Source: Urban et al. [151]/with permission of AAAS.

properties that are favorable for self-healing. This was demonstrated in a study that utilized weak van der Waal interactions to enable self-healing of simple commodity polymers [151]. Although individual van der Waals interaction is extremely weak, the collective force of many of these bonds significantly affects the viscoelasticity of the bulk polymer, similar to small molecules for which intermolecular forces determine many of their physical properties. In this study, commodity copolymers of methyl methacrylate and *n*-butyl acrylate were shown to have favorable interchain van der Waals forces that form key-and-lock interchain junctions within certain compositional range (Figure 1.19). The interaction was attributed to the polarity of the poly(butyl acrylate) pendant groups and a favorable steric environment formed by copolymerizing with the methyl methacrylate. As a result, cohesive energy density increased, and the material gained sufficient elasticity so that chains underwent elastic deformation instead of slippages during mechanical damage, storing the mechanical energy. Upon force removal, the cut then closed on its own by entropy-driven elastic recovery. Copolymer chains returned to their initial conformations in a spring-like manner. Meanwhile, fast dynamics of the van der Waals forces allowed chains to randomize at the interface on a longer time scale and reform key-and-lock junctions.

In this example, the collective force of noncovalent interactions seems to inhibit bond breaking during damage, contradicting the common conception that their breaking and reforming dynamics are responsible for healing. However, it is the favorable thermodynamic and kinetic characteristics within the material that drive the self-healing processes. The change of chain morphologies, increased entanglements, force transduction, and viscoelasticity affected by noncovalent interactions should be taken into consideration in addition to the bond dynamics when considering the self-healing mechanisms. Covalent bonds may also break during damage unless the chains slip out before breaking. Mending of covalent bonds is done through thermal or photo-reversible reactions. Despite the importance of mechanochemistry in self-healing materials, many studies do not offer evidence as to what bonds break at the site of damage and how they contribute to the recovery. However, the amount of chemistries that have been utilized in mechanoresponsive self-healing materials is tremendous and can be found in many comprehensive reviews [152–156].

There have been increasing numbers of studies that utilize noncovalent mechanochemistry to develop functional materials beyond self-healing in recent years. One of such applications is drug activation by ultrasound-induced scission of mechanosensitive noncovalent bonds [70]. When a drug is complexed to a substrate via noncovalent interactions, it is deactivated. The hypothesis is thus that if ultrasound can break the interactions, the drug can then be freed from the substrates and become activated to bind to the target. In the work by Huo et al., two systems were developed based on this hypothesis. One relies on the strong binding of RNA aptamers to their targets (Figure 1.20a). The RNA aptamer used here is R23, which selectively binds to aminoglycoside antibiotics, including neomycin B and paromomycin via non-covalent interactions, such as hydrogen bonds or electrostatic interactions. Once bound to the aptamer, these drugs are deactivated. To make the complex mechanically cleavable, the authors incorporated R23 aptamer units into high molar mass RNA strands followed by loading of the antibiotics. It was found that ultrasonication destroyed the noncovalent interactions between the drug and the aptamer. Additionally, covalent bond scission of the phosphodiester RNA backbone was observed. Ultimately, antimicrobial properties of the drugs were activated by 30 minutes of ultrasound treatment. In comparison, the same drugs loaded to just the aptamers, which are oligonucleotide sequences, did not respond to ultrasonication because of the short-chain length. Another approach developed by the authors was based on force-sensitive prodrug assemblies. The supramolecular binding between vancomycin (Van) with its H-bond-complementary peptide target sequence Cys-Lys-Lys(Ac)-d-Ala-d-Ala (DADA) was used as a proof of concept. As shown in Figure 1.20b, to obtain sufficient force sensitivity, this supramolecular motif was incorporated between poly(oligo(ethylene glycol) methyl ether methacrylate) (POEGMEMA) polymer chains and gold nanoparticles (AuNPs), or between two AuNPs. For both systems, selective mechanochemical scission of the H-bonded motif occurred, leading to successful activation of the antimicrobial drug Van toward *Staphylococcus aureus*. The concept was also successfully extended to ultrasound-induced activation of enzyme activity [157].

Host–guest interactions are highly dynamic in aqueous environments. Therefore, they have been widely used in stimuli-responsive hydrogels [158, 159]. Recently, the use of β-cyclodextrin (host) and adamantane (guest) complex to develop mechanoresponsive hydrogels has been explored. An approach by Hippler et al. utilized a small-molecule competitive guest, adamantanecarboxylic acid, to control the material's force responses [160]. Both host and guest moieties were attached to acrylamide backbone that formed a rigid hydrogel following host–guest crosslinking. Addition of a solution containing the competitive guest decreased the crosslink density as the competitive guest began to occupy host sites. This resulted in a softer hydrogel. Under the same applied force, the competitive guest containing hydrogel was reported to experience a greater deformation than the hydrogel without the competitive guest. A different approach by Uyama and

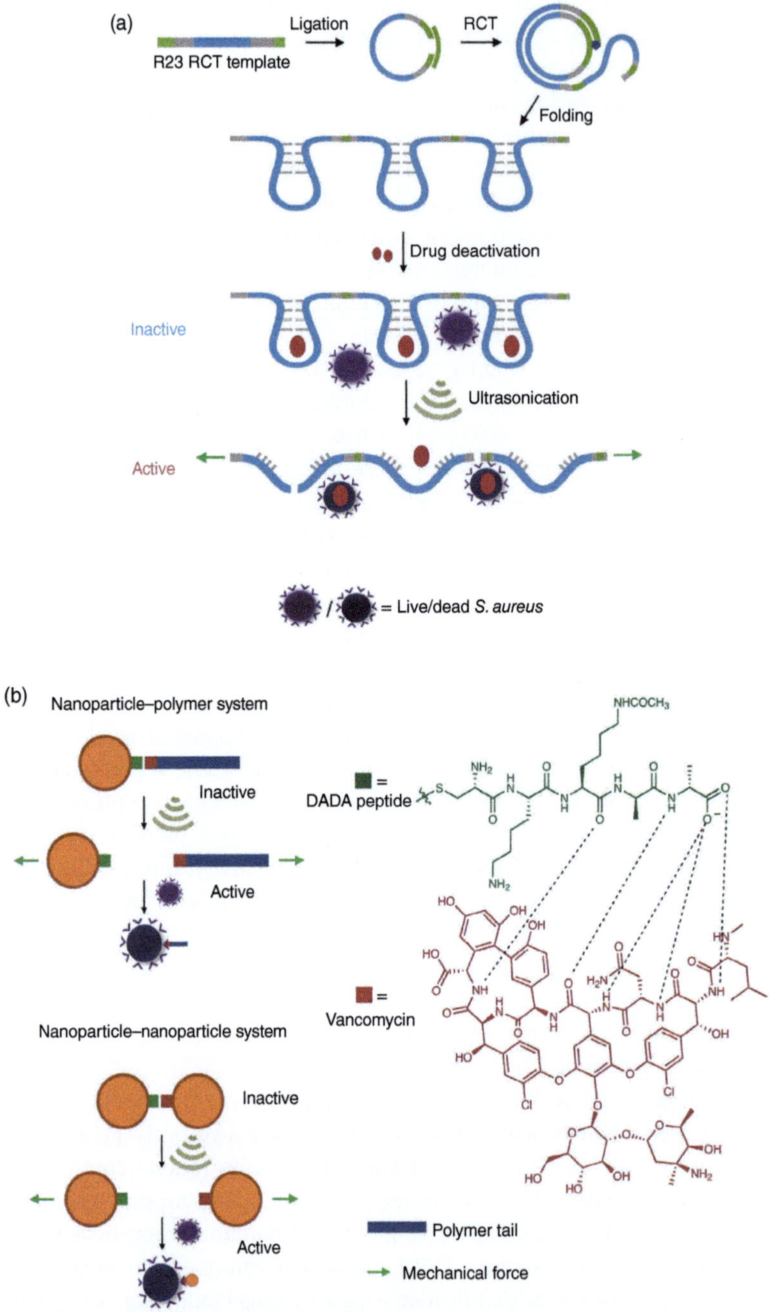

Figure 1.20 Design principle for drug activation by ultrasound scission of noncovalent bonds. (a) Synthesis of polyaptamers R23 via rolling-circle transcription (RCT), drug loading, and subsequent activation and release of the drug by ultrasound to kill *S. aureus* bacteria. (b) Mechanoactivation of nanoparticle–polymer and nanoparticle–nanoparticle assemblies via breaking the H-bonding between DADA–vancomycin pairs to activate the antibiotic properties of the vancomycin drug. Source: Huo et al. [70]/with permission of Springer Nature.

coworkers utilized a hydrogel comprising an adamantane containing thermoresponsive polymer and β-cyclodextrin containing non-thermoresponsive network to afford a stimuli-responsive phase transition [161]. The phase transition of the poly(*N*-isopropylacrylamide) thermoresponsive polymer was restricted even above its lower critical solution temperature (LCST) because of its host–guest connection to the non-thermoresponsive network. Following applied mechanical stress, the host–guest interactions dissociate, and the coil–globule transition typical of the thermoresponsive polymer above its LCST occurred. The hydrogel became opaque following compression, quantified by a decrease in transmittance. Cooling the hydrogel below LCST redistributed the thermoresponsive polymer within the network, regaining original transparency.

1.6 Conclusions

Mechanochemistry plays an indispensable role in many complicated biological processes from cellular growth to repair; however, the design of comparable abiotic mechanical systems that are as sophisticated as nature remains challenging. Over the past decade, synthetic mechanoresponsive materials have attracted immense attention for their numerous practical applications and unique reactivity that is simply not feasible with thermo- or photochemistry. The emergence of new tools to gain more insight into mechanisms of mechanochemical reactions has sparked increasing curiosity in the field. In this chapter, the unique mechanisms of mechanochemistry are discussed followed by introduction of various covalent and noncovalent mechanoactive bonds. By incorporating these bonds into polymers, materials with functionalities, including stress sensing, self-healing, release of small molecules, such as catalysts or drugs, stress mitigation, creation of tougher and stronger materials, and mechanoactivation of smart functionalities, can be created. The ubiquitous energy difference between covalent and noncovalent bonds allows for a wide variety of molecular compositions to experience selective force-induced bond dissociations. Covalent mechanochemistry has undergone more extensive studies, while noncovalent interactions will continue to play an important role in synthesizing functional mechanopolymers. Polymer materials functionalized with mechanophores that are based on drug molecules will find applications in the biomedical field which are shown in recent studies using covalent mechanophores or supramolecular assemblies [69, 162]. Future developments will take inspiration from new chemistries to improve on current strength, stretchability, and self-healing to exhibit properties that emulate biological tissues [163–166]. When specifically considering muscle tissue, adding electrically conductive functionality to mechanopolymers will be the next step in achieving materials that closely match complete tissue function [145, 167–170]. Creating sensing and signaling pathways similar to bio mechanotransduction will allow future materials to be adaptive to the mechanical environment to perform vital functionalities.

References

1 Takacs, L. (2013). The historical development of mechanochemistry. *Chem. Soc. Rev.* 42 (18): 7649–7659.

2 Takacs, L.M. (2004). Carey Lea, the first mechanochemist. *J. Mater. Sci.* 39 (16): 4987–4993.

3 Ostwald, W. (1919). Die Chemische Literatur Und Die Organisation Der Wissenschaft. In: *Handbuch Der Allgemeinen Chemie*. Gesel.: Akad. Verlag. m. b. H. 1–120.

4 IUPAC In Compendium of Chemical Terminology (1997). 2nd ed. (the "Gold Book"). Compiled by A.D. McNaught and A. Wilkinson. Oxford: Blackwell Scientific Publications. Online version (2019) created by S.J. Chalk. ISBN 0-9678550-9-8. https://doi.org/10.1351/goldbook.

5 Baláž, P., Achimovičová, M., Baláž, M. et al. (2013). Hallmarks of mechanochemistry: from nanoparticles to technology. *Chem. Soc. Rev.* 42 (18): 7571–7637.

6 Akbulatov, S. and Boulatov, R. (2017). Experimental polymer mechanochemistry and its interpretational frameworks. *ChemPhysChem* 18 (11): 1422–1450.

7 Willis-Fox, N., Rognin, E., Aljohani, T.A., and Daly, R. (2018). Polymer mechanochemistry: manufacturing is now a force to be reckoned with. *Chem* 4 (11): 2499–2537.

8 Hu, H., Ma, Z., and Jia, X. (2020). Reaction cascades in polymer mechanochemistry. *Mater. Chem. Front.* 4 (11): 3115–3129.

9 Izak-Nau, E., Campagna, D., Baumann, C., and Göstl, R. (2020). Polymer mechanochemistry-enabled pericyclic reactions. *Polym. Chem.* 11 (13): 2274–2299.

10 Abi Ghanem, M., Basu, A., Behrou, R. et al. (2020). The role of polymer mechanochemistry in responsive materials and additive manufacturing. *Nat. Rev. Mater.* 6: 84–98.

11 O, Neill, R.T. and Boulatov, R. (2021). The many flavours of mechanochemistry and its plausible conceptual underpinnings. *Nat. Rev. Chem.* 5 (3): 148–167.

12 Chen, Y., Mellot, G., Van Luijk, D. et al. (2021). Mechanochemical tools for polymer materials. *Chem. Soc. Rev.* 50 (6): 4100–4140.

13 Tsimbouri, P. (2015). Adult stem cell responses to nanostimuli. *J. Funct. Biomater.* 6 (3): 598–622.

14 McNamara, L.E., Burchmore, R., Riehle, M.O. et al. (2012). The role of microtopography in cellular mechanotransduction. *Biomaterials* 33 (10): 2835–2847.

15 Majkut, S., Dingal, P.C.D.P., and Discher, D.E. (2014). Stress sensitivity and mechanotransduction during heart development. *Curr. Biol.* 24 (10): R495–R501.

16 Ucar, H. and Kasai, H. (2021). Forceful synapses reveal mechanical interactions in the brain. *Nature* https://doi.org/10.1038/d41586-021-03516-0.

17 Lavalle, P., Boulmedais, F., Schaaf, P., and Jierry, L. (2016). Soft-mechanochemistry: mechanochemistry inspired by nature. *Langmuir* 32 (29): 7265–7276.

18 Collier, J.H., Rudra, J.S., Gasiorowski, J.Z., and Jung, J.P. (2010). Multi-component extracellular matrices based on peptide self-assembly. *Chem. Soc. Rev.* 39 (9): 3413–3424.

19 Chidanguro, T., Weng, W., and Simon, Y.C. (2018). Mechanochemistry: inspiration from biology. In: *RSC Polymer Chemistry Series*, Chapter 1, vol. 2018 (ed. Y.C. Simon and S.L. Craig), 1–35. Royal Society of Chemistry.

20 Okhota, S., Melnikov, I., Avtaeva, Y. et al. (2020). Shear stress-induced activation of von Willebrand factor and cardiovascular pathology. *Int. J. Mol. Sci.* 21 (20): 7804.

21 Crawley, J.T.B., De Groot, R., Xiang, Y. et al. (2011). Unraveling the scissile bond: how ADAMTS13 recognizes and cleaves von Willebrand factor. *Blood* 118 (12): 3212–3221.

22 Fogelson, A.L. and Neeves, K.B. (2015). Fluid mechanics of blood clot formation. *Annu. Rev. Fluid Mech.* 47: 377–403.

23 Zhang, Q., Zhou, Y.F., Zhang, C.Z. et al. (2009). Structural specializations of A2, a force-sensing domain in the ultralarge vascular protein von Willebrand factor. *Proc. Natl. Acad. Sci. U.S.A.* 106 (23): 9226–9231.

24 Holten-Andersen, N., Harrington, M.J., Birkedal, H. et al. (2011). PH-induced metal–ligand cross-links inspired by mussel yield self-healing polymer networks with near-covalent elastic moduli. *Proc. Natl. Acad. Sci. U.S.A.* 108 (7): 2651–2655.

25 Springer, T.A. (2011). Biology and physics of von Willebrand factor concatamers. *J. Thromb. Haemost.* 9: 130–143.

26 Harrington, M.J., Masic, A., Holten-Andersen, N. et al. (2010). Iron-clad fibers: a metal-based biological strategy for hard flexible coatings. *Science* 328 (5975): 216–220.

27 Hoffmann, T., Tych, K.M., Hughes, M.L. et al. (2013). Towards design principles for determining the mechanical stability of proteins. *Phys. Chem. Chem. Phys.* 15: 15767–15780.

28 Vélez-Ortega, A.C., Freeman, M.J., Indzhykulian, A.A. et al. (2017). Mechanotransduction current is essential for stability of the transducing stereocilia in mammalian auditory hair cells. *Elife* 6: e24661.

29 Lumpkin, E.A. and Caterina, M.J. (2007). Mechanisms of sensory transduction in the skin. *Nature* 445: 858–865.

30 Walsh, C.M., Bautista, D.M., and Lumpkin, E.A. (2015). Mammalian touch catches up. *Curr. Opin. Neurobiol.* 34: 133–139.

31 Hermans, T.M., Frauenrath, H., and Stellacci, F. (2013). Droplets out of equilibrium. *Science* 341 (6143): 243–244.

32 Hickenboth, C.R., Moore, J.S., White, S.R. et al. (2007). Biasing reaction pathways with mechanical force. *Nature* 446 (7134): 423–427.

33 Lenhardt, J.M., Black, A.L., and Craig, S.L. (2009). *gem*-Dichlorocyclopropanes as abundant and efficient mechanophores in polybutadiene copolymers under mechanical stress. *J. Am. Chem. Soc.* 131 (31): 10818–10819.

34 Lenhardt, J.M., Ong, M.T., Choe, R. et al. (2010). Trapping a diradical transition state by mechanochemical polymer extension. *Science* 329 (5995): 1057–1060.

35 Akbulatov, S., Tian, Y., and Boulatov, R. (2012). Force–reactivity property of a single monomer is sufficient to predict the micromechanical behavior of its polymer. *J. Am. Chem. Soc.* 134 (18): 7620–7623.

36 Ong, M.T., Leiding, J., Tao, H. et al. (2009). First principles dynamics and minimum energy pathways for mechanochemical ring opening of cyclobutene. *J. Am. Chem. Soc.* 131 (18): 6377–6379.

37 Kochhar, G.S., Bailey, A., and Mosey, N.J. (2010). Competition between orbitals and stress in mechanochemistry. *Angew. Chem. Int. Ed.* 49 (41): 7452–7455.

38 Wollenhaupt, M., Krupička, M., and Marx, D. (2015). Should the Woodward–Hoffmann rules be applied to mechanochemical reactions? *ChemPhysChem* 16 (8): 1593–1597.

39 Ribas-Arino, J., Shiga, M., and Marx, D. (2009). Understanding covalent mechanochemistry. *Angew. Chem. Int. Ed.* 48 (23): 4190–4193.

40 Bell, G.I. (1978). Models for the specific adhesion of cells to cells. *Science* 200 (4342): 618–627.

41 Hess, H. (2012). Optimal loading of molecular bonds. *Nano Lett.* 12 (11): 5813–5814.

42 Stauch, T. and Dreuw, A. (2016). Advances in quantum mechanochemistry: electronic structure methods and force analysis. *Chem. Rev.* 116 (22): 14137–14180.

43 Dammer, U., Hegner, M., Anselmetti, D. et al. (1996). Specific antigen/antibody interactions measured by force microscopy. *Biophys. J.* 70 (5): 2437–2441.

44 Smith, S.B., Cui, Y., and Bustamante, C. (1996). Overstretching B-DNA: the elastic response of individual double-stranded and single-stranded DNA molecules. *Science* 271 (5250): 795–799.

45 Hinterdorfer, P., Baumgartner, W., Gruber, H.J. et al. (1996). Detection and localization of individual antibody–antigen recognition events by atomic force microscopy. *Proc. Natl. Acad. Sci. U.S.A.* 93 (8): 3477–3481.

46 Oberhauser, A.F., Marszalek, P.E., Erickson, H.P., and Fernandez, J.M. (1998). The molecular elasticity of the extracellular matrix protein tenascin. *Nature* 393 (6681): 181–185.

47 Marszalek, P.E., Oberhauser, A.F., Pang, Y.-P., and Fernandez, J.M. (1998). Polysaccharide elasticity governed by chair–boat transitions of the glucopyranose ring. *Nature* 396 (6712): 661–664.

48 Grandbois, M., Beyer, M., Rief, M. et al. (1999). How strong is a covalent bond? *Science* 283 (5408): 1727–1730.

49 Wang, J., Kouznetsova, T.B., Niu, Z. et al. (2015). Inducing and quantifying forbidden reactivity with single-molecule polymer mechanochemistry. *Nat. Chem.* 7 (4): 323.

50 Wiita, A.P., Ainavarapu, S.R.K., Huang, H.H., and Fernandez, J.M. (2006). Force-dependent chemical kinetics of disulfide bond reduction observed with single-molecule techniques. *Proc. Natl. Acad. Sci. U.S.A.* 103 (19): 7222–7227.

51 Beedle, A.E.M., Mora, M., Davis, C.T. et al. (2018). Forcing the reversibility of a mechanochemical reaction. *Nat. Commun.* 9: 3155.

52 Pobelov, I.V., Lauritzen, K.P., Yoshida, K. et al. (2017). Dynamic breaking of a single gold bond. *Nat. Commun.* 8: 15931.

53 Evans, E. and Ritchie, K. (1997). Dynamic strength of molecular adhesion bonds. *Biophys. J.* 72 (4): 1541–1555.

54 Schönherr, H., Beulen, M.W.J., Bügler, J. et al. (2000). Individual supramolecular host–guest interactions studied by dynamic single molecule force spectroscopy. *J. Am. Chem. Soc.* 122 (20): 4963–4967.

55 Merkel, R., Nassoy, P., Leung, A. et al. (1999). Energy landscapes of receptor–ligand bonds explored with dynamic force spectroscopy. *Nature* 397 (6714): 50–53.

56 Friddle, R.W., Podsiadlo, P., Artyukhin, A.B., and Noy, A. (2008). Near-equilibrium chemical force microscopy. *J. Phys. Chem. C* 112 (13): 4986–4990.

57 Huang, Z., Xu, B., Chen, Y. et al. (2006). Measurement of current-induced local heating in a single molecule junction. *Nano Lett.* 6 (6): 1240–1244.

58 Ribas-Arino, J. and Marx, D. (2012). Covalent mechanochemistry: theoretical concepts and computational tools with applications to molecular nanomechanics. *Chem. Rev.* 112 (10): 5412–5487.

59 Beyer, M.K. (2000). The mechanical strength of a covalent bond calculated by density functional theory. *J. Chem. Phys.* 112 (17): 7307–7312.

60 Kean, Z.S., Gossweiler, G.R., Kouznetsova, T.B. et al. (2015). A coumarin dimer probe of mechanochemical scission efficiency in the sonochemical activation of chain-centered mechanophore polymers. *Chem. Commun.* 51 (44): 9157–9160.

61 Saitta, A.M., Soper, P.D., Wasserman, E., and Klein, M.L. (1999). Influence of a knot on the strength of a polymer strand. *Nature* 399 (6731): 46–48.

62 Stauch, T. and Dreuw, A. (2016). Knots "choke off" polymers upon stretching. *Angew. Chem. Int. Ed.* 55 (2): 811–814.

63 Patil, V.P., Sandt, J.D., Kolle, M., and Dunkel, J. (2020). Topological mechanics of knots and tangles. *Science* 367 (6473): 71–75.

64 Zhang, H., Chen, Y., Lin, Y. et al. (2014). Spiropyran as a mechanochromic probe in dual cross-linked elastomers. *Macromolecules* 47 (19): 6783–6790.

65 Celestine, A.D.N., Beiermann, B.A., May, P.A. et al. (2014). Fracture-induced activation in mechanophore-linked, rubber toughened PMMA. *Polymer (Guildf)* 55 (16): 4164–4171.

66 Clough, J.M., van der Gucht, J., Kodger, T.E., and Sprakel, J. (2020). Cephalopod-inspired high dynamic range mechano-imaging in polymeric materials. *Adv. Funct. Mater.* 30 (38), 2002716.

67 Chen, Y., Joshua Yeh, C., Qi, Y. et al. (2020). From force-responsive molecules to quantifying and mapping stresses in soft materials. *Sci. Adv.* 6 (20): 1–9.

68 Matsuda, T., Kawakami, R., Namba, R. et al. (2019). Mechanoresponsive self-growing hydrogels inspired by muscle training. *Science* 363 (6426): 504–508.

69 Shi, Z., Wu, J., Song, Q. et al. (2020). Toward drug release using polymer mechanochemical disulfide scission. *J. Am. Chem. Soc.* 142 (34): 14725–14732.

70 Huo, S., Zhao, P., Shi, Z. et al. (2021). Mechanochemical bond scission for the activation of drugs. *Nat. Chem.* 13 (2): 131–139.

71 Klein, I.M., Husic, C.C., Kovács, D.P. et al. (2020). Validation of the CoGEF method as a predictive tool for polymer mechanochemistry. *J. Am. Chem. Soc.* 142 (38): 16364–16381.

72 Davis, D.A., Hamilton, A., Yang, J. et al. (2009). Force-induced activation of covalent bonds in mechanoresponsive polymeric materials. *Nature* 459 (7243): 68–72.

73 Potisek, S.L., Davis, D.A., Sottos, N.R. et al. (2007). Mechanophore-linked addition polymers. *J. Am. Chem. Soc.* 129 (45): 13808–13809.

74 Kingsbury, C.M., May, P.A., Davis, D.A. et al. (2011). Shear activation of mechanophore-crosslinked polymers. *J. Mater. Chem.* 21 (23): 8381–8388.

75 Gossweiler, G.R., Hewage, G.B., Soriano, G. et al. (2014). Mechanochemical activation of covalent bonds in polymers with full and repeatable macroscopic shape recovery. *ACS Macro Lett.* 3 (3): 216–219.

76 Degen, C.M., May, P.A., Moore, J.S. et al. (2013). Time-dependent mechanochemical response of SP-cross-linked PMMA. *Macromolecules* 46 (22): 8917–8921.

77 Gossweiler, G.R., Kouznetsova, T.B., and Craig, S.L. (2015). Force-rate characterization of two spiropyran-based molecular force probes. *J. Am. Chem. Soc.* 137 (19): 6148–6151.

78 Beiermann, B.A., Davis, D.A., Kramer, S.L.B. et al. (2011). Environmental effects on mechanochemical activation of spiropyran in linear PMMA. *J. Mater. Chem.* 21 (23): 8443–8447.

79 Barbee, M.H., Kouznetsova, T., Barrett, S.L. et al. (2018). Substituent effects and mechanism in a mechanochemical reaction. *J. Am. Chem. Soc.* 140 (40): 12746–12750.

80 Zhang, H., Gao, F., Cao, X. et al. (2016). Mechanochromism and mechanical-force-triggered cross-linking from a single reactive moiety incorporated into polymer chains. *Angew. Chem. Int. Ed.* 55 (9): 3040–3044.

81 Versaw, B.A., McFadden, M.E., Husic, C.C., and Robb, M.J. (2020). Designing naphthopyran mechanophores with tunable mechanochromic behavior. *Chem. Sci.* 11 (17): 4525–4530.

82 Qian, H., Purwanto, N.S., Ivanoff, D.G. et al. (2021). Fast, reversible mechanochromism of regioisomeric oxazine mechanophores: developing in situ responsive force probes for polymeric materials. *Chem* 7 (4): 1080–1091.

83 McFadden, M.E. and Robb, M.J. (2019). Force-dependent multicolor mechanochromism from a single mechanophore. *J. Am. Chem. Soc.* 141 (29): 11388–11392.

84 Song, Y.K., Lee, K.H., Hong, W.S. et al. (2012). Fluorescence sensing of microcracks based on cycloreversion of a dimeric anthracene moiety. *J. Mater. Chem.* 22 (4): 1380–1386.

85 Kabb, C.P., O, Bryan, C.S., Morley, C.D. et al. (2019). Anthracene-based mechanophores for compression-activated fluorescence in polymeric networks. *Chem. Sci.* 10 (33): 7702–7708.

86 Konda, S.S.M., Brantley, J.N., Varghese, B.T. et al. (2013). Molecular catch bonds and the anti-hammond effect in polymer mechanochemistry. *J. Am. Chem. Soc.* 135 (34): 12722–12729.

87 Peterson, G.I., Lee, J., and Choi, T.L. (2019). Multimechanophore graft polymers: mechanochemical reactions at backbone-arm junctions. *Macromolecules* 52 (24): 9561–9568.

88 Sulkanen, A.R., Sung, J., Robb, M.J. et al. (2019). Spatially selective and density-controlled activation of interfacial mechanophores. *J. Am. Chem. Soc.* 141 (9): 4080–4085.

89 Göstl, R. and Sijbesma, R.P. (2016). π-Extended anthracenes as sensitive probes for mechanical stress. *Chem. Sci.* 7 (1): 370–375.

90 Li, J., Shiraki, T., Hu, B. et al. (2014). Mechanophore activation at heterointerfaces. *J. Am. Chem. Soc.* 136 (45): 15925–15928.

91 Li, J., Hu, B., Yang, K. et al. (2016). Effect of polymer grafting density on mechanophore activation at heterointerfaces. *ACS Macro Lett.* 5 (7): 819–822.

92 Stratigaki, M., Baumann, C., Van Breemen, L.C.A. et al. (2020). Fractography of poly(*N*-isopropylacrylamide) hydrogel networks crosslinked with mechanofluorophores using confocal laser scanning microscopy. *Polym. Chem.* 11 (2): 358–366.

93 Larsen, M.B. and Boydston, A.J. (2013). "Flex-activated" mechanophores: using polymer mechanochemistry to direct bond bending activation. *J. Am. Chem. Soc.* 135 (22): 8189–8192.

94 Larsen, M.B. and Boydston, A.J. (2014). Successive mechanochemical activation and small molecule release in an elastomeric material. *J. Am. Chem. Soc.* 136 (4): 1276–1279.

95 Hu, X., Zeng, T., Husic, C.C., and Robb, M.J. (2019). Mechanically triggered small molecule release from a masked furfuryl carbonate. *J. Am. Chem. Soc.* 141 (38): 15018–15023.

96 Akbulatov, S., Tian, Y., Huang, Z. et al. (2017). Experimentally realized mechanochemistry distinct from force-accelerated scission of loaded bonds. *Science* 357 (6348): 299–303.

97 Wang, J., Kouznetsova, T.B., Boulatov, R., and Craig, S.L. (2016). Mechanical gating of a mechanochemical reaction cascade. *Nat. Commun.* 7: 13433.

98 Hu, X., McFadden, M.E., Barber, R.W., and Robb, M.J. (2018). Mechanochemical regulation of a photochemical reaction. *J. Am. Chem. Soc.* 140 (43): 14073–14077.

99 Hsu, T.G., Zhou, J., Su, H.W. et al. (2020). A polymer with "locked" degradability: superior backbone stability and accessible degradability enabled by mechanophore installation. *J. Am. Chem. Soc.* 142 (5): 2100–2104.

100 Lin, Y., Kouznetsova, T.B., and Craig, S.L. (2020). Mechanically gated degradable polymers. *J. Am. Chem. Soc.* 142 (5): 2105–2109.

101 Barbee, M.H., Wang, J., Kouznetsova, T. et al. (2019). Mechanochemical ring-opening of allylic epoxides. *Macromolecules* 52 (16): 6234–6240.

102 Chen, Z., Mercer, J.A.M., Zhu, X. et al. (2017). Mechanochemical unzipping of insulating polyladderene to semiconducting polyacetylene. *Science* 357 (6350): 475–479.

103 Boswell, B.R., Mansson, C.M.F., Cox, J.M. et al. (2021). Mechanochemical synthesis of an elusive fluorinated polyacetylene. *Nat. Chem.* 13 (1): 41–46.

104 Mercer, J.A.M., Cohen, C.M., Shuken, S.R. et al. (2016). Chemical synthesis and self-assembly of a ladderane phospholipid. *J. Am. Chem. Soc.* 138 (49): 15845–15848.

105 Yang, J., Horst, M., Romaniuk, J.A.H. et al. (2019). Benzoladderene mechanophores: synthesis, polymerization, and mechanochemical transformation. *J. Am. Chem. Soc.* 141 (16): 6479–6483.

106 Yang, J., Horst, M., Werby, S.H. et al. (2020). Bicyclohexene- peri-naphthalenes: scalable synthesis, diverse functionalization, efficient polymerization, and facile mechanoactivation of their polymers. *J. Am. Chem. Soc.* 142 (34): 14619–14626.

107 Wang, F., Burck, M., and Diesendruck, C.E. (2017). Following homolytic mechanochemical kinetics with a pyrenyl nitrone spin trap. *ACS Macro Lett.* 6 (1): 42–45.

108 Ghosh, B. and Urban, M.W. (2009). Self-repairing oxetane-substituted chitosan polyurethane networks. *Science* 323: 1458–1460.

109 Amamoto, Y., Akamada, A., Otsuka, H. et al. (2011). Repeatable photoinduced self-healing of covalently cross-linked polymers through reshuffling of trithiocarbonate units. *Angew. Chem. Int. Ed.* 50: 1660–1663.

110 Yuan, C., Zhi Rong, M., Qiu Zhang, M. et al. (2011). Self-healing of polymers via synchronous covalent bond fission/radical recombination. *Chem. Mater.* 23 (22): 5076–5081.

111 Imato, K., Nishihara, M., Kanehara, T. et al. (2012). Self-healing of chemical gels cross-linked by diarylbibenzofuranone-based trigger-free dynamic covalent bonds at room temperature. *Angew. Chem.Int. Ed.* 51 (5): 1138–1142.

112 Frenette, M., Aliaga, C., Font-Sanchis, E., and Scaiano, J.C. (2004). Bond dissociation energies for radical dimers derived from highly stabilized carbon-centered radicals. *Org. Lett.* 6 (15): 2579–2582.

113 Imato, K., Kanehara, T., Ohishi, T. et al. (2015). Mechanochromic dynamic covalent elastomers: quantitative stress evaluation and autonomous recovery. *ACS Macro Lett.* 4 (11): 1307–1311.

114 Sumi, T., Goseki, R., and Otsuka, H. (2017). Tetraarylsuccinonitriles as mechanochromophores to generate highly stable luminescent carbon-centered radicals. *Chem. Commun.* 53 (87): 11885–11888.

115 Ishizuki, K., Oka, H., Aoki, D. et al. (2018). Mechanochromic polymers that turn green upon the dissociation of diarylbibenzothiophenonyl: the missing piece toward rainbow mechanochromism. *Chem. Eur. J.* 24 (13): 3170–3173.

116 Mori, Y., Yamada, N., Kanazawa, M. et al. (1996). Radical dissociation of 2,2′-bis(2-aryl-3-benzothiophenonyl)s by mechanical energy. *Bull. Chem. Soc. Jpn.* 69 (8): 2355–2359.

117 Sakai, H., Sumi, T., Aoki, D. et al. (2018). Thermally stable radical-type mechanochromic polymers based on difluorenylsuccinonitrile. *ACS Macro Lett.* 7 (11): 1359–1363.

118 Verstraeten, F., Göstl, R., and Sijbesma, R.P. (2016). Stress-induced colouration and crosslinking of polymeric materials by mechanochemical formation of triphenylimidazolyl radicals. *Chem. Commun.* 52 (55): 8608–8611.

119 Wu, Q., Yuan, Y., Chen, F. et al. (2020). Diselenide-linked polymers under sonication. *ACS Macro Lett.* 9 (11): 1547–1551.

120 Xia, J., Zhao, P., Pan, S., and Xu, H. (2019). Diselenide-containing polymeric vesicles with osmotic pressure response. *ACS Macro Lett.* 8 (6): 629–633.

121 Wang, Z., Ma, Z., Wang, Y. et al. (2015). A novel mechanochromic and photochromic polymer film: when rhodamine joins polyurethane. *Adv. Mater.* 27 (41): 6469–6474.

122 Karman, M., Verde-Sesto, E., and Weder, C. (2018). Mechanochemical activation of polymer-embedded photoluminescent benzoxazole moieties. *ACS Macro Lett.* 7 (8): 1028–1033.

123 Chen, Y., Spiering, A.J.H., Karthikeyan, S. et al. (2012). Mechanically induced chemiluminescence from polymers incorporating a 1,2-dioxetane unit in the main chain. *Nat. Chem.* 4 (7): 559–562.

124 Yang, F., Yuan, Y., Sijbesma, R.P., and Chen, Y. (2020). Sensitized mechanoluminescence design toward mechanically induced intense red emission from transparent polymer films. *Macromolecules* 53 (3): 905–912.

125 Ducrot, E., Chen, Y., Bulters, M. et al. (2014). Toughening elastomers with sacrificial bonds and watching them break. *Science* 344 (6180): 186–189.

126 Encina, M.V., Lissi, E., Sarasua, M. et al. (1980). Ultrasonic degradation of polyvinylpyrrolidone: effect of peroxide linkages. *J. Polym. Sci.: Polym. Lett. Ed.* 18 (12): 757–760.

127 Lin, Y., Barbee, M.H., Chang, C.-C., and Craig, S.L. (2018). Regiochemical effects on mechanophore activation in bulk materials. *J. Am. Chem. Soc.* 140 (46): 15969–15975.

128 Huheey, J. and Cottrell, T. (1958). *The Strengths of Chemical Bonds*. Waltham, MA: Butterworths.

129 Kersey, F.R., Yount, W.C., and Craig, S.L. (2006). Single-molecule force spectroscopy of bimolecular reactions: system homology in the mechanical activation of ligand substitution reactions. *J. Am. Chem. Soc.* 128 (12): 3886–3887.

130 Rodgers, M.T. and Armentrout, P.B. (2000). Noncovalent metal–ligand bond energies as studied by threshold collision-induced dissociation. *Mass Spectrom. Rev.* 19 (4): 215–247.

131 Skinner, H.A. and Connor, J.A. (1985). Metal–ligand bond-energies in organometallic compounds. *Pure Appl. Chem.* 57 (1): 79–88.

132 Sha, Y., Zhang, Y., Xu, E. et al. (2018). Quantitative and mechanistic mechanochemistry in ferrocene dissociation. *ACS Macro Lett.* 7 (10): 1174–1179.

133 Di Giannantonio, M., Ayer, M.A., Verde-Sesto, E. et al. (2018). Triggered metal ion release and oxidation: ferrocene as a mechanophore in polymers. *Angew. Chem. Int. Ed.* 57 (35): 11445–11450.

134 Zhang, Y., Wang, Z., Kouznetsova, T.B. et al. (2021). Distal conformational locks on ferrocene mechanophores guide reaction pathways for increased mechanochemical reactivity. *Nat. Chem.* 13 (1): 56–62.

135 Monsaert, S., Vila, A.L., Drozdzak, R. et al. (2009). Latent olefin metathesis catalysts. *Chem. Soc. Rev.* 38 (12): 3360–3372.

136 Piermattei, A., Karthikeyan, S., and Sijbesma, R.P. (2009). Activating catalysts with mechanical force. *Nat. Chem.* 1 (2): 133–137.

137 Ledoux, N., Allaert, B., Linden, A. et al. (2007). Bis-coordination of *N*-(alkyl)-*N′*-(2,6-diisopropylphenyl) heterocyclic carbenes to grubbs catalysts. *Organometallics* 26 (4): 1052–1056.

138 Ulman, M. and Grubbs, R.H. (1999). Ruthenium carbene-based olefin metathesis initiators: catalyst decomposition and longevity. *J. Org. Chem.* 64 (19): 7202–7207.

139 Küng, R., Pausch, T., Rasch, D. et al. (2021). Mechanochemical release of non-covalently bound guests from a polymer-decorated supramolecular cage. *Angew. Chem. Int. Ed.* 60 (24): 13626–13630.

140 Li, C.-H., Wang, C., Keplinger, C. et al. (2016). A highly stretchable autonomous self-healing elastomer. *Nat. Chem.* 8 (6): 618–624.

141 Brown, I.D. (2009). Recent developments in the methods and applications of the bond valence model. *Chem. Rev.* 109 (12): 6858–6919.

142 Yan, T., Schröter, K., Herbst, F. et al. (2017). What controls the structure and the linear and nonlinear rheological properties of dense, dynamic supramolecular polymer networks? *Macromolecules* 50 (7): 2973–2985.

143 Zhang, Q., Zhu, X., Li, C.-H. et al. (2019). Disassociation and reformation under strain in polymer with dynamic metal–ligand coordination cross-linking. *Macromolecules* 52 (2): 660–668.

144 Chen, Y., Kushner, A.M., Williams, G.A., and Guan, Z. (2012). Multiphase design of autonomic self-healing thermoplastic elastomers. *Nat. Chem.* 4 (6): 467.

145 Zhao, P.-C., Li, W., Huang, W., and Li, C.-H. (2020). A self-healing polymer with fast elastic recovery upon stretching. *Molecules* 25 (3): 597.

146 Kang, J., Son, D., Wang, G.N. et al. (2018). Tough and water-insensitive self-healing elastomer for robust electronic skin. *Adv. Mater.* 30 (13): 1706846.

147 Mozhdehi, D., Neal, J.A., Grindy, S.C. et al. (2016). Tuning dynamic mechanical response in metallopolymer networks through simultaneous control of structural and temporal properties of the networks. *Macromolecules* 49 (17): 6310–6321.

148 Sun, J.-Y., Zhao, X., Illeperuma, W.R.K. et al. (2012). Highly stretchable and tough hydrogels. *Nature* 489 (7414): 133–136.

149 Morelle, X.P., Illeperuma, W.R., Tian, K. et al. (2018). Highly stretchable and tough hydrogels below water freezing temperature. *Adv. Mater.* 30 (35): 1801541.

150 Guo, H., Han, Y., Zhao, W. et al. (2020). Universally autonomous self-healing elastomer with high stretchability. *Nat. Commun.* 11: 2037.

151 Urban, M.W., Davydovich, D., Yang, Y. et al. (2018). Key-and-lock commodity self-healing copolymers. *Science* 362 (6411): 220–225.

152 Wang, S. and Urban, M.W. (2020). Self-healing polymers. *Nat. Rev. Mater.* 5 (8): 562–583.

153 Yang, Y. and Urban, M.W. (2013). Self-healing polymeric materials. *Chem. Soc. Rev.* 42 (17): 7446–7467.

154 Zheng, N., Xu, Y., Zhao, Q., and Xie, T. (2021). Dynamic covalent polymer networks: a molecular platform for designing functions beyond chemical recycling and self-healing. *Chem. Rev.* 121 (3): 1716–1745.

155 Roy, N., Bruchmann, B., and Lehn, J.M. (2015). DYNAMERS: dynamic polymers as self-healing materials. *Chem. Soc. Rev.* 44 (11): 3786–3807.

156 Wojtecki, R.J., Meador, M.A., and Rowan, S.J. (2011). Using the dynamic bond to access macroscopically responsive structurally dynamic polymers. *Nat. Mater.* 10 (1): 14–27.

157 Zhao, P., Huo, S., Fan, J. et al. (2021). Activation of the catalytic activity of thrombin for fibrin formation by ultrasound. *Angew. Chem. Int. Ed.* 60 (26): 14707–14714.

158 Zhang, M., Xu, D., Yan, X. et al. (2012). Self-healing supramolecular gels formed by crown ether based host–guest interactions. *Angew. Chem.* 124 (28): 7117–7121.

159 Miyamae, K., Nakahata, M., Takashima, Y., and Harada, A. (2015). Self-healing, expansion–contraction, and shape-memory properties of a preorganized supramolecular hydrogel through host–guest interactions. *Angew. Chem. Int. Ed.* 54 (31): 8984–8987.

160 Hippler, M., Weißenbruch, K., Richler, K. et al. (2020). Mechanical stimulation of single cells by reversible host–guest interactions in 3D microscaffolds. *Sci. Adv.* 6 (39): eabc2648.

161 Sugawara, A., Asoh, T.-A., Takashima, Y. et al. (2021). Mechano-responsive hydrogels driven by the dissociation of a host–guest complex. *ACS Macro Lett.* 10 (7): 971–977.

162 Bhangu, S.K., Bocchinfuso, G., Ashokkumar, M., and Cavalieri, F. (2020). Sound-driven dissipative self-assembly of aromatic biomolecules into functional nanoparticles. *Nanoscale Horiz.* 5 (3): 553–563.

163 Self, J.L., Sample, C.S., Levi, A.E. et al. (2020). Dynamic bottlebrush polymer networks: self-healing in super-soft materials. *J. Am. Chem. Soc.* 142 (16): 7567–7573.

164 Jo, Y.H., Li, S., Zuo, C. et al. (2020). Self-healing solid polymer electrolyte facilitated by a dynamic cross-linked polymer matrix for lithium–ion batteries. *Macromolecules* 53 (3): 1024–1032.

165 Pan, Y., Zhang, H., Xu, P. et al. (2020). A mechanochemical reaction cascade for controlling load-strengthening of a mechanochromic polymer. *Angew. Chem. Int. Ed.* 59 (49): 21980–21985.

166 Wang, Z., Wang, J., Ayarza, J. et al. (2021). Bio-inspired mechanically adaptive materials through vibration-induced crosslinking. *Nat. Mater.* 20: 869–874.

167 Gao, Q., Kopera, B.A.F., Zhu, J. et al. (2020). Breathable and flexible polymer membranes with mechanoresponsive electric resistance. *Adv. Funct. Mater.* 30 (26): 1907555.

168 Ashizawa, M., Zheng, Y., Tran, H., and Bao, Z. (2020). Intrinsically stretchable conjugated polymer semiconductors in field effect transistors. *Prog. Polym. Sci.* 100: 101181.

169 Yao, B., Hong, W., Chen, T. et al. (2020). Highly stretchable polymer composite with strain-enhanced electromagnetic interference shielding effectiveness. *Adv. Mater.* 32 (14): 1907499.

170 Kim, D.C., Shim, H.J., Lee, W. et al. (2020). Material-based approaches for the fabrication of stretchable electronics. *Adv. Mater.* 32 (15): 1902743.

2

Photoresponsive Polymers

Hossein Roghani-Mamaqani and Zeinab Tajmoradi

Sahand University of Technology, Faculty of Polymer Engineering, P.O. Box: 51335-1996, Tabriz, Iran

2.1 Introduction

Smart polymers change their physical or chemical characteristics by induction of different physical (temperature, light, electrical potential, mechanical stress, and magnetite field), chemical (pH, redox, ion, and gas), and biological (glucose and enzyme) stimuli. Inherent smart polymers respond to a stimulus without incorporation of external moieties in their matrix. In this context, temperature-responsive polymers show variation in their solubility by phase behavior upon variation of temperature. In addition, pH-responsive polymers, which contain either weakly acidic or basic groups in their molecular structure, show variation in the charge, hydrophilicity/hydrophobicity, and volume in response to changes in the pH of the surrounding medium. Another class of smart polymers is prepared by incorporation of stimuli-responsive compounds into the polymer matrix. Most of the light-responsive polymers are prepared by introduction of photoresponsive moieties to the polymer matrices. Incorporation of light-responsive moieties in inherent stimuli-responsive polymers yields dual- or multi-responsive polymers.

Light-responsive polymers are among the main categories of smart materials, which have received large attention due to specific physical characteristics of light. As a nondestructive stimulus, light has low cost, quick responsivity, high accessibility, diffusibility in most of the media, and high controllability out of the system, which can be localized in time and space. Induction of light could be resulted in regulation of different molecular characteristics, including isomerization, conformation, polarity, amphiphilicity, optical chirality, and conjugation, which finally caused macroscopic variation of polymer characteristics, including shape, wettability, solubility, optical properties, conductivity, and adhesion. Light-responsive polymers are divided into different categories based on the type of light-sensitive moieties incorporated into the polymer matrix. Photoresponsive polymer composites are prepared by the addition of light-responsive organic/inorganic particles, such as carbon dots (CDs) into the polymeric matrix [1, 2]. Polymer dots (PDs), as organic semiconductor nanomaterials, consist of semiconducting π-conjugated

Smart Stimuli-Responsive Polymers, Films, and Gels, First Edition.
Edited by Liang Hu, Yongfeng Gao, and Michael J. Serpe.

organic polymers with sizes in the range of 1–100 nm [3, 4]. Photoresponsive polymers are prepared by chemical incorporation of organic photoresponsive moieties to the polymeric backbone or as a pendant group [5–9]. Photoresponsive polymers can show substantial changes in their chemical and/or physical properties, such as conformation, polarity, charge, amphiphilicity, conjugation, or optical chirality in response to light irradiation in the solution or solid states [10].

Depending on the type of the photosensitive moiety incorporated into the polymer structure, responsivity of these polymers could be either reversible, irreversible, or dynamic. Therefore, the kind of photosensitive moiety introduced in the polymer and its response to light irradiation could be highly important [7, 9]. Reversible photoresponsive behavior, including cis–trans isomerization, ionic dissociation, hydrogen transfer, or ring formation/cleavage, could generally be observed in polymers having reversible molecular switches, such as spiropyran (SP), spirooxazine, azobenzene, coumarin, diarylethene, oxazine, fulgide, and stilbene. Spiropyran and spirooxazine are highly attractive photoresponsive compounds because of their facile synthesis methods and unique properties, such as photostability and high reversibility, compatibility with polymer matrices, and multi-responsivity. Such isomerizations could be accompanied by variation in the physical or chemical characteristics of the photoresponsive compounds and consequently the polymer chain, including dipole moment or geometrical structural changes. Upon light exposure, the irreversible photoresponsive materials, such as the photolabile *ortho*-nitrobenzyl ester (ONB), coumarin-4-ylmethyl derivatives, pyrenylmethyl, phenacyl esters, cinnamate derivatives, and 2-naphthoquinone-3-methides groups, are cleaved from the polymer structure into a more polar species through an irreversible transformation [9]. Incorporation of photoinduced dynamic covalent bonds into polymers backbone also results in smart polymers with stimuli-controlled structural transformation and self-healing properties [11, 12]. Another type of photoresponsive behavior in smart polymers is reversible or irreversible photocross-linking reactions based on [4π+4π] or [2π+2π] cycloaddition mechanisms, such as photodimerization of coumarin, anthracene, cinnamic acid, stilbene, and thymine [9, 13, 14]. The reversible, irreversible, and dynamic photoresponsive compounds are mainly applied in various applications, such as smart polymeric inks [15, 16], chemical or mechanical sensors [17–20], photopatterning [21–23], high-resolution bioimaging [24–26], surface-active agents [27–31], viscosity modifiers [32–37], self-healing polymers [38–40], shape-memory polymers (SMPs) [41–46], photoconductive polymers [47–49], switchable fluorescent polymer particles [50, 51], drug-delivery systems [52–56], membranes [57, 58], films [59, 60], and textile [61, 62].

Photoresponsive polymers and their applications have widely been studied in different views. Aguilar et al. described different photoresponsive polymers, their light-induced molecular responses, and also their applications [9]. Photocross-linking by formation of cyclobutane ring, dimerization, and ionic mechanism, photoisomerization reaction in polymeric materials, and photoconducting polymers were also discussed by A. Ravve [13]. Bertand and Gohy discussed photoresponsive polymers, the synthesis methods, and their applications [5]. Preparation, properties, and applications of photochromic materials were also described by Tian and Zhang [63]. The progress in the field of synthetic smart polymers combining light and temperature responsivities was investigated by Theato and coworker [64]. Applications of photoresponsive polymers in membrane [14, 65], micro and nanocapsules [66], surface modification and fluorescence modulation [67], smart coating [68], and smart textile [69] were also reported. Light-responsive polymers were also discussed in different views [66, 70–74]. Abdollahi et al. highlighted light-controlling of temperature-responsivity in stimuli-responsive polymers [75], the advances of stimuli-chromism of photoswitches in smart polymers [7], and photoluminescent and chromic nanomaterials in anticounterfeiting technologies [8]. The present chapter introduces the main types of photoresponsive polymers and the mechanism of their photoreactivity. The photoresponsive polymers that incorporate organic photoresponsive molecules are discussed in Section 2.2, and the main applications of these systems are presented in Section 2.3. Finally, critical insights in this field and its future advances are discussed in Section 2.4.

2.2 Photoresponsive Polymers

2.2.1 Photoluminescent Polymers

Photoluminescence is the emission of light radiation from photoactive materials, which arises from a radiative transition between a higher electronic excited state and a lower energy state that is induced by the absorption of a photon of light. After the initial photon absorption, the photoluminescent molecule undergoes internal energy redistribution, and the emitted photon (light) is of longer wavelength or lower energy than the initially absorbed photon. Depending on the nature of the excited state, photoluminescent materials are formally divided into fluorescence and phosphorescence categories. In both fluorescence and phosphorescence, the emitted photon (light) has lower energy than the absorbed photon and emission occurs at a longer wavelength than the incident light. The main difference is the duration for each manner to occur after the initial absorption of light. Fluorescence occurs much more quickly than phosphorescence and does not change the spin direction of the electrons. The glow of fluorescence emission is switched off when the source of excitatory radiation is turned off. On the other hand, an afterglow with durations of fractions of a second up to hours could be observed for the phosphorescence process [76–79]. Fluorescent and phosphorescent polymers are a class of photoluminescent materials that have high importance in many applications, such as anticounterfeiting and authentication technologies, sensor development, security inks document (invisible marking on postage stamps and currency), cell labeling, drug delivery, and most importantly high-resolution bioimaging [8]. The fluorescent polymers incorporating an organic fluorophore could immediately show bright emission in various colors followed by absorption of electromagnetic radiation (light of current wavelength, or a group of wavelengths), and the glow of fluorescence emission is switched off by turning the light source off. Similar to the fluorescent polymers, the phosphorescent polymers incorporating organic phosphorescent compounds could immediately show bright emissions in different colors followed by absorption of electromagnetic radiation and have a wide range of applications in anticounterfeiting/authentication technologies with high security and also sensors [8, 80]. This section is an attempt to present some of these luminescent materials and also understand the discussions on photoluminescent polymers.

2.2.1.1 Fluorescent Polymers

Fluorescence emission process requires a conjugated chemical structure substituted by electron-donating and electron-withdrawing functionalities, electron resonance, and suitable electron density. Meanwhile, chemical or physical parameters, such as light intensity and wavelength, nature of the substituted groups, conjugation length in the chemical structure, particle size in the case of colloids, transparency or turbidity of the solutions, and concentration and polarity of the surrounding media (interactions between the fluorescent molecules with solvent molecules or substrate) could affect the intensity and wavelength of the fluorescence emission. The fluorescent materials were classified into six categories of upconverting nanoparticles (UCNPs), quantum dots (QDs), luminescent CDs, PDs, organic fluorescent dyes, supramolecular structures, and rare-earth metal complexes [8]. UCNPs are inorganic crystalline photoluminescent nanoscale particles (diameter 1–100 nm) composed of lanthanide or actinide ion-doped transition metals, which could absorb two or more incident photons of relatively low energy and convert them into one emitted photon with higher energy. Generally, absorption occurs in the near infrared (NIR), while upconverting/emission occurs in the visible or ultraviolet wavelength regions of the electromagnetic spectrum. QDs are zero-dimensional crystalline semiconducting fluorescent nanoparticles with a small size in the range of 2–10 nm (10–50 atoms), which exhibit size and composition-dependent optical and electronic (optoelectronic) properties. In the presence of UV light irradiation, an electron in the semiconductor QD can be excited to a state of higher energy corresponding to the transition of an electron from the valence band to the conductance band. The excited electron can be emitting its energy in the form of fluorescence as the electron reverts from the conductance band to the valence band. QDs are mainly applied to display and illumination systems because of their unique properties, such as high quantum yield and also narrow and tunable emitting spectra [81–84]. Luminescent CDs are novel zero-dimensional carbon-based fluorescent nanomaterials in the size range of a few tens of nanometers, which can be doped with N, S, P, and B heteroatoms. They have attracted broad research interest because of their diverse physicochemical properties, such as unique optical properties, small size, relatively strong fluorescence characteristics, high quantum yield, electron mobility, photobleaching resistance, biocompatibility, high water solubility, low cost, eco-friendliness, abundant functional groups (e.g. amino, hydroxyl, carboxyl, and aldehyde), and high stability toward environmental degradation. When the CDs are illuminated by UV light, an electron in the CD can be excited to a state of higher energy and then moved between the highest occupied molecular orbital (HOMO) and lowest unoccupied molecular orbital (LUMO) accompanied by fluorescence emission. The energy gap between the HOMO and LUMO orbitals strongly depends on the incorporated oxygen content on the CDs surface indicating the excitation and emission wavelengths and corresponding color of the fluorescence emission [85–87]. PDs are novel organic semiconductor nanomaterials containing semiconducting π-conjugated organic polymers with sizes in the range of 1–100 nm. Excitation and emission wavelengths and corresponding color of the fluorescence emission depend on chemical structure of the conjugated polymer. Consequently, PDs have been combined with other QDs and nanoparticles to form advanced hybrid nanomaterials. Compared with the traditional organic small molecules, semiconducting QDs, organic dyes, and organic–inorganic rare-earth metal complexes, the PDs have significant advantages, such as highly bright photoluminescence, stable optical properties, high stability toward environmental degradation, improved photostability, easy surface modification, high biocompatibility, and low toxicity, which exhibit significant potential applications in biological imaging, sensing and detection, and drug-delivery applications [3, 4]. Organic dyes (organic fluorescent dyes) are a significant class of the photoluminescent materials, because of their distinct characteristics, such as conjugated chemical structure, electron-donating or electron-withdrawing substituted groups, and electron resonance. The fluorescence emission of organic dyes (fluorophore) typically originates either from intramolecular charge-transfer transitions (refer to as CT dyes) or from an optical transition delocalized over the whole fluorophore (refer to as resonant dyes because of their resonance emission). CT dyes, such as coumarins, have well-separated, broader, and structureless absorption and emission bands in polar solvents, and show a strong polarity dependence on their spectroscopic properties. Resonant dyes, such as fluorescein, rhodamine, and most cyanines, are characterized by slightly structured and comparatively narrow absorption and emission bands that often mirror each other. In addition, their molar absorption coefficients, and in most cases also their fluorescence quantum yields, are generally higher than those of dyes with intramolecular charge-transfer transitions [8, 88]. The optical properties and the color of fluorescence emission of the fluorescent compound are dependent on the change in their distinct characteristics. As shown in the Jablonski diagram for the organic molecules (Figure 2.1), when a certain compound absorbs a photon of light, a series of photophysical events occur, such as excitation of an electron to a higher vibrational energy state, loss of energy in the absence of light emission (internal conversion or vibrational relaxation), singlet–triplet intersystem crossing (ISC) (from the lowest excited singlet state (S1) to a triplet state (T*n*)), which forms the basis for phosphorescence of organic luminophores, loss of energy again until the lowest energy level of the first exited state, and the excited electron relaxes to its lowest energy excited vibrational state (emits the energy in the form of fluorescence or phosphorescence depending upon the path required to relax) [90–95]. In the case of fluorophores, the electron is excited from the lowest vibrational level in its ground state (its most stable state) to a higher vibrational energy state. Generally, this process occurs with a $\pi \rightarrow \pi^*$ electronic transition within a time scale of about 10^{-15} seconds (femtosecond). From there, the fluorophore compound emits its energy in the form of fluorescence as the electron reverts from the first excited singlet state (S1) to various vibrational levels of the ground state (S0), and fluorescence emission occurs within a time frame of 10^{-7}–10^{-9} seconds

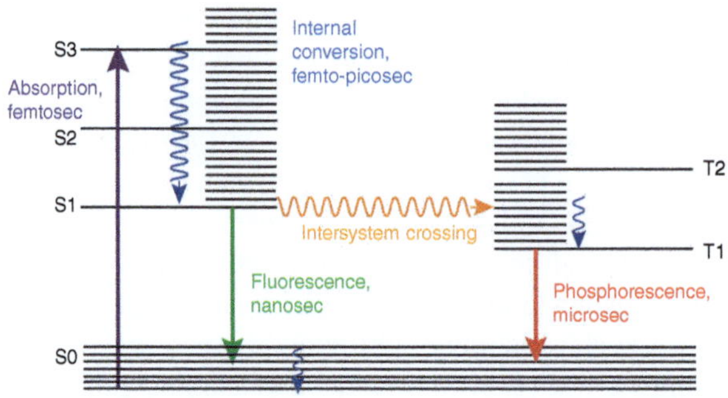

Figure 2.1 Jablonski diagram of photosynthetic process leading to fluorescence emission [89]. Source: Reproduced with permission from Berezin and Achilefu [89]. Copyright 2010 Royal Society of Chemistry (RSC).

(nanoseconds). From this level, the energy gap between the HOMO and the LUMO indicates the excitation and emission wavelengths and corresponding color of the fluorescence emission due to light absorption in terms of the absorption wavelength [77, 78, 89, 96].

The fluorescent polymers containing organic fluorophore species could immediately show bright emissions in various colors followed by absorption of electromagnetic radiation, and the glow of fluorescence emission is switched off by turning the light source off. The fluorescence emission of these types of polymers and their color is dependent on the absorbed energy or light photons from the source of excitatory radiation, which is a function of electron resonance and conjugation. These types of smart polymers have a wide range of applications in cell labeling, anticounterfeiting, rewritable patterning, drug delivery, sensors, fingerprint identification, and most importantly high-resolution bioimaging [8, 96, 97]. Some of the most general organic fluorescent compounds (fluorophores), such as coumarin derivatives, cyanine derivatives, perylene and its derivatives, fluorescein, rhodamine, pyrene, naphthalimide, and nitrobenzoxadiazole, which are commonly used in this class of fluorescent polymers, are presented in Table 2.1 [8, 138, 139]. Physical and chemical characteristics of these fluorescent dyes are dependent on their intermolecular or intramolecular interactions and their interactions with the solvent molecules or the surrounding media. The visual effect provided by a fluorophore, whether in solution or in a polymer matrix, is due to the combination of two special effects: (i) the base color originates from the wavelengths of the light transmitted or reflected as a result of selective light absorption and (ii) the color assigned to the wavelengths of the emitted light by fluorescence [97].

2.2.1.2 Phosphorescent Polymers

Phosphorescence is an important type of photoluminescence phenomena with an emission mechanism different than fluorescence. The difference was made by visual observation in response to light irradiation. Fluorescent compounds stopped emitting immediately when the excitatory radiation was turned off, while phosphorescent compounds had a lasting visible afterglow emission for

Table 2.1 Chemical structure and spectroscopic characteristics for some of the most general organic fluorescent compounds.

Name	Chemical structure	Absorbance wavelength (nm)	Emission wavelength (nm)	References
4-Hydroxycoumarin	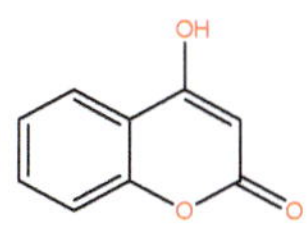	325	386	[98, 99]
7-Methoxycoumarin	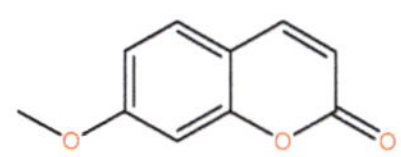	360	410	[98, 100]
7-Hydroxy-4-methyl coumarin	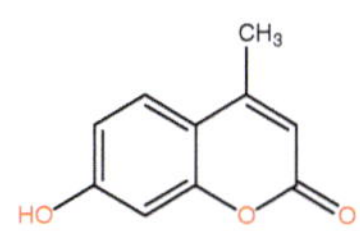	310	450	[101–108]
4-Aminocoumarin	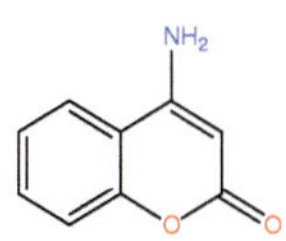	350	445	[109, 110]
Pyrene	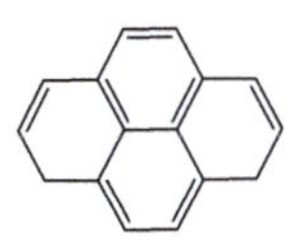	490	375, 385	[111, 112]
Nitrobenzoxadiazole (NBD)	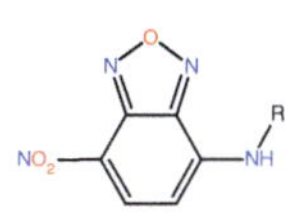	450	530	[113–115]
Cyanine 3	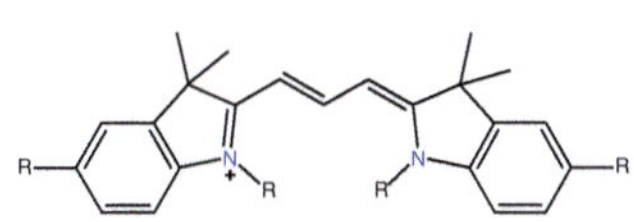	550	570	[116–118]
Cyanine 5	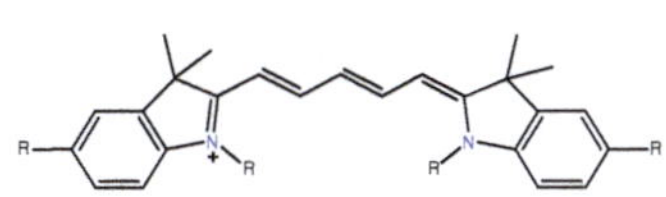	650	670	[119, 120]
Borondipyrromethene (BODIPY)	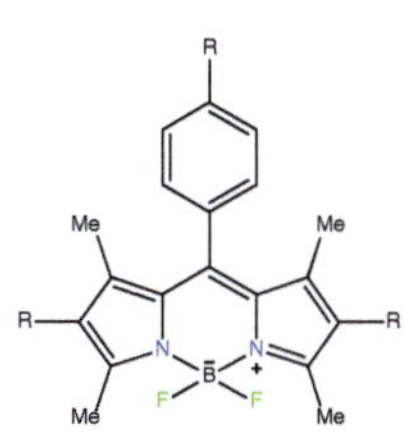	520	550	[121–124]
Pyrylenediimide	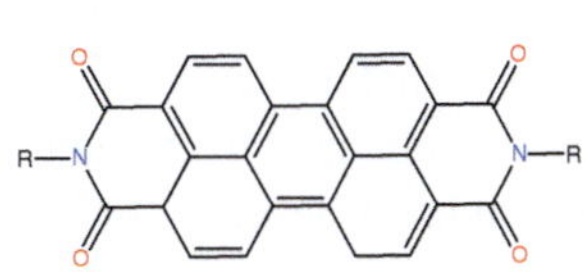	450	535	[125–128]

(Continued)

Table 2.1 (Continued)

Name	Chemical structure	Absorbance wavelength (nm)	Emission wavelength (nm)	References
Fluorescein	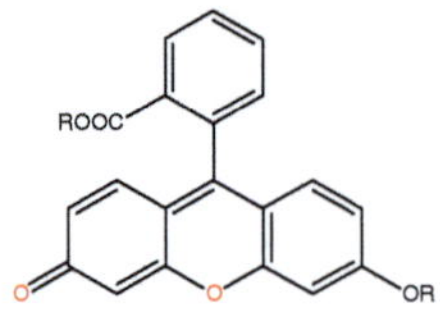	460	530	[129, 130]
Rhodamine	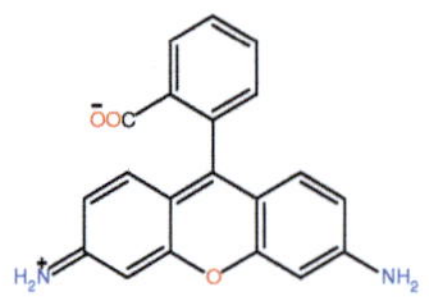	550	580	[131, 132]
Naphthalimide	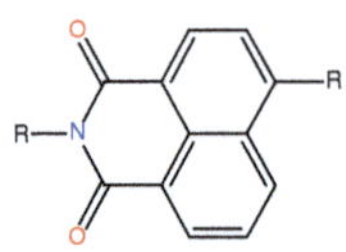	420	495	[133–136]
Azocarbazole	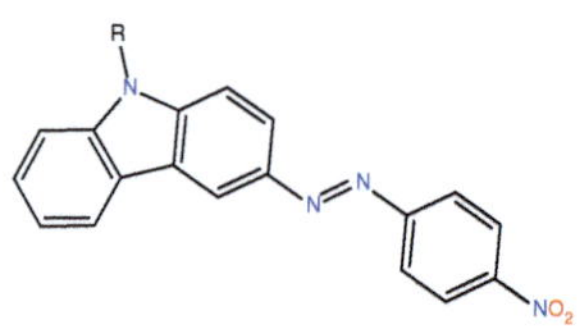	430–440	530	[137]

seconds, minutes, or hours after excitation attributed to the forbidden energy state transitions in quantum mechanics [76, 79]. However, to produce phosphorescence emission, it is essential to have an effective nonradiative singlet-triplet ISC between the isoenergetic excited singlet state (S1) and triplet state (T*n*). Phosphorescent materials with long-persistent luminescence (afterglow) have gained increasing interest due to their wide applications in sensing, bioimaging, and data encryption [140–142]. The phosphorescence phenomena could be observed for all of the fluorescent materials discussed in the Section 2.2.1.1, such as organic compounds, PDs, QDs, UCNPs, and organic–inorganic complexes [8]. In general, production of the phosphorescence phenomena includes two critical processes of ISC from the lowest excited energy state (usually S1) to another state with a higher spin multiplicity (usually T*n* ($n > 1$)) and radiative transition from the lowest excited T1 to the ground state (S0) (Figure 2.1). Nevertheless, the emission from an excited T1 could easily be quenched at room temperature under ambient conditions. To date, most phosphorescent materials are limited to inorganic and organometallic complexes, such as Pt^{2+}, Ir^{3+}, or Au^{+} compounds, which inevitably lead to high cost, high toxicity, limited resources, and difficulty of preparation owing to the existence of noble metals. Therefore, it is highly desirable to develop metal-free organic materials pure-organic room-temperature phosphorescence luminophores, including persulfurated derivatives, dibenzothiophene and related sulfur-containing heteroaromatic derivatives, carbazole and related nitrogen-containing heteroaromatic derivatives, fluorene derivatives, sulfur–nitrogen-containing heteroaromatic derivatives, dibenzothiophene and related sulfur-containing heteroaromatic derivatives, borate derivatives, and polyacid derivatives with ultralong phosphorescence due to their advantages, such as electroconductivity, diverse molecular architectures, low cost, relatively environment-friendly, and also easy synthesis and modification. There is still a major challenge to suppress nonradiative transition derived from molecular motions and enhance ISC from the lowest excited S1 to the excited T*n* for developing metal-free organic phosphorescence with long-lived emission under ambient conditions. Therefore, to obtain room-temperature phosphorescence with long-persistent luminescence (afterglow), a set of possible approaches, such as crystallization engineering, doping in rigid matrices, CDs formation, metal–organic framework (MOF) construction, H-aggregation, and many other strategies have been offered [95, 143]. These room-temperature phosphorescent materials were mostly limited to crystalline systems or doped composites, which significantly hindered their applications that is mainly due to the stringent requirement of the formation of crystalline state and poor processability for small organic compounds or unavoidable phase separation in host–guest systems. To improve this limitation and also to reduce nonradiative transitions from the chromophore motions and restrict deactivation of triplet excitations under ambient conditions (oxygen, humidity, and thermal motion), polymer-based phosphorescent materials have been offered. These polymeric systems have long-lived phosphorescence under ambient conditions (in air and at room temperature) owing to their intriguing features, such as good processability, high-thermal stability, self-standing transparent property, flexibility, and stretchability, which have great potential in flexible lighting and displays [9, 93, 143, 144]. Generally, there are two ways to achieve this, which are (i) noncovalent incorporating phosphor molecules into polymer matrices via physical encapsulation and (ii) covalent connecting potential phosphor groups into polymers through chemical modification. In the first way, small purely organic phosphors are embedded in the rigid environment of a polymer matrix with high transparency and strong hydrogen bonding capability, such as poly(lactic acid), poly(vinyl alcohol), and poly(methyl methacrylate). In a second way, functionalized organic phosphors are coupled with polymer chains to produce room-temperature phosphorescence emission, in which the phosphor groups could act as side chains, main chains, or terminal groups [95, 144, 145]. Phosphorescent polymers incorporating the metal-free organic phosphors could immediately show bright emissions in various colors followed by absorption of electromagnetic radiation. They have played a distinctive role in the field of organic flexible electronics, such as solar cells, organic light-emitting diodes (LEDs), memory devices, and field-effect transistors. They have a wide range of applications in anticounterfeiting/authentication technologies with high security, which are mainly due to their unique advantages, including easy processing, low cost, high flexibility and stretchability, and even fantastic electron mobility and thermal conductivity.

Table 2.2 Chemical structure and some of the most general organic phosphorescent compounds.

Phosphorescent group	Chemical structure	References
Difluoroboron dibenzoylmethane	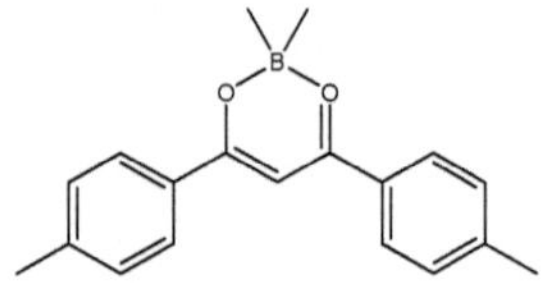	[95]
2-Bromo-5-hydroxybenzaldehyde derivative	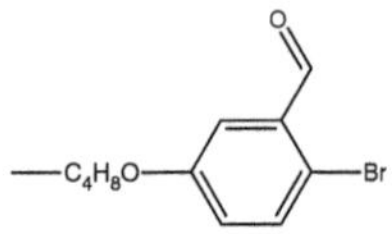	[95, 146]
4-Bromo-1,8-naphthalic anhydride	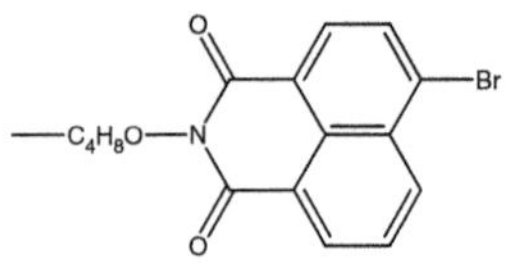	[93, 95, 147]
α-Bromonaphthalene derivative	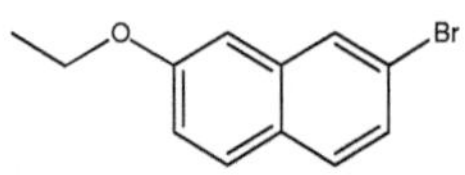	[93, 95, 147]
2,5-Dihexyloxy-4-bromobenzaldehyde (Br6A)	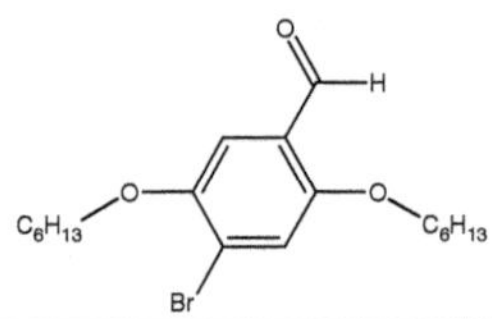	[92, 93, 95, 147]

Table 2.2 shows some classical phosphor groups incorporated into polymers, such as difluoroboron dibenzoylmethane, naphthalene imide, 4-bromobenzaldehyde, 2-bromo-5-hydroxybenzaldehyde derivative, 4-bromo-1,8-naphthalic anhydride, and α-bromonaphthalene derivatives [92, 93, 95, 146, 148]. These phosphor groups act as guests, main chains, side chains, or terminal groups of the resultant polymers, and their intramolecular motions could also be controlled positively and the room-temperature phosphorescence effect could effectively be realized [92, 95].

2.2.2 Photochromic Polymers

The word "photochromic" or "photochromism" has been entered in numerous dictionaries, which originates from the two Greek words of "photos" and "chroma" meaning light and color, respectively. Photochromism is an optical phenomenon in photoresponsive materials related to light-induced reversible color change in photochromic molecule due to the light and heat effects, where the photochromic molecule undergoes isomerization between two chemical structures which have different absorption spectra [7, 8, 149, 150]. The photoswitchable isomerization of these photochromic materials is accompanied by a change in the conformation, leading to molecular changes in physical properties, such as polarity as well as a color change in the chromophoric units, viscosity, chemical structure, conjugation, absorbance and fluorescence emission, and macroscopic changes in the material properties, such as wettability, thickness, and stability. Such phenomena can be observed in chemical compounds, such as spiropyran, spirooxazine, spirothiopyran, azobenzene, benzopyran, naphthopyran, oxazine, oxazolidine, diarylethene, fulgide, stilbene, hemithioindigo, viologen, chromene, salicylideneaniline, dinitrobenzylpyridine (DNBP), triphenylimidazole, and over-crowed

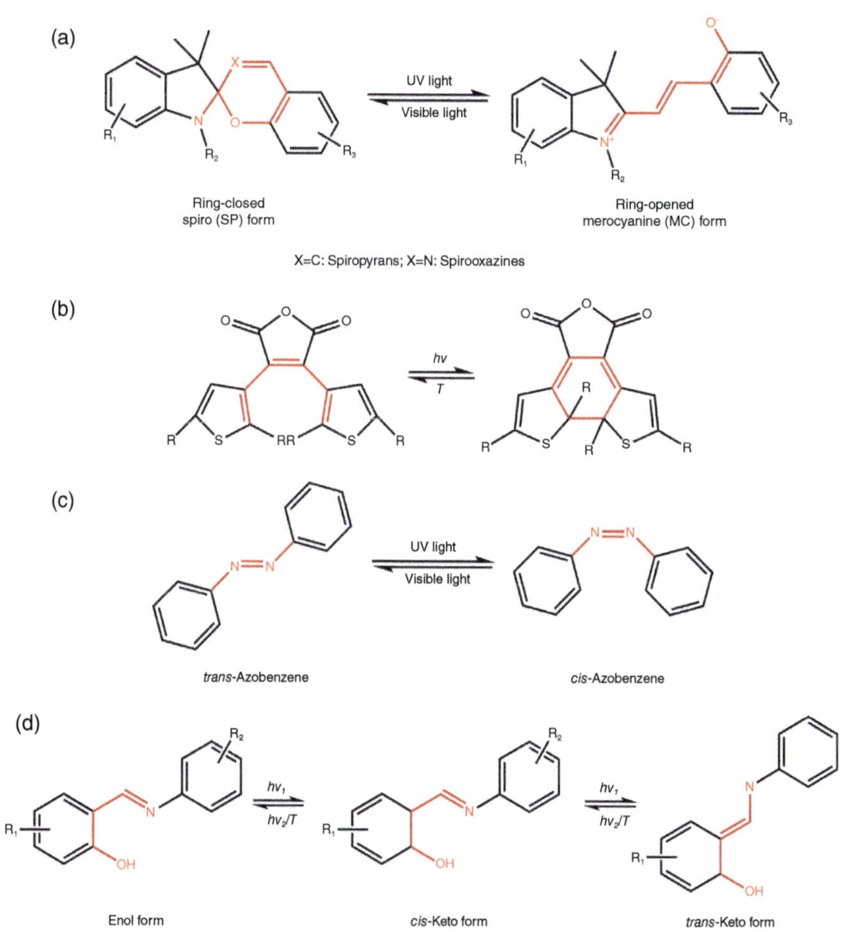

Figure 2.2 (a) Ring formation and cleavage of spiropyran, (b) photochromism of diarylethenes by closing the six-membered ring in the core of the antiparallel rotamer, (c) photoisomerization of azobenzene, (d) photochromism (phototautomerism) of anils, the enol-to-keto reaction via intramolecular proton transfer, (e) photoreaction of dinitrobenzylpyridine (DNBP) between the CH and NH forms, involving the OH species through intramolecular proton-transfer reaction, (f) redox in viologen, and (g) photochromism of hexaarylbiimidazole (HABI) through hemolytic cleavage reaction.

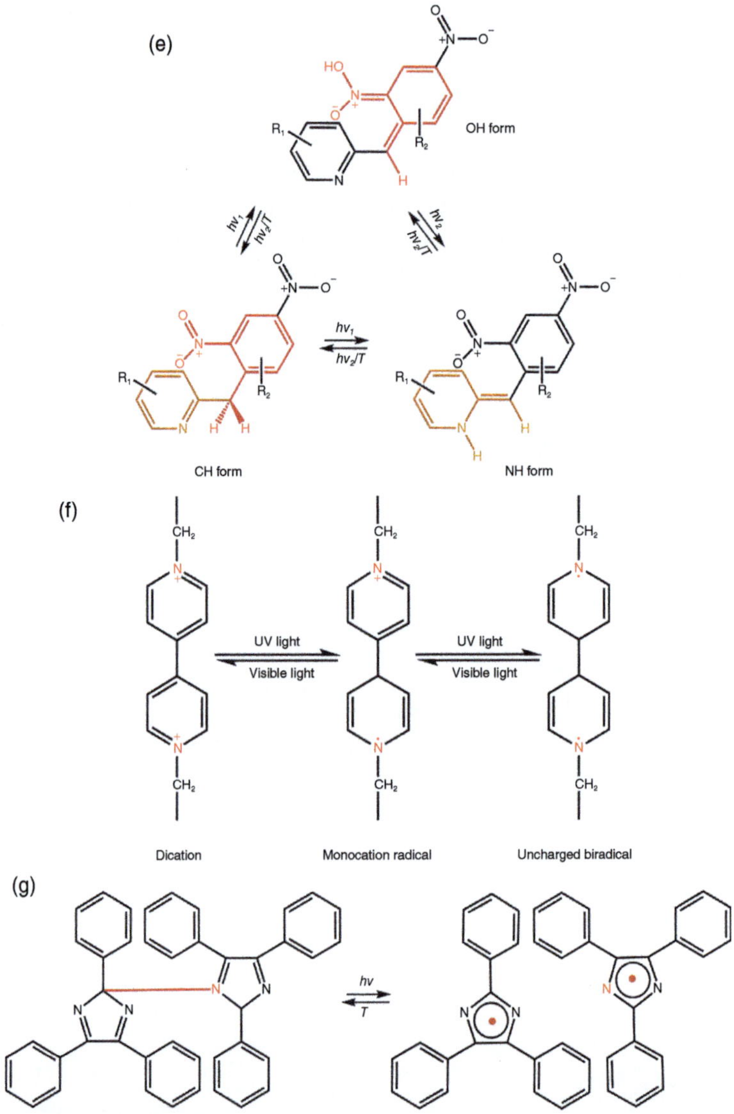

Figure 2.2 (*Continued*)

alkene, as presented in Table 2.3 [8, 63, 66]. Depending on the structure of the chromophores, the light-induced molecular change is classified into various categories, i.e. cyclization reaction, cis–trans isomerization, redox reaction, ionic dissociation, intramolecular hydrogen transfer (proton transfer) or group transfer, and radical formation (homolytic cleavage) [63, 67]. In some of these photochromic systems, the reaction reverts from products back to the starting reactants through a thermally induced reaction (named *T-type*), such as in spiropyran, spirooxazine, azobenzene, stilbene, chromene, salicylideneaniline, triphenylimidazole, and fulgide. However, other systems, including diarylethenes and furylfugides or indolinofulgides, show no change in the absence of light (named *P-type*) [210–212].

As an example of photocyclization reaction, spiropyrans show chromism and fluorescence emission upon light irradiation by cleavage of the spiro C—O bond via photoisomerization of the nonpolar, nonconjugated, ring-closed, discolored, and nonfluorescent spiropyran (SP) form to the zwitterionic, conjugated, ring-opened, colored and fluorescent merocyanine (MC) form, as shown in Figure 2.2a. In this process, the hydrophobic spiro-heterocycle isomer is converted into the hydrophilic zwitterionic merocyanine upon UV light irradiation, while the reverse process is triggered by visible light irradiation (620 nm) [21, 22]. In the case of spiropyran and other organic photochromic compounds with a similar chemical structure, responsivity, and photoisomerization mechanism, such as spirooxazines, their color and fluorescence emission could considerably be affected by characteristics of the surrounding media, owing to the zwitterionic nature of the MC molecule [7, 8]. By incorporation into polymer chains in the backbone or as the side chain, spiropyran, spirooxazine, spirothiopyran, and the related compounds and derivatives can enable switching of wettability, optical properties, including absorption and emission spectra, solubility, and ion recognition in polymer materials. In the case of diarylethenes and fugides, as shown in Figure 2.2b for diarylethenes, they exist as either antiparallel or parallel rotamers. By light irradiation, the six-membered ring in the core of the antiparallel rotamer is closed, which could cause extension of conjugation structure when incorporated into a polymer structure. In the case of fulgides, such a ring-closing reaction results in the formation of thermally irreversible colored isomers. In summary, in the case of spiropyrans, spirooxazines, chromenes, fulgides, fulgimides, diarylethenes, and the related families, the photochromism reaction is based on six π electrons delocalized over six different atoms, which undergo cyclization. Azobenzene chromophores are another promising photoisomerizable groups, which undergo a reversible cis–trans isomerization upon light irradiation, similar to the stilbene chromophores. They show photoisomerization from a nonpolar trans isomer to a polar cis one upon UV irradiation. Figure 2.2c shows the photoisomerization of azobenzene chromophore under UV and visible light irradiation. Irradiation of visible light caused to reverse isomerization of the chromophore from cis-to-trans, and also exchanges the polarity from a polar to a nonpolar structure [213–215]. By incorporation into polymer chains in the backbone or as the side chain, azobenzene could result in switching of hydrophilicity/hydrophobicity, optical properties, including absorption and emission spectra, and coordinative interaction in polymer materials.

For the photochromic molecules involving an intramolecular hydrogen transfer (or a proton transfer), derivatives of salicylideneaniline have probably received more attention and higher study. They are mainly substituted salicylideneanilines and variants, where aniline is replaced by different groups, such as aminothiophene or aminopyridine, and they are also called anils. However, photochromism of anils can be accomplished in a variety of environments, from solution to inclusions or encapsulated media, and even in some cases in the solid state, with a media-dependent behavior. Photoreaction of anils is shown in Figure 2.2d, where in the first step of the

Table 2.3 Typical examples of reversible photochromic compounds.

Photochromic materials	Visible light	UV-irradiation	References
Spiropyran	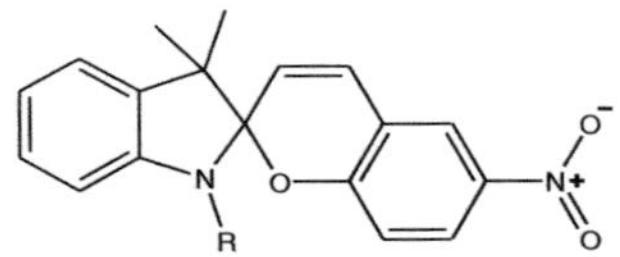	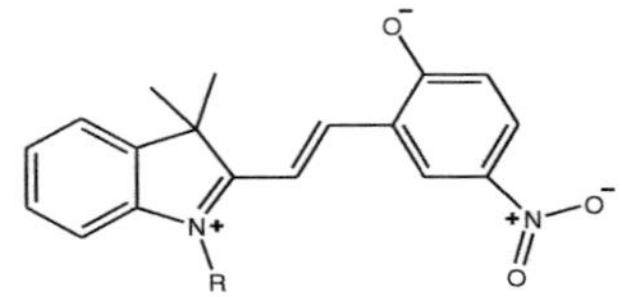	[15, 16, 21, 22, 151–154]
Spirooxazine	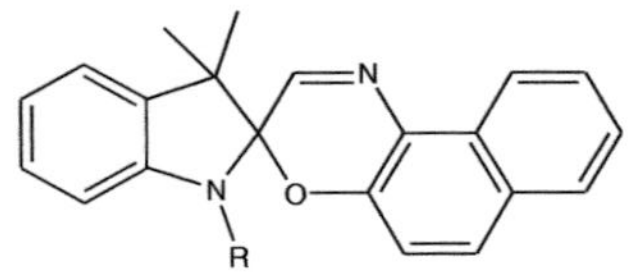	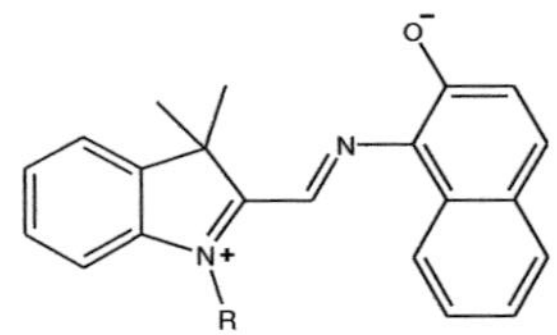	[155, 156]
Spirothiopyran	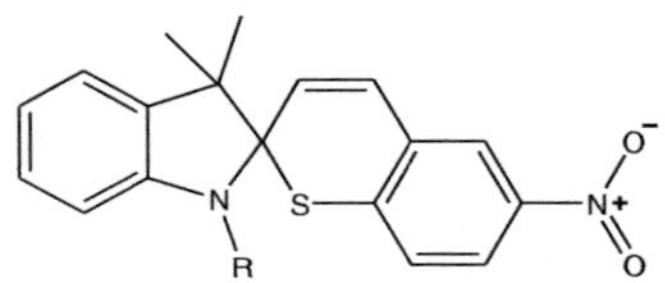	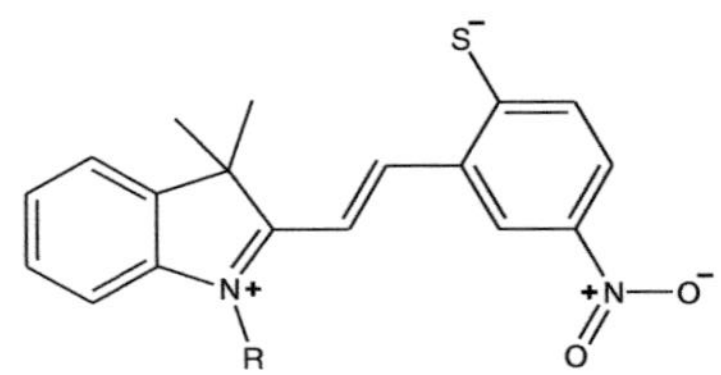	[157–159]
Azobenzene	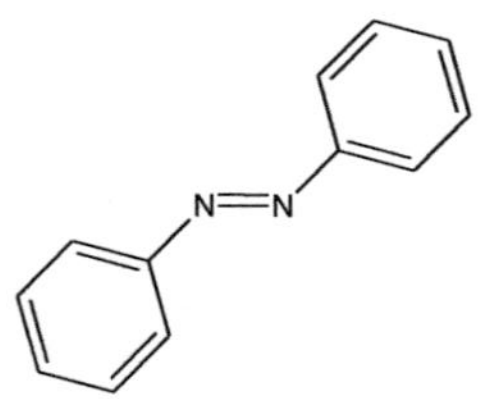	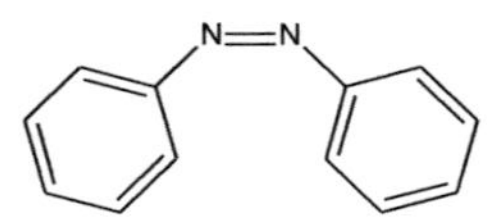	[155–160]
Benzopyran	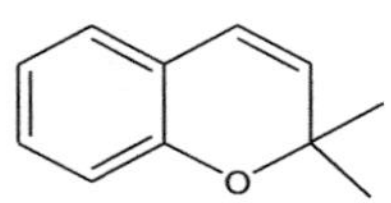	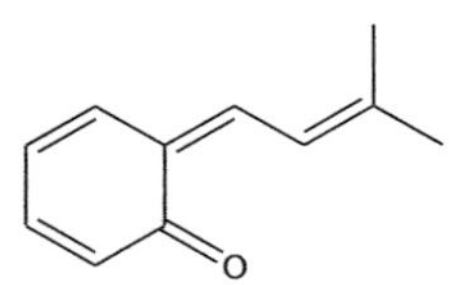	[161–166]

Naphthopyran	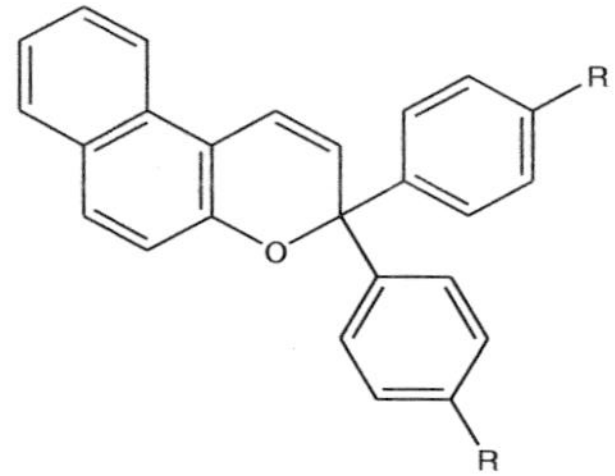	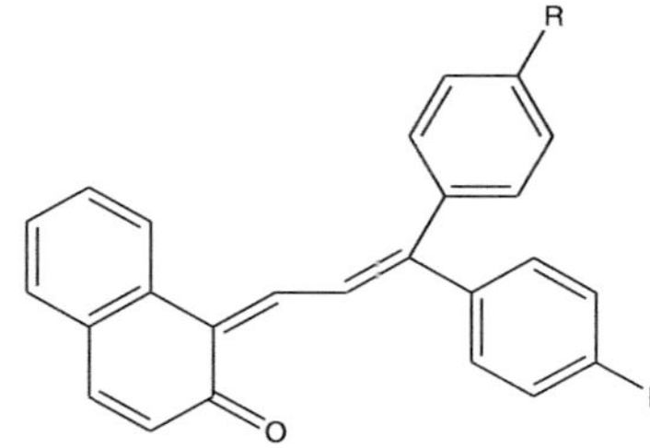	[167–171]
Oxazine	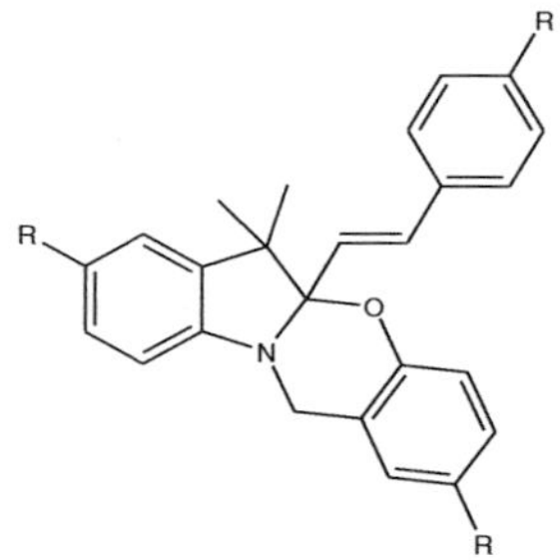	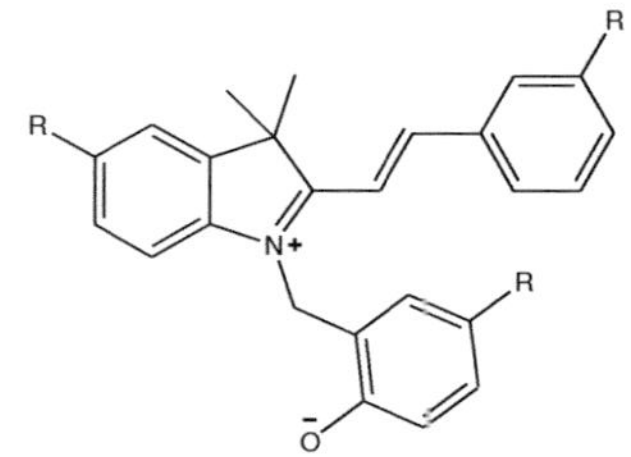	[172–179]
Oxazolidine	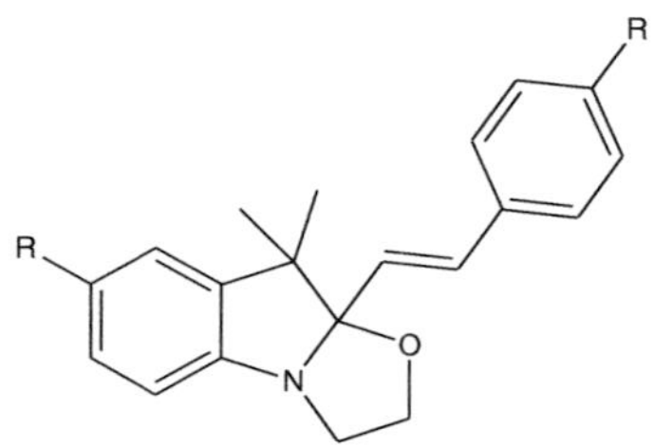	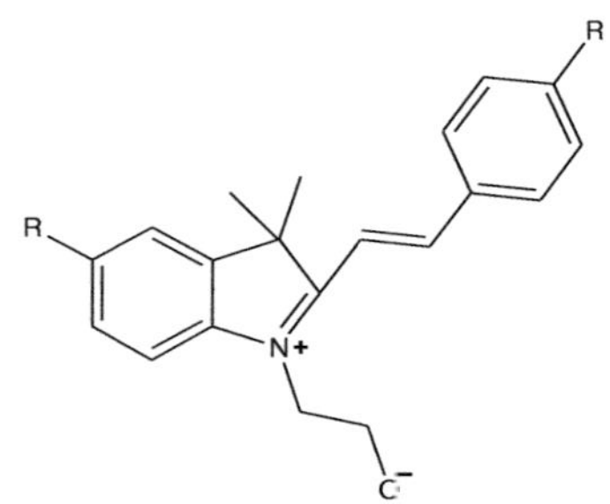	[156, 180–182]
Diarylethene	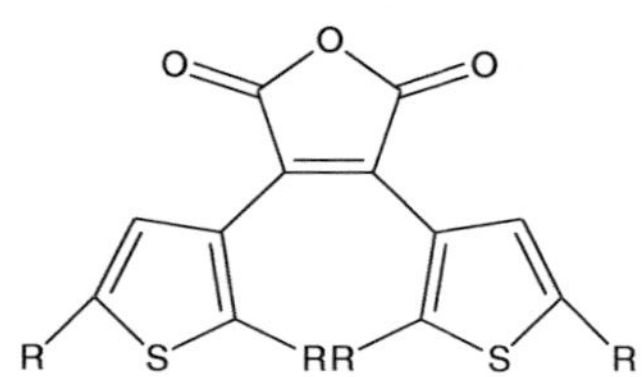	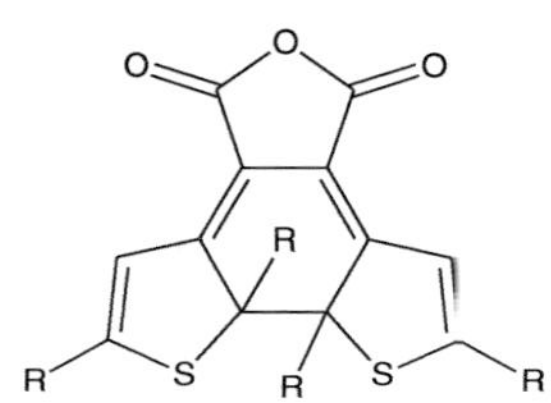	[183–185]

(Continued)

Table 2.3 (Continued)

Photochromic materials	Visible light	UV-irradiation	References
Fulgide	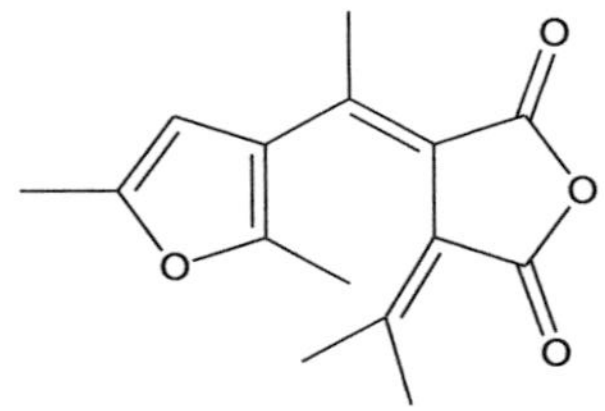	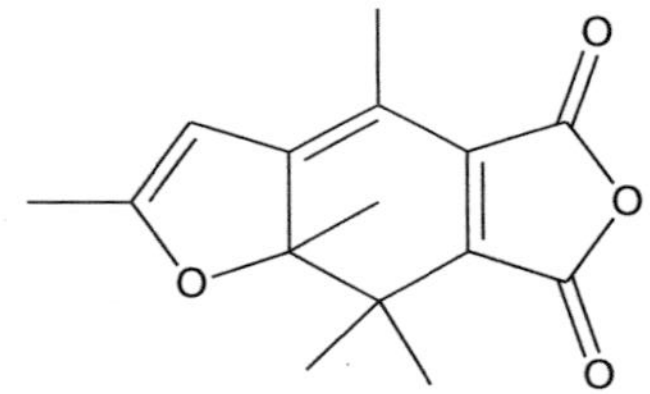	[186–188]
Stilbene	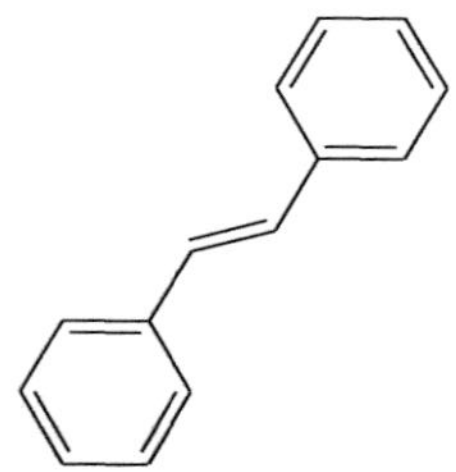	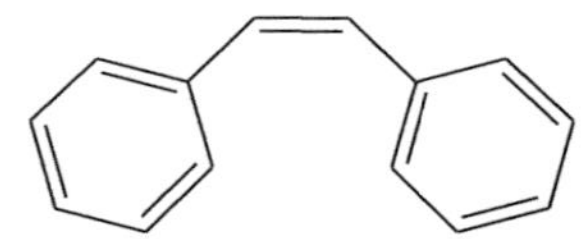	[189, 190]
Hemithioindigo	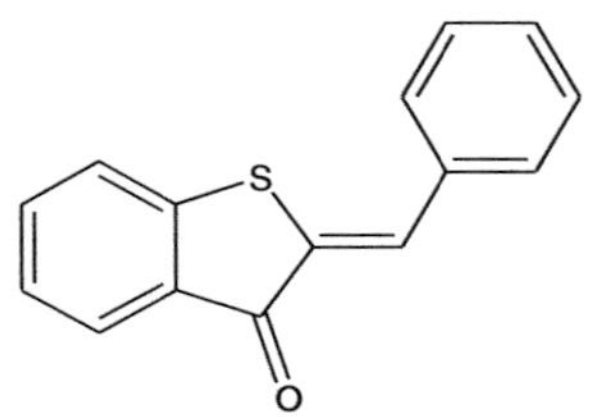	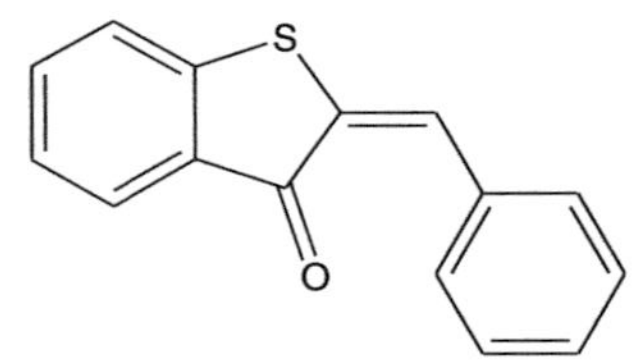	[191–193]
Viologen	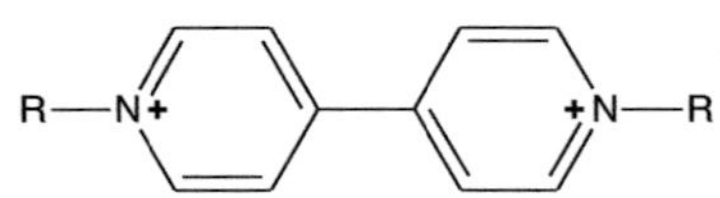	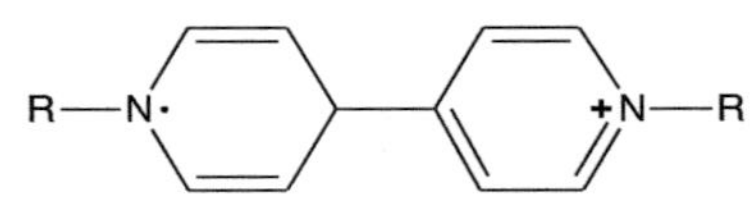	[194–196]
Chromene	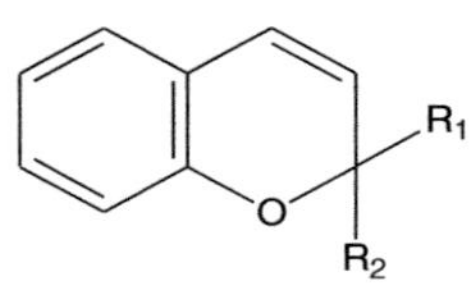	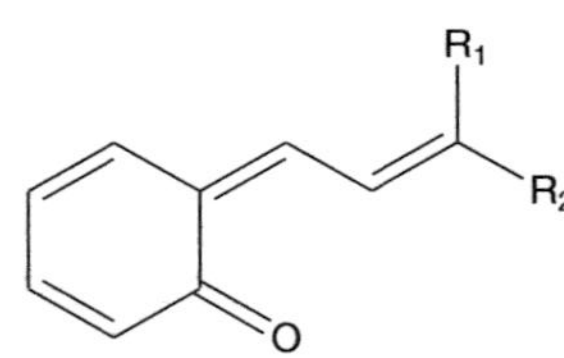	[155, 167]

Salicylideneaniline	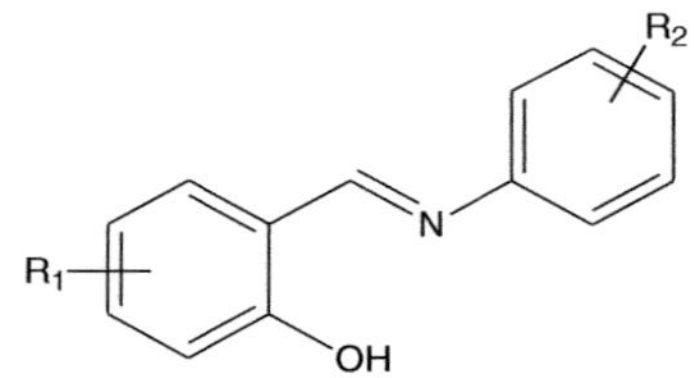	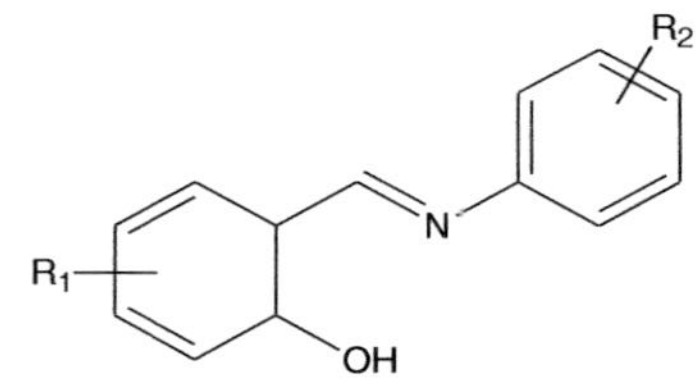	[196–201]
Dinitrobenzylpyridine	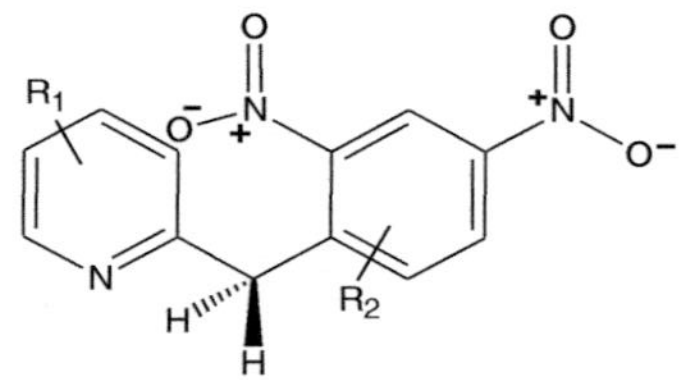	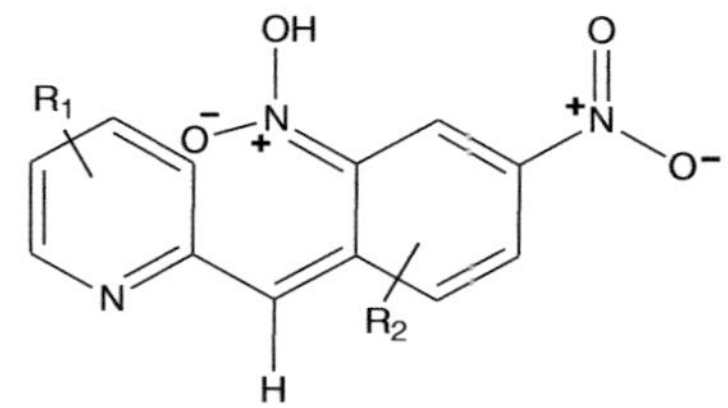	[202–205]
Triphenylimidazolyl	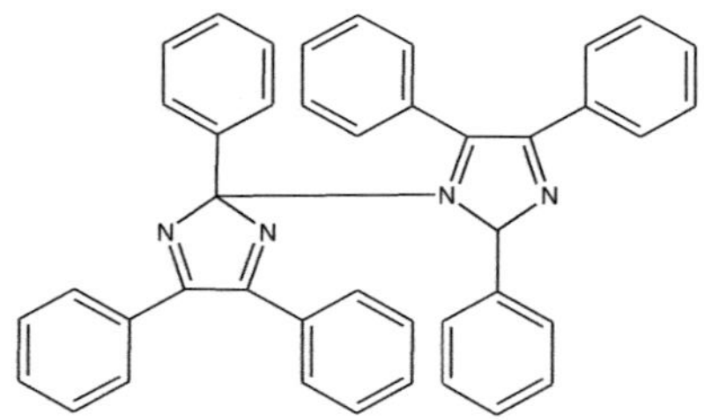	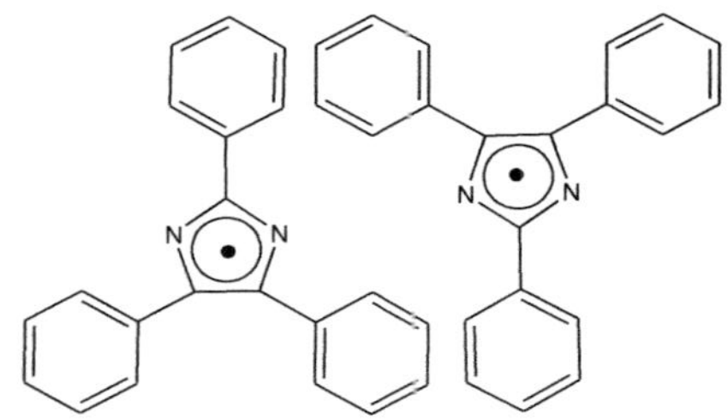	[206, 207]
Over-crowed alkene	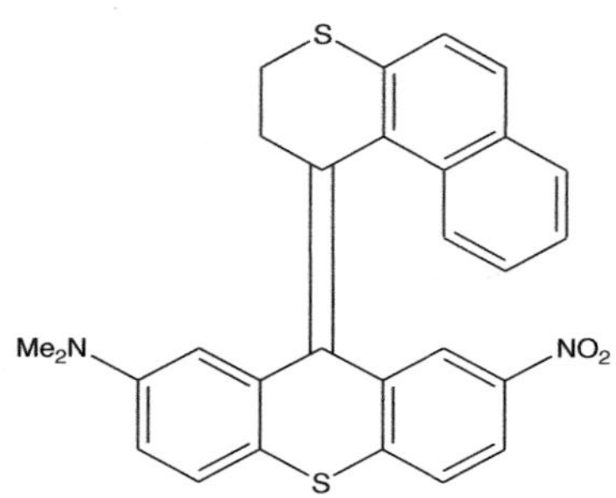	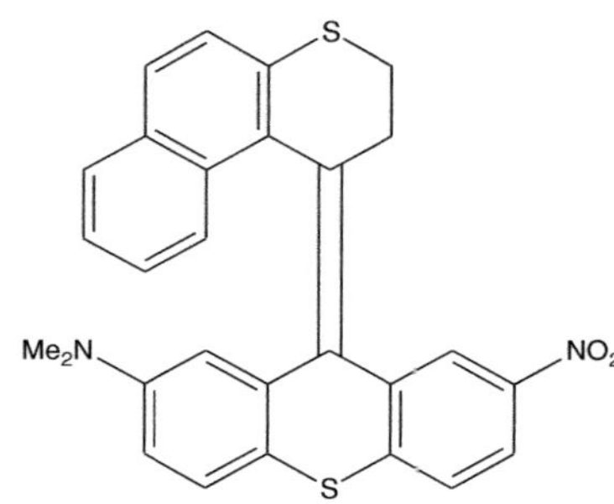	[208, 209]

reaction, an intramolecular proton transfer is induced and allows the passage from the enol (in fact a phenol) form to the *cis*-keto (ketone) form in the excited state called excited-state intramolecular proton transfer (ESIPT). In the next step, this transfer is quickly followed by cis–trans photoisomerization at an excited state to achieve a *trans*-keto form. The process may also be called keto–enol phototautomerism. The tautomerism process occurs within a few picoseconds in solution and a few hundred picoseconds in a solid state. The keto photoproduct reverts back to the enol very quickly in solution state, typically within a few milliseconds, whereas it may vary from a few seconds to several months in the solid state, thus the photochromism of anils can barely be observed in the solution state [216–222]. The second most studied family of photochromic molecules with an intramolecular proton transfer is DNBP, which can be prepared by nitration in HNO_3/H_2SO_4 mixture from benzylpyridine. Under UV excitation, the colorless CH form becomes the colored OH and NH forms. The CH form is the most stable, while the OH form is the least stable and spontaneously converts to its isomers (Figure 2.2e) [63, 202]. In the case of redox reactions, viologens, such as 1,10-dimethyl-4,40-bipyridinium dichloride, are typical photochromic compounds based on a photoinduced electron transfer (PET) that change their color reversibly upon reduction and oxidation reactions [14, 69]. Under UV light irradiation, they are reduced to monocation radical and then an uncharged compound in a reversible way, as shown in Figure 2.2f. Other chromophores with the same photochromism mechanism are thiazine derivatives and tetraphenylhydrazine. The other light-induced molecular change in photochromic compound is based on hemolytic cleavage, which leads to the formation of two radicals, as depicted in Figure 2.2g. The homolytic cleavage of the C—N bond between the two imidazole rings of the photochromic compound, triphenylimidazolyl dimer (TPID), can be induced by light irradiation, heating, or pressure, which could be resulted in formation of two triphenylimidazolyl radicals (TPIR). The reverse reaction is carried out thermally and driven by radical diffusion, leading to re-formation of the starting imidazole dimer. TPIR has a large absorption wavelength in the visible light region, whereas TPID absorbs wavelengths only in the UV light region and therefore, it is colorless. Moreover, the cleavage of photochromic compound of this family is induced upon UV light irradiation within less than 100 fs, while recombination can take up to few minutes at room temperature [206, 223, 224].

Incorporation of photochromic materials into polymer backbone or side chains can significantly affect various properties of the polymer matrix and also their surrounding environment, through reversible photoisomerization between nonpolar and polar molecular structures. Moreover, chemical bonding between the photochromic compounds and polymer results in increasing their chemical stability, switchability, and photostability. Consequently, to protect the photochromic molecular species against environmental degradation, dye aggregation, negative photochromism, and as well as increase their photoswitchability, photostability, and reversible response, they must be chemically bonded in weakly polar polymer matrices [15, 16, 21, 22, 105, 156]. Hence, recognition of photochromic compound characteristics and the polymer matrix are determinative factors in the preparation of photochromic polymers. Generally, photochromic polymers are synthesized through free radical polymerization, controlled/living radical polymerization (CRP), and rarely poly-condensation methods. Photoswitchable isomerization of these photochromic polymers is usually accompanied by reversible changes in their physical and chemical properties, such as polymer chain conformation, pH, polarity and solubility, sol–gel transition temperature, surface wettability, shape of polymer gels, membrane potential, membrane permeability, and phase separation temperature of the polymer blends [63]. In these type of smart polymers, the fascinating color change by incorporated photochromic molecules in response to specific wavelengths of light produces number of applications, such as anticounterfeiting and authentication technologies [15, 151, 179, 225, 226], rewritable patterning [16, 21, 22, 156], polymer sensors [100, 184, 227, 228], smart surface-active agents [213, 229–231], self-healing polymers [28, 232–234], shape-changing polymers [39, 235–237], photoconductive polymers [13, 238, 239], drug-delivery systems [146, 240–242], and membranes [14, 65]. In the case of organic photochromic compounds, such as spiropyran and diarylethene families (two of the important photochromic compounds) and their related polymers, general cleavage or formation of a specific chemical bond upon light irradiation, electrical potential, or mechanical forces leads to an extensive electron resonance in the conjugated aromatic structures, and also cause to development of the color changes [147, 243]. Also, these photochromic compounds and their related polymers can be used in chemical sensing of other stimulants, such as polarity, pH, heat, stress, and electrical potential [63, 244–246], which will be discussed in Section 2.3. Furthermore, combination of photochromic materials with the other photoactive compounds, such as fluorescent materials, results in polymers with application in multiple security inks which will be discussed.

2.2.3 Photocleavable Polymers

Polymers with photodegradable moieties have recently received large attention due to their interesting characteristics. These types of polymers known as photodegradable polymers have found numerous applications in photopatterning technology, hydrophilicity/hydrophobicity transition, micellization and dissociation transitions, etc. [9, 10, 14]. Photolabile protecting groups (PPGs) can be removed upon application of an appropriate wavelength of light, which include three main categories of 2-nitrobenzyl (NB) series, carbonyl-based PPGs, and benzyl-based PPGs. In general, photocleavage of PPGs occurs under mild conditions and does not require any chemical reagents. Therefore, this type of PPGs makes it possible to deal with very sensitive molecules, in spite of their incompatibility with acids or bases [247, 248]. Light-sensitive molecules, including benzyl, 2-nitrobenzyl, benzoin, cinnamyl, phenacyl, vinyl silane PPGs, and their derivatives and also polycyclic aromatics, such as pyren-1-ylmethyl, pyren-1-ylmethoxycarbonyl, anthraquinon-2-ylethyl-1′,2′-diol, anthraquinon-2-ylmethoxycarbonyl, oxobenzopyrans (coumarins), anthracene-9-methanol, and phenanthren-9-ylmethoxycarbonyl, have been developed and applied in the protection of alcohols, carboxylic acids, amines, ketones, aldehydes, and phosphates [249]. However, ONB derivatives, coumarin-4-ylmethyl derivatives, phenacyl esters [250–253], cinnamate derivatives, pyrenylmethyl, and anthracene-9-methanol [254] are the well-known PPGs used in polymeric systems, which can cleave themselves into two or more groups with the resulting degradation of polymer structure upon light exposure. Within this class of photolabile groups, the ONB alcohol derivatives and coumarin groups, which undergo selective bond cleavage upon irradiation with UV light, are the most applicable photolabile groups in polymeric structures [9, 75]. The first photodegradable polymer incorporating chromophore groups in its structure was synthesized by Barltrop et al. in 1966. They used ONB compounds as photocleavable groups, which undergo photocleavage and isomerization from ONB to *ortho*-nitrozobenzaldehyde groups upon UV irradiation at wavelength of about 300–365 nm via a Norrish II type intramolecular rearrangement [255]. Photoisomerization of ONB derivatives is based on the fast intramolecular H-atom transfer affording *aci*-nitro tautomer that can occur both in the solid and solution states [247, 256, 257]. As shown in Table 2.4, ONB is cleaved to a light-sensitive water-insoluble *ortho*-nitrosobenzaldehyde, and releases a carboxylic acid group through a n–π transition upon UV exposure with wavelengths between 300 and 365 nm [3, 87, 288]. The light-induced cleavage of this chemical is accompanied by an irreversible phase transition from hydrophobic to hydrophilic states [9, 64, 65, 75]. Consequently, this photoinduced phase transition could be used for variation of solubility and tuning critical solution temperature (LCST and UCST) of the temperature-responsive polymers in aqueous solution by photocleavage of the light-sensitive molecules [258, 259, 289]. The carbonyl-based PPGs can be divided into the phenacyl family and also 7-methoxycoumarinyl-4-methyl derivatives. The benzyl-based PPGs can be divided into the (2-naphthyl)methyl, anthracen-9-ylmethyl, (phenanthren-9-yl)methyl, pyrenylmethyl, and perylen-3-ylmethyl derivatives [248]. As an example, pyrenylmethyl esters undergo a photolysis reaction in the presence of water or a protonic solvent, where the main by-product, *para*-hydroxyphenylacetic acid, is water-soluble, which could be its advantage over the ONB derivatives.

Depending on the type and position of these photolabile compounds in the polymer chain, different applications could be obtained from the photocleavage reactions in polymers, including charge generation in the side groups, hydrophilicity/hydrophobicity transition, depolymerization and chain breakage, activation of catalyst and "click" reactant, formation of active groups, and carboxylic acid functionalization of the polymer chain for coupling reactions. The ONB derivatives release carboxylic acid functionality on polymer chain upon intramolecular oxidation by light irradiation. Similarly, phenacyl esters could be used in carboxylic acid functionalization of polymer chains. They undergo a homolytic cleavage of the C—O bond to give an acryloxy radical and a phenacyl derivative radical, which are converted to the carboxylic acid group by protonation of the acryloxy radical and yielding *para*-methoxyacetophenone byproduct. In addition, highly reactive radicals generated from light irradiation of 2-naphthoquinone-3-methides could further react with vinyl compounds. Incorporation of these photolabile groups into side chains results in photoinduced reactivity of the resulted polymer chains, which is highly applicable in photolithography applications [247, 256, 257, 290, 292].

Table 2.4 Typical examples of irreversible photolabile protecting groups.

Irreversible photoresponsive groups	Light-induced responsivity	References
ortho-Nitrobenzyl ester	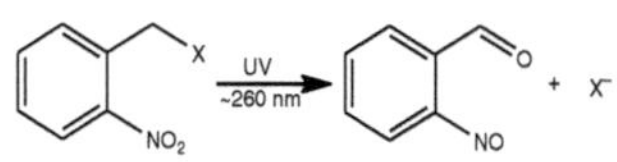	[23, 27, 258–273]
Coumarin-4-ylmethyl	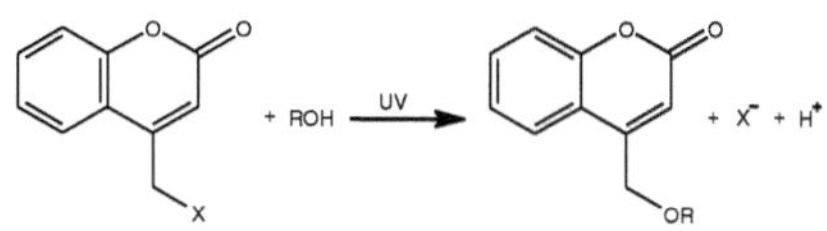	[272, 273]
para-Methoxy-phenacyl ester	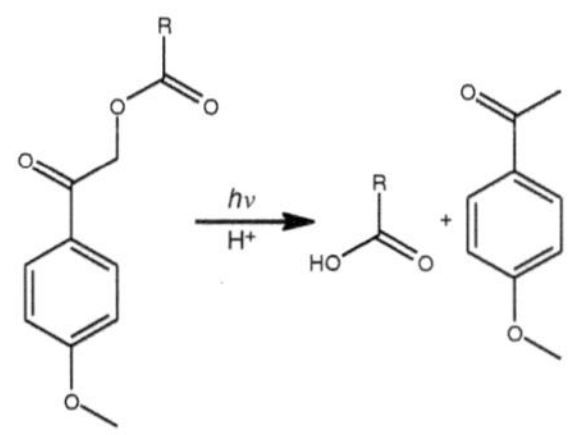	[272, 273]
Pyrenylmethyl ester	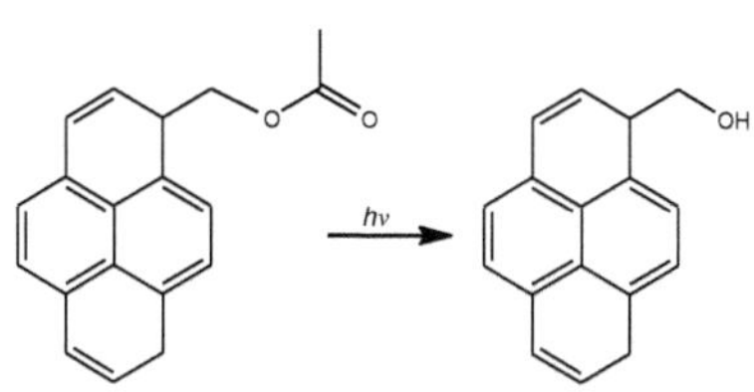	[274–276]
Anthracene-9-methanol	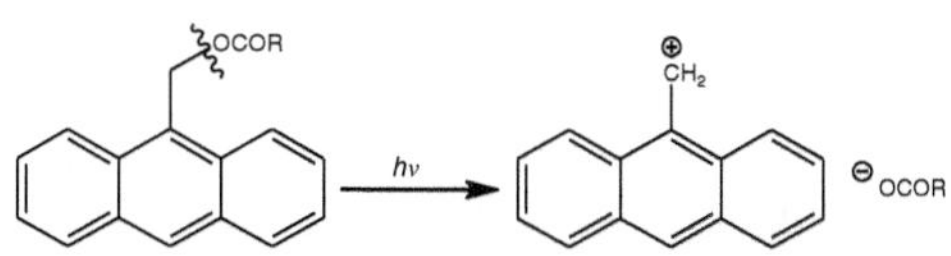	[254, 277]
Coumarinyl ester	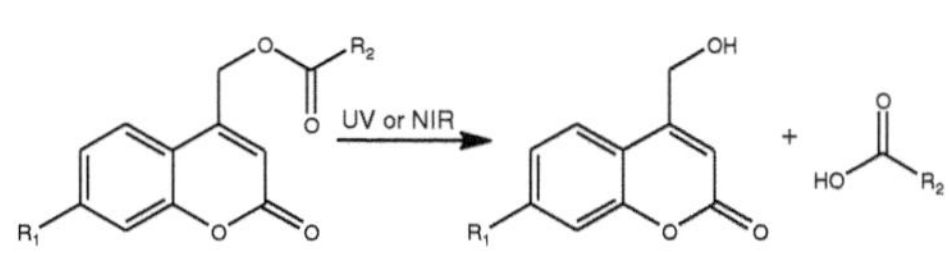	[5]
Truxylic acid	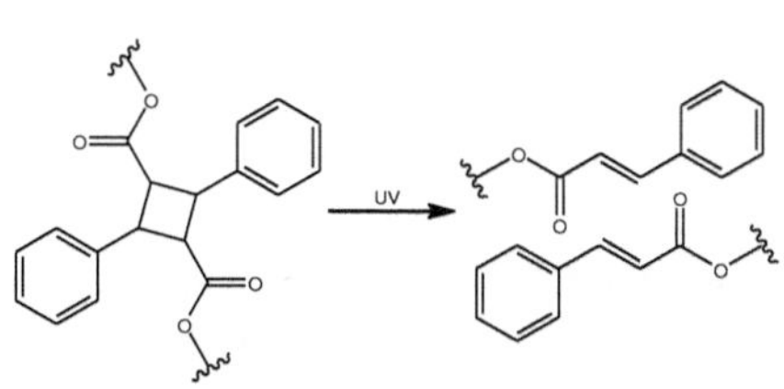	[5, 278]

Table 2.4 (Continued)

Irreversible photoresponsive groups	Light-induced responsivity	References
Diphenyliodonium-2-carboxylate monohydrate	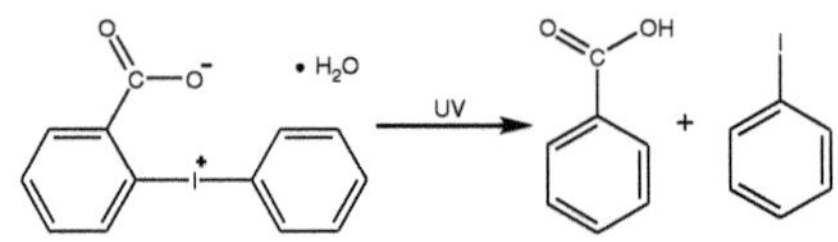	[279, 280]
Stilbazolium	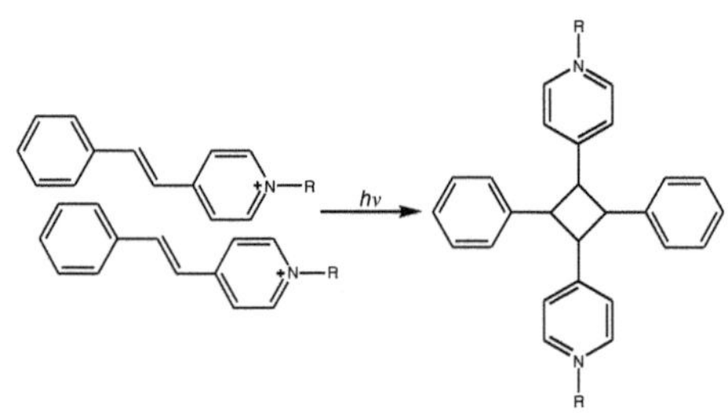	[281–287]

2.2.4 Photodimerizable Polymers

Light-responsive cross-linkable compounds have also been used for the development of photodimerizable polymeric systems. The dimerization reactions, namely intermolecular cyclization, are common photoinduced reactions of organic chemistry predicted by the Woodward–Hoffmann rules [293]. Photoinduced cyclization (refer as photocross-linking reaction) takes place between two reactive species, with one of them electronically excited and so provides the possibility to reversibly create and break bonds between the photosensitive molecules upon application of an appropriate wavelength of light. Incorporation of photodimerizable moieties into polymer structures allows fabrication of various photocross-linkable polymeric systems. Increasing attention has recently been paid to development of various photochemical-induced reversible cross-linkable polymeric materials due to their significant potential applications in photoresists, organic LEDs, and printing inks [16, 294–298]. The photocross-linking reactions occur through different pathways, such as cycloaddition reactions or dimerization of chromophores and ionic mechanisms. Many of the photocross-linkable compounds react by either a [2π+2π] or [4π+4π] cycloaddition mechanism with the formation of the cyclobutane rings that can be reversed upon application of an appropriate wavelength of light. The overall mechanism of a [2π+2π] cycloaddition is shown in Figure 2.3. Accordingly, an alkene absorbs a photon in the UV region to excite one of its electrons from the c1 orbital in the ground state of the HOMO to the c2* orbital (known as the excited-state HOMO). Likewise, [2π+2π] cycloaddition occurs when the excited-state HOMO of one alkene interacts with the LUMO of the second alkene. The symmetry interaction between the terminal π lobes, allowing interaction between lobes on the same face of one alkene and lobes on the same face of a second

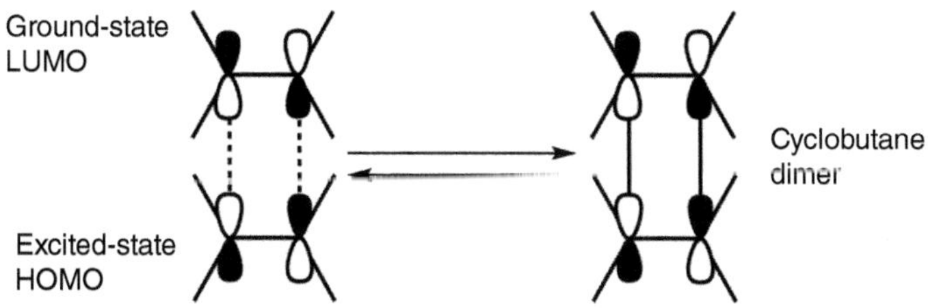

Figure 2.3 Orbital summary required for [2π+2π] cycloaddition of two general alkenes [72]. Source: Reproduced with permission from Kaur et al. [72]. Copyright 2014 Royal Society of Chemistry (RSC).

alkene, caused to photocycloaddition phenomenon in these systems [72, 280]. However, reversible photodimerization reactions (refer as photocycloaddition reactions or photocross-linking reactions) in polymeric systems are limited to a few molecules, including benzothiophene oxide, coumarin, dibenzazepine, stilbene, thymine, and cinnamic acid for [2π+2π] cycloadditions and anthracene for [4π+4π] cycloaddition (Table 2.5) [66, 72, 105]. For both of these cycloaddition reactions, the photocross-linking or photodimerization process can be either reversible or irreversible, depending on the incoming light and the cross-linking reagent [65, 72]. For best understanding, the reversible photodimerization reaction of coumarin based on [2π+2π] cycloaddition is illustrated in Figure 2.4. When the coumarin functionalities are exposed to UV irradiation with wavelength of higher than 350 nm, the cyclobutane rings are formed through [2π+2π] cycloaddition reaction. On the other hand, the reverse reaction is performed at a wavelength shorter than 260 nm [65]. Researches showed formation of a dimer is classified into three typical products depending on the combination of coumarin concentration, dose, and solvent – the anti-head-to-head dimer (Figure 2.4a), the syn head-to-head dimer (Figure 2.4b), and the syn head-to-tail dimer (Figure 2.4c). At low concentrations, the triplet coumarin is produced and the anti-head-to-head dimer is formed due to inefficient ISC, while at high concentrations the syn head-to-head dimer is formed due to reaction of the coumarin singlet with ground-state coumarin reference [3, 71].

In the case of polymers incorporated photocross-linkable compounds, a strong UV absorption at certain wavelengths (typically around the 260–300 nm regions) is observed due to the electronic conjugation of their incorporated photoreactive molecules. The conjugation is then destroyed through dimerization of the molecules, and intensity of the absorbance band of the irradiated sample is reduced in the stated wavelength range. In another hand, the retro-[2π+2π] cycloaddition, the ring-breaking reaction of the dimer linkages (i.e. cyclobutane cleavage), caused to restore conjugation in the molecule and absorption again increases in the 270–300 nm region [72, 321, 322]. On the basis of the fact that dimerization of the mentioned pendant chromophores, such as coumarin moieties in a polymer matrix can remarkably change their physical and chemical properties, such as solubility, density, viscosity, and also mechanical strength, they can also display significant role for application of stable light-induced changes in physical and

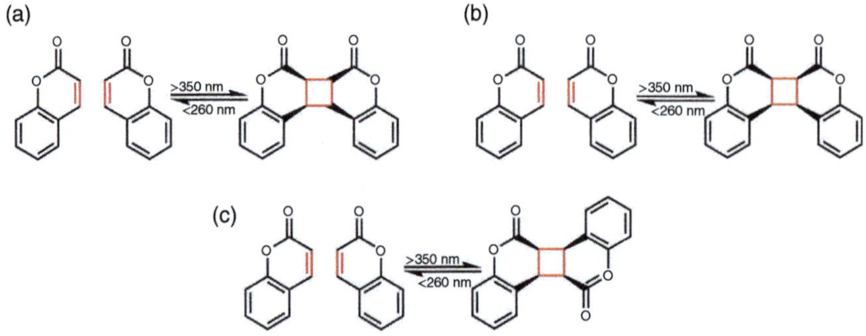

Figure 2.4 Typical isomers resulting from the photodimerization of coumarin molecules using different conditions: (a) anti-head-to-head dimer, (b) syn head-to-head dimer, and (c) syn head-to-tail dimer [71]. Source: Reproduced with permission from Trenor et al. [71]. Copyright 2004 American Chemical Society (ACS).

chemical properties of the polymeric materials [105]. Furthermore, this reversible coupling/decoupling responsivity of the photodimerizable compounds has potential applications in light-induced cross-linking of various polymer systems, photoresist polymers, self-healing materials, shell and core cross-linking in self-assembled polymer particles, photoresponsive rheological modifiers, and also photodegradable hydrogels, some of which will be discussed in detail in the Section 2.3.

2.2.5 Photoadaptable Polymers

Dynamic bonds are capable of being reversibly broken and reformed under certain conditions, and are widely used in stimuli-responsive systems, dynamic combinatorial chemistry, and dynamic materials [323–326]. These dynamic bonds are classified into two categories of supramolecular or noncovalent interactions and dynamic covalent bonds, as shown in Figure 2.5a. Rapid exchange of noncovalent supramolecular interactions facilitates dynamic properties of materials, but they are often unstable and prone to creep. On the other hand, dynamic covalent bonds are typically static under environmental conditions and dynamic under stimuli, such as light, pH or temperature, and stress [91, 236, 329–335]. Introduction of dynamic covalent bonds into a polymeric system (refer as dynamic covalent polymer) provides it with unprecedented properties, such as shape-memory properties, increased toughness, self-healing ability, ability to relax stresses, malleability, and adaptability [327, 336–339]. Depending on their introduced dynamic covalent bond, the exchange in these polymers can occur in two associative and dissociative pathways. In associative pathway, cleavage of the exiting linkage and formation of a new one happen concurrently; however, an existing bond breaks first and then a new bond is formed in the dissociative pathway. In an associative polymeric system, there is minimal change in its structure due to an approximately constant bond density even as bonds are activated. While in a dissociative polymeric system, due to a decrease in the number of linkages, as bonds are activated, a dramatic change in polymer structure occurs during the exchange process (Figure 2.5b,c) [91, 328, 335].

Table 2.5 Typical moieties capable to give reversible cycloaddition reactions under UV light irradiation.

Photocross-linkable groups	Reversible cycloaddition reaction	References
Anthracene	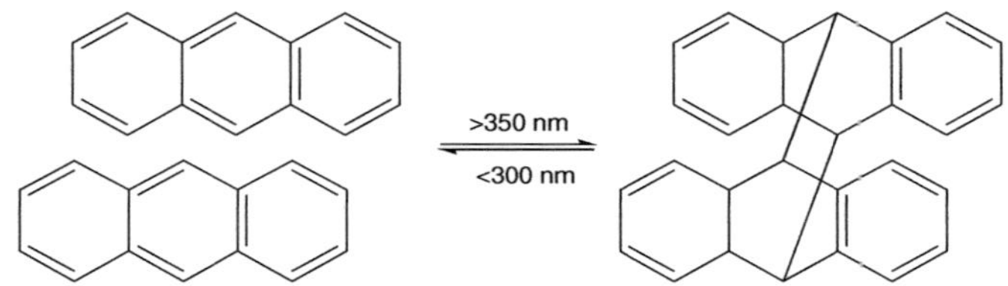	[299–304]
Cinnamic acid	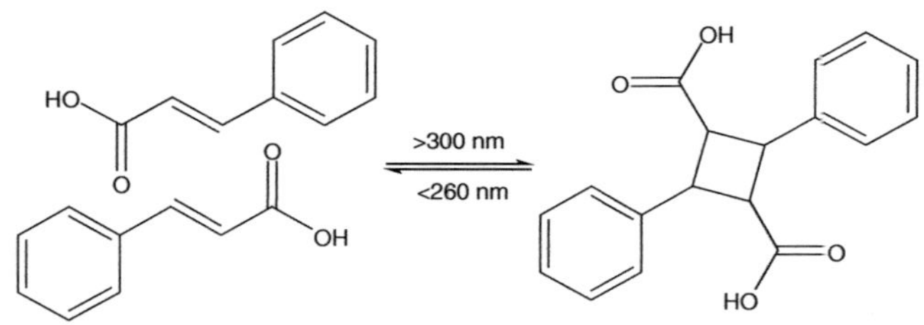	[305–307]
Nitrocinnamate	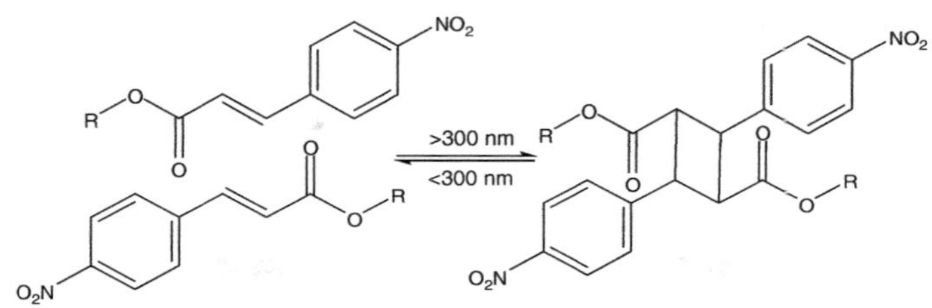	[308–311]

Coumarin		[38, 103, 105, 106, 312]
Stilbene	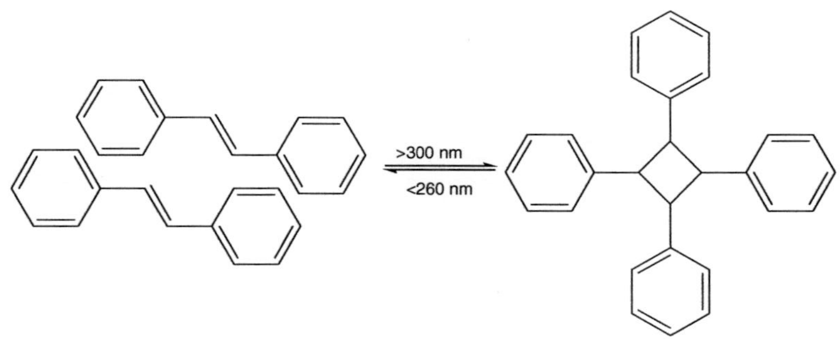	[313–317]
Tymine	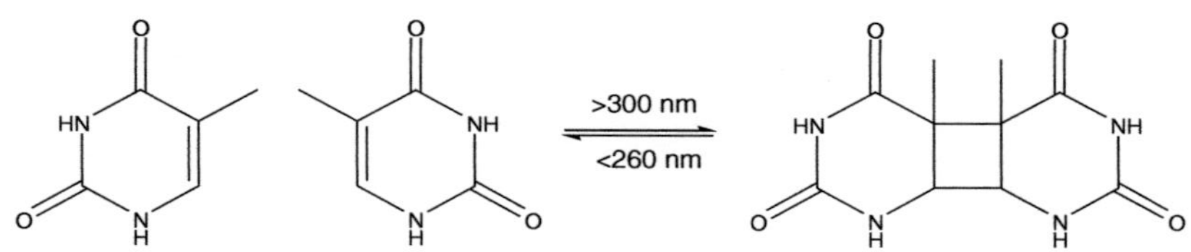	[318–320]

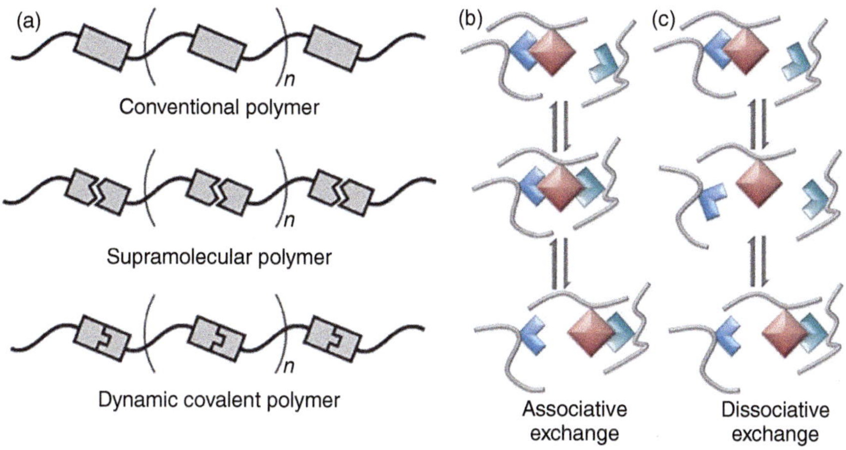

Figure 2.5 Schematic illustration of (a) conventional polymer, supramolecular polymer, and dynamic covalent polymer [327]. Source: Reproduced with permission from Liu et al. [327]. Copyright 2009 Elsevier Ltd.; and (b) associative exchange and (c) dissociative exchange in dynamic covalent bonds [328]. Source: Reproduced with permission from Perera et al. [328]. Copyright 2019 John Wiley & Sons Inc.

Photodynamic covalent polymers (photodynamers) are polymeric structures that exhibit dynamic features originating in the reversibility of photodynamic covalent bonds. Disulfides and diselenides are the most important dynamic covalent bonds used in polymeric systems that provide possibility to exchange the dynamic properties upon application of an appropriate wavelength of light [325, 340–345]. Trithiocarbonate, allyl sulfide, thiuram disulfide, 1,2-dithiolane, seleno-sulfide, and hexaarylbiimidazole exchange reactions are the other dynamic covalent bonds used in polymeric systems [320–358]. Exchange reactions in these dynamic covalent bonds are based on free radical recombination and addition-fragmentation chain-transfer reactions. The main types of reversible reactions used for the construction of dynamic polymers are shown in Table 2.6 [11, 325, 340, 366, 367, 369]. Disulfide-exchange reaction occurs by recombination of the sulfenyl radicals formed from homolytic dissociation of disulfide bonds. In the case of allyl sulfides, the reaction between thiyl radicals and allyl sulfide forms radical intermediate that undergoes beta scission and forms regenerates a thiyl radical and allyl sulfide. Thiuram disulfide undergoes photocontrolled reshuffling reaction through homolytic dissociation of dithiocarbamate central bond under visible light irradiation. 1,2-Dithiolanes are five-membered cyclic disulfides that form disulfide bonds by thiol-initiated ring-opening reaction in the presence of small-molecule thiols. Diselenide dynamic covalent bonds are known as redox-responsive linkages, where the seleno-sulfide groups follow the same reaction. Hexaarylbiimidazole derivatives yield radicals with low reactivity upon homolytic breakage of C—N bonds by UV or visible light irradiation. The inherent dynamic nature of the dynamic covalent bonds used in polymers and their position in the polymer structure (dynamic covalent bond is used in the main chain of the polymer or it is used to allow side chain modification of the polymer) are two

Table 2.6 Representative exchange reactions used in dynamic chemistry.

Dynamic covalent group	Exchange reactions in dynamic chemistry	References
Disulfide	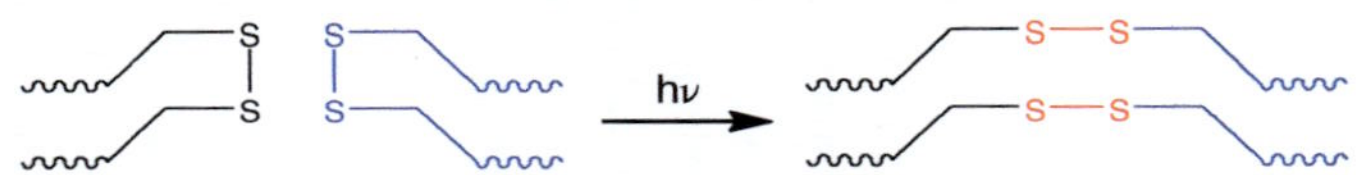	[359–365]
Trithiocarbonates	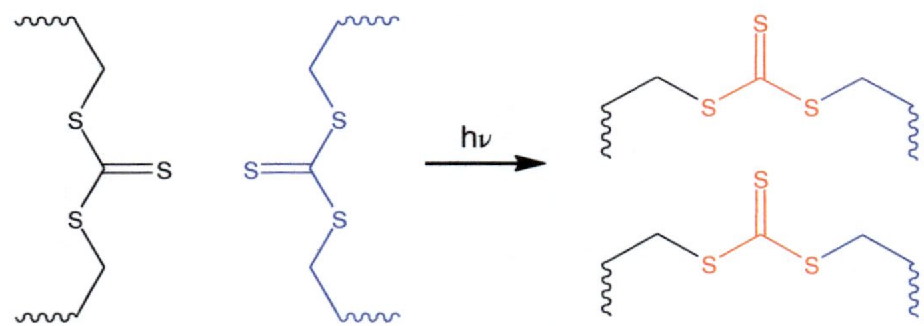	[357, 358]
Allyl sulfide	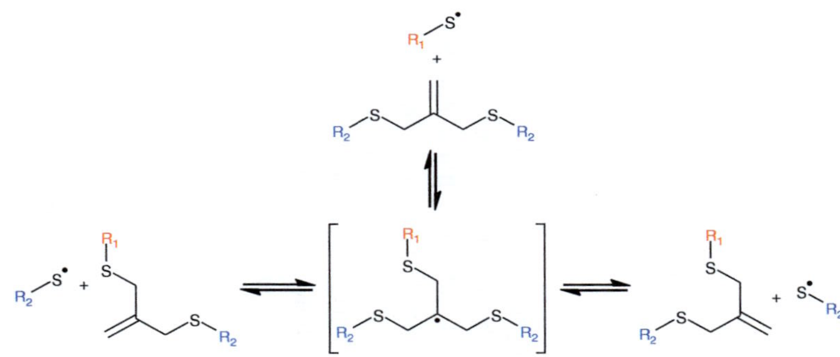	[351–354]

Diselenide	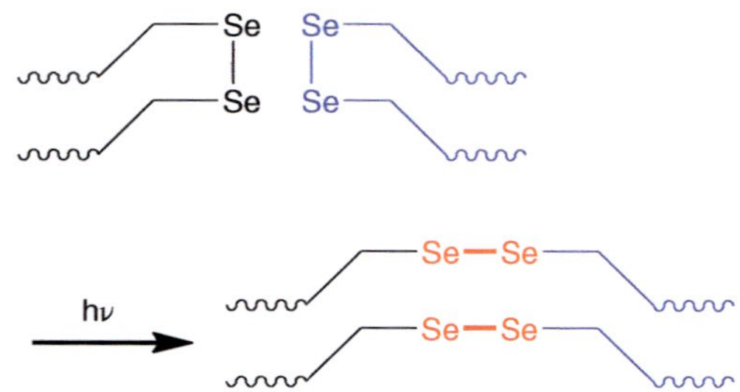	[366, 367]
Thiuram disulfide	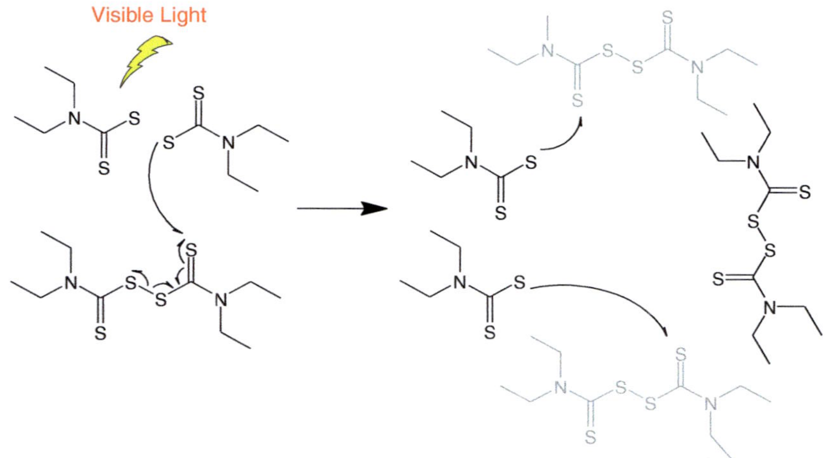	[350–356]
1,2-Dithiolane	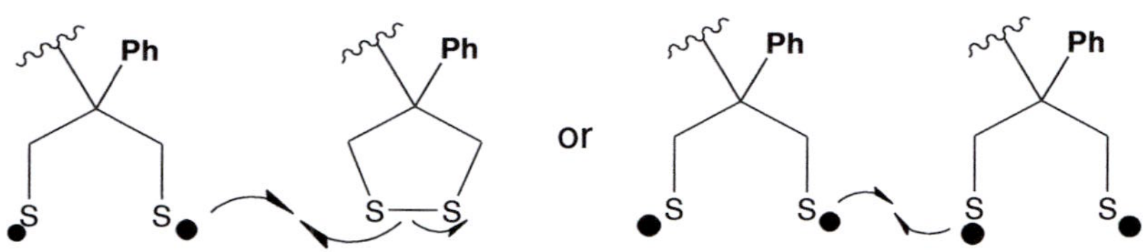	[368]

(Continued)

Table 2.6 (Continued)

Dynamic covalent group	Exchange reactions in dynamic chemistry	References
Seleno-sulfide	S—S Se—Se $\xrightarrow{h\nu}$ S—Se	[355]
Hexaarylbiimidazole	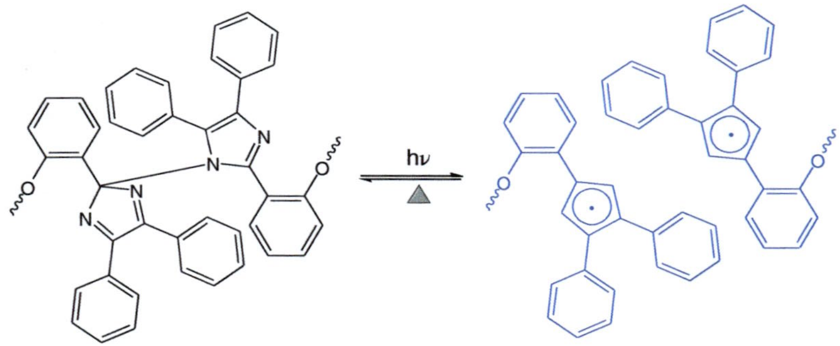	[346, 347, 349]

important points in the preparation of these photodynamic polymers. Based on the dynamic covalent bond and its position, photodynamic covalent polymers show many unique properties, such as self-healing ability, recyclability, degradability, and shape-memory behavior [335, 369], some of which will be discussed in the Section 2.3.

2.3 Applications of Photoresponsive Polymers

Light-induced variations in polarity of polymer chain structures, viscosity of polymers, color change and conjugation of chromophoric units and the resulted polymers, and also UV absorbance and fluorescence/phosphorescence emission of the polymers in solution or solid state can be reflected in the macroscopic level, which has been used in different applications, such as surfactants and variation of critical micelle concentration (CMC), smart inks, optical sensors, optical patterning, viscosity modifiers, drug delivery, self-healing, self-repairing, shape-changing, and also smart membranes, films, and textiles. These smart polymers, their synthesis, and also their applications are presented in the following.

2.3.1 Smart Polymeric Inks

Smart polymeric inks have gained increasing interest, due to their anticounterfeiting, authentication, and encoding applications in banknotes, money papers, passports, national cards, certificates. Such inks should show photostability, reversible photoluminescence, resistance to environmental degradations, fast authentication, high photofatigue resistance, and also high stability of printing on confidential documents. Different printing methods, such as handwriting, stamping, inkjet printing, and screen printing, as well as anticounterfeiting labels, are used in application of different anticounterfeiting inks. Anticounterfeiting and authentication technologies with high securities have commonly been on the basis of chromism and luminescence phenomena. The photochromic and photoluminescent (fluorescent and phosphorescent) polymers have been the most applicable materials in such applications with a high security and efficiency level and fast authentication. Polymers in the solution, colloidal, solid, or fibrous states could be used, if they show a high brightness, intensity, and even emission with different colors. Application of solvent-based polymeric inks containing different additives in security documents with permanent contact with human bodies may cause serious health problems. Therefore, water-based colloidal polymers were developed, in which the formulated anticounterfeiting/authentication inks are composed of high content of water. The polymer particles contain photoactive compounds could be dispersed in aqueous media and formulated for ink applications in this method [8, 63]. Therefore, most of the recent studies have focused on development of water-based polymeric inks with low prices, high photochromism and photoluminescence intensities, and also high lifetimes. Polymer particles containing organic or inorganic luminescent materials and chromic species are commonly used for this purpose. However, printing issues of the colloidal polymer particles are the main challenge for production of these smart inks in large quantities. Therefore, other types of polymer matrices, including fibrous or solid labels, are also applicable. Chemical or physical incorporation of photoresponsive materials, including different dyes, QDs, CDs, and UCNPs, are considered in the preparation of smart inks. Emission color, brightness, and intensity are the most significant factors in application of different fluorescent, phosphorescent, and chromic materials in anticounterfeiting inks. The fluorescent and phosphorescent materials could only induce one level of security to the polymer inks by their fluorescence emission; however, photochromic compounds could induce both of the fluorescence and photochromism characteristics under UV irradiation. In all these systems, highly important role of polymer matrices on lifetime, photostability, and fluorescence emission or chromism of these materials should be studied in detail.

The fluorescent and phosphorescent materials have extensively been studied for authentication, encoding, and anticounterfeiting applications. Figure 2.6 shows several examples of some of the fluorescent and phosphorescent materials incorporated in polymer matrices and used as smart inks. Coumarin-containing polymers were used for security making of confidential documents. Application of coumarin derivatives was studied in the preparation of fluorescent offset security ink on cellulosic paper [51]. The security ink was synthesized by mixing coumarin with a mixture of a very long alkyl resin, phenolic varnish, and maleic varnish. The fluorescent ink pattern with highly bright green fluorescence emission was observed under UV light irradiation with a wavelength of 365 nm. Figure 2.6a shows functional polymer nanoparticles prepared by surfactant-free emulsion polymerization with blue fluorescence emission due to physical/chemical modification of the latex nanoparticles with coumarin derivatives [105]. Epoxy-functionalized fluorescent polymer nanoparticles synthesized through emulsion copolymerization of methyl methacrylate (MMA), glycidyl methacrylate, and 7-acryloxycoumarin were also used for the preparation of photoluminescent cellulosic papers, which showed intensified fluorescence emission with potential applications as anticounterfeiting inks in security documents [372]. Figure 2.6b shows different colors for the fluorescent polymer nanoparticles prepared by physical post-polymerization modification methods with two different oxazolidine derivatives as highly fluorescent chromophores [156]. These multicolor fluorescent nanoparticles were applied as security inks for stamping on cellulosic papers that displayed highly bright fluorescence emissions in different colors. Anticounterfeiting fluorescent inks based on epoxy-functionalized polyacrylic nanoparticles modified with Rhodamine B were also synthesized for cellulosic substrates [370], as shown in Figure 2.6c. The pattern planned by this ink was immediately developed and showed brilliant red–pink fluorescence emission upon UV irradiation. An amorphous copolymer composed of 2-bromocarbazole phosphor and acrylamide on its side chains was synthesized and used as a phosphorescent polymer ink, which can engender blue–purple phosphorescence emission with high quantum yield at room temperature [373]. The polymer matrix and the hydrogen bonding between the polymer chains can effectively help to strengthen the phosphorescent emission. Three different phosphorescent

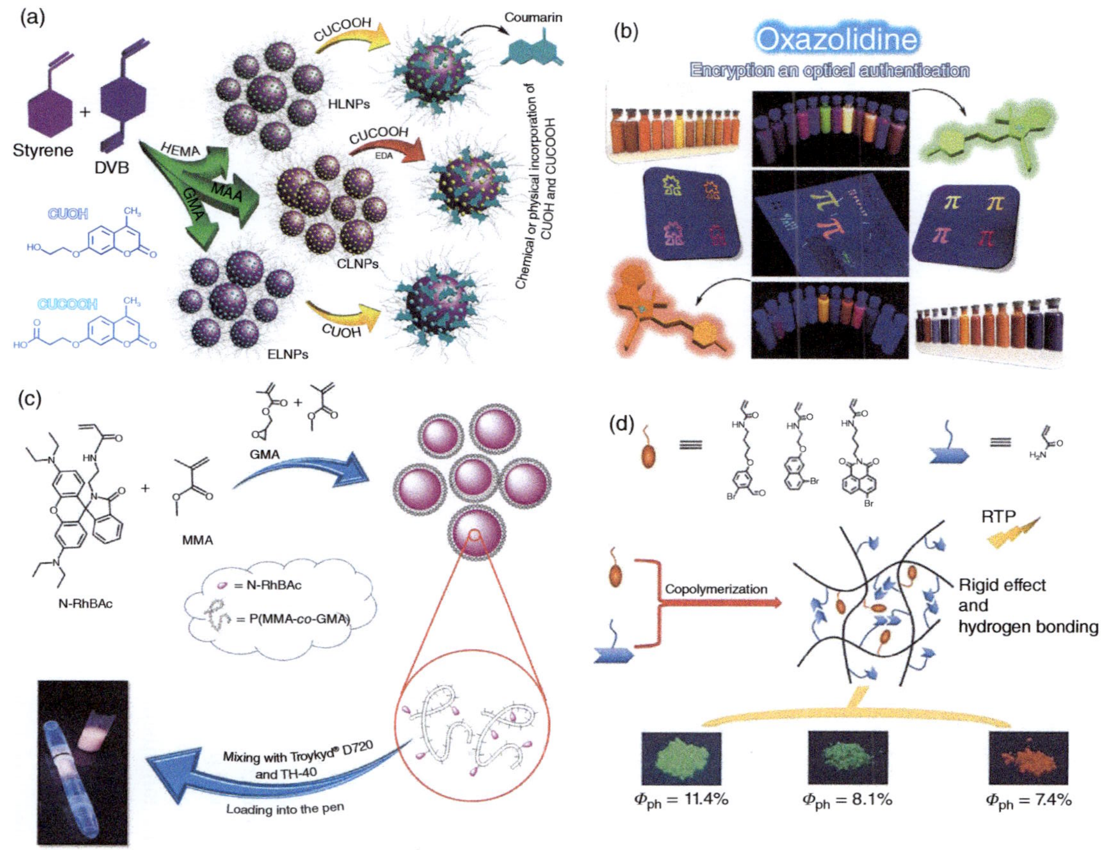

Figure 2.6 (a) Synthesis and application of polymer particles containing coumarin in the preparation of anticounterfeiting inks [105]. Source: Reproduced with permission from Abdollahi et al. [105]. Copyright 2020 Royal Society of Chemistry (RSC), (b) synthesis and application of polymers containing oxazolidine derivatives in the preparation of anticounterfeiting inks [156]. Source: Reproduced with permission from Abdollahi et al. [156]/Elsevier., (c) synthesis and application of polymer structures containing Rhodamine B in the preparation of anticounterfeiting inks [370]. Source: Reproduced with permission from Sun et al. [370]/Elsevier., and (d) synthesis and application of polymers containing 2-bromo-5-hydroxybenzaldehyde, α-bromonaphthalene, and 4-bromo-1,8-naphthalic anhydride derivatives in the preparation of anticounterfeiting inks [371]. Source: Reproduced with permission from Zhu et al. [371].

polymers were prepared by radical copolymerization, of acrylamide and phosphors containing a 2-bromo-5-hydroxybenzaldehyde, an α-bromonaphthalene, and a 4-bromo-1,8-naphthalic anhydride derivative, respectively [371]. These amorphous polymers showed strong phosphorescence with high quantum yields at room temperature, as shown in Figure 2.6d. Other fluorescent and phosphorescent compounds could also easily be incorporated into polymer solution, polymer particles, polymer fibers, and also solid polymer films.

Photochromic polymer inks show dual-mode authentication by color change in response to UV light irradiation and also fluorescence emission under UV excitation. Therefore, photochromic polymer inks are highly important in the development of anticounterfeiting inks. Anticounterfeiting polymeric inks based on spiropyran photochromic compounds have received much more attention [374–378]. Copolymerization or post-polymerization modification reactions are commonly used for incorporation of spiropyran into the polymer structures. The latter is highly efficient in development of photochromic inks due to surface decoration of polymer particles, which let the photochromic compounds absorb higher amount of UV irradiation. Figure 2.7 presents some typical examples for rewritable anticounterfeiting polymer inks based on photochromic compounds. Figure 2.7a shows functionalized stimuli-responsive latex particles containing spiropyran photoswitch with applications for marking on security documents and also photopatterning on cellulosic papers [21]. The prepared photoresponsive latex nanoparticles displayed fluorescence emission in different colors and photochromic coloration under UV light and after excitation with UV light, respectively. Such different colorations are function of polarity of the polymer media, type of its surface functional groups, and the small-content morphology of the particles. Water-based polymer inks with highly intense red, pink, and orange colors containing two different spiropyran derivatives were also developed as optical anticounterfeiting and authentication inks for security-marking of model security documents [151]. For this purpose, MMA and styrene nanoparticles were synthesized by emulsifier-free emulsion polymerization, and loaded to stamps with different patterns (quick response code, barcode, fingerprint, and other marks) for security marking of cellulosic papers. This work also shows that the color and fluorescence emission intensity of the patterns vary depending on the polarity of the polymer matrix and its functional groups, as shown in Figure 2.7b. Preparation of photoswitchable polyacrylic nanofibers and films doped with spiropyran by means of electrospinning and drop-casting techniques were also reported. Optical properties of the prepared photoresponsive acrylic nanofibers and films were investigated upon illumination with UV and visible lights [379], as shown in Figure 2.7c. Photochromic styrene and spiropyran copolymers with carboxylic acid chain end groups were synthesized by solution polymerization, and electrospun to yield photochromic nanofibers with well-defined morphology [380], as shown in Figure 2.7d. These nanofibers were applied on cellulosic papers and banknotes for induction of dual-mode security. Both the photochromic coloration and fluorescence emission showed high intensities, the security marks were rapidly authenticated upon UV irradiation,

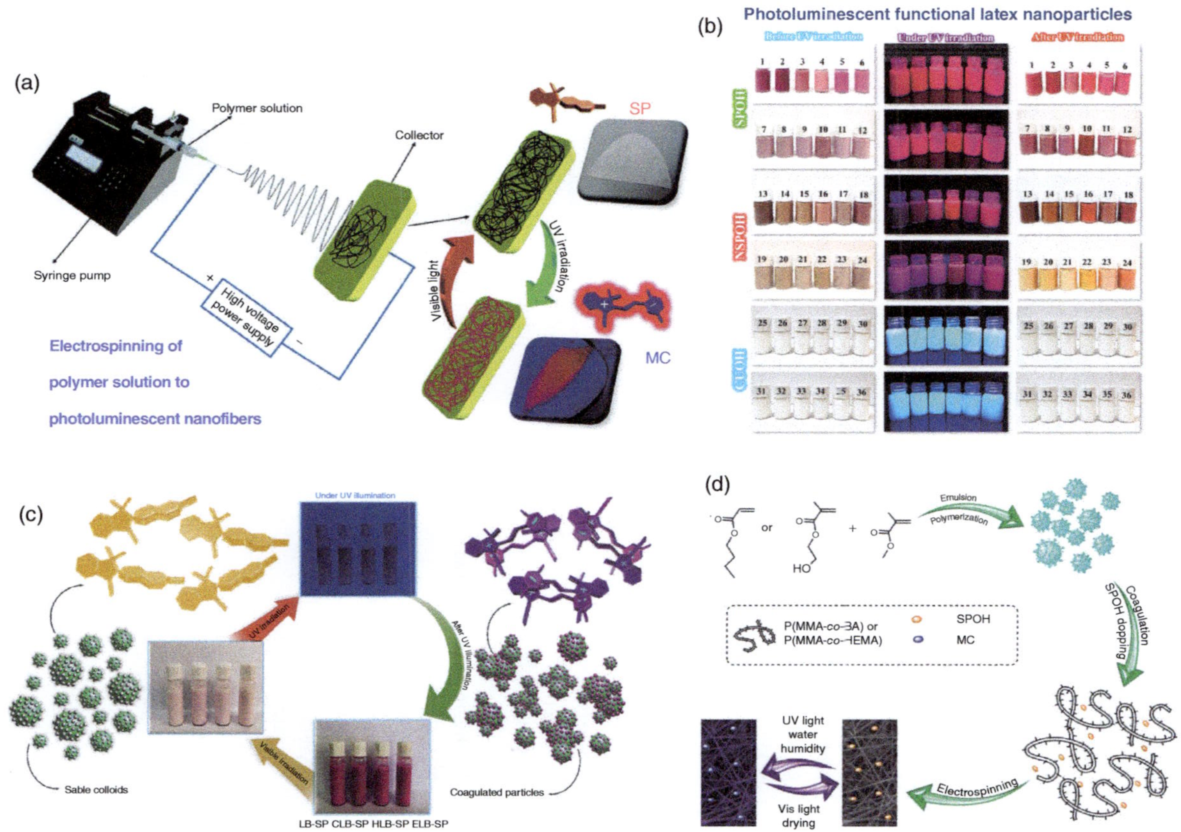

Figure 2.7 Synthesis and application of polymer structures containing different functional groups in the preparation of anticounterfeiting inks: (a) spiropyran-containing functional latex particles [21]. Source: Reproduced with permission from Abdollahi et al. [21]/Royal Society of Chemistry, (b) functional polymer particles with two spiropyran derivatives [151]. Source: Reproduced with permission from Abdollahi et al. [151]/Elsevier., (c) photoswitchable polyacrylic nanofibers and films doped with spiropyran [379]. Source: Reproduced with permission from Yang et al. [379]/American Chemical Society., and (d) polystyrene nanofibers containing chemically copolymerized spiropyran moieties [380]. Source: Reproduced with permission from Wang et al. [380]/Elsevier.

and this process was fully reversible. Two different commercial naphthopyran- and spirooxazine-based photochromic dyes were applied on PET fabric by inkjet printing and UV-curing. The photochromic behavior of the textile was also evaluated from the viewpoints of the controls of color intensities, switching speeds, and material durability [381]. Application of some different luminescent and photochromic compounds in polymer inks is summarized in Table 2.7.

Table 2.7 Application of some different luminescent and photochromic compounds in smart polymer inks.

Name	Polymer matrix	Polymer state	Printing method	References
Coumarin	Styrene and acrylic acid copolymer	Particle	Patterning	[381, 382]
	Methyl methacrylate and glycidyl methacrylate copolymer	Particle	Coating and handwriting	[380]
BODIPY	*tert*-Butyl methacrylate and methacrylated BODIPY random copolymer	Solution	Coating, handwriting, patterning	[383]
Fluorescein	Surfactant-free thiol-ene ink	Foam	Bubble writing	[129]
	Poly(lactic acid)/poly(glycidyl methacrylate) hybrid	Nanofiber mat	Silk-screen printing	[130]
Rhodamine	Methyl methacrylate and glycidyl methacrylate copolymer	Particle	Handwriting	[370]
Spiropyran	Styrene and *n*-butyl acrylate copolymer	Particle, film	Screen-printed textiles	[382]
	Styrene and *n*-butyl acrylate copolymer	Fiber	Electrospinning	[384]
	Polystyrene	Solution, fiber	Electrospinning	[380]
	PMMA	Particle	Patterning, stamping	[385]
	Styrene and acrylamide copolymer	Particle	Patterning	[376]
	MMA and *n*-butyl acrylate copolymer	Particle	Patterning	[16]
Oxazolidine	Copolymer of MMA with HEMA, GMA, AAM, and DMAEMA Copolymer of styrene with HEMA, GMA, AAM, and DMAEMA	Particle	Patterning	[156]

2.3.2 Polymer Sensors

Design of fluorescent and photochromic polymeric chemosensors have recently attracted attentions of chemists, clinical biochemists, biologists, and environmentalists because of their simplicity, high selectivity, and sensitivity in color and fluorescent assays. In fluorescent chemosensors, variation of physicochemical properties upon interaction with different chemicals result in change in fluorescence characteristics. Fluorescent chemosensors are composed of a signaling fluorophore and a receptor with a recognition capability that is connected by a spacer. By binding guest species to the receptor, fluorescence intensity, emission wavelength, or fluorescence lifetime of the fluorophore are changed via different mechanisms [97]. Conventional sensing mechanisms are PET, intramolecular charge transfer (ICT), metal–ligand charge transfer (MLCT), twisted intramolecular charge transfer (TICT), electronic energy transfer (EET), fluorescence resonance energy transfer (FRET), aggregation-induced emission (AIE), ESIPT, and excimer/exciplex formation [386, 387]. The conventional fluorescein, rhodamine, and cyanine fluorophores have bright emissions only in dilute solutions, and are quenched at high concentrations. AIE emission fluorogens, such as tetraphenylethene (TPE), tetraphenylpyrazine (TPP), silole, quinoline-malononitrile (QM), cyanostilbene, 9,10-distrylanthracene (DSA), and organo-boron complexes, have highly intense emission upon aggregate formation [97]. Incorporation of these fluorophores in the matrix of smart polymers may lead to preparation of highly sensitive polymeric chemosensors. Polymeric fluorescence chemosensors have commonly been used for detection of metal cations, anions, neutral molecules, biomolecules, and also pH of aqueous media [386]. Conjugated polymer sensors, polymers containing pendant ligand groups, dendrimer-based sensors, and imprinted polymer-based sensors are the main categories of the polymer fluorescence chemosensors. Figure 2.8a shows novel NIR fluorescence silver ion sensors of a dendrimer containing a perylenediimide core as the energy acceptor and the 1,8-naphthalimide derivative as the energy donor [388]. They observed that the probe showed selective fluorescence responses to protons and silver ions. The probe exhibited rapid, sensitive, and selective dual-channel fluorescence responses (at 480 and 784 nm) toward silver ions and protons based on its special structure that includes FRET, ICT, and PET. In addition, in response to silver ions, the probe exhibited a color change visible by the naked eye. A fluorescent poly(amidoamine)-based wavelength-shifting bichromophore was synthesized and labeled with yellow-green emitting 4-(*N*-piperazinyl)-1,8-naphthalimide "donor" units capable of absorbing light and efficiently transferring the energy to a focal Rhodamine 6G "acceptor" [385]. Due to the PET in the periphery and pH-dependent spirolactam to ring-opening amide equilibrium of rhodamine core, novel bichromophoric system acted as a selective ratiometric pH fluorescence probe, as shown in Figure 2.8b. A "turn-on" thiol-responsive fluorescence probe was into polymeric nanoparticles was used for detection of intracellular thiols with a photoinduced electron-transfer process in the off state of the probe, which is terminated upon the reaction with thiol compounds [377], as shown in Figure 2.8c.

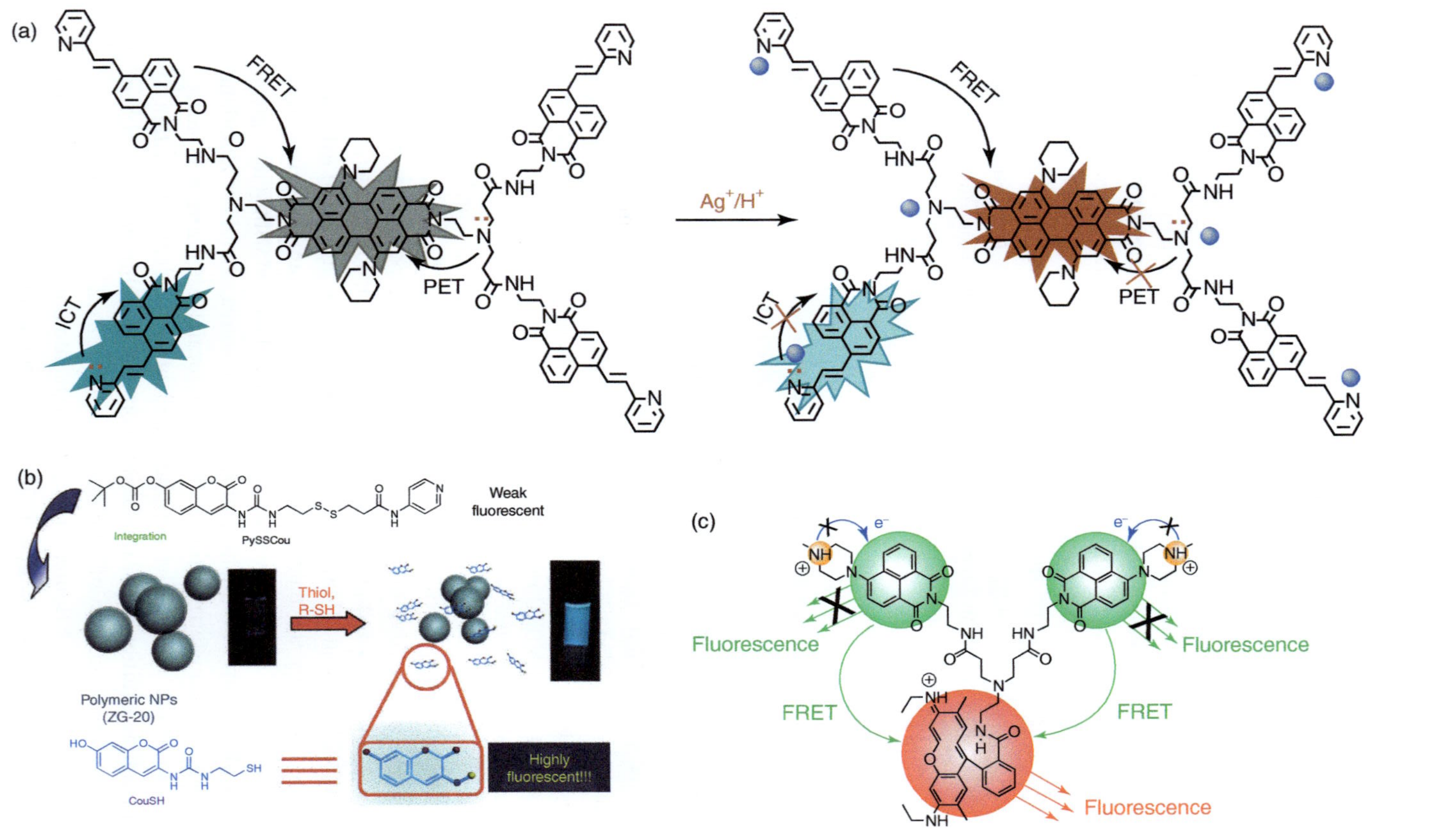

Figure 2.8 (a) Possible mechanism of the near-infrared star-shaped probe dendrimer with silver cation and proton [388]. Source: Reproduced with permission from Zhu et al. [388]. Copyright 2019 John Wiley & Sons Inc., (b) the mechanism for pH detection by poly(amidoamine) (PAMAM) dendrimer decorated with donor and acceptor molecules [385]. Source: Reproduced with permission from Cheng et al. [385]/Springer Nature., and (c) the thiol-responsive fluorescence probe in polymeric nanoparticles for sensing intracellular thiols [377]. Source: Reproduced with permission from Nikolai I. Georgiev et al.,(2014) [286] / Elsevier.

Photochromic sensors, unlike single-state sensors, have numerous advantages. The diagnostic forms of photochromic sensors can be altered by different exciting lights, resulting in more sensing information for analysis and multiple signal outputs. Multiple external stimulations can be monitored and detected by analysis of the changes in signal outputs. The photoinduced process can produce isomers with several diagnostic units that can be attached to different analytes. The isomerization of the photochromic sensors can be adjusted not only with the light, but also with pH values [378], temperature [379], electric fields [389], and ion species [390]. This type of sensor is applicable in different fields, i.e. for specific detection (pH, solvent temperature, metal ions, anions, gas and vapor, and biomolecules) and multianalyte discrimination. Spiropyran and its derivatives have shown to be an excellent example of photochromic compounds as efficient sensors for multiassay detection. Based on the solvatochromism of spiropyran, fused silica microcapillaries functionalized with spiropyran-polymer brushes and synthesized through surface-initiated ring-opening metathesis polymerization were used to identify six different solvents with different polarities [228]. In various solvents, spiropyran showed different colors and absorption spectra under UV excitation; an increase in polarity leads to change of absorption and color from blue to red. The solvent-dependent color of spiropyran resulted in detecting capability of the SP polymeric brush-coated capillaries. Various solvents were passed through the capillaries upon UV light irradiation, and then the hypsochromic shifts of the colors from highly polar solvents (such as methanol) to highly nonpolar solvents (such as toluene) were observed. The colorless spiropyran polymer coating can also be recovered under white-light irradiation, provided that the polymer has potential for fabrication of reusable miniaturized sensing platforms, as shown in Figure 2.9a. Shiraishi et al. designed a copolymer that consisted of poly(*N*-isopropylacrylamide) (PNIPAAm) and spiropyran units to create a colorimetric thermometer [227], as shown in Figure 2.9b. The copolymer behaves as a colorimetric thermometer exhibiting linear and reversible bathochromic/hypsochromic shift of the absorption spectrum in a wide-temperature range (10–34 °C), and increase in temperature led to change in the color from orange (10 °C) to pink and then to purple (34 °C, $\lambda_{max} = 546$ nm). The bathochromic shift of the absorption spectra and color changes are due to the conversion of MC from the zwitterionic to quinoidal form, which was induced by the coil-to-globule phase transition of PNIPAAm through increasing temperature. When the temperature reaches above 34 °C, the thermal reversion of MC to SP form led to the color fade. Therefore, the copolymer could be used as a temperature indicator in the range of 10–34 °C. A reversible ion sensor based on spiropyran-containing polymer brushes was prepared using surface-initiated copolymerization of a spiropyran methacrylate derivative and MMA. The confined microenvironment provides a colorimetric response that is sensitive and selective for different metal ion complexation [391], as shown in Figure 2.9c. Application of some different luminescent and photochromic compounds in polymer sensors is summarized in Table 2.8.

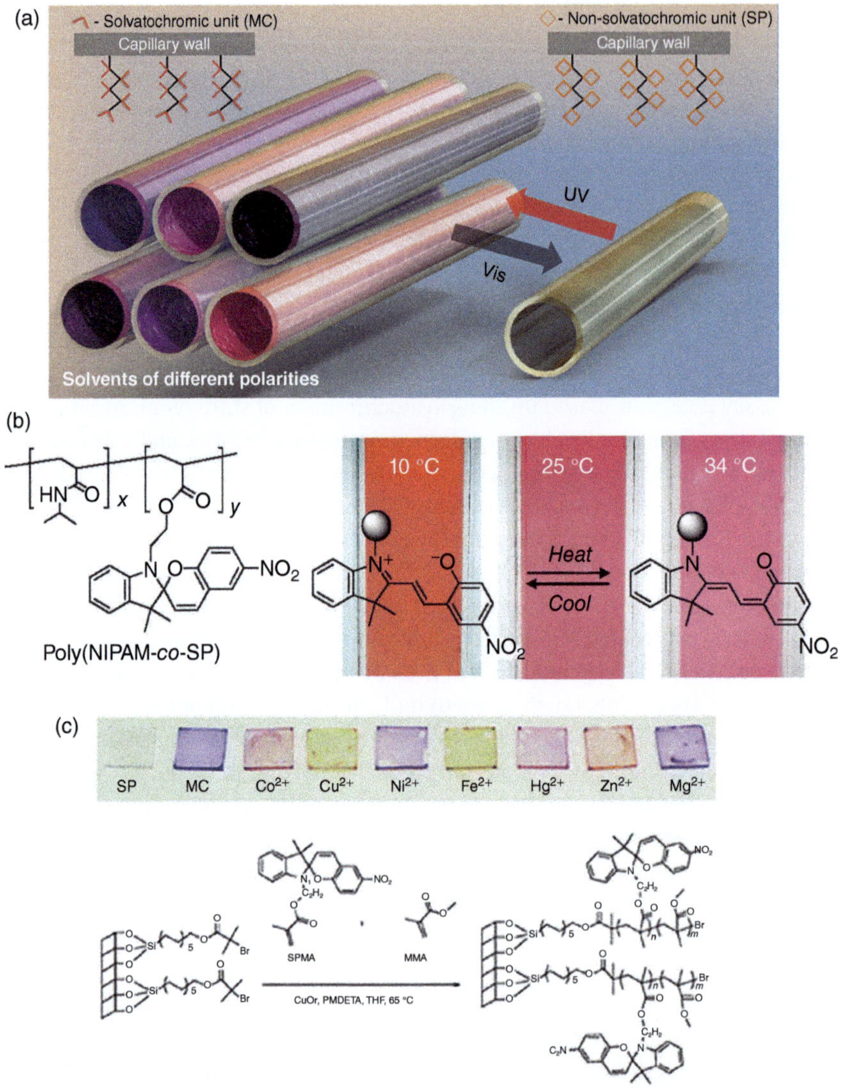

Figure 2.9 (a) Response of spiropyran-containing polymer brush coated microcapillaries to different polarized solvents passing through [228]. Source: Reproduced with permission from Florea et al. [228]/American Chemical Society, (b) temperature colorimetric sensing by *N*-isopropylacrylamide and spiropyran copolymer [227]. Source: Reproduced with permission from Shiraishi et al. [227]/American Chemical Society, and (c) selectivity of spiropyran-containing polymer brushes to different ions [391]. Source: Reproduced with permission from Fries et al. [391]/Royal Society of Chemistry.

Table 2.8 Application of some different luminescent and photochromic compounds in polymer sensors.

Name	Polymer matrix and shape	Analyte	Mechanism	References
Coumarin	Styrene and DMAEMA copolymer nanoparticles	CO_2	Fluorometry	[17]
	Grounded methacrylic acid and coumarin copolymer	pH	Fluorometry based on PET	[392]
	Coumarin–triazole polymer	Hg^{2+}	Ratiometry based on ESIPT	[18]
	Polyethylene film	Cu^{2+} and Zn^{2+}	Colorimetry based on PET	[393]
Fluorescein	Poly(ethylene glycol) diacrylate hydrogel fiber	Iodine ion	Fluorescence quenching	[19]
	Polystyrene particle	pH	Ratiometry	[394]
Spiropyran	Cellulosic papers modified by spiropyran-acrylic copolymer	pH	Photochromism	[20]
	Polydimethylsiloxane composite	HCl vapor	Photochromism	[395]
	PNIPAAm	Cyanide anion	Photochromism	[396]
	PMMA	Mechanical force	FRET	[397]

2.3.3 Photolithography

The stimuli-responsive photoresist polymers, which are effectively degraded upon exposure to UV light, are significantly interested in preparation of photopatterns. Photolithography is one of the methods used to transfer a pattern onto a substrate by means of an irradiation source and a photoresponsive polymer. Smart photoresist polymers, which are degraded upon UV light irradiation, have significantly been interested in the preparation of photopatterns with application in biotechnologies [290, 398], sensor design [16, 21, 22, 399], microelectronics [400], and microfluidics [401]. Figure 2.10 shows the process of light-induced patterning using photo-cleavable polymers. PNIPAAm-based photoresists with temperature-triggered development are shown in Figure 2.10a, in which the patterns could irreversibly be transferred on substrate [290]. Temperature and photoresponsive random copolymers of *N*-isopropylacrylamide and ONB were synthesized in different molar ratios of *o*-nitrobenzyl acrylate to *N*-isopropylacrylamide. The prepared copolymer was used in photolithography applications managed by altering lower critical solution temperature (LCST) by the photocleavage reaction. These copolymers are soluble in an aqueous media below their LCST. After exposure to UV illumination, the photocleavable groups are detached from the side chain of copolymer resulting in an increased LCST due to conversion of the hydrophobic copolymer to hydrophilic

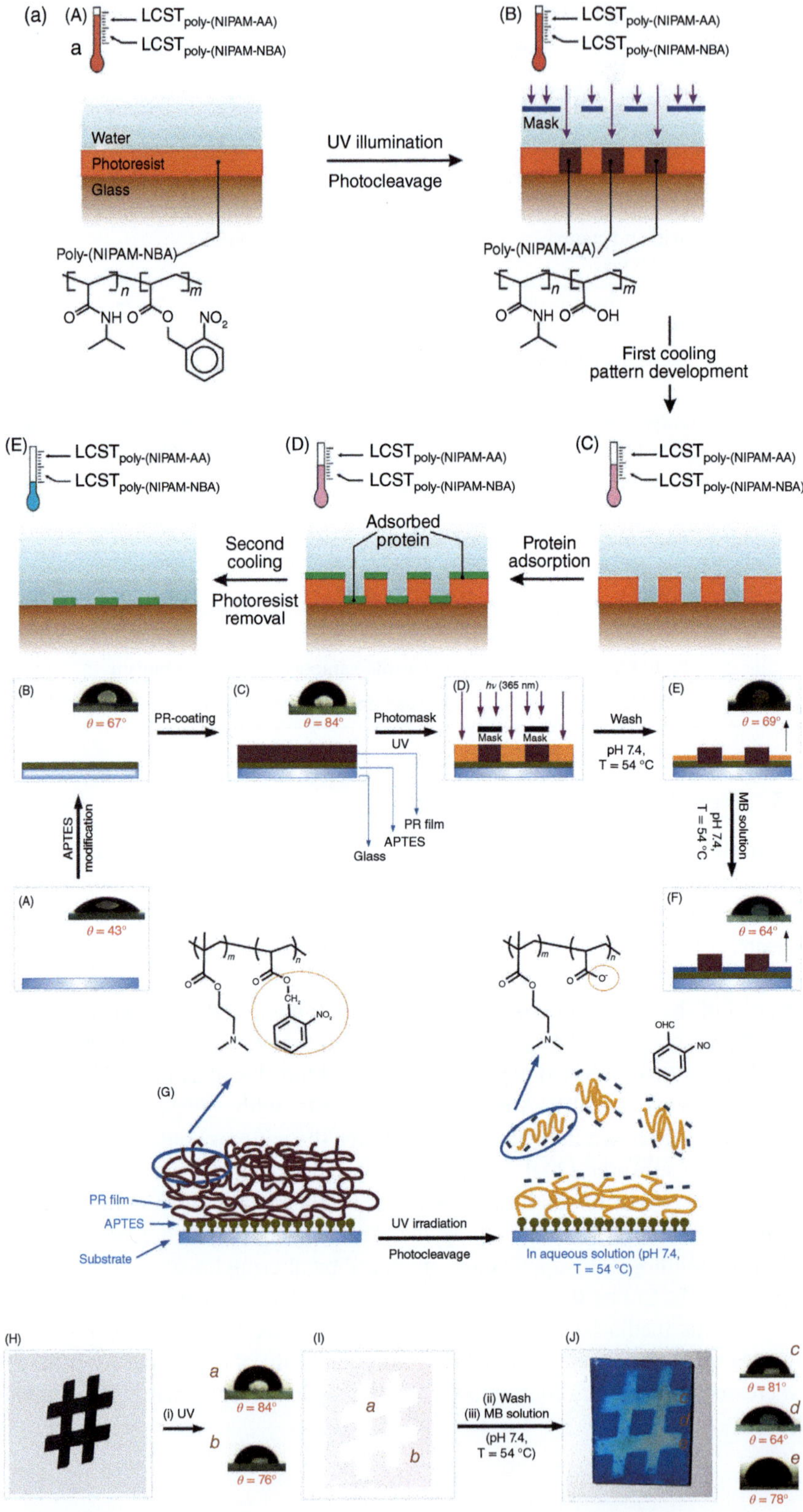
(a)
(A)
a
LCST poly-(NIPAM-AA)
LCST poly-(NIPAM-NBA)
Water
Photoresist
Glass
Poly-(NIPAM-NBA)
UV illumination
Photocleavage
(B)
Mask
Poly-(NIPAM-AA)
First cooling
pattern development
(C)
Protein adsorption
(D)
Adsorbed protein
Second cooling
Photoresist removal
(E)
(B)
θ = 67°
PR-coating
(C)
θ = 84°
Photomask
UV
(D)
hν (365 nm)
Mask
Wash
pH 7.4, T = 54 °C
(E)
θ = 69°
PR film
APTES
Glass
APTES modification
MB solution
pH 7.4, T = 54 °C
(A)
θ = 43°
(F)
θ = 64°
(G)
PR film
APTES
Substrate
UV irradiation
Photocleavage
In aqueous solution (pH 7.4, T = 54 °C)
(H)
(i) UV
θ = 84°
θ = 76°
(I)
(ii) Wash
(iii) MB solution
(pH 7.4, T = 54 °C)
(J)
θ = 81°
θ = 64°
θ = 78°

Figure 2.10 Schematic illustration for (a) photopatterning using PNIPAAm-based photoresists with temperature-triggered development with the deposition of thermos-responsive poly(2-nitrobenzyl acrylate-*co*-*N*-isopropylacrylamide) photoresist onto a substrate, UV illumination of the photoresist through a photomask followed by photocleavage of ONB segments and increasing the LCST, development of the photoresist pattern at slightly lower temperature, proteins adsorption, and complete removal of the photoresist in aqueous environment at low temperatures [290]. Source: Reproduced with permission from Ionov and Diez [290]. Copyright 2009 American Chemical Society (ACS), (b) the procedure to transfer methylene blue dye onto a glass-based smart surface with pH- and temperature-triggered development via photolithography technique [23]. Source: Reproduced with permission from Tajmoradi et al. [23]/Elsevier.

one. Consequently, the illuminated parts of spin-coated copolymer layers dissolve at higher temperatures than the LCST, leading to pattern development. The stimuli-responsive copolymers of *N*,*N*′-dimethylaminoethyl methacrylate and ONB acrylate prepared by reversible addition-fragmentation chain transfer (RAFT) polymerization were used as a water-soluble photoresist to transfer a methylene blue (MB) pattern onto glass surface by using photolithography technique [23], as shown in Figure 2.10b. The procedure was followed by copolymer film deposition on the 3-aminopropyltriethoxysilane-modified glass substrate via deep coating, illumination of the film through a photomask, rinsing with water (pH 7.4, $T = 54\,°\mathrm{C}$), and immersing in MB solution (pH 7.4, $T = 54\,°\mathrm{C}$). Responsivity of the photoresist polymer-modified glass to UV light, masking substrate, the illuminated substrate through a photomask, and MB-patterned glass surface are also shown in Figure 2.10b.

2.3.4 Surface Active Agents

Photoresponsive polymeric surfactants have gained considerable attention in the preparation of heterogeneous polymer systems, such as dyestuff, paper coatings, inks, pharmaceuticals, and personal care and cosmetic products [402]. One of the most significant applications of photoresponsive polymeric surfactants is decreasing interfacial tension between oil and water and also stabilization of oil-in-water (O/W) and water-in-oil (W/O) droplets in heterogeneous media. Hydrophilic/hydrophobic block copolymers and graft copolymers are the most appropriate photoresponsive polymeric surfactants. In the case of emulsion stabilization by reversible chromophore-containing polymeric surfactants, such as azobenzene, coumarin, and spiropyran compounds, the emulsification/demulsification process is reversible upon UV/visible light irradiation due to the hydrophobic/hydrophilic transition of the surfactant under light irradiation switches [228, 403–407]. In the case of the photoresponsive azobenzene [213, 230, 231], the chromophore changes its conformation from an apolar trans isomer to a polar cis one, reversibly, when illumination is switched between UV and visible light. Effect of light-responsivity was investigated on the stability and controlling foamability of an emulsion prepared using a temperature and photoresponsive polymeric surfactant containing an azobenzene chromophore [408]. Photoresponsive O/W emulsions were prepared using *n*-dodecane and aqueous solution of polymer in trans state of azobenzene. The advantage of this emulsion system is reversible control of the emulsion stability by switching light irradiation wavelength (UV/blue or dark) at temperatures above the phase-inversion temperature. However, photoisomerization of the trans-to-cis isomer under UV light irradiation caused a reduction of molecular area of the polymeric surfactant at the O/W interface, and then decreasing stability of the emulsion was led to demulsification on the basis of coalescence of the oil droplets. In other words, no foam could be created upon UV light irradiation. Reversible transition between emulsification and demulsification of a light-responsive *n*-octane in water Pickering emulsion was also studied by using azobenzene ionic liquid surfactant. Adsorption of azobenzene ionic liquid surfactant on the surface of the Pd-supported silica nanoparticles and the light-responsive activity of ionic liquid surfactant caused such a behavior [290]. Effect of illumination was studied on stability of the emulsions prepared using a photoresponsive surfactant containing a coumarin chromophore as a cross-linker [407], as shown in Figure 2.11a. Dual-light and pH-responsive linear amphiphilic random poly(methacrylic acid-*co*-methyl methacrylate-*co*-7-(4-vinylbenzyloxyl)-4-methylcoumarin) self-assembled into micellar structure in dimethylformamide/H_2O solvent was used as a particulate emulsifier to stabilize an O/W system. The swelling behavior and emulsification performance of the resultants stimuli-responsive micelles are investigated by the dosage of UV irradiation and pH variation. The photocross-linking of coumarin units under alkaline conditions and UV irradiation of $\lambda > 310$ nm results in interchain cross-linking and interparticle cross-linking in the micelle emulsifiers causing to increase in the size and polydispersity index of the micelles. The photocross-linked micelle emulsifiers became more rigid with increasing pH due to the increasing interchain cross-linking and electrostatic repulsion. These factors make them difficult to adsorb at the paraffin oil/water interface, displaying lower emulsification efficiency and lower stability of the emulsions over time. Effect of photocross-linking was also studied on emulsifying performance of the self-assemblies of photo and pH-responsive poly(7-(4-vinylbenzyloxyl)-4-methylcoumarin-*co*-acrylic acid) [406], in which structure of the obtained polymeric micelles could be simply controlled by the amphiphilicity of the copolymer. As the amount of shrinkage degree was increased by photocross-linking, size of the emulsion droplets was significantly increased and the micelles showed lower emulsification efficiency. Kim et al. prepared block copolymer particles switching their shape and color in response to light by using a spiropyran-dodecyltrimethylammoium bromide surfactant that changes its amphiphilicity upon photoisomerization. The hydrophilic merocyanine form of the surfactant affords the formation of spherical and onion-like particles. However, the ring-closed form of the surfactants yields prolate or oblate ellipsoids with axially stacked nanostructures [409], as shown in Figure 2.11b.

Emulsions prepared by irreversible photoresponsive surfactants can irreversibly be emulsified or demulsified upon irradiation with an appropriate wavelength of light. Photoresponsive surfactants based on irreversible photoresponsive ONB and coumarin chromophores are the most important types of this category [23, 27, 108]. Hydrophobic/hydrophilic transition in stimuli-responsive copolymers of *N*,*N*′-dimethylaminoethyl methacrylate and ONB acrylate prepared by RAFT polymerization was also studied. The smart transition was controlled by the block lengths, light irradiation, pH and temperature variation, and CO_2 purging. CMC of the copolymers was determined by turbidity measurement and fluorescence spectroscopy using coumarin as the fluorescence probe. This copolymer was also used as an emulsifier to prepare stable O/W emulsions and subsequent synthesis of colloidal polystyrene latex by emulsion polymerization [23], as shown in Figure 2.12a,b. Stimuli-responsive block copolymer-grafted CNC was also prepared via surface-initiated RAFT polymerization, and used as a smart

(a)

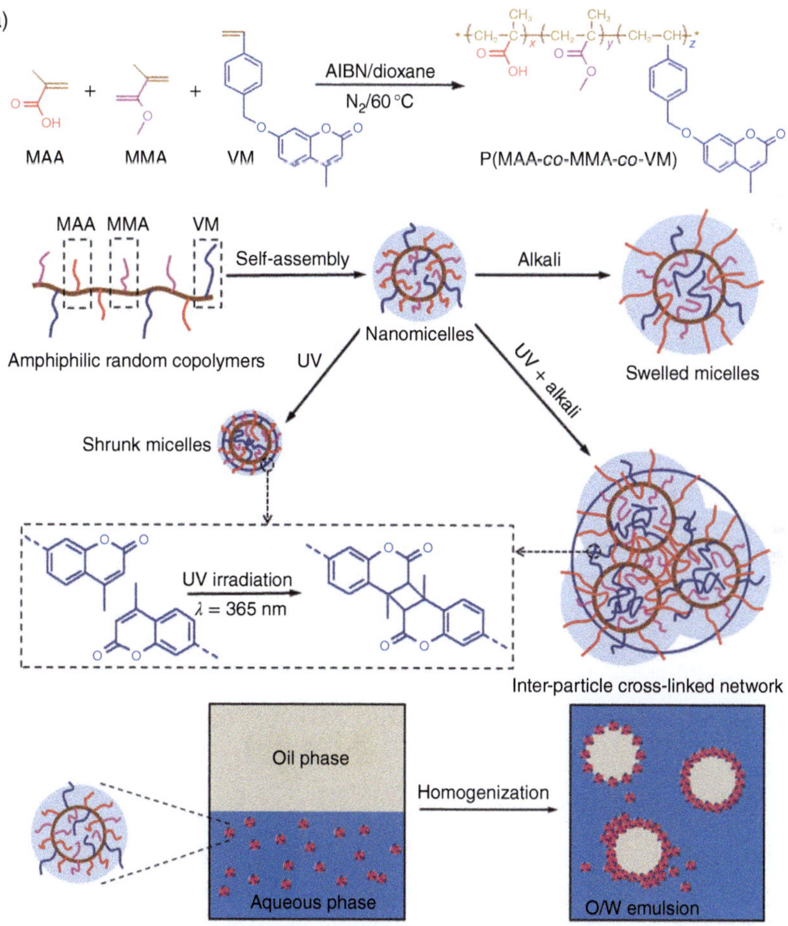

Figure 2.11 (a) Schematic illustration for the synthesis of random copolymer, its self-assembly and responsivity behavior of the obtained micelles, and finally, Pickering emulsions stabilized with the micelles as particulate emulsifiers [407]. Source: Reproduced with permission from Wang et al. [407]. Copyright 2018 American Chemical Society (ACS) and (b) synthesis of the spiropyran-dodecyltrimethylammoium bromide surfactant and light-responsivity of the block copolymer particles changing their shape through wavelength-selective photoisomerization of the surfactants [409]. Source: Reproduced with permission from Kim et al. [409]. Copyright 2021 American Chemical Society (ACS).

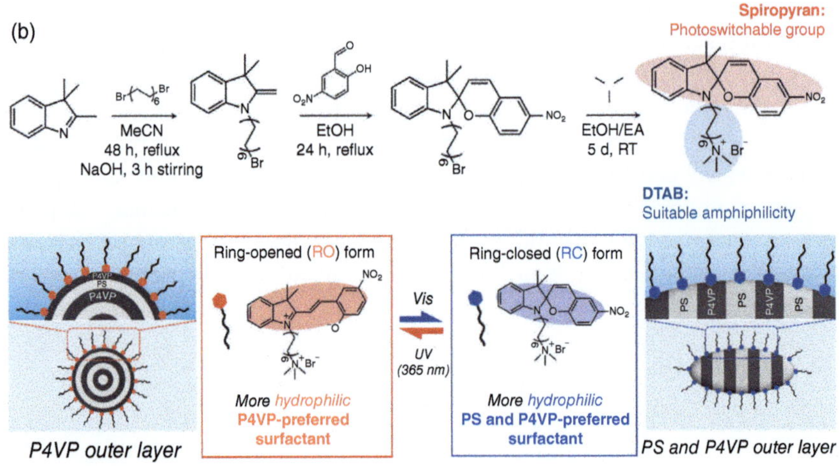

Figure 2.11 (*Continued*)

surfactant to run Pickering emulsion polymerization of styrene. Stability and emulsification behavior of styrene-in-water droplets under UV light irradiation was also investigated, as shown in Figure 2.12d. Irreversible demulsification was observed by separation of the surfactant from styrene and water interface under UV irradiation, due to increasing hydrophilicity of the surfactant [27]. Emulsions prepared by surfactants on the basis of host–guest modulation can be emulsified or demulsified upon irradiation with an appropriate wavelength of light. Application of azobenzene photoisomerization to control the host–guest interaction of an azobenzene-surfactant and cyclodextrin and tune molecular amphiphilicity was studied. The azobenzene-surfactant formed vesicle-like aggregates in aqueous solution. Binding α-CD to the azobenzene enhanced water solubility and disrupted the vesicles. However, upon photoirradiation with UV-light, α-CD cannot bind with the azobenzene anymore and slides over the alkyl chain, which resulted in the reformation of vesicles. This process could be reversed by applying visible light and recovering azobenzene into its trans form [410], as shown in Figure 2.12d.

2.3.5 Photorheological Polymers

Photorheological fluids can modify their rheological properties, most importantly viscosity, by light irradiation. Controlling viscosity of fluids by light has potential applications in valves, sensors, microfluidic devices, shock absorbers, etc. [33, 411–416]. In photorheological polymers, the mechanism for viscosity modulation is mainly based on physical interactions between the light-responsive materials incorporated into the chain structures. Such physical interchain interactions are mainly reversible and based on the electrostatic interactions, hydrogen bonding, van der Waals forces, and hydrophobic interactions. These reversible photoisomerization reactions under light irradiation result in viscosity variations in reversible and repeatable cycles by different mechanisms of polymer self-assembly structure

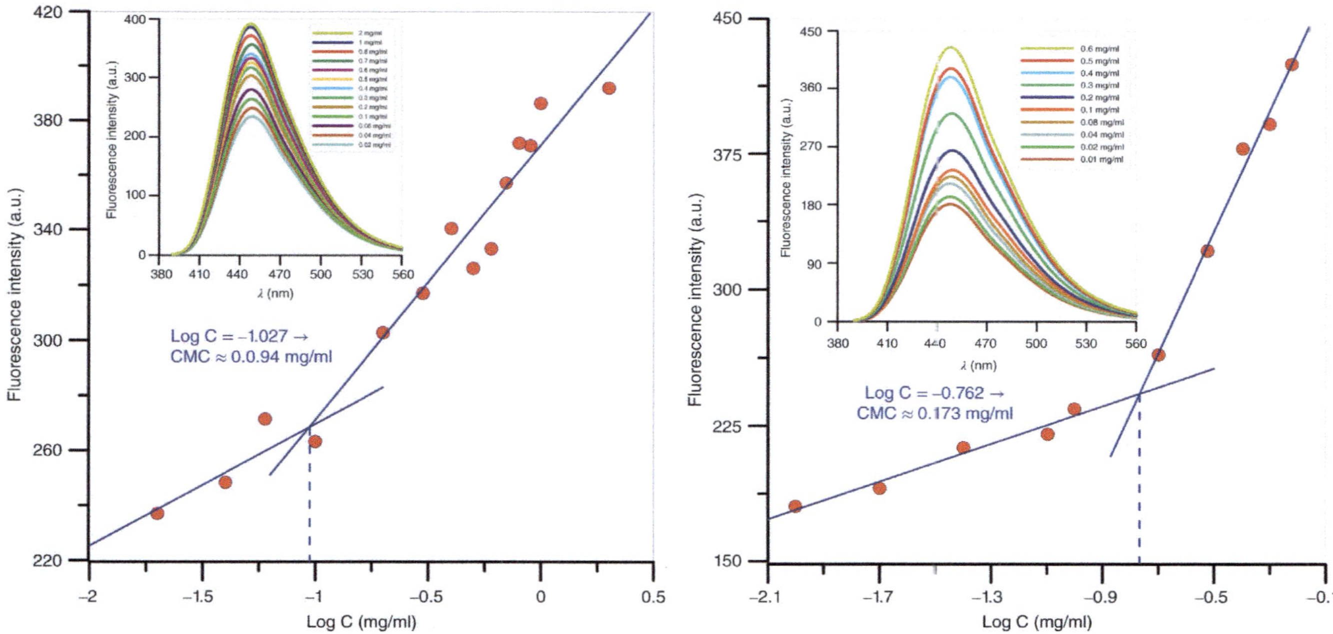

Figure 2.12 (a, b) Determination of CMC values for the copolymer before and after cleavage of ONB groups, respectively, by plotting emission intensity at 450 nm versus the log of micelle concentration [23]. Source: Reproduced with permission from Tajmoradi et al. [23]/Elsevier, (c) photographs of the emulsions stabilized with the copolymer after cleavage of ONB groups and responsive behavior of the emulsions under different stimuli [27]. Source: Reproduced with permission from Tajmoradi et al. [27]/Elsevier, and (d) emulsions prepared by surfactants on the basis of host–guest modulation [410]. Source: Reproduced with permission from Chen et al. [410]. Copyright 2003 Royal Society of Chemistry.

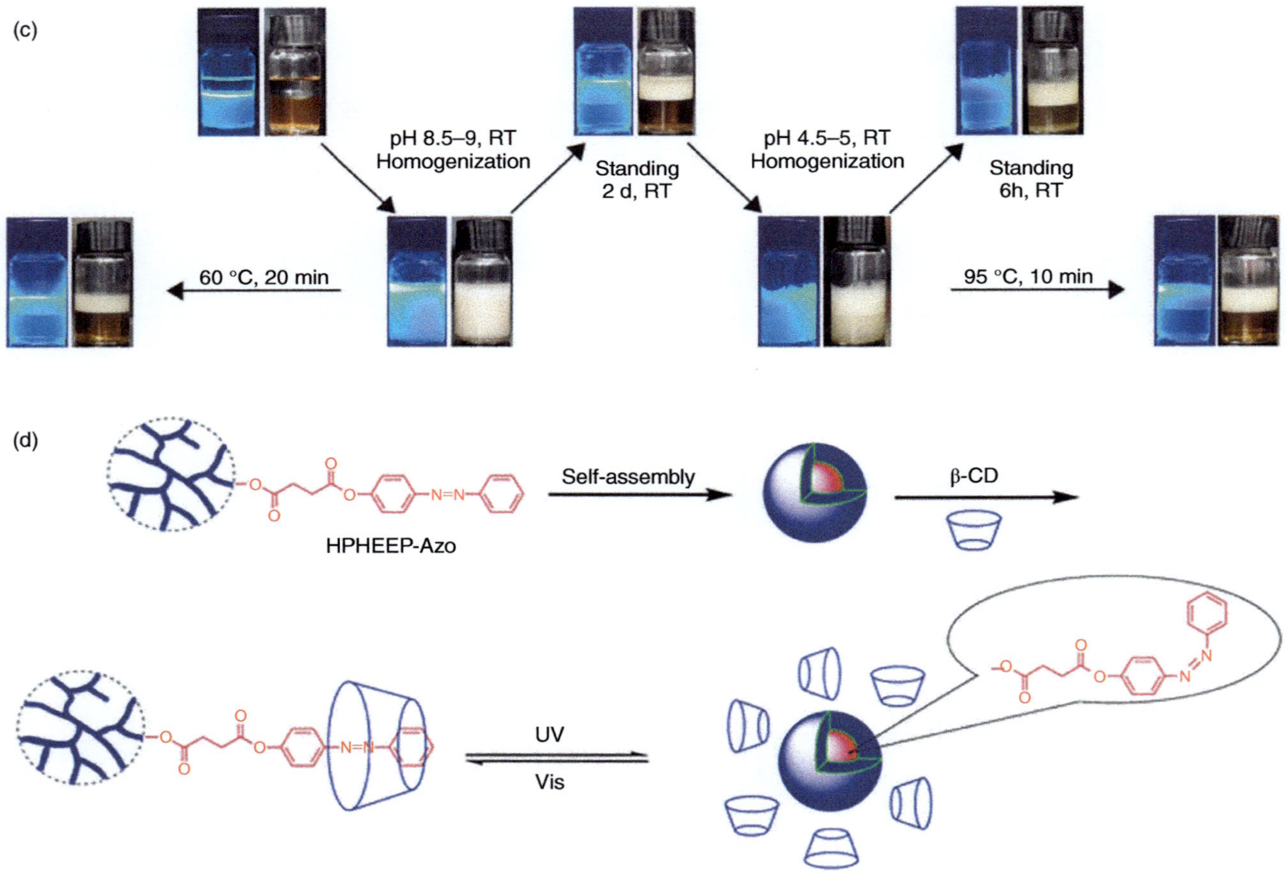

Figure 2.12 (*Continued*)

variation or interchain interactions [34–36]. The irreversible light-responsive compounds can also be used for modification of viscosity in polymer systems. Covalent bond cleavage or formation mechanisms result in variation of polymer chain structure, length, hydrophilicity/hydrophobicity, and self-assembly in media, which can result in irreversible viscosity variation in polymer systems.

Variation of rheological properties under light irradiation in fluorescent materials can be accompanied by changes in fluorescence characteristics of the fluorescent materials. Du et al. prepared a coumarin-functionalized telechelic associative model polymer and used it to study the effect of light-induced coumarin dimerization on the polymer structure and solution rheological properties. They studied effects of light intensity and its irradiation time, polymer concentration, and temperature on the transient network and its rheological behavior in aqueous solution. A novel transient network model composed of train loops and bridges covalently linked to train loops was proposed to describe the unique solution properties [417], as shown in Figure 2.13a. Reinicke and coworkers used anthracene for reversible tuning of rheological properties in a 9-anthracene ester-terminated polydimethylsiloxane. Reversible dimerization of anthracene esters is used to cross-link polydimethylsiloxane chains by UV radiation at the wavelength of 365 nm or cleave the chains by UV radiation at the wavelength of 254 nm or elevated temperatures (>130 °C) [418], as shown in Figure 2.13b. Variation of rheological properties under light irradiation in photochromic materials may also be accompanied by color changes or other photochromic characteristics. Azobenzene-containing polymers are also highly important photorheological polymers that show variation in viscosity by cis–trans isomerization of the azobenzene units [32, 33]. For a spiropyran-containing polymer, viscosity modification of the polymer solution by light irradiation can be achieved because of changing interactions between the polymer chains after transformation of the spiropyran to merocyanine form upon light irradiation [37, 419–421]. In such smart systems, viscosity variation is precisely controllable by the intensity and wavelength of the irradiated light. Peng and coworkers synthesized ethoxylated telechelic polyurethane chains containing spiropyran end groups, which were self-assembled into micellar structures in aqueous media. They showed reversible hydrophilic–lipophilic balance variation of polymer chains by conversion of the spiropyran to merocyanine form upon alternating UV and visible light irradiation. Subsequently, reversible rearrangement of micellar junctions through bridge-to-loop or loop-to-bridge transitions resulted in reversible variation in their photorheological properties [37], as shown in Figure 2.13c. Dou et al. prepared a block copolymer containing bifunctional spiropyran, poly(ethylene glycol) (PEG), and poly(propylene glycol) (PPG). They reported that spyropiran and PPG induce increase of viscosity by their isomerization to the MC form and in response to temperature increase, respectively. The main reason for increase of viscosity by light irradiation is reported as the π–π stacking interactions of the MC forms in the adjacent polymer chains [413]. Variation of rheological properties under light irradiation in irreversible light-responsive materials could be followed by different

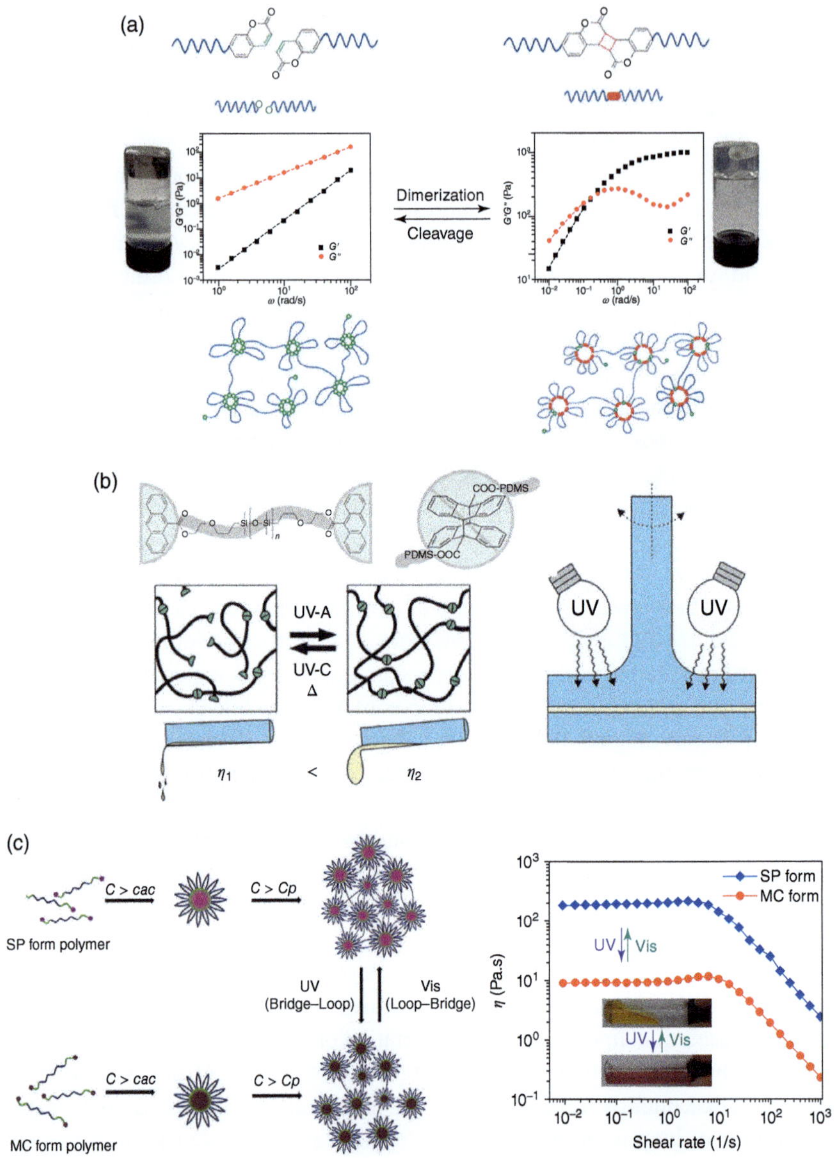

Figure 2.13 (a) Preparation of a coumarin-functionalized telechelic associative model polymer and its application in study of the effect of light-induced coumarin dimerization on the polymer structure and solution rheological properties [417]. Source: Reproduced with permission from Du et al. [417]/American Chemical Society, (b) Chemical structure of the spiropyran-containing polyurethane, its self-assembly and morphological changes upon UV irradiation, and variation of rheological properties [418]. Source: Reproduced with permission from Gäbert et al. [418]. Copyright 2017 American Chemical Society, and (c) reversible tuning of rheological properties in 9-anthracene ester-terminated polydimethylsiloxane [38]. Source: Reproduced with permission from Ling et al. [38]/The Society of Rheology.

evidence than the fluorescence and chromism phenomena. For this purpose, ONB moieties can be used for irreversible variation of hydrophilicity/hydrophobicity and consequent morphology transition and viscosity changes.

2.3.6 Self-Healing Polymers

Breaking of chemical bonds or polymer chains by heat, light, and external mechanical forces caused damage inside a polymer or its surfaces, such as in the formation and propagation of a crack or fracture. The polymers with self-healable properties, self-healing polymers can repair cracks or fractures by themselves. With the exception of a few self-healing polymer systems whose healing process is truly autonomous, i.e. without any assistance and in environmental conditions, most polymers require external stimuli, such as light, heat, magnetic, laser beam, chemical, and mechanical, to trigger the self-healing in a nonautonomous way. Two main methods have been developed for preparation of self-healing polymers by the nonautonomous insight. The nonautonomous external systems include introduction of catalysts in the form of capsules or fibers into polymer matrix, where the catalyst is activated at the crack aria and forms the new polymer network. The nonautonomous internal systems are based on implementation of dynamic reversible bonds in a polymer matrix, where the dynamic bond is re-formed by a stimulus under ambient conditions. Light-induced healable polymers represent fascinating stimuli-responsive self-healing characteristics, as the healing process can remotely be controlled by turning off or on the excitation light.

Photoinduced healing processes have been developed in three styles based on the nature of the photoreactive segments used in polymer – (i) reversible photocross-linking reaction based on the reversible photodimerization and photocleavage reactions between the photoreactive moieties resulting in linkage of polymer chains located on both sides of the crack surfaces, (ii) phototriggered metathesis or "reshuffling" reaction across the crack surfaces of opposite sides resulting in linkage of polymer chains located on both sides of the crack, and (iii) dissociation of the noncovalent chemical bonds on either side of the crack upon photothermal reaction and then, formation of new noncovalent bonds across the crack surfaces of opposite sides upon cooling after turning off the light [11, 12, 422–424]. In the field of light-induced self-healing polymers incorporated chromophores or dynamic covalent bonds, a number of studies have exploited the use of reversible photocross-linking reactions based on the reversible photodimerization and photocleavage of chromophores, such as cinnamoyl, coumarin, azobenzenes, spiropyrans (SP), oxetanes, and anthracene [72, 356, 357, 417, 421–442] and also metathesis reactions of particularly interesting dynamic covalent bonds, including disulfide [356, 425, 443, 444], allyl sulfide [445, 446], and trithiocarbonates [357]. As presented in Figure 2.14a, an effective self-healing can be achieved by light exposure to the damage if the crack surfaces contain a sufficient amount of such photoreactive moieties that can react to form new photoinduced chemical bonds, either covalent or noncovalent, across the crack surfaces to merge them. These photoreactive groups should be part of the polymer, and polymer chains must be sufficiently mobile to penetrate into the crack and bring photoreactive moieties on the crack surface of opposite sides closer for formation of new chemical bonds [11, 12].

Yang et al. designed a self-healing hydrogel that possesses light-switchable host–guest complex between azobenzene acrylamide and poly(β-cyclodextrin) as the macro-cross-linker, and pH-sensitive boronate ester to develop light- and

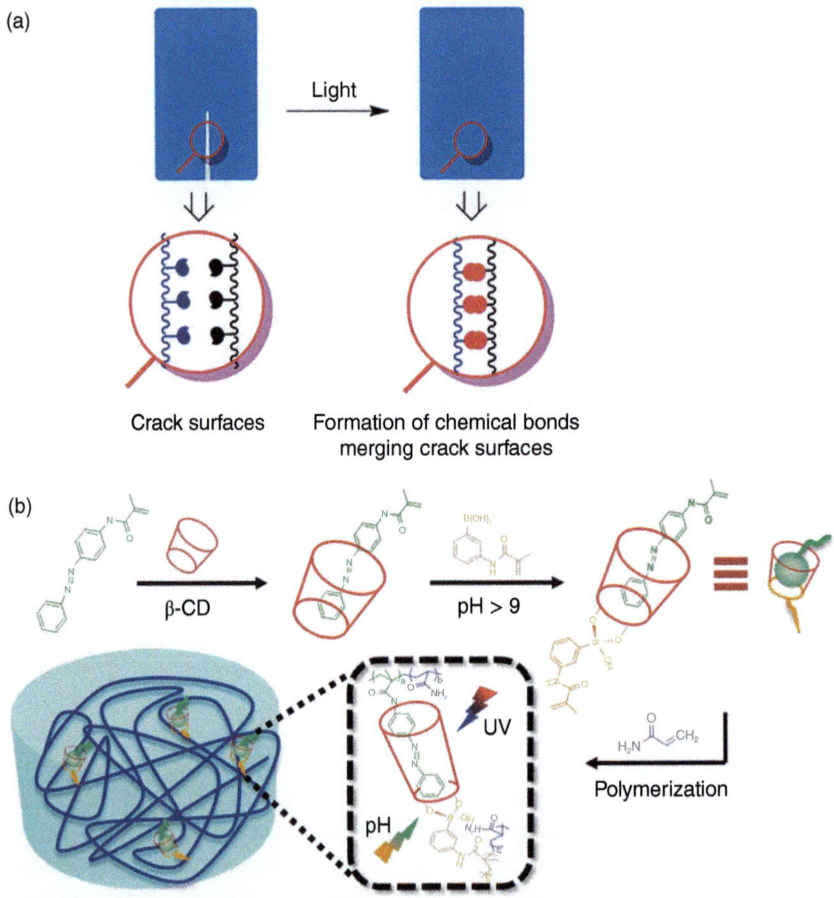

Figure 2.14 Schematic illustration for (a) light-induced self-healing of a crack [11]. Source: Reproduced with permission from Habault et al. [11]. Copyright 2012 Royal Society of Chemistry (RSC), (b) synthesis of light- and pH-responsive self-healing hydrogel [231]. Source: Reproduced with permission from Long et al. [231]. Copyright 2019 Springer, (c, d) chemical structure of coumarin functionalized single-chain polymers, intra and interchain collapse by means of photodimerization of coumarin and self-healing pathway with corresponding optical microscopy images depicting autonomic and nonautonomic route of self-healing process [28]. Source: Reproduced with permission from Joseph et al. [28]/Royal Society of Chemistry, and (e) light-induced healable epoxy polymer networks via anthracene dimer scission of the diamine cross-linker [232]. Source: Reproduced with permission from Hughes et al. [232]/American Chemical Society.

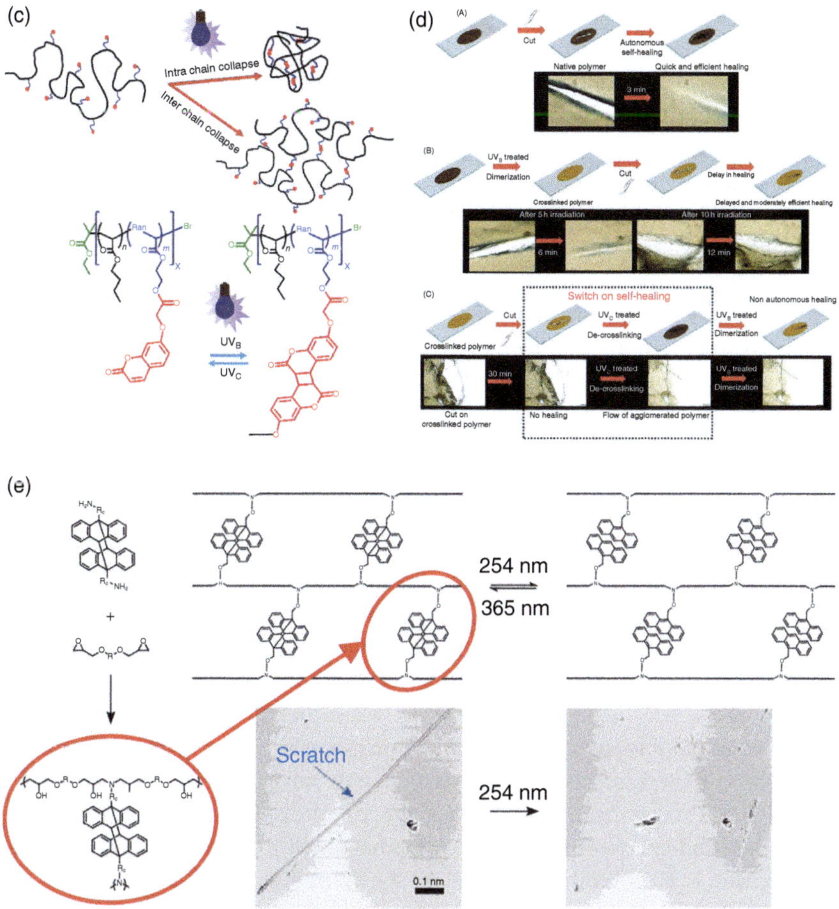

Figure 2.14 (*Continued*)

pH-responsive hydrogel [234], as shown in Figure 2.14b. They concluded that design of the combined host–guest inclusion complexations and boronate esters on the hydrogel showed an advantage for improving self-healing, mechanical, and recovery properties simultaneously. They found that the self-healing ability of the hydrogel could be switched off and on with UV and visible light due to transition of azobenzene segments from trans-to-cis states. They also investigated the influence of photoisomerization of azobenzene segments on the switchable self-healing ability of the hydrogel. Chemical and light-responsive self-healing polymeric hydrogels with switchable stiffness transitions were fabricated via functionalization of carboxymethyl cellulose chains with photoisomerizable arylazopyrazole (*trans*-AAP), β-cyclodextrin (β-CD), and nucleic acid tethers [233]. The control over the light-induced switchable stiffness properties of the hydrogel was used to develop shape-memory and self-healing matrices. A hydrogel of high stiffness was obtained upon cross-linking of tethered nucleic acids and the host–guest *trans*-AAP functionalized linker. In the presence of UV illumination, the *trans*-AAP was converted to *cis*-AAP, resulting in the partial separation of the host–guest cross-linking units and yielding a low-stiffness hydrogel.

Autonomous and nonautonomous self-healing functional polymers were developed by incorporating photodimerizable coumarin moieties in a linear chain of butyl acrylate-based graft copolymers that undergo intrachain or interchain collapse with UV irradiation to form diverse nanostructures [28], as shown in Figure 2.14c. Such coumarin incorporated polymers synthesized by atom transfer radical polymerization (ATRP) followed by post-functionalization with reactive functional units were undergoing photocross-linking reactions upon light irradiation in the wavelength range of 280–350 nm. Figure 2.14d shows an autonomic and intrinsic nature of the self-healing behavior, where the photoinduced healable polymer exhibited quick and efficient healing of the scratch within a few minutes without any outside intervention. Increasing the photodimerization percent of the photocross-linked polymer film resulted in an increase in the T_g values, which caused a significant delay in the healing process and inefficient autonomic healing of a cut on the film (up to 30 minutes) owing to the rigidification in the hard segments. In the third condition, nonautonomic healing of the cut in the cross-linked polymer film could be obtained through phototriggered using UV irradiation with the wavelengths of 200–280 nm that was further covalently fixed by UV irradiation with the wavelengths of 280–320 nm. Anthracene is another important chromophore with dynamic cycloaddition reaction that has been utilized for light-induced self-healing of polymers. In this context, Hughes et al. fabricated highly photoreversible cross-linked epoxy polymers, in which anthracene-based diamine cross-linkers act as a photocross-linkable curing agent. Such polymers change their structure from a hard and rigid cross-linked structure to a soft and flowable state to heal damage and then were hardened to a cross-linked structure in the response to UV illumination [232], as shown in Figure 2.14e.

2.3.7 Shape-Changing Polymers

Light-actuating shape-changing polymer systems are an exciting class of intelligent materials that can be deformed into a provisional shape under light stimuli and recover their initial shape once the stimuli are terminated. In shape-changing polymers, shape transformation is based on the incorporation of photosensitive moieties or based on implementation of dynamic bonds into a tailored polymer structure that acts as photoswitches. Triggering these switches through the use of an appropriate wavelength of light leads to reversible shape changes in the polymer systems. The most common photoactive moieties that can be used in light-actuation shape-changing polymers as light-responsive switches include azobenzene, spiropyran, cinnamic moieties, anthracene, coumarin, diarylethene, and imine chromophores and also ester, disulfide, allyl sulfide, hydrazone, and acylhydrazone exchangeable bonds. Light-actuation shape-changing polymers are used in various polymers, including hydrogels, liquid crystalline polymeric (LCP) materials, and SMPs, which can be either amorphous or semicrystalline. In addition, common starting shapes for the shape-changing polymers are in the form of films, fibers, and tubes. A different shape-changing mechanism is characteristic of SMPs that can be deformed into a temporary shape by application of external stress and return to the initial shape under suitable stimuli. They can be fixed in deformed shapes and their initial shapes can be recovered by applying an external stimulus [11, 447–449]. Azobenzenes are the most widely used class of photoresponsive molecules for light-actuated shape-changing materials because of a large shape change in the molecule upon trans–cis photoisomerization could lead to large shape deformations in materials. Azobenzene-containing microactuators with various shapes (straight, "Y"-shaped, serpentine, and helical) were prepared using a linear LCP that showed asymmetric geometry change upon 470 nm light irradiation with an intensity gradient along the tubular microactuator [237]. LCPs were prepared through ring-opening metathesis polymerization in the form of mechanically robust high-molecular-weight polyolefins. These microactuators comprised a long alkyl backbone-containing double bonds and azobenzene moieties in side chains acting as both mesogens and photoresponsive groups, which are able to apply photocontrol of a wide diversity of liquids over a long distance with controllable velocity and direction, and hence to combine liquids, and to mix multiphase liquids even to make liquids run uphill. Furthermore, this photodeformable tubular microactuator could be used in micro-reactors, micro-optomechanical systems, and laboratory-on-a-chip settings, as shown in Figure 2.15a. As the performance of photochromic switches (such as azobenzene) is sensitive to the viscosity of the polymer matrix, T_g can play an important role in the efficient photomechanical activation of light-actuated shape-changing LCP materials. Figure 2.15b,c show photoactuator polymer films with reversible photoswitchable T_g at ambient conditions and photomechanical actuation upon illumination of UV/visible light developed by incorporating a "photomelting" azobenzene monomer cross-linked into liquid crystalline (LC) networks [236]. The reversible trans-to-cis photoisomerization of azobenzene chromophores results in a change in T_g of the cross-linked LC polymer networks. The polymer network had a T_g higher than the room temperature in the trans-form upon exposure to visible light and lower than the room temperature in the cis-form after UV light irradiation.

In the case of hydrogels' photoresponsive reversible expansion–contraction, molecular actuators built from rotaxane-based compounds known as [c2]daisy chains were prepared via polycondensation between four-armed poly(ethylene glycol) and a [c2]daisy chain based on α-cyclodextrin as the host component and azobenzene as a photoresponsive guest component [235], as shown in Figure 2.16a. Light-induced trans-to-cis transition of azobenzene caused to sliding motion of the [c2]daisy chain unit and also this azo hydrogel is reversibly exchanged between the contracted and bend states before and after UV irradiation. The azobenzene in its trans conformation forms a stable host–guest complex with α-CD under visible light and caused to increase in the number density of cross-links in the material. The trans-to-cis isomerization upon light irradiation allows the α-CD to freely slide along the polymer chain, reducing the elastic modulus of the material and causing it to shrink, as shown in Figure 2.16b.

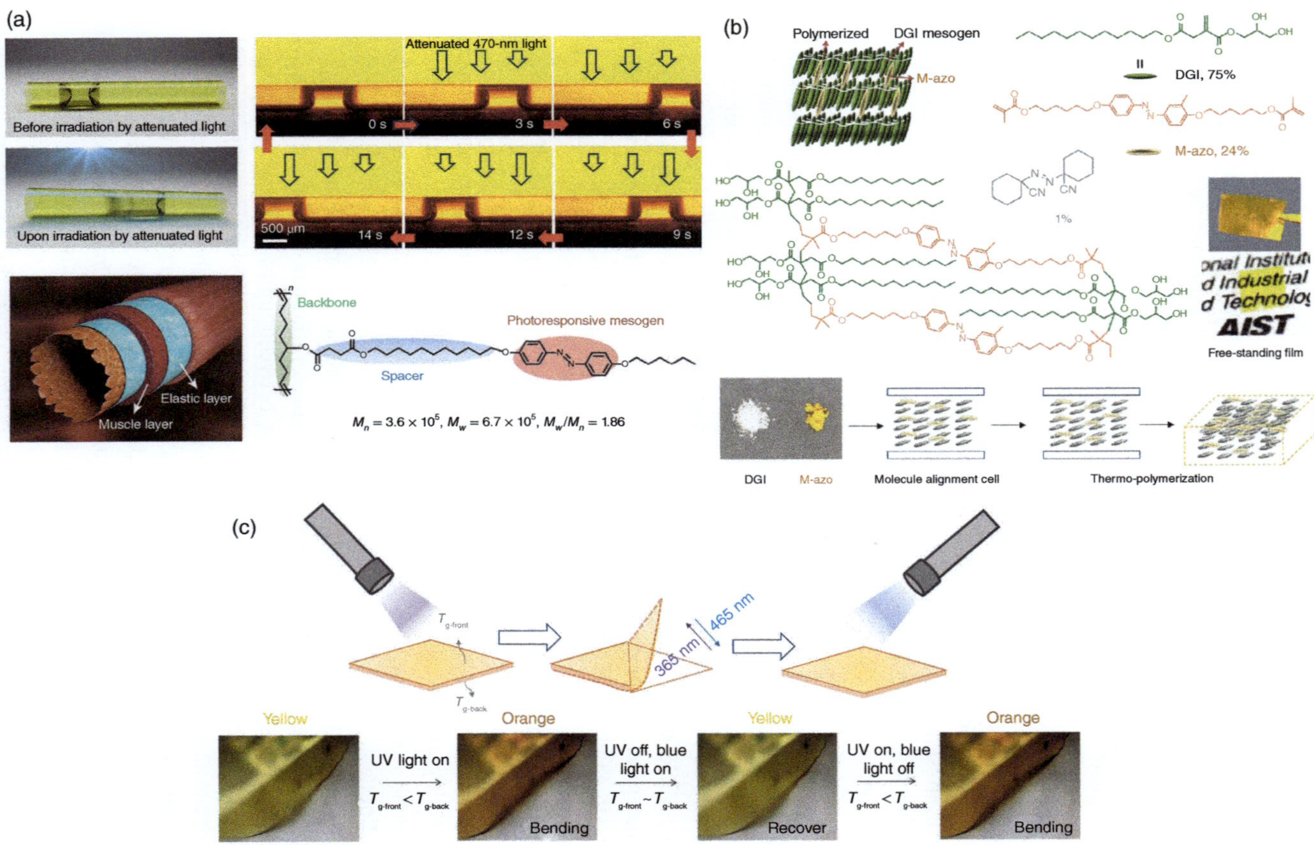

Figure 2.15 Photocontrol of fluid slugs in liquid crystal azobenzene-containing polymer microactuators [237]. Source: Reproduced with permission from Lv et al. [237]/Springer Nature and (b, c) the cross-linked polymer network structure in the film, photoswitchable T_g, and the continuous photomechanical response of the polymer film under light irradiation [236]. Source: Reproduced with permission from Yue et al. [236]/Springer Nature.

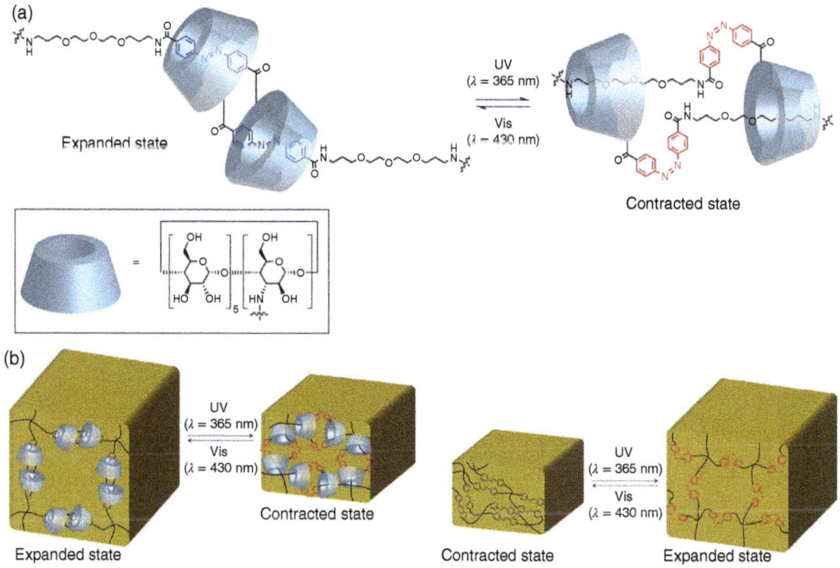

Figure 2.16 Schematic illustration for (a) chemical structure and photoisomerization of an α-CD-based [c2]daisy chain with an azobenzene derivative as the axis and (b) expansion–contraction of the hydrogel upon photoirradiation [235]. Source: Reproduced with permission from Iwaso et al. [235]. Copyright 2016 Nature Groups.

Figure 2.17a,b shows a light- and thermal-responsive SMP composed of 4,4-azodibenzoic acid and hexamethylene diisocyanate (HDI) unit in their hard segments and *N*-methyldiethanolamine and HDI unit in their soft segments that is capable of shape deformation under UV light and shape fixation in visible light irradiation, named as staging-responsive SMP. This system undergoes shape deformation with irradiation by UV light and shape fixation under visible light. Unlike common SMPs, which are only temporarily deformed under UV irradiation without a fixed temporary shape, the staging-responsive SMP developed in this work achieves a spontaneous deformation using light irradiation and remained fixed and kept the temporary shapes under ambient conditions. Shape recovery was controlled using a suitable temperature and facilitated by "hot programming," as hydrogen-bond interactions were weakened at high temperatures [39].

2.3.8 Photoconductive Polymers

Photoconductivity is an optical and electrical phenomenon, in which a material becomes more electrically conductive due to the absorption of specific light wavelengths. Some of polymers, including poly(*N*-vinyl carbazole), poly(vinyl fluoride), poly(vinyl alcohol), or poly(vinyl acetate), show measurable increase in conductivity, when irradiated with light. These polymeric materials are produced charged species in response to light, where the charged species can transfer under an electric field and then conduct electricity. When they are also doped with photoresponsive compounds that form charge-transfer complexes, the photoresponsivity

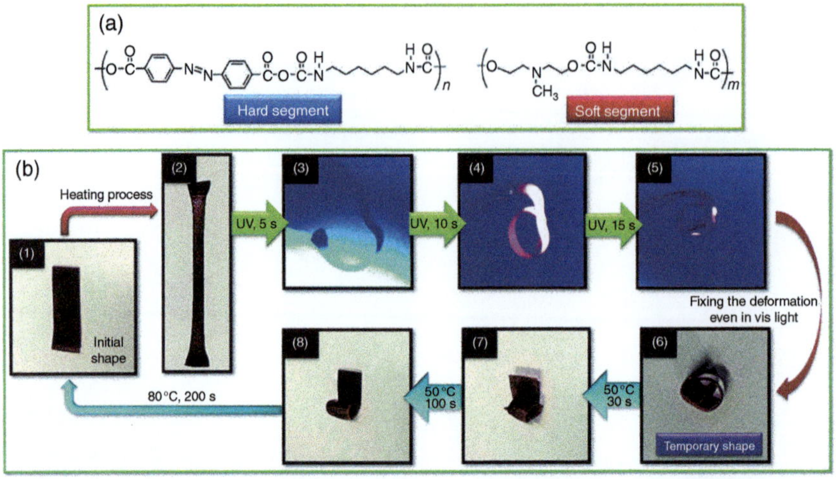

Figure 2.17 Schematic illustration for (a) molecular structures of the stimuli-responsive shape memory polymers, (b) shape deformation with irradiation by UV light, and shape fixation in visible light [39]. Source: Reproduced with permission from Ban et al. [39]/Royal Society of Chemistry.

can be increased and even extended into the visible region of the spectrum of light. Photoconductive polymers are categorized into three classes – n-type (electron-transporting), p-type (hole-transporting), or bipolar (capable of transporting both holes and electrons). Vinyl derivatives of polynuclear aromatic polymers, such as poly(vinyl pyrene) or poly(vinyl carbazole), absorb ultraviolet light in the wavelength of 360 nm and develop an exciton that ionizes in the electric field, where the exciting form is not conductive. Absorption of this polymer shifts into the visible region due to the formation of charge-transfer states through addition of an equivalent amount of an electron acceptor, such as trinitrofluorenone derivative. They can be classified into five categories based on their structures and modes of photoconduction, including polymers with pendant groups, molecularly doped polymers, backbone-conjugated polymers, liquid-crystalline systems, and nanoclusters/polymer composites [13, 239, 450]. Reversible switching of conductivity within diarylethene-based redox-active polymer films is shown in Figure 2.18a. The polymer comprises a central switch unit as a core and polymerizable bithiophene units, in which the polymerizable bithiophene units are separated from the central photochromic unit by phenyl spacers. Upon light irradiation, the conductivity of the polymer films at potentials below 0.8 V decreases due to switching of the polymer to the open form [238]. Optical and charge-transport properties of a photoconductive low-bandgap conjugated polymer, poly[2,6-(4,4-bis-(2-ethylhexyl)-4H-cyclopenta [2,1-b;3,4-b′]dithiophene)-alt-4,7-(2,1,3-benzothiadiazole)] (PCPDTBT), with an optical energy gap of $E_g \sim 1.5$ eV, have been studied. The polymer displays optical absorption and photoconductive response over a wide range of wavelengths from 850 to 300 nm. Hence, due to these properties, PCPDTBT has a good advantage for efficient light-harvesting and charge-carrier extraction in polymer-based

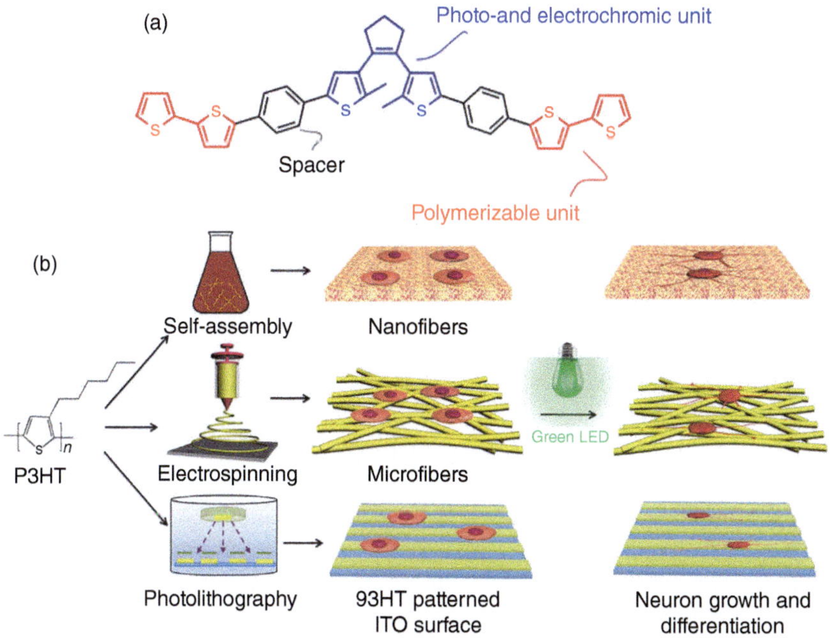

Figure 2.18 Schematic representation of (a) bifunctional polymerizable dithienylethene switch [238]. Source: Reproduced with permission from Logtenberg et al. [238]. Copyright 2012 American Chemical Society (ACS) and (b) preparation of P3HT micro/nanofibers and P3HT-patterned surface for PC12 cell culture [451]. Source: Reproduced with permission from Wu et al. [451]. Copyright 2019 American Chemical Society (ACS).

photovoltaic cells [450]. Multiscale-structured fibers and patterned films, based on a semiconducting polymer, poly(3-hexylthiophene) (P3HT), were synthesized as photoconductive biointerfaces to promote neuronal stimulation upon light irradiation, as shown in Figure 2.18b. In this work, the rat pheochromocytoma (PC12) was selected as an example of a primary neuron to investigate its responsivity behavior upon light irradiation, on the surface of P3HT with different structures and dimensions. All the results show that neurons cultured on the P3HT substrate have longer neuritis outgrowth under LED light irradiation compared to those grown on a blank substrate, which could be mainly attributed to the photoconductive effect of P3HT under illumination that dramatically promotes neurite outgrowth [451]. Other photoconductive polymers based on perylene, thiophene, carbazole, and triphenylamine derivatives with application in organic LEDs, smart thin films, and high-performance photorefractive devices were also investigated [450, 452–456].

2.3.9 Drug Delivery

Significant studies have led to the development of several novel photoresponsive polymers for drug-delivery systems, as light is nondestructive and allows remote localized activation. Photoresponsive polymer micelles or vesicles formed through self-assembly of photoresponsive block copolymers are the most applicable systems as carriers for controlled drug-delivery applications. Reversible photoresponsive chromophores, such as spiropyran and azobenzene, can reversibly be isomerized between two different hydrophobic and hydrophilic forms upon UV/visible light irradiation. This is the main reason for development of smart drug-delivery systems based on self-assembled photoresponsive copolymers. Reversible self-assembly of light-responsive amphiphilic copolymers based on spiropyran moieties is used for this purpose, as shown in Figure 2.19a. Polymersomes based on poly(ethylene oxide)-*b*-PSPA (PEO-*b*-PSPA) were synthesized as a novel amphiphilic diblock copolymer, where SPA is a derivative of spiropyran containing a unique carbamate linkage. These block copolymers were self-assembled to polymersomes in aqueous media with a hydrophobic PSPA core and a hydrophilic PEO shell. The formed self-assembled polymersomes were disrupted by isomerization of the nonpolar SP to the highly polar forms upon UV irradiation. This work developed a significant method to prepare light-controllable drug-delivery systems, where drug release can be switched on and off by alternating UV/visible irradiation [242]. Figure 2.19b shows a typical example of irreversible photoresponsive polymeric micelle as a drug carrier. The diblock copolymers were self-assembled into micellar structures with the hydrophobic ONB-based block in the core and hydrophilic block in the shell. Under UV irradiation, the hydrophobic moiety of chromophore is detached from its structure under photocleavage reaction, which enhances solubility of polymer chains and, consequently, the micelles disrupt. When the micelles were loaded with a hydrophobic drug, UV exposure allowed the irreversible disruption of micelles and burst release of the drug. In this type of irreversible photoresponsive drug-delivery system, drug release reaches its maximum value due to change in the hydrophilicity of the core and disruption of the structure of micelles [258]. Figure 2.19c shows a typical example of application of a coumarin derivative in the preparation of photoresponsive polymers to control drug delivery. Coumarin was incorporated into linear dendritic copolymers to prepare photocross-linkable nanocarriers for efficient *in vivo* tumor-targeted drug delivery. The prepared nanocarriers can be degraded upon light irradiation and release drug molecules *in vivo* by taking advantage of the reversible photodimerization of coumarin groups [241]. Furthermore, azobenzene containing-polymers are another good host–guest candidate for the preparation of self-assembled stable vesicles and reversible rapid releasing of anticancer drugs [146, 240]. Some examples of photoresponsive polymer systems in drug-delivery applications are summarized in Table 2.9.

2.3.10 Membranes, Films, and Textiles

Membranes are selective barriers that can separate components with different physical/chemical properties. The membrane selectivity is affected by affinity between the pore surface and the substances. They are playing paramount roles in mass transfer and separation because of many advantages, such as no phase change, low-energy consumption, and no additives in the membrane processes. However, pore size and surface properties of traditional porous membranes usually remain unalterable

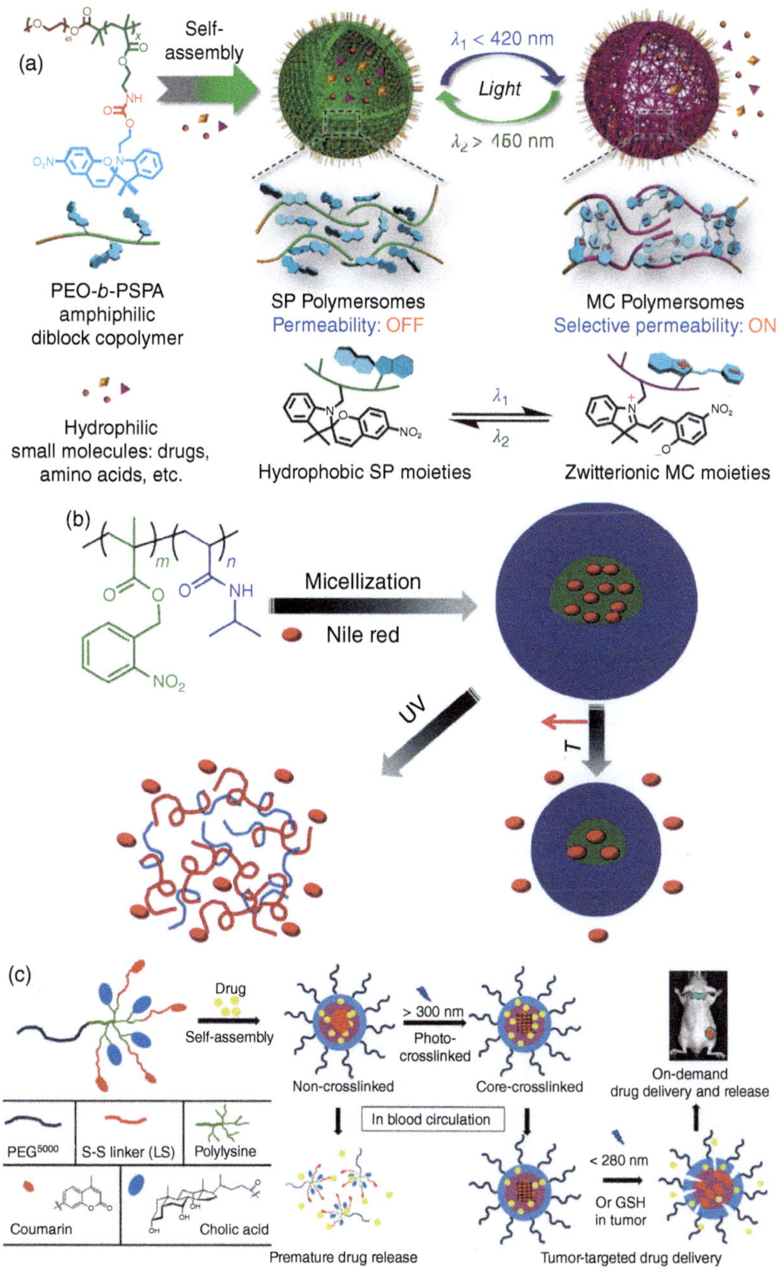

Figure 2.19 (a) Chemical structure of the photochromic polymersomes with reversible light-switchable on/off permeability and its light-controllable drug-release behavior [242]. Source: Reproduced with permission from Wang et al. [242]. Copyright 2015 American Chemical Society (ACS), (b) chemical structure of PNIPAAm-*b*-PNBM and morphological changes of the self-assembled micellar nanoparticle under photo and temperature stimulation [258]. Source: Reproduced with permission from Yang et al. [258]. Copyright 2013 Royal Society of Chemistry (RSC), and (c) schematic representation of the cross-linking and de-cross-linking processes of the coumarin-containing photosensitive phase-segregated micelle nanocarriers [241]. Source: Reproduced with permission from Shao et al. [241]/American Chemical Society.

Table 2.9 Photoresponsive polymer systems in drug-delivery applications.

Photoresponsive agent	Polymer structure	Drug	References
Coumarin	PAMAM dendrimer	5-Fluorouracil	[54]
	Polycaprolactone nanoparticle	DOX and chlorambucil (Cbl)	[55]
	PAMAM dendrimer	DOX	[103, 457]
	PEG-coumarin nanoaggregates	Curcumin (CUR)	[458]
	Poly(2-hydroxyethyl methacrylate)	DOX	[459]
	2-Hydroxyethyl methacrylate and coumarin copolymer	DOX	[460]
	Poly(ethylene glycol)-*b*-poly(L-lysine)	DOX	[461]
	Hyaluronic acid-based nanogels	Paclitaxel	[462]
Azobenzene	Carboxymethyl cellulose hydrogels	Naproxen	[56]
Spiropyran	interpenetrating polymer network	Doxycycline hyclate, dopamine, levodopa, prednisone, CUR	[463]
	Homo and copolymers of DMAEMA and MMA	DOX	[53]
	Homo and copolymers of NIPAAm and MMA	DOX	[52]
	Poly(ε-caprolactone)-*block*-poly (methacrylic acid-*co*-spiropyran methacrylate)	DOX	[464]

due to their unchangeable physical/chemical structures. Such unalterable pore size and surface properties of traditional membranes could restrict their efficient applications in extended fields. To provide advanced functions and enhanced performances of traditional membranes, stimuli-responsive gating membranes, such as photoresponsive membranes, are established by chemical/physical incorporation of stimuli-responsive compounds, such as chromophores as functional gates into traditional porous membranes. Photoresponsive chromophores, such as spiropyrans, diarylethenes, and azobenzenes, can be immobilized by coating or grafting onto the surface and/or pores of such membranes. They can also be chemically bounded or physically mixed into a membrane bulk structure. Specific wavelengths can cause the transition of a property at the microscopic level in responsive sites, or/and heating of the surrounding environment leading to macroscopic and, often, reversible changes in the membrane performance [14, 65, 465]. Photoresponsive bactericidal and bacteria-resistant membranes based on nitrobenzyl derivatives were synthesized via the integration of ATRP and click chemistry reaction [466]

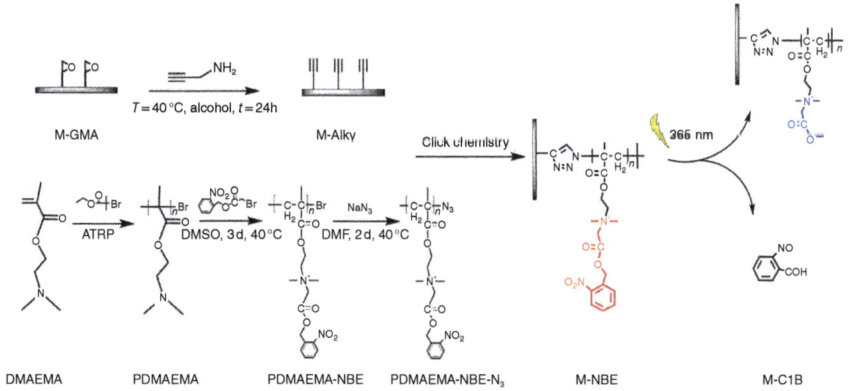

Figure 2.20 Schematic illustration for the preparation of a photoresponsive membrane surface [466]. Source: Reproduced with permission from Xie et al. [466]. Copyright 2017 Elsevier Ltd.

to obtain alkynyl-functionalized poly(ethersulfone) (PES) membrane. To endow the membrane surface with photoresponsive property, azide-functionalized polycationic poly (*N*,*N*-dimethyl-*N*-(2-(methacryloyloxy)ethyl)-*N*-((2-nitrobenzyl)oxy)-2-oxoethanaminium bromide) (PDMAEMA-NBE) was firstly synthesized by ATRP. Then, the PDMAEMA-NBE was grafted onto the alkynyl-functionalized PES membrane surface through click chemistry to get a charged surface. Upon UV light irradiation, the modified membrane surface could switch the surface charge, and the hydrophilic/hydrophobic property would affect the interaction of protein and surface to cause some bioresponses. Also, the surface would switch from bactericidal to bacteria-resistant surface by photoswitching of the polymer membrane from polycationic to polyzwitterionic form, as shown in Figure 2.20.

As organic photochromic materials, spiropyrans, azobenzenes, and viologens have widely been used in textiles [69, 466]. Smart polymers used in textiles usually appear in various forms, such as solution, microcapsule, gel, film/foam, fiber, and nonwoven, to meet different processing requirements in textiles. The potential of digital inkjet printing of solvent-based ink to produce photochromic textiles was investigated. A spiro-based UV-responsive smart textile through inkjet printing and UV-LED curing of a specifically designed photochromic ink on PET fabric was also reported [455]. The work focused on tuning and characterization of the color performance of a photochromic dye embedded in a UV-curable ink resin and showed during fabrication of textile, the rigidity of the UV-resin, the dye-polymer matrix interaction, and the degree of freedom of the dye to switch between isomers in the matrix was affected by curing intensity of the photochromic ink.

2.4 Summary and the Future

Light-responsive materials show different responses to light irradiation, including luminescence, chromism, covalent bond cleavage, covalent bond formation, and covalent bond exchange, which could be reversible or irreversible. These unique characteristics make light-responsive polymers as an important candidate in high-tech applications, including smart sensors, inks, surfactants, self-healing, self-cleaning, self-repairing, drug and gene delivery, membranes, films, fibers, and many other instances. Combination of other stimuli-responsive characteristics of smart polymers with such properties yields dual and multi-responsive polymers with lots of unique properties.

2.4.1 Water Contact Angle Variation

Photoresponsive polymers could respond to light irradiation by variation of hydrophilicity/hydrophobicity and consequently water contact angle. Such behavior is applicable in the preparation of photoresponsive polymeric films, membranes, surface-active agents, ion absorbents, and photorheological polymers. In addition, it has applications in drug-delivery systems by formation of different copolymer assemblies in aqueous media or even disruption of the previously formed self-assembled structures after light irradiation. Disruption of micellar and vesicular structures could be used in delivery of different drugs by the light irradiation, which could also be targeted by local light irradiation. Photolithography is also highly applicable in these categories of smart polymers after cleavage of a moiety and variation of the polymers amphiphilicity. In such applications, LCST of the dual photo and thermo-responsive polymers is varied and used as a trigger for dissolution of the irradiated or nonirradiated sections by photomasking. In this field, polymer coatings, fibers, and films are the most commonly used smart polymeric products. Self-cleaning applications, morphological transitions, tuning LCST or phase characteristics of polymer solutions by light irradiation, smart emulsifications and demulsifications, smart drying systems, smart foaming and defoamers, and many other applications based on these characteristics of photoresponsive polymers could be investigated in the future. It seems that combination of light-responsivity with pH, redox, and temperature responsivities might lead to even more high-tech applications.

2.4.2 Viscosity Variation

Variation of viscosity could occur with breakage of covalent bonds in the backbone and the consequent polymer chain sessions. As an example, it is observable in ONB-based polymers and hydrogels when ONB has been inserted in the junctions or the polymer backbone. The opposite trend could also be observed when the coupling reaction in polymers containing light-responsive materials is capable of cycloaddition reactions, as in the case of coumarin. In such systems, increase of polymer molecular weight or network formation is possible by light irradiation depending on the location and content of coumarin molecules. Physical interactions can also alter viscosity of polymers in different states after light irradiations. When ONB is located in the side chains, the polymers viscosity could be varied after light irradiation by polar interactions between the polymer chains. In spiropyran and similar materials, ionic interaction causes viscosity variation when incorporated into the backbone or side chains. Variation of viscosity may be accompanied by color change or fluorescence emission variations in these materials, which could be a source of sensor production for sensing viscosity by spectroscopic investigations. Gelochromism phenomenon is one of the most important chromism phenomena related to this category. Valves, sensors, microfluidic devices, shock absorbers, and many other high-tech applications could use these characteristics in the future to prepare smart materials for different applications. In such applications, control of viscosity of polymer solution or its modulus in the bulk state could be very important. Such control over viscosity and modulus is achieved by the light intensity and wavelength and also the time of light irradiation. Combination of these characteristics with other polymers with different mechanisms of responsivity toward various stimuli can result in new characteristics of such dual- and multi-stimuli-responsive polymers in the future.

2.4.3 Color Change and Emission

Variation of color or fluorescence emission by light irradiation is related to the stimuli-chromic and luminescent materials. Visual and spectroscopic sensations of these variations help to prepare chemosensors of ions, pH, polarity of solvent, gas, and redox molecules by colorimetric and fluorometric chemosensors. Anticounterfeiting inks, latent fingerprints, security labels, security marking, and many other similar applications could be designed by these materials. Visual or spectroscopic detection of fluorophores in smart surfactants, cell recognition, photolithography systems, self-healed area, drug-delivery area, and ion-adsorbed area in membranes are other important applications in this category. Chromism could also be observed by other stimuli, which could be an interesting topic in the future of this category. Chromism in response to base, acid, water, ion, solvent, mechanical stress, and pressure could be used as a source of sensing these analytes by chromism in visual or spectroscopic methods.

2.4.4 Sol–Gel Transition

Light-induced sol–gel transition is highly similar to the viscosity variation with a difference that gel formation is the final point in such materials. This could be reversible as in bifunctional light-responsive coumarin or irreversible as in bifunctional ONB. It could also be physical as in spiropyran-containing polymers or chemicals in the bifunctional coumarin-containing polymers. If coumarin or similar molecules are located in the side chains, the cycloaddition reaction may result in gel formation. Gelochromism is the most important phenomenon in this category. Combination of different sol–gel transitions by other stimuli, such as temperature, redox, and pH, with light-responsive sol–gel transition reactions may result in dual- and multi-stimuli-driven sol–gel transitions in a reversible or irreversible manner, which could be highly interesting in injectable hydrogels in drug delivery and tissue engineering, wound healing, self-healing, and shape-recovery applications in the future.

Abbreviations

AIE	aggregation-induced emission
ATRP	atom transfer radical polymerization
CDs	carbon dots
CMC	critical micelle concentration
CRP	controlled/living radical polymerization
DNBP	dinitrobenzylpyridine
DSA	9,10-distrylanthracene
EET	electronic energy transfer
ESIPT	excited-state intramolecular proton transfer
FRET	fluorescence resonance energy transfer
HABI	hexaarylbiimidazole
HDI	hexamethylene diisocyanate
HOMO	highest occupied molecular orbital
LCP	liquid-crystalline polymeric
ICT	intramolecular charge transfer
LED	light-emitting diode
ISC	intersystem crossing
LCST	lower critical solution temperature
LUMO	lowest unoccupied molecular orbital
MB	methylene blue
MC	merocyanine
MLCT	metal–ligand charge transfer
MMA	methyl methacrylate
MOF	metal–organic framework
NIR	near infrared
ONB	*ortho*-nitrobenzyl ester
O/W and W/O	oil-in-water and water-in-oil droplets
P3HT	poly(3-hexylthiophene)
PC12	pheochromocytoma
PCPDTBT	poly[2,6-(4,4-bis-(2-ethylhexyl)-4*H*-cyclopenta[2,1-*b*;3,4-*b*′] dithiophene)-*alt*-4,7-(2,1,3-benzothiadiazole)]
PDs	polymer dots
PEG	poly(ethylene glycol)
PES	poly(ethersulfone)
PET	photoinduced electron transfer
PNIPAAm	poly(*N*-isopropylacrylamide)
PPGs	photolabile protecting groups
QDs	quantum dots
QM	quinoline-malononitrile
RAFT	reversible addition-fragmentation chain transfer
S1 and T*n*	the lowest excited singlet state and a triplet state
SMPs	shape-memory polymers
SP	spiropyran

TICT	twisted intramolecular charge transfer
TPE	tetraphenylethene
TPID	triphenylimidazolyl dimer
TPIR	triphenylimidazolyl radicals
TPP	tetraphenylpyrazine
UCNPs	upconverting nanoparticles
UCST	upper critical solution temperature

References

1 Shi, Y. and Chen, Z. (2018). *J. Mater. Chem. C* 6: 11817–11834.
2 Hribar, K.C., Lee, M.H., Lee, D., and Burdick, J.A. (2011). *ACS Nano* 4: 2948–2956.
3 Guo, L., Ge, J., and Wang, P. (2018). *Photochem. Photobiol.* 94: 916–934.
4 Wu, C. and Chiu, D.T. (2013). *Angew. Chem. Int. Ed.* 52: 3086–3109.
5 Bertrand, O. and Gohy, J.F. (2017). *Polym. Chem.* 8: 52–73.
6 Irie, M. and Kunwatchakun, D. (1986). *Macromolecules* 19: 2476–2480.
7 Abdollahi, A., Roghani-Mamaqani, H., and Razavi, B. (2019). *Prog. Polym. Sci.* 98: 101149.
8 Abdollahi, A., Roghani-Mamaqani, H., Razavi, B., and Salami-Kalajahi, M. (2020). *ACS Nano* 14: 14417–14492.
9 Aguilar, M.R., Elvira, C., Gallardo, A. et al. (2007). *Top. Tissue Eng.* 3: 1–27.
10 Shahi, S., Roghani-Mamaqani, H., Talebi, S., and Mardani, H. (2022). *Polym. Chem.* 13: 161–192.
11 Habault, D., Zhang, H., and Zhao, Y. (2013). *Chem. Soc. Rev.* 42: 7244–7256.
12 Pasparakis, G. (2017). *Polym. Chem.* 8: 6464–6484.
13 Ravve, A. (2006). *Light-Associated Reactions of Synthetic Polymers*. New York: Springer.
14 Nicoletta, F.P., Cupelli, D., Formoso, P. et al. (2012). *Membranes* 2: 134–197.
15 Abdollahi, A., Herizchi, A., Roghani-Mamaqani, H., and Alidaei-Sharif, H.L.P. (2019). *Carbohydr. Polym.* 230: 115603.
16 Abdollahi, A., Alidaei-Sharif, H., Roghani-Mamaqani, H., and Herizchi, A. (2020). *J. Mater. Chem. C* 8: 5476–5493.
17 Asadi-Zaki, N., Mardani, H., Roghani-Mamaqani, H., and Shahi, S. (2021). *Polymer* 228: 123899.
18 Ngororabanga, J.M.V., Tshentu, Z.R., and Mama, N. (2019). *New J. Chem.* 43: 12168–12177.
19 Zhang, L., Bian, Z., Hu, A. et al. (2020). *Opt. Commun.* 475: 126225.
20 Abdollahi, A., Mouraki, A., Sharifian, M.H., and Mahdavian, A.R. (2018). *Carbohydr. Polym.* 200: 583–594.
21 Abdollahi, A., Sahandi-Zangabad, K., and Roghani-Mamaqani, H. (2018). *ACS Appl. Mater. Interfaces* 10: 39279–39292.
22 Abdollahi, A., Sahandi-Zangabad, K., and Roghani-Mamaqani, H. (2018). *Langmuir* 34: 13910–13923.

23 Tajmoradi, Z., Roghani-Mamaqani, H., and Salami-Kalajahi, M. (2020). *Polymer* 205: 122859.
24 Chen, J., Zhong, W., Tang, Y. et al. (2015). *ACS Macromol.* 48: 3500–3508.
25 Sharker, S.M., Jeong, C.J., Kim, S.M. et al. (2014). *Chem. Asian J.* 9: 2921–2927.
26 Zhu, M., Zhang, G., Hu, Z. et al. (2014). *Macromolecules* 47: 1543–1552.
27 Tajmoradi, Z., Roghani-Mamaqani, H., and Salami-Kalajahi, M. (2020). *Chem. Eng. J.* 417: 128005.
28 Joseph, J.P., Miglani, C., Singh, A. et al. (2020). *Soft Matter* 16: 2506–2515.
29 McCoy, T.M., Liu, A.C., and Tabor, R.F. (2016). *Nanoscale* 8: 6969–6974.
30 Takahashi, Y., Koizumi, N., and Kondo, Y. (2015). *Langmuir* 32: 683–688.
31 Ruchmann, J., Fouilloux, S., and Tribet, C. (2008). *Soft Matter* 10: 2098–2108.
32 Irie, M., Hirano, Y., Hashimoto, S., and Hayashi, K. (1981). *Macromolecules* 14: 262–267.
33 Butler, C.S.G., King, J.P., Giles, L.W. et al. (2021). *J. Colloid Interface Sci.* 594: 669–680.
34 Oh, H., Ketner, A.M., Heymann, R. et al. (2013). *Soft Matter* 9: 5025–5033.
35 Li, J., Zhao, M., Zhou, H. et al. (2012). *Soft Matter* 8: 7858–7864.
36 Chang, D., Yan, W., Yang, Y. et al. (2016). *Dyes Pigm.* 134: 186–189.
37 Dong, R., Peng, J., Chang, X. et al. (2017). *J. Rheol.* 61: 107–116.
38 Ling, J., Rong, M.Z., and Zhang, M.Q. (2014). *Chin. J. Polym. Sci.* 32: 1286–1297.
39 Ban, J., Mu, L., Yang, J. et al. (2017). *J. Mater. Chem. A* 5: 14514–14518.
40 Peng, B., Li, H., Li, Y. et al. (2020). *Chem. Eng. J.* 395: 125079.
41 Nagata, M. and Yamamoto, Y.U. (2009). *J. Polym. Sci., Part A: Polym. Chem.* 47: 2422–2433.
42 Xie, H., Cheng, C., Deng, X. et al. (2017). *Macromolecules* 50: 5155–5164.
43 Xie, H., He, M.J., Deng, X.Y. et al. (2016). *ACS Appl. Mater. Interfaces* 8: 9431–9439.
44 Elastomers, S., Li, J., Lewis, C.L. et al. (2011). *Macromolecules* 44: 5336–5343.
45 Wang, L., Yang, X., Chen, H. et al. (2013). *ACS Appl. Mater. Interfaces* 5: 10520–10528.
46 Jiang, H.Y., Kelch, S., and Lendlein, A. (2006). *Adv. Mater.* 18: 1471–1475.
47 Sakata, T., Kajiya, D., and Saitow, K. (2020). *ACS Appl. Mater. Interfaces* 12: 46598–46608.
48 Liang, H.Y., Cao, W.L., Du, M. et al. (2006). *Chem. Phys. Lett.* 419: 292–296.
49 Kim, H., Choi, K., Mun, J. et al. (2002). *Opt. Mater.* 21: 657–662.
50 Chen, Q., Liang, X., Du, J. et al. (2019). *J. Mater. Chem. C* 7: 8045–8052.
51 Talebnia, F., Nourmohammadian, F., and Bastani, S. (2020). *Prog. Org. Coat.* 77: 1351–1359.
52 Razavi, B., Abdollahi, A., Roghani-Mamaqani, H., and Salami-Kalajahi, M. (2020). *Polymer* 187: 122046.
53 Razavi, B., Abdollahi, A., Roghani-Mamaqani, H., and Salami-Kalajahi, M. (2020). *Mater. Sci. Eng., C* 109: 110524.
54 Wang, H., Miao, W., Wang, F., and Cheng, Y. (2018). *Biomacromolecules* 19: 2194–2201.

55 Karthik, S., Jana, A., Selvakumar, M. et al. (2017). *J. Mater. Chem. B* 5: 1734–1741.

56 Kim, Y., Jeong, D., Shinde, V.V. et al. (2020). *Int. J. Biol. Macromol.* 163: 824–832.

57 Weh, K., Noack, M., Ruhmann, R. et al. (1998). *J. Chem. Eng. Technol.* 21: 408–412.

58 Kaner, P., Hu, X., Thomas, S.W. III, and Asatekin, A. (2017). *ACS Appl. Mater. Interfaces* 9: 13619–13631.

59 Kim, S., Yu, C., Shin, C., and Choi, M. (2007). *Dyes Pigm.* 75: 250–252.

60 Seki, T. (2007). *Curr. Opin. Solid State Mater. Sci.* 10: 241–248.

61 Little, A.F. and Christie, R.M. (2010). *Color. Technol.* 126: 157–163.

62 Little, A.F. and Christie, R.M. (2010). *Color. Technol.* 126: 164–170.

63 Tian, H. and Zhang, J. (ed.) (2016). *Photochromic Materials: Preparation, Properties and Applications*. Wiley.

64 Jochum, F.D. and Theato, P. (2013). *Chem. Soc. Rev.* 42: 7468–7483.

65 Pantuso, E., De Filpo, G., and Nicoletta, F.P. (2019). *Adv. Opt. Mater.* 1900252, 1–35.

66 Marturano, V., Cerruti, P., Giamberini, M. et al. (2017). *Polymers* 9: 8.

67 Ercole, F., Davis, T.P., and Evans, R.A. (2010). *Polym. Chem.* 1: 37–54.

68 Stumpel, J.E., Broer, D.J., and Schenning, A.P.H. (2014). *J. Chem. Commun.* 50: 15839–15848.

69 Hu, J., Meng, H., Li, G., and Ibekwe, S.I. (2012). *Smart Mater. Struct.* 21: 053001.

70 Dugave, C. and Demange, L. (2003). *Chem. Rev.* 103: 2475–2532.

71 Trenor, S.R., Shultz, A.R., Love, B.J., and Long, T.E. (2004). *Chem. Rev.* 104: 3059–3077.

72 Kaur, G., Johnston, P., and Saito, K. (2014). *Polym. Chem.* 5: 2171–2186.

73 Feringa, B.L. and Browne, W.R. (ed.) (2011). *Molecular Switches*, 2 Volume Set. Wiley.

74 Narayan, K.S. and Kumar, N. (2001). *Appl. Phys. Lett.* 79: 1891–1893.

75 Abdollahi, A., Roghani-Mamaqani, H., Salami-Kalajahi, M., and Razavi, B. (2019). *Polym. Chem.* 10: 5686–5720.

76 Shinde, K.N., Dhoble, S.J., Swart, H.C., and Park, K. (2012). *Phosphate Phosphors for Solid-State Lighting*. Springer Science & Business Media.

77 Marsden, E. (2018). *Photoluminescence Advances in Research and Applications-Physics Research and Technology*. Nova Science Publishers, Inc.

78 Kitai, A. (ed.) (2008). *Luminescent Materials and Applications*, vol. 25. Wiley.

79 Bebb, H.B. and Williams, E.W. (1972). *Semiconductors and Semimetals, Chapter 4 Photoluminescence I: Theory*, vol. 8, 181–320. Elsevier.

80 Tokito, S., Suzuki, M., and Sato, F. (2003). *Org. Electron.* 4: 105–111.

81 Li, X., Rui, M., Song, J. et al. (2015). *Adv. Funct. Matter* 25: 4929–4947.

82 Park, Y.S., Lim, J., and Klimov, V.I. (2019). *Nat. Mater.* 18: 249–255.

83 Xu, G., Zeng, S., Zhang, B. et al. (2016). *Chem. Rev.* 116: 12234–12327.

84 Emin, S., Singh, S.P., Han, L. et al. (2011). *Sol. Energy* 85: 1264–1282.

85 Tao, S., Zhu, S., Feng, T. et al. (2017). *Mater. Today Chem.* 6: 13–25.

86 Liu, J., Li, R., and Yang, B. (2020). *ACS Cent. Sci.* 6 (12): 2179–2195.
87 Sagbas, S. and Sahiner, N. (2019). *Nanocarbon and its Composites Preparation, Properties and Applications*, 651–676. Woodhead Publishing.
88 Resch-genger, U., Grabolle, M., Cavaliere-jaricot, S. et al. (2008). *Nat. Methods* 5: 763–775.
89 Berezin, M.Y. and Achilefu, S. (2010). *Chem. Rev.* 110: 2641–2684.
90 Gong, S., Yang, C., and Qin, J. (2012). *Chem. Soc. Rev.* 41: 4797–4807.
91 Kuila, S. and George, S.J. (2020). *Angew. Chem. Int. Ed.* 132: 9479–9483.
92 Chen, C. and Liu, B. (2019). *Nat. Commun.* 10: 1–15.
93 Gan, N., Shi, H., An, Z., and Huang, W. (2018). *Adv. Funct. Mater.* 28: 1802657.
94 Zhao, W. (2020). *Nat. Rev. Mater.* 5: 869–885.
95 Fang, M., Yang, J., and Li, Z. (2019). *Chin. J. Polym. Sci.* 37: 383–393.
96 Christie, R.M. (2011). *Handbook of Textile and Industrial Dyeing Principles, Processes and Types of Dyes: 17-Fluorescent dyes*, 562–587. Woodhead Publishing Limited.
97 Wu, J., Liu, W., Ge, J. et al. (2011). *Chem. Soc. Rev.* 40: 3483–3495.
98 Wagner, B.D. (2009). *Molecules* 14: 210–237.
99 Manolov, I., Maichle-moessmer, C., and Danchev, N. (2006). *Eur. J. Med. Chem.* 41: 882–890.
100 Lim, B. and Lee, J. (2016). *Molecules* 21: 339.
101 Hiremani, V.D., Anandalli, M.H., Gasti, T. et al. (2021). *J. Polym. Res.* 28: 1–12.
102 Georgieva, I., Trendafilova, N., Aquino, A., and Lischka, H. (2005). *J. Phys. Chem. A* 109: 11860–11869.
103 Razavi, B., Abbaszadeh, R., Salami-Kalajahi, M., and Roghani-Mamaqani, H. (2020). *J. Mol. Liq.* 319: 114138.
104 Rostami-Tapeh-Esmail, E., Golshan, M., Salami-Kalajahi, M. et al. (2020). *J. Mol. Liq.* 313: 113504.
105 Abdollahi, A., Roghani-Mamaqani, H., Herizchi, A. et al. (2020). *Polym. Chem.* 11: 2053–2069.
106 Rostami-Tapeh-Esmail, E., Golshan, M., Salami-kalajahi, M., and Roghani-Mamaqani, H. (2020). *Mater. Sci. Eng., C* 110: 110745.
107 Abousalman-Rezvani, Z., Eskandari, P., Roghani-Mamaqani, H., and Salami-Kalajahi, M. (2019). *Carbohydr. Polym.* 225: 115247.
108 Abousalman-Rezvani, Z., Eskandari, P., Roghani-Mamaqani, H. et al. (2019). *Polymer* 182: 121830.
109 Guha, S., Lohar, S., Bolte, M. et al. (2012). *Spectrosc. Lett.* 45: 225–235.
110 Stamboliyska, B., Janevska, V., Shivachev, B., and Nikolova, R.P. (2010). *Arkivoc* 10: 62–76.
111 Irmler, P., Gogesch, F.S., Larsen, C.B. et al. (2019). *Dalton Trans.* 48: 1171–1174.
112 Li, Y., Huang, W., Yong, J. et al. (2018). *New J. Chem.* 42: 12644–12648.
113 Yin, J., Hu, H., Wu, Y., and Liu, S. (2011). *Polym. Chem.* 2: 363–371.
114 Yamaguchi, T., Asanuma, M., Nakanishi, S. et al. (2014). *Chem. Sci.* 5: 1021–1029.
115 Kim, H.W., Choi, M.G., Park, H. et al. (2015). *RSC Adv.* 5: 4623–4627.

116 Rashid, F., Raducanu, V., Zaher, M.S., and Hamdan, S.M. (2019). *Nat. Commun.* 10: 1–14.
117 Klohs, J., Schering, B., and Ag, P. (2008). *Basic Res. Cardiol.* 103: 144–151.
118 Guillaume, M., Zutterman, F., and Lie, V. (2007). *Chem. Phys. Lett.* 446: 165–169.
119 Haidekker, M.A., Nipper, M., Mustafic, A. et al. (2010). *Advanced Fluorescence Reporters in Chemistry and Biology I*, 267–308. Berlin, Heidelberg: Springer-Verlag.
120 Nogues, I.F.T., Goutayer, M., Da Silva, A. et al. (2009). *J. Biomed. Opt.* 14: 054005.
121 Tian, D., Qi, F., Ma, H. et al. (2018). *Nat. Commun.* 9: 1–14.
122 Cao, J., Zhao, C., Feng, P. et al. (2012). *RSC Adv.* 2: 418–420.
123 Rao, M.R., Mobin, S.M., and Ravikanth, M. (2010). *Tetrahedron* 66: 1728–1734.
124 Wei, X., Bu, L., Li, X. et al. (2017). *Dyes Pigm.* 136: 480–487.
125 Rostami-Tapeh-Esmail, E., Golshan, M., Salami-Kalajahi, M., and Roghani-Mamaqani, H. (2020). *Dyes Pigm.* 180: 108488.
126 Goodson, F.S., Panda, D.K., Ray, S. et al. (2013). *Org. Biomol. Chem.* 11: 4797–4803.
127 Cheng, H. and Qian, Y. (2015). *Dyes Pigm.* 112: 317–326.
128 Golshan, M., Rostami-Tapeh-Esmail, E., and Salami-Kalajahi, M. (2020). *Eur. Polym. J.* 137: 109933.
129 Amato, D.N., Amato, D.V., Sandoz, M. et al. (2020). *ACS Appl. Mater. Interfaces* 12: 42048–42055.
130 Wei, J., Yang, S., Wang, L. et al. (2013). *RSC Adv.* 3: 19403–19408.
131 Qi, Q., Huang, L., Yang, R. et al. (2019). *Chem. Commun.* 55: 1446–1449.
132 Stobiecka, M. and Hepel, M. (2011). *Phys. Chem. Chem. Phys.* 13: 1131–1139.
133 Li, D., Jiang, J., Huang, Q. et al. (2016). *Polym. Chem.* 7: 3444–3450.
134 Wang, C., Zhang, D., Huang, X. et al. (2014). *Sens. Actuators, B* 198: 33–40.
135 Kawai, K., Kawabata, K., Tojo, S., and Majima, T. (2002). *Bioorg. Med. Chem. Lett.* 12: 2363–2366.
136 Li, D., Munyentwali, A., Wang, G. et al. (2015). *Dyes Pigm.* 117: 92–99.
137 Rad, J.K., Mahdavian, A.R., Salehi-Mobarakeh, H., and Abdollahi, A. (2016). *Macromolecules* 49: 141–152.
138 Sabnis, R.W. (2015). *Handbook of Fluorescent Dyes and Probes*. Wiley.
139 Parker, C.A. and Rees, W.T. (1962). *Analyst* 87: 83–111.
140 Zhen, X., Tao, Y., An, Z. et al. (2017). *Adv. Mater.* 29: 1606665.
141 Xu, S., Chen, R., Zheng, C., and Huang, W. (2016). *Adv. Mater.* 28: 9920–9940.
142 Kabe, R., Notsuka, N., Yoshida, K., and Adachi, C. (2016). *Adv. Mater.* 28: 655–660.
143 Cai, S., Ma, H., Shi, H. et al. (2019). *Nat. Commun.* 10: 1–8.
144 Wu, Q., Xiong, H., Zhu, Y. et al. (2020). *ACS Appl. Mater. Interfaces* 2: 669–705.
145 Lee, D., Bolton, O., Kim, B.C. et al. (2013). *J. Am. Chem. Soc.* 135 (16): 6325–6329.
146 Xia, D., Yu, G., Li, J., and Huang, F. (2014). *Chem. Commun.* 50: 3606–3608.
147 Hirata, S. (2017). *Adv. Opt. Mater.* 5: 1–50.

148 Hirata, S. (2017). *Adv. Opt. Mater.* 5: 1700116.

149 Helmy, S. and de Alaniz, J.R. (2015). *Adv. Heterocycl. Chem.* 117: 131–177.

150 Bouas-Laurent, H. and Durr, H. (2001). *Pure Appl. Chem.* 73: 639–665.

151 Abdollahi, A., Roghani-Mamaqani, H., Salami-Kalajahi, M., and Razavi, B. (2020). *Carbohydr. Polym.* 245: 116507.

152 Achilleos, D.S. and Vamvakaki, M. (2010). *Macromolecules* 43 (17): 7073–7081.

153 Schiphorst, J., Van Den Broek, M., De Koning, T. et al. (2016). *J. Mater. Chem. A* 4: 8676–8681.

154 Chen, S., Jiang, F., Cao, Z. et al. (2015). *Chem. Commun.* 51: 12633–12636.

155 Paramonov, S.V., Lokshin, V., and Fedorova, O.A. (2011). *J. Photochem. Photobiol., C* 12: 209–236.

156 Abdollahi, A., Roghani-Mamaqani, H., and Salami-Kalajahi, M. (2020). *J. Colloid Interface Sci.* 580: 192–210.

157 Mueller, P., Hammer, L., Mueller, R. et al. (2018). Conference on Lasers and Electro-Optics, pp. 1–2.

158 Vijayamohanan, H., Palermo, E.F., and Ullal, C.K. (2017). *Chem. Mater.* 29: 4754–4760.

159 Mueller, P., Hammer, L., Mueller, R. et al. (2019). *Chem. Mater.* 31: 1966–1972.

160 Zhang, Y., Chen, S., Pang, M., and Zhang, W. (2016). *Polym. Chem.* 7: 6880–6884.

161 Peng, S., Guo, Q., Hartley, P.G., and Hughes, T.C. (2014). *J. Mater. Chem. C* 39: 8303–8312.

162 Ravi, P., Sin, S.L., Gan, L.H. et al. (2005). *Polymer* 46: 137–146.

163 Wang, G., Tong, X., and Zhao, Y. (2004). *Macromolecules* 37: 8911–8917.

164 Zhao, Y. and Stoddart, J.F. (2009). *Langmuir* 25: 8442–8446.

165 Jochum, F.D. and Theato, P. (2009). *Polymer* 50: 3079–3085.

166 Jochum, F.D., Borg, L., Roth, P.J., and Theato, P. (2009). *Macromolecules* 42: 7854–7862.

167 Van Gemert, B. *Benzo and Naphthopyrans (Chromenes)*, 111–140. Boston, MA: Springer.

168 Rosaria, M., Gentili, P.L., Romani, A., and Favaro, G. (2010). *J. Phys. Chem. C* 114: 6123–6131.

169 Nunzio, M.R., Gentili, P.L., Romani, A., and Favaro, G. (2008). *ChemPhysChem* 9: 768–775.

170 Zayat, M., Levy, D., De Ciencia, I., and De Madrid, D.M. (2003). *J. Mater. Chem.* 13: 727–730.

171 Van Gemert, B., Kumar, A., and Knowles, D.B. (2006). *J. Mater. Chem.* 13: 37–41.

172 Zhang, Y., Garcia-Amorós, J., Captain, B., and Raymo, F.M. (2016). *J. Mater. Chem. C* 4: 2744–2747.

173 Tomasulo, M., Sortino, S., White, A.J.P. et al. (2005). *J. Org. Chem.* 70: 8180–8189.

174 Garcia-amorós, J., Swaminathan, S., and Sortino, S. (2014). *Chem. Eur. J.* 20: 10276–10284.

175 Deniz, E., Tomasulo, M., Cusido, J. et al. (2011). *Langmuir* 27: 11773–11783.

176 Beaujean, P., Bondu, F., Garcia-amoro, J. et al. (2016). *J. Am. Chem. Soc.* 138: 5052–5062.

177 Minkin, V.I. (2004). *Chem. Rev.* 104: 2751–2776.

178 Sheng, L., Li, M., Zhu, S. et al. (2014). *Nat. Commun.* 5: 1–8.

179 Garcia-amoros, J., Swaminathan, S., and Raymo, F.M. (2014). *Dyes Pigm.* 106: 71–73.

180 Chen, Q., Sheng, L., Du, J. et al. (2018). *Chem. Commun.* 54: 5094–5097.

181 Na, S. and Kim, H. (2016). *Dyes Pigm.* 134: 526–530.

182 Guerrin, C., Szaloki, G., Berthet, J. et al. (2018). *J. Org. Chem.* 83: 10409–10419.

183 Stellacci, B.F., Bertarelli, C., Toscano, F. et al. (1999). *Adv. Mater.* 11: 292–295.

184 Pu, S., Sun, Q., and Zheng, C. (2017). *Mater. Sci. Eng.* 222: 012019.

185 Zou, Q., Li, X., Zhang, J. et al. (2012). *Chem. Commun.* 48: 2095–2097.

186 Fan, M.G., Yu, L., and Zhao, W. (2002). *Fulgide Family Compounds: Synthesis, Photochromism, and Applications*. Boston, MA: Springer.

187 Tomasello, G., Bearpark, M.J., Robb, M.A. et al. (2010). *Angew. Chem. Int. Ed.* 49: 2913–2916.

188 Chen, Y., Wang, C., Fan, M. et al. (2004). *Opt. Mater.* 26: 75–77.

189 Zaplo, O., Stumpe, J., Seeboth, A., and Hermel, H. (1992). *Mol. Cryst. Liq. Cryst.* 213: 153–161.

190 Görner, H. and Kuhn, H.J. (1994). *Adv. Photochem.* 19: 1–117.

191 Kitzig, S., Thilemann, M., Cordes, T., and Rück-Braun, K. (2016). *Chem. Phys. Chem.* 17: 1252–1263.

192 Eggers, K., Fyles, T.M., and Montoya-pelaez, P.J. (2001). *J. Org. Chem.* 66: 2966–2977.

193 Steinle, W. and Rück-Braun, K. (2003). *Org. Lett.* 5: 141–144.

194 Lin, R.G., Xu, G., Wang, M.S. et al. (2013). *Inorg. Chem.* 52: 1199–1205.

195 Horhold, H., Rathe, H., Helbig, M., and Opfermann, J. (1987). *Macromol. Chem. Phys.* 188: 2083–2104.

196 Sata, T., Matsuo, Y., Yamaguchi, T., and Matsusaki, K. (1997). *J. Chem. Soc.* 93: 2553–2560.

197 Hadjoudis, E. (1995). *Mol. Eng.* 5: 301–337.

198 Thompson, B.C., Abboud, K.A., Reynolds, J.R. et al. (2005). *New J. Chem.* 29: 1128–1134.

199 Hutchins, K.M., Dutta, S., Loren, B.P., and Macgillivray, L.R. (2014). *Chem. Mater.* 26: 3042–3044.

200 Ohshima, A., Momotake, A., and Arai, T. (2004). *J. Photochem. Photobiol.* 162: 473–479.

201 Zgierski, M.Z., Grabowska, A., and Zgierski, M.Z. (2010). *J. Chem. Phys.* 112: 6329–6337.

202 Naumov, P., Sakurai, K., Ishikawa, T. et al. (2005). *J. Phys. Chem. A* 109: 7264–7275.

203 Corval, A., Kuldova, K., Eichen, Y. et al. (1996). *J. Chem. Phys.* 100: 19315–19320.

204 Casalegno, R., Kuldovb, K., and Trommsdorff, H.P. (1997). *J. Lumin.* 72: 78–80.

205 Ikeda, Y., Takeda, J., and Kurita, S. (2000). *Mol. Cryst. Liq. Cryst. Sci. Technol., Sect. A* 345: 197–202.
206 Edkins, R.M., Probert, M.R., Fucke, K. et al. (2013). *Phys. Chem. Chem. Phys.* 15: 7848–7853.
207 Kishimoto, Y. and Abe, J. (2009). *J. Am. Chem. Soc.* 4876: 4227–4229.
208 Castiglioni, F., Danowski, W., Perego, J. et al. (2020). *Nat. Chem.* 12: 595–602.
209 Stuckhardt, C., Roke, D., Danowski, W. et al. (2019). *Beilstein J. Org. Chem.* 15: 2767–2773.
210 Sakata, Y., Fukushima, S., Akine, S., and Setsune, J.I. (2016). *Chem. Commun.* 52: 1278–1281.
211 Corns, S.N., Partington, S.M., and Towns, A.D. (2009). *Color. Technol.* 125: 249–261.
212 Kobatake, S. and Terakawa, Y. (2007). *Chem. Commun.* 17: 1698–1700.
213 Schimka, S., Lomadze, N., Rabe, M. et al. (2016). *Phys. Chem. Chem. Phys.* 19: 108–117.
214 Tabor, R.F., McCoy, T.M., Hu, Y., and Wilkinson, B.L. (2018). *Bull. Chem. Soc. Jpn.* 91: 932–939.
215 Arya, P., Jelken, J., Lomadze, N. et al. (2020). *J. Chem. Phys.* 152: 024904.
216 Avadanei, M., Tigoianu, R., Serpa, C. et al. (2017). *J. Photochem. Photobiol., A* 332: 475–486.
217 Jacquemin, P.L., Robeyns, K., Devillers, M., and Garcia, Y. (2014). *Chem. Commun.* 50: 649–651.
218 Mitra, S. and Tamai, N. (2003). *Phys. Chem. Chem. Phys.* 5: 4647–4652.
219 Sliwa, M., Mouton, N., Ruckebusch, C. et al. (2009). *J. Phys. Chem. C* 27: 11959–11968.
220 Ziółek, M., Kubicki, J., Maciejewski, A. et al. (2004). *Phys. Chem. Chem. Phys.* 6: 4682–4689.
221 Hadjoudis, E., Chatziefthimiou, S.D., and Mavridis, I.M. (2009). *Curr. Org. Chem.* 13: 269–286.
222 Hadjoudis, E., Mavridis, I.M., Box, P.O. et al. (2004). *Chem. Soc. Rev.* 33: 579–588.
223 Ito, H., Mutoh, K., and Abe, J. (2020). *Org. Lett.* 22: 5680–5684.
224 Maeda, K. and Hayashi, T. (1969). *Bull. Chem. Soc. Jpn.* 42: 3509–3514.
225 Zhu, M., Xu, B., and Tian, W. (2017). *J. Am. Chem. Soc.* 139: 16036–16039.
226 Tian, W., Qi, L., Chao, X. et al. (2018). *Compos. Struct.* 189: 1–8.
227 Shiraishi, Y., Miyamoto, R., and Hirai, T. (2009). *Org. Lett.* 11: 1571–1574.
228 Florea, L., Mckeon, A., Diamond, D., and Benito-lopez, F. (2013). *Langmuir* 29: 2790–2797.
229 Takahashi, Y., Fukuyasu, K., Horiuchi, T. et al. (2014). *Langmuir* 30: 41–47.
230 Shang, T., Smith, K.A., and Hatton, T.A. (2003). *Langmuir* 19: 10764–10773.
231 Long, J., Tian, S., Niu, Y. et al. (2014). *Colloids Surf., A* 454: 172–179.
232 Hughes, T., Simon, G.P., and Saito, K. (2019). *ACS Appl. Mater. Interfaces* 11: 19429–19443.
233 Davidson-Rozenfeld, G., Stricker, L., Simke, J. et al. (2019). *Polym. Chem.* 10: 4106–4115.

234 Yang, M., Wang, L., Cheng, Y. et al. *J. Mater. Sci.* 54: 9983–9994.
235 Iwaso, K., Takashima, Y., and Harada, A. (2016). *Nat. Chem.* 8: 625–632.
236 Yue, Y., Norikane, Y., Azumi, R., and Koyama, E. (2018). *Nat. Commun.* 9: 1–8.
237 Lv, J.A., Liu, Y., Wei, J. et al. (2016). *Nature* 537: 179–184.
238 Logtenberg, H., van der Velde, J.H., de Mendoza, P. et al. (2012). *J. Phys. Chem. C* 116: 24136–24142.
239 Pearson, J.M. (1977). *Photochemical Processes in Polymer Chemistry – 2*, 463–477. Pergamon.
240 Zhang, H., Tian, W., Suo, R. et al. (2015). *J. Mater. Chem. B* 3: 8528–8536.
241 Shao, Y., Shi, C., Xu, G. et al. (2014). *ACS Appl. Mater. Interfaces* 6: 10381–10392.
242 Wang, X., Hu, J., Liu, G. et al. (2015). *J. Am. Chem. Soc.* 137: 15262–15275.
243 Utsumi, H., Nagahama, D., Nakano, H., and Shirota, Y. (2002). *J. Mater. Chem.* 12: 2612–2619.
244 Scarmagnani, S., Walsh, Z., Alhashimy, N., Radu, A., Paull, B., Macka, M.; Diamond, D. (2007). 29th Annual International Conference of the IEEE Engineering in Medicine and Biology Society, 4096–4097.
245 Miyatake, T., Takamori, Y., and Yamaguchi, K.J. (2015). *Photochem. Photobiol. A: Chem.* 313: 36–43.
246 Hur, D.Y. and Shin, E.J. (2015). *Bull. Korean Chem. Soc.* 36: 104–110.
247 Bochet, C.G. (2002). *J. Chem. Soc., Perkin Transactions* 1: 125–142.
248 Wang, P. (2013). *Asian J. Organ. Chem.* 2: 452–464.
249 Fernandes, M.J.G., Gonc, M.S.T., and Costa, S.P.G. (2008). *Tetrahedron* 64: 3032–3038.
250 Inomata, K., Kawasaki, S., Kameyama, A., and Nishikubo, T. (2000). *React. Funct. Polym.* 45: 1–9.
251 Bertrand, O., Gohy, J.F., and Fustin, C.A. (2011). *Polym. Chem.* 2: 2284–2292.
252 Okamoto, A., Tanabe, K., Inasaki, T., and Saito, I. (2003). *Angew. Chem.* 42: 2502–2504.
253 Orth, R. and Sieber, S.A. (2009). *J. Org. Chem.* 74: 8476–8479.
254 Singh, A.K. and Khade, P.K. (2005). *Tetrahedron Lett.* 46: 5563–5566.
255 Barltrop, J.A., Plant, P.J., and Schofield, P. (1966). *Chem. Commun. (London)* 22: 822–823.
256 Pelliccioli, A.P., Wirz, J.J., and Wirz, J.J. (2002). *Photochem. Photobiol. Sci.* 1: 441–458.
257 Hellrung, B., Kamdzhilov, Y., Schwo, M., and Wirz, J. (2005). *J. Am. Chem. Soc.* 127: 8934–8935.
258 Yang, F., Cao, Z., and Wang, G. (2015). *Polym. Chem.* 6: 7995–8002.
259 He, L., Vibhagool, S., Zhao, H. et al. (2018). *Macromol. Chem. Phys.* 219: 1–5.
260 Shen, H., Xia, Y., Qin, Z. et al. (2017). *Polym. Chem.* 55: 1–11.
261 Jana, S., Bose, A., Saha, A., and Mandal, T.K. (2017). *J. Polym. Sci., Part A: Polym. Chem.* 55: 1714–1729.
262 Doh, J. and Irvine, D.J. (2004). *J. Am. Chem. Soc.* 126: 9170–9171.
263 Wu, H., Dong, J., Zhan, X. et al. (2014). *RSC Adv.* 4: 35757–35761.
264 Doh, J. and Irvine, D.J. (2006). *Proc. Natl. Acad. Sci. U.S.A.* 103: 5700–5705.

265 Wang, X., Jiang, G., Li, X. et al. (2013). *Polym. Chem.* 4: 4574–4577.
266 Cao, Z., Wu, H., Dong, J., and Wang, G. (2014). *Macromolecules* 47: 8777–8783.
267 Lai, J., Mu, X., Xu, Y. et al. (2010). *Chem. Commun.* 46: 7370–7372.
268 Wu, H., Dong, J., Li, C. et al. (2013). *Chem. Commun.* 49: 3516–3518.
269 Woodcock, J.W., Wright, R.A.E., Jiang, X. et al. (2010). *Soft Matter* 6: 3325–3336.
270 Zhang, J., Shi, N., Zhang, J. et al. (2016). *Macromol. Rapid Commun.* 38: 1–7.
271 Yao, C., Wang, X., Liu, G. et al. (2016). *Macromolecules* 49: 8282–8295.
272 Goeldner, M. and Givens, R. (ed.) (2006). *Dynamic Studies in Biology: Phototriggers, Photoswitches and Caged Biomolecules.* Wiley.
273 Yu, H., Li, J., Wu, D. et al. (2010). *Chem. Soc. Rev.* 39: 464–473.
274 Ai, X., Mu, J., and Xing, B. (2016). *Theranostics* 6: 2439–2457.
275 Liu, Z., Shi, J., Wang, Y. et al. (2019). *Eng. Asp.* 568: 436–444.
276 Jiang, J., Tong, X., and Zhao, Y. (2005). *J. Am. Chem. Soc.* 127: 8290–8291.
277 Tanaka, K., Takahashi, Y., Isobe, T. et al. (2005). *J. Photochem. Photobiol. A: Chem.* 174: 130–137.
278 Gohy, J.F. and Zhao, Y. (2013). *Chem. Soc. Rev.* 42: 7117–7129.
279 Sun, K., Kumar, R., Falvey, D.E., and Raghavan, S.R. (2009). *J. Am. Chem. Soc.* 131: 7135–7141.
280 Griesbeck, A.G. and Mattay, J. (2004). *Synthetic Organic Photochemistry*, 13–22. CRC Press.
281 Kashida, H., Doi, T., Sakakibara, T. et al. (2013). *J. Am. Chem. Soc.* 135: 7960–7966.
282 Wolff, T. and Nees, D. (1998). *Prog. Colloid Polym. Sci.* 111: 113–116.
283 Itoh, K., Yamamoto, M., and Wajima, T.J. (1993). *Mol. Struct.* 293: 311–314.
284 Takagi, K., Suddaby, B.R., Vadas, S.L., and Backer, C.A. (1986). *J. Am. Chem. Soc.* 108: 7865–7867.
285 Ramamurthy, V. and Parthasarathy, A. (2011). *Isr. J. Chem.* 51: 817–829.
286 Kaliappan, R., Maddipatla, M.V.S.N., Kaanumalle, L.S., and Ramamurthy, V. (2007). *Photobiol. Sci.* 6: 737–740.
287 Mondal, B., Zhang, T., Prabhakar, R. et al. (2014). *Photobiol. Sci.* 13: 1509–1520.
288 Bryan, G.O., Wong, B.M., and Mcelhanon, J.R. (2010). *ACS Appl. Mater. Interfaces* 2: 1594–1600.
289 Jiang, X., Lavender, C.A., Woodcock, J.W., and Zhao, B. (2008). *Macromolecules* 41: 2632–2643.
290 Ionov, L.; Diez, S. 2009, 34, 13315–13319.
291 Il'ichev, Y.V., Schwo, M.A., and Wirz, J. (2004). *J. Am. Chem. Soc.* 126: 4581–4595.
292 Zhao, H., Sterner, E.S., Coughlin, E.B., and Theato, P. (2012). *Macromolecules* 45: 1723–1736.
293 Woodward, R.B. and Hoffmann, R. (1969). *Angew. Chem. Int. Ed. Engl.* 8: 781.
294 Chen, J., Chang, C., and Road, K. (2014). *Materials* 7: 805–875.
295 Shahi, S., Roghani-Mamaqani, H., Talebi, S., and Mardani, H. (2022). *Coord. Chem. Rev.* 455: 214368.
296 Shirai, M. (2014). *Polym. J.* 12: 1–7.

297 Zhang, Y., Hreha, R.D., Jabbour, G.E. et al. (2002). *J. Mater. Chem.* 12: 1703–1708.

298 Chandross, E.A. (1965). *J. Chem. Phys.* 43: 4175–4176.

299 Tamaki, T., Kokubu, T., and Ichimura, K. (1987). *Tetrahedron* 43: 1485–1494.

300 Van Damme, J. and Du Prez, F. (2018). *Prog. Polym. Sci.* 82: 92–119.

301 Rameshbabu, K., Kim, Y., Kwon, T. et al. (2007). *Tetrahedron Lett.* 48: 4755–4760.

302 Kan, L., Cheng, H., Li, B. et al. (2019). *New J. Chem.* 43: 2658–2664.

303 Wei, P., Yan, X., and Huang, F. (2014). *Chem. Commun.* 50: 14105–14108.

304 Fang, Y., Du, X., Du, Z. et al. (2017). *J. Mater. Chem. A* 5: 8010–8017.

305 Li, M.H. and Keller, P. (2009). *Soft Matter* 5: 927–937.

306 Inkinen, J., Niskanen, J., Talka, T. et al. (2015). *Sci. Rep.* 5: 1–8.

307 Ding, L., Liu, L., Chen, Y.F. et al. (2019). *Chem. Eng. J.* 374: 1317–1325.

308 Yuan, X., Fischer, K., and Schärtl, W. (2004). *Adv. Funct. Mater.* 14: 457–463.

309 Ishigami, T., Murata, T., and Endo, T. (1976). *Bull. Chem. Soc. Jpn.* 49: 3578–3583.

310 Gattás-Asfura, K.M., Weisman, E., Andreopoulos, F.M. et al. (2005). *Biomacromolecules* 6: 1503–1509.

311 Kuang, G.C., Ji, Y., Jia, X.R. et al. (2009). *Tetrahedron* 65: 3496–3501.

312 Zhang, W., Hong, C., and Pan, C. (2017). *Biomacromolecules* 4: 1210–1217.

313 Sinnwell, M.A., Ingenthron, B.J., Groeneman, R.H., and MacGillivray, L.R. (2016). *J. Fluorine Chem.* 188: 5–9.

314 Kang, B., Lee, J.Y., and Kang, S. (2018). *Bull. Korean Chem. Soc.* 39: 1283–1286.

315 Liao, L.Y., Li, Y.B., Zhang, X.M. et al. (2014). *J. Phys. Chem. C* 118: 15963–15969.

316 Serhan, M., Sprowls, M., Jackemeyer, D. et al. (2019). *AIChE Annual Meeting Conference Proceedings American Institute of Chemical Engineers.*

317 Vicinelli, V., Ceroni, P., Maestri, M. et al. (2004). *Org. Biomol. Chem.* 2: 2207–2213.

318 Hirano, S., Toyota, S., and Toda, F. (2005). *Chem. Commun.* 5: 643–644.

319 Finch, A.S., Davis, W.B., and Rokita, S.E. (2013). *Photochem. Photobiol. Sci.* 12: 1474–1482.

320 Neelakandan, P.P., Pan, Z., Hariharan, M., and Lewis, F.D. (2012). *Photochem. Photobiol. Sci.* 11: 889–892.

321 Maddipatla, M.V.S.N., Wehrung, D., Tang, C. et al. (2013). *Macromolecules* 13: 5133–5140.

322 Ling, J., Zhi, M., and Qiu, M. (2012). *Polymer* 53: 2691–2698.

323 Minkenberg, C.B., Florusse, L., Eelkema, R. et al. (2009). *J. Am. Chem. Soc.* 131: 11274–11275.

324 Shahi, S., Roghani-Mamaqani, H., Hoogenboom, R. et al. (2020). *Chem. Mater.* 34, 468–498.

325 Spitzer, D., Rodrigues, L.L., Straßburger, D. et al. (2017). *Angew. Chem.* 56: 15461–15465.

326 Schattling, P., Jochum, D., and Theato, P. (2014). *Polym. Chem. Talents* 5: 25–36.

327 Liu, C., Fan, Z., Tan, Y. et al. (2020). *Adv. Mater.* 32: 1907569.

328 Perera, M.M., Chimala, P., Elhusain-elnegres, A. et al. (2020). *ACS Macro Lett.* 9: 1552–1557.

329 Guo, W. and Lei, Z. (2015). *J. Mater. Res.* 30: 3201–3210.

330 Gröger, G., Meyer-Zaika, W., Böttcher, C. et al. (2011). *J. Am. Chem. Soc.* 133: 8961–8971.

331 Doncom, K.E., Hansell, C.F., Theato, P., and O'Reilly, R.K. (2012). *Polym. Chem.* 3: 3007–3015.

332 Azcune, I. and Odriozola, I. (2016). *Eur. Polym. J.* 84: 147–160.

333 AbdolahZadeh, M., Van Der Zwaag, S., and Garcia, S.J. (2016). *Self-healing Materials. Advances in Polymer Science*, vol. 273, 185–218. Cham: Springer.

334 Yang, Y. and Urban, M.W. (2013). *Chem. Soc. Rev.* 42: 7446–7467.

335 Wojtecki, R.J., Meador, M.A., and Rowan, S.J. (2010). *Nat. Publ. Gr.* 10: 14–27.

336 Kloxin, C.J. and Bowman, C.N. (2013). *Chem. Soc. Rev.* 42: 7161–7173.

337 Maeda, T., Otsuka, H., and Takahara, A. (2009). *Polymer* 34: 581–604.

338 Chakma, P. and Konkolewicz, D. (2019). *Angew. Chem.* 131: 9784–9797.

339 García, F. and Smulders, M.M. (2016). *J. Polym. Sci., Part A: Polym. Chem.* 54: 3551–3577.

340 Fan, F., Ji, S., Sun, C. et al. (2018). *Angew. Chem. Int. Ed.* 57: 16426–16430.

341 Wagner, M., Krieger, A., Minameyer, M. et al. (2021). *Macromolecules* 54: 2899–2911.

342 Otsuka, H., Nagano, S., Kobashi, Y. et al. (2010). *Chem. Commun.* 46: 1150–1152.

343 Miyasaka, H., Satoh, Y., Ishibashi, Y. et al. (2009). *J. Am. Chem. Soc.* 131: 7256–7263.

344 Liu, Y., Yang, Y., Shi, D. et al. (2019). *Adv. Mater.* 31: 1–7.

345 Fujita, K., Hatano, S., Kato, D., and Abe, J. (2008). *Org. Lett.* 10: 3105–3108.

346 Qin, X.Z., Liu, A., Trifunac, A.D., and Krongauz, V.V. (1991). *J. Phys. Chem.* 95: 5822–5826.

347 Li, W., Lu, S., Zhao, M. et al. (2018). *Polymers* 10: 1392.

348 McBride, M.K., Hendrikx, M., Liu, D. et al. (2017). *Adv. Mater.* 29: 1606509.

349 Gandavarapu, N.R., Azagarsamy, M.A., and Anseth, K.S. (2014). *Adv. Mater.* 26: 2521–2526.

350 Wu, Z., Cheng, P., Zhao, W. et al. (2020). *New J. Chem.* 44: 10902–10910.

351 Meng, Y., Fenoli, C.R., Aguirre-Soto, A. et al. (2014). *Adv. Mater.* 26: 6497–6502.

352 Zhao, P., Xia, J., Cao, M., and Xu, H. (2020). *ACS Macro Lett.* 9: 163–168.

353 Amamoto, Y., Otsuka, H., Takahara, A., and Matyjaszewski, K. (2012). *Adv. Mater.* 24: 3975–3980.

354 Amamoto, Y., Kamada, J., Otsuka, H. et al. (2011). *Angew. Chem. Int. Ed.* 50: 1660–1663.

355 Dong, P., Cui, K., Xu, F. et al. (2018). *Polym. Int.* 67: 868–873.

356 Ji, S., Cao, W., Yu, Y., and Xu, H. (2014). *Angew. Chem. Int. Ed.* 53: 6781–6785.

357 Ji, S., Xia, J., and Xu, H. (2016). *ACS Macro Lett.* 5: 78–82.

358 Zhang, Y., Qi, Y., Ulrich, S. et al. (2019). *Mater. Chem. Front.* 4: 489–506.

359 Du, X., Li, J., Welle, A. et al. (2015). *Adv. Mater.* 27: 5000–5001.

360 Fortman, D.J., Snyder, R.L., Sheppard, D.T., and Dichtel, W.R. (2018). *ACS Macro Lett.* 7: 1226–1231.

361 Nevejans, S., Ballard, N., Miranda, J.I. et al. (2016). *Phys. Chem. Chem. Phys.* 18: 27577–27583.

362 Klepel, F. and Ravoo, B.J. (2017). *Org. Biomol. Chem.* 15: 3840–3842.

363 Zhang, X. and Waymouth, R.M. (2017). *J. Am. Chem. Soc.* 139: 3822–3833.

364 Delavari, S., Ziadzade, S., Keyvan Rad, J. et al. (2020). *Carbohydr. Polym.* 247: 116756.

365 Hajiali, M., Keyvan Rad, J., Ghezelsefloo, S., and Mahdavian, A.R. (2020). *J. Ind. Eng. Chem.* 92: 287–296.

366 Anderson, L., Nehrlich, S., and Champigny, M.L. (1978). *Plant Physiol.* 61: 601–605.

367 Zhao, D., Du, Z., Liu, S. et al. (2019). *ACS Appl. Polym. Mater.* 1: 2951–2960.

368 Zhang, T., Chen, H., Ma, X., and Tian, H. (2017). *Ind. Eng. Chem. Res.* 56: 3123–3128.

369 Huang, L., Ding, Y., Ma, Y. et al. (2019). *Macromolecules* 52: 4703–4712.

370 Sun, B., He, Z., Hou, Q. et al. (2013). *Carbohydr. Polym.* 95: 598–605.

371 Zhu, M.Q., Zhu, L., Han, J.J. et al. (2006). *J. Am. Chem. Soc.* 128: 4303–4309.

372 Chen, H., Yao, X., Ma, X., and Tian, H. (2016). *Adv. Opt. Mater.* 4: 1397–1401.

373 Brown, C.L., Barbee, M.H., Ko, J.H. et al. (2017). *J. Chem. Educ.* 94: 1752–1755.

374 Zhu, M.Q., Zhang, G.F., Li, C. et al. (2011). *J. Am. Chem. Soc.* 133: 365–372.

375 Sun, B., Hou, Q., He, Z. et al. (2014). *Carbohydr. Polym.* 111: 419–424.

376 Karimipour, K., Keyvan Rad, J., Ghomi, A.R. et al. (2020). *Dyes Pigm.* 175: 108185.

377 Azimi, R. and Abdollahi, A. (2021). *J. Mater. Chem. C* 9: 9571–9583.

378 Seipel, S., Yu, J., Viková, M. et al. (2019). *Fibers Polym.* 20: 1424–1435.

379 Yang, Y., Li, M., and Fu, S. (2021). *Prog. Org. Coat.* 158: 106348.

380 Wang, J., Jin, B., Wang, N. et al. (2017). *Macromolecules* 50: 4629–4638.

381 Nezhadghaffar-Borhani, E., Abdollahi, A., Roghani-Mamaqani, H., and Salami-Kalajahi, M. (2021). *J. Colloid Interface Sci.* 593: 67–78.

382 Ang, C.Y., Tan, S.Y., Lu, Y. et al. (2014). *Sci. Rep.* 4: 1–8.

383 Kim, H.N., Guo, Z., Zhu, W. et al. (2011). *Chem. Soc. Rev.* 40: 79–93.

384 Wu, D., Sedgwick, A.C., Gunnlaugsson, T. et al. (2017). *Chem. Soc. Rev.* 46: 7105–7123.

385 Cheng, H.-R., Yang, L., Zhang, S. et al. (2019). *Luminescence* 34: 615–622.

386 Georgiev, N.I., Asiri, A.M., Qusti, A.H. et al. (2014). *Dyes Pigm.* 102: 35–45.

387 Jiang, G., Wang, S., Yuan, W. et al. (2007). *Eur. J. Org. Chem.* 13: 2064–2067.

388 Zhu, L., Ma, X., Ji, F. et al. (2007). *Chem. Eur. J.* 13: 9216–9222.

389 Wagner, K., Byrne, R., Zanoni, M. et al. (2011). *J. Am. Chem. Soc.* 133: 5453–5462.

390 Kopelman, R.A., Snyder, S.M., and Frank, N.L. (2003). *J. Am. Chem. Soc.* 125: 13684–13685.

391 Fries, K., Samanta, S., Orski, S., and Locklin, J. (2008). *Chem. Commun.* 6288–6290.

392 Nguyen, T.H., Sun, T., and Grattan, K.T.V. (2020). *Dyes Pigm.* 177: 108312.

393 Vonlanthen, M., Rojas-Montoya, S.M., Cuétara-Guadarrama, F. et al. (2020). *Macromol. Chem. Phys.* 221: 1–9.
394 Grazon, C., Si, Y., Placial, J.P. et al. (2019). *Photochem. Photobiol. Sci.* 18: 1156–1165.
395 Guo, J., Wei, X., Fang, X. et al. (2021). *Sens. Actuators, B* 347: 130623.
396 Shiraishi, Y., Sumiya, S., Manabe, K., and Hirai, T.T. (2011). *ACS Appl. Mater. Interfaces* 3: 4649–4656.
397 Jia, Y., Wang, S., Wang, W.J. et al. (2019). *Macromolecules* 52: 7920–7928.
398 Ionov, B.L., Synytska, A., and Diez, S. (2008). *Adv. Funct. Mater.* 1501–1508.
399 Ionov, L., Minko, S., Stamm, M., and Je, R. (2003). *J. Am. Chem. Soc.* 125: 8302–8306.
400 Weng, B., Beh, S., Kim, I.T. et al. (1999). *Adv. Mater.* 11: 1038–1041.
401 Zhao, B., Moore, J.S., and Beebe, D.J. (2001). *Science* 291: 1023–1026.
402 Tadros, T.F. (2006). *Applied Surfactants: Principles and Applications*. Wiley.
403 Li, S., Wang, F., and Liu, H. (2019). *ChemistrySelect* 4: 7404–7412.
404 Perrin, P., Porcar, I., and Tribet, C. (2003). *Polym. Int.* 470: 465–470.
405 Wang, F. and Liu, H. (2018). *J. Phys. Chem. C* 122: 3434–3442.
406 Yi, C., Sun, J., Zhao, D. et al. (2014). *Langmuir* 30: 6669–6677.
407 Wang, F., Yu, X., Yang, Z. et al. (2018). *J. Phys. Chem. C* 122: 18995–19003.
408 Salonen, A., Langevin, D., and Perrin, P. (2010). *Soft Matter* 6: 5308–5311.
409 Kim, J., Yun, H., Lee, Y.J. et al. (2021). *J. Am. Chem. Soc.* 143: 13333–13341.
410 Chen, C.J., Liu, G.Y., Liu, X.S. et al. (2012). *New J. Chem.* 36: 694–701.
411 Kelly, E.A., Willis-Fox, N., Houston, J.E. et al. (2020). *Nanoscale* 12: 6300–6306.
412 Sakai, H., Orihara, Y., Kodashima, H. et al. (2005). *J. Am. Chem. Soc.* 127: 13454–13455.
413 Dou, Q., Liow, S.S., Weng, W., and Loh, X.J. (2016). *J. Polym. Sci., Part A: Polym. Chem.* 54: 2837–2844.
414 Tan, C.S.Y., Del Barrio, J., Liu, J., and Scherman, O.A. (2015). *Polym. Chem.* 6: 7652–7657.
415 Ketner, A.M., Kumar, R., Davies, T.S. et al. (2007). *J. Am. Chem. Soc.* 129: 1553–1559.
416 Cho, M.Y., Kim, J.S., Choi, H.J. et al. (2017). *Smart Mater. Struct.* 26: 054007.
417 Du, Z., Yan, X., Dong, R. et al. (2018). *Macromolecules* 51: 1518–1528.
418 Gäbert, C., Rosenstingl, T., Linsler, D. et al. (2020). *ACS Appl. Polym. Mater.* 2: 5460–5468.
419 Irie, M., Hayashi, K., and Menju, A. (1981). *Polym. Photochem.* 1: 233–242.
420 Sugiura, S., Sumaru, K., Ohi, K. et al. (2007). *Sens. Actuators, A* 140: 176–184.
421 Menju, A., Hayashi, K., and Irie, M. (1981). *Macromolecules* 14: 755–758.
422 Mphahlele, K., Ray, S.S., and Kolesnikov, A. (2017). *Polymers (Basel)* 9: 1–22.
423 Wypych, G. (2017). *Self-Healing Materials: Principles and Technology*, 1e, vol. 1. ChemTec Publishing.
424 Ghosh, S.K. (ed.) (2009). Weinheim: Wiley-VCH, 138–217.
425 Zhang, H., Fortin, D., Xia, H., and Zhao, Y. (2013). *Macromol. Rapid Commun.* 34: 1742–1746.
426 Burnworth, M., Tang, L., Kumpfer, J.R. et al. (2011). *Nature* 472: 334–337.

427 Liao, X., Chen, G., Liu, X. et al. (2010). *Angew. Chem.* 122: 4511–4515.

428 Chen, H., Ma, X., Wu, S., and Tian, H. (2014). *Angew. Chem.* 126: 14373–14376.

429 Gordon, M.B., French, J.M., Wagner, N.J., and Kloxin, C. (2015). *J. Adv. Mater.* 27: 8007–8010.

430 Kunz, W. and Kellermeier, M. (2009). *Science* 323: 344–345.

431 Chujo, Y., Sada, K., Nomura, R. et al. (1993). *Macromolecules* 26: 5611–5614.

432 Abdallh, M., Hearn, M.T.W., Simon, G.P., and Saito, K. (2017). *Polym. Chem.* 8: 5875–5883.

433 Chung, C.M., Roh, Y.S., Cho, S.Y., and Kim, J.G. (2004). *Chem. Mater.* 16: 3982–3984.

434 Concellón, A., Schenning, A.P.H.J., Romero, P. et al. (2018). *Macromolecules* 51: 2349–2358.

435 Joseph, J.P., Singh, A., Gupta, D. et al. (2019). *ACS Appl. Mater. Interfaces* 11: 28213–28220.

436 Ling, J., Rong, M.Z., and Zhang, M.Q. (2011). *J. Mater. Chem.* 21: 18373–18380.

437 Froimowicz, P., Frey, H., and Landfester, K. (2011). *Macromol. Rapid Commun.* 32: 468–473.

438 Ota, M., Sasamori, T., Tokitoh, N. et al. (2015). *J. Org. Chem.* 80: 5687–5695.

439 Trenor, S.R., Shultz, A.R., Love, B.J., and Long, T.E. (2004). *Chem. Rev.* 104: 3059–3078.

440 He, J., Tremblay, L., Lacelle, S., and Zhao, Y. (2011). *Soft Matter* 7: 2380–2386.

441 Altintas, O., Willenbacher, J., Wuest, K.N.R. et al. (2013). *Macromolecules* 46: 8092–8101.

442 Zhang, M.Q. and Rong, M.Z. (2013). *Polym. Chem.* 4: 4878–4884.

443 Canadell, J., Goossens, H., and Klumperman, B. (2011). *Macromolecules* 44: 2536–2541.

444 Fairbanks, B.D., Singh, S.P., Bowman, C.N., and Anseth, K.S. (2011). *Macromolecules* 44: 2444–2450.

445 Scott, T.F., Schneider, A.D., Cook, W.D., and Bowman, C.N. (2005). *Science* 308: 1615–1617.

446 Scott, T.F., Draughon, R.B., and Bowman, C.N. (2006). *Adv. Mater.* 18: 2128–2132.

447 Li, W., Liu, Y., and Leng, J. (2018). *Compos. Part A: Appl. Sci. Manuf.* 110: 70–75.

448 Stoychev, G., Kirillova, A., and Ionov, L. (2019). *Adv. Opt. Mater.* 7: 1–30.

449 Wei, Y., Ebendorff-Heidepriem, H., and Zhao, J. (2019). *Adv. Opt. Mater.* 7 (21): 1900702.

450 Hwang, I.W., Soci, C., Moses, D. et al. (2007). *J. Adv. Mater.* 19: 2307–2312.

451 Wu, Y., Peng, Y., Bohra, H. et al. (2019). *ACS Appl. Mater. Interfaces* 11: 4833–4841.

452 Nian, L., Zhang, W., Zhu, N. et al. (2015). *J. Am. Chem. Soc.* 137: 6995–6998.

453 Liguori, R., Rubino, A., Botta, A., and Pragliola, S. (2017). *International Semiconductor Conference (CAS)*, pp. 79–82.

454 Giang, H.N., Kinashi, K., Sakai, W., and Tsutsumi, N. (2014). *J. Photochem. Photobiol., A* 291: 26–33.

455 Jiang, H., Hershtig, G., Richter, S., and Jelinek, R. (2016). *J. Phys. Chem. Lett.* 7: 1628–1631.

456 Wu, Y., Wang, K., Huang, S. et al. (2017). *ACS Appl. Mater. Interfaces* 9: 13602–13610.

457 Pashaei-Sarnaghi, R., Najafi, F., Taghavi-Kahagh, A. et al. (2021). *Eur. Polym. J.* 158: 110686.

458 Behl, G., Kumar, P., Sikka, M. et al. (2018). *J. Biomater. Sci., Polym. Ed.* 29: 360–375.

459 Arjmand, F., Salami-Kalajahi, M., and Roghani-Mamaqani, H. (2021). *J. Phys. Chem. Solids* 154: 110102.

460 Samanta, P., Kapat, K., Maiti, S. et al. (2019). *J. Colloid Interface Sci.* 555: 132–144.

461 Zhang, L., Pu, Y., Li, J. et al. (2020). *Drug Deliv. Sci. Technol.* 58: 101789.

462 Stefanello, T.F., Couturaud, B., Szarpak-Jankowska, A. et al. (2017). *Nanoscale* 9: 12150–12162.

463 Ghani, M., Heiskanen, A., Kajtez, J. et al. (2021). *ACS Appl. Mater. Interfaces* 13: 3591–3604.

464 Yuan, W., Gao, X., Pei, E., and Li, Z. (2018). *Polym. Chem.* 9: 3651–3661.

465 Liu, Z., Wang, W., Xie, R. et al. (2016). *Chem. Soc. Rev.* 45: 460–475.

466 Xie, Y., Chen, S., Qian, Y. et al. (2017). *Mater. Sci. Eng., C* 84: 52–59.

3

Polymer Systems for Ionizing Radiation Dosimetry and Radiotherapy

Li Jiang, Chengfang Zhang, Rensheng Wang, and Liang Hu

Soochow University, State Key Laboratory of Radiation Medicine and Protection, Collaborative Innovation Center of Radiological Medicine of Jiangsu Higher Education Institutions, School for Radiological and Interdisciplinary Sciences (RAD-X), 199 Ren'ai Road, Suzhou 215000, China

3.1 Introduction

A variety of radioactive nuclides are known to exist in nature that decay other nuclides with emission of various kinds of radiation. In 1895, Roentgen observed X-rays studying cathode rays in a gas discharge tube. In 1898, the Curies discovered the radioactivity of polonium and radium. Later, Becquerel revealed that uranium decays emitting β-radiation. In 1911, Rutherford proposed a planetary model of atom according to which an atom consists of dense positively charged nucleus and negatively charged electrons orbiting around it. Radiation can be classified into charged particles (i.e. α-, β-particles, and protons) and neutral radiation (i.e. photons and neutrons). Since the nineteenth century, the interaction of radiation with matter, in particular, with human beings, was widely studied. The radiation absorbed by our body causes radiobiological effects on the cells.

The ionizing radiation is widely used in medical diagnosis and treatment, food processing, nuclear power generation, military, and other fields [1–6]. The radiation dose is defined as the amount of energy deposited per unit mass (J/kg or Gy). The real-time/cumulative measurement of the radiation dose is beneficial for assessing the health status of the relevant personnel [7]. In the field of radiotherapy, monitoring the radiation dose can deposit the maximum dose in the tumor and minimize the adverse effects on the healthy tissues, thereby significantly improving the effectiveness of the radiotherapy.

Meanwhile, stimuli-responsive polymers have been widely studied for decades owing to the changes in their transformations in response to external stimuli, such as temperature, pH, and light [8]. Among the various stimuli, the ionizing radiation responsive polymers are rare, although these exhibit a high potential in the combined radiotherapy and chemotherapy [9–11]. The polymers containing the S—S (bond energy: 240 KJ/mol), Se—Se (bond energy: 172 KJ/mol), and Te—Te (126 KJ/mol) bonds are reported to be ideal for generating the ionizing radiation-responsive polymers [12–15].

Smart Stimuli-Responsive Polymers, Films, and Gels, First Edition.
Edited by Liang Hu, Yongfeng Gao, and Michael J. Serpe.

Hence, in this chapter, we first introduce the mechanisms of interaction between different kinds of radiation with matter. Then, focused on the polymer-based dosimeters, including polymer/dye, polymer fluorescent, and polymer/noble metal nanoparticles dosimeters. We finally move to stimuli-responsive polymers and their use in the combined radiotherapy and chemotherapy processes.

3.2 Interaction of Radiation with Matter

Understanding radiation–matter interaction enhances our capability to utilize radioactive emission for medical imaging, nuclear medicine applications, and radiotherapy at the same time also avoiding its toxic action to the highest possible extent [16].

3.2.1 α-Particles

α-Particles, which are the most common heavily charged particles observed in radioactive radiation, are helium nuclei composed of two protons and two neutrons. Their emission usually accompanies decay of heavy nuclei. Normally, α-particles emitted by natural radionuclides have a kinetic energy of 4–10 MeV. On penetrating the material, they have a negligible probability of interaction with its nuclei. Instead, α-particles lose almost all their energy by interacting with electrons, which results in the ionization or excitation of atoms in the matter (Figure 3.1). Due to the larger mass of α-particles as compared to that of electrons, they move on straight trajectories in the material while losing their energy. Provided that the kinetic energy of α-particles is sufficient for ionization, the electrons are stripped off neutral atoms, yielding formation of pairs of electrons and positively charged ions referred as the electron–ion pairs (or ion pairs). The average energy required to produce an ion pair is called the average ionization energy (W_e), which is fully determined by the chemical composition of the material and the nature of radiation. According to the Bethe–Bloch formula (Eq. (3.1)), the average energy loss of incident α-particles per unit path length ($-dE/dx$) is proportional to the square of particle charge ($z = 2$) and inversely proportional to the square of its velocity (v), which is expressed as follows:

$$-\frac{dE}{dx} = \frac{4\pi z^2 e^4}{m_0 v^2}\rho\frac{Z}{A}\left[\ln\left(\frac{2m_0v^2}{I}\right) - \ln\left(1-\left(\frac{v}{c}\right)^2\right) - \left(\frac{v}{c}\right)^2\right] \tag{3.1}$$

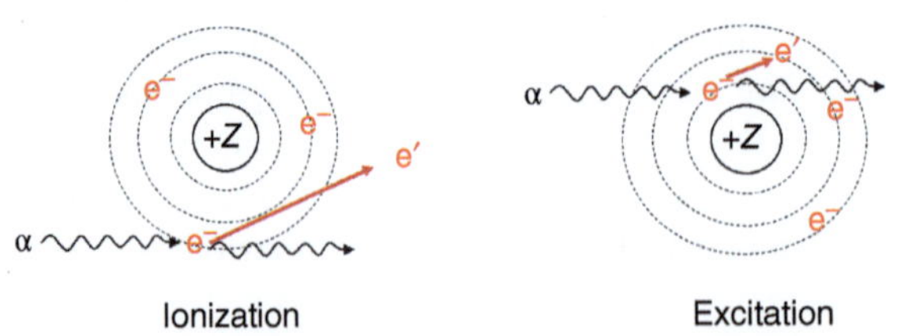

Figure 3.1 Interaction of α-particles with matter.

where m_0 is the mass of an α-particle, ρ, Z, and A are the density, atomic number, and mass number of an irradiated substance respectively, I is the average ionization potential, and c is the speed of light.

According to the Bethe–Bloch formula, the $-dE/dx$ rapidly increases when α-particles stop ($v \to 0$) at the end of their trajectory in the matter. That is how, in all the rest of particles kinetic energy is released in a very short path, which leads to appearance of a Bragg peak. Therefore, by adjusting the incident particle velocity, the position of the Bragg peak can be controlled. Such control underlies the principle of proton and heavy-ion radiotherapy.

When the kinetic energy of an α-particle is not sufficient to ionize an atom but is enough to excite an electron to a higher energy level, this process is referred to as excitation. The electrons in the excited state are unstable and return to the lower energy levels accompanied by emission of photons.

3.2.2 Electrons

Electrons are negatively charged particles with small masses (511 keV), such that they are significantly scattered during propagation throughout the matter. Similar to α-particles, they can ionize or excite the atoms (Figure 3.2). When the interaction of an electron with an atom leads to the ejection of an atomic electron from inner orbital, the atom turns to be in an excited state with an electron vacancy. This vacancy is filled by an electron from outer orbital, at this time, characteristic X-ray emission is produced. Electrons can also interact with nuclei. During this process, the velocity of an electron rapidly decreases with energy release inducing bremsstrahlung (emission of a high-energy photon). The fraction of energy loss due to bremsstrahlung increases with the increase in the electron energy and the atomic number of the material. The bremsstrahlung process is the physical mechanism of X-ray generation and is used in medical radiotherapy, digital radiography, and computed tomography.

3.2.3 Photons

γ- and X-rays generally refer to electromagnetic radiation emitted in the de-excitation processes of nuclei or by electron transition/bremsstrahlung. By nature, both γ- and X-rays are electromagnetic waves, i.e. photons. Unlike α-particles, when a photon

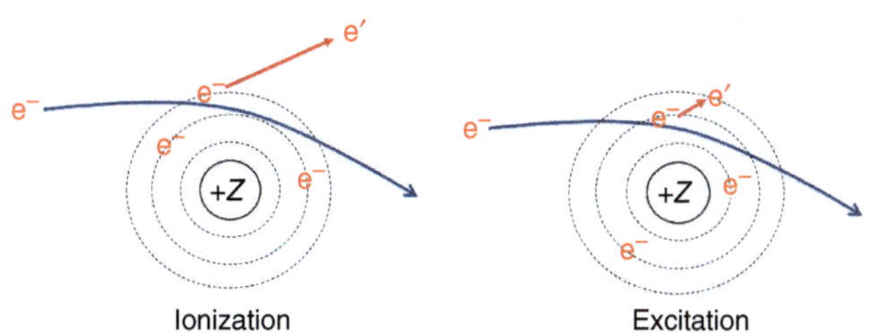

Figure 3.2 Interaction of electrons with matter.

passes through the matter, the decrease in the intensity of electromagnetic radiation propagating in the matter is described according to the concept of attenuation probability. This decrease is expressed by the famous exponential attenuation law as follows:

$$I(d) = I_0 e^{-\mu d} \tag{3.2}$$

where I_0 is the incident radiation intensity, and d and μ are the material thickness and absorption coefficient, respectively.

The main interactions between photons and matter include coherent scattering, photoelectric effect, Compton scattering, and pair production. Coherent scattering occurs at low photon energies and does not lead to a change of the wavelength of scattered photons with respect to the incident ones. In this section, we focus on the other three kinds of photon–matter interaction (Figure 3.3).

Photoelectric effect is the process, in which a photon interacts with an atom transferring all its energy to a bound electron. As a result, a free electron called photoelectron is generated. The photoelectron energy (E_e) is equal to $E_\gamma - E_B$, where E_γ is the photon energy and E_B is the electron binding energy, respectively. After the emission of a photoelectron, a vacancy in the electron shell of the atom is created. This vacancy can be filled by an electron from an outer shell, resulting in the characteristic X-ray emission. Due to low energies, both photoelectron and X-ray photon will most likely be absorbed by the material substance. Therefore, a photon transfers all its energy to the matter in the photoelectric effect. The probability of this effect is proportional to Z and inversely proportional to the photon energy. This means that the photoelectric effect is mostly observed in the case of low-energy photons or high-Z materials.

Compton effect refers to a scattering process of a photon on collision with an electron in an atom. The Compton effect results in generation of a scattered free electron with the energy received from the photon, a positively charged atom, and the photon with the lower energy as compared to that of the initial one. The Compton effect is directly related to the electron density of matter. Similar to the photoelectric effect, the probability of Compton scattering is inversely proportional to the photon energy. According to the conservation of energy and momentum, the maximum energy transferred by the photon to the electron (E_{max}) is calculated as follows:

$$E_{max} = \frac{E_\gamma}{1 + m_e c^2/(2E_\gamma)} \tag{3.3}$$

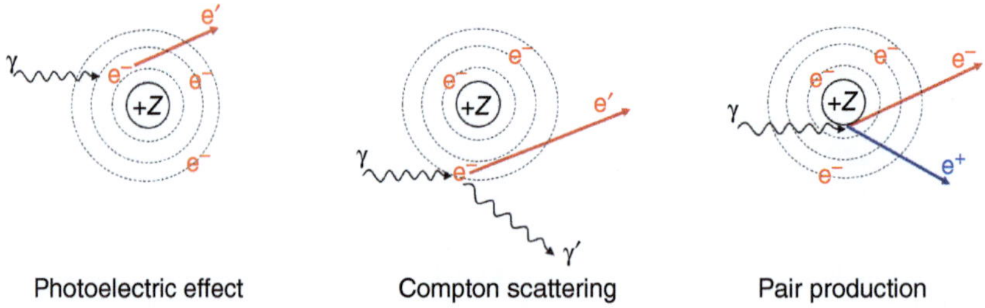

Figure 3.3 Interaction of photons with matter.

where E_γ is the initial photon energy, m_e is the electron mass, and c is the speed of light, respectively.

It can be seen from Eq. (3.3) that E_{max} is always smaller than E_γ. This means that a photon cannot transfer all its energy to the material in a single Compton scattering event. This leads to appearance of a Compton plateau in the energy spectrum where photon energies are measured by a finite-size detector.

Pair production refers to a transformation of a high-energy photon into an electron–positron pair as a result of interaction with an atomic nucleus. The photon energy threshold of this reaction is 1.02 MeV corresponding to the double rest mass of each particle. After generation, positron and electron gradually lose their kinetic energy in various processes (such as ionization and excitation), as discussed in the section describing charged particles. It is interesting to notice that a slowly moving positron annihilates with an electron in the material, yielding two 0.51 MeV photons. The probability of electron–positron pair generation is proportional to the photon energy and Z^2 of the matter. Therefore, pair production is the primary interaction of high-energy photons with the matter, especially for high-Z materials.

3.3 Polymer Systems for Ionizing Radiation Dosimetry

3.3.1 Polymer-Based Dosimeters

The dosimeters that can produce physical, chemical, and/or biological effects caused by the ionizing radiation are able to measure the ionizing radiation dose. The typical radiation dosimeters include semiconductors, film, thermoluminescence, and hydrogel dosimeters [17–20]. The semiconductor dosimeters are affected by temperature, incident angle, dose rate, and energy [21, 22]. The film dosimeters are affected by temperature and humidity [23]. The dose signals cannot be read out immediately after exposure to the ionizing radiation. The thermoluminescence dosimeters need annealing treatment prior to use and suffer from the sample-to-sample deviation [24, 25]. In this section, we review polymer/dye, polymer fluorescent, and polymer/noble metal nanoparticle dosimeters.

3.3.2 Polymer/Dye Dosimeters

The introduction of the photochromic compounds (e.g. organic dyes) in the polymer matrices is a common strategy for constructing the polymer-based dosimeters [26, 27]. The organic dyes include spiropyran (SP), azobenzene, methylene blue (MB), methyl orange (MO) leucomalachite green (LMG), and nitro blue tetrazolium (NBT, Figure 3.4) [28–31]. For instance, MB and MO decompose on exposure to the ionizing radiation (i.e. γ rays, electrons, and heavy ions), thus, resulting in a gradual color fading [29, 32, 33]. The dose–response ranges of MB and MO have been reported to be 0–40 and 0–300 kGy, respectively [32]. Alqathami et al. prepared a series of PRESAGE® dosimeters (polyurethane [PU] resin) with LMG. The authors observed that the radiation dose sensitivity of the as-prepared dosimeters was

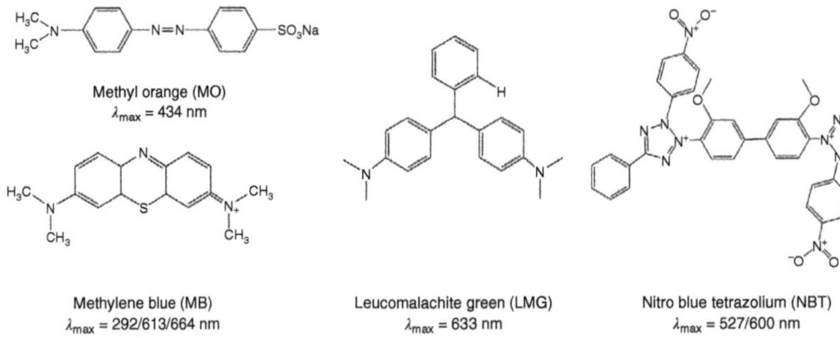

Figure 3.4 Representative radiochromic dyes.

dependent on the type of the LMG derivatives. The bromo-substituted derivative exhibited the best sensitivity, followed by the chloro- and methoxy-substituted derivatives [31].

The polymeric hydrogels are crosslinked by using the hydrophilic macromolecules and feature a three-dimensional network structure [8, 34, 35]. The hydrogel-based dosimeters have superior tissue equivalence and biocompatibility. Therefore, the polymer gel-based dosimeters have attracted widespread research attention [36–38]. The mechanism of the hydrogel-based dosimeters is primarily based on dose-dependent radiation-induced polymerization. The absorbed dose can be read out using magnetic resonance imaging (MRI) as the amount of the radiation-induced polymerization is directly proportional to the spin–spin relaxation rate ($R_2 = 1/T_2$) of the hydrogen nucleus [39]. However, the MRI process is time-consuming and is easily affected by the magnetic field temperature, along with requiring the professionals for operation. Recently, Abdek-Fattach et al. reported a NBT-gelation system for dosimetry. The unirradiated NBT gel demonstrated a pale-yellow color without any absorption band in the 430–750 nm range. As the irradiation dose was increased to 1000 Gy, the color of the NBT gel changed to pale violet, followed by blue-violet. Accordingly, the NBT gel exhibited a significant absorption band with a maximum at 527 nm (due to the mono-formazan form) and 600 nm (due to the di-formazan form) upon radiation exposure. The authors attributed the increment in the absorption band at 527 nm to the amount of the reduced NBT by the hydrated electrons [30].

As a typical organic photochromic compound, spiropyran can undergo a phase transition from the colorless closed spiropyran structure (SP) to the colored open merocyanine structure (MC) under UV light stimulation [40]. In addition, after exposure to the 1–20 kGy ionizing radiation dose, the SP structure transitions to the MC structure by the absorption of the emitted photons or excitation via nonradiative energy transfer [41]. Thus, spiropyran can be used for measuring the ionizing radiation dose. To reduce the lower limit of dose (LOD) detection, Kinashi et al. reported the polystyrene (PS)-based composite microfibers, composed of a spiropyran dye (1,3,3-trimethylindolino-6′-nitrobenzopyrylospiran, 6-nitro BIPS) and a photostimulable phosphor (europium-doped barium fluorochloride, $BaFCl:Eu^{2+}$) [42]. Under X-ray irradiation, the $BaFCl:Eu^{2+}$ particles yielded a strong

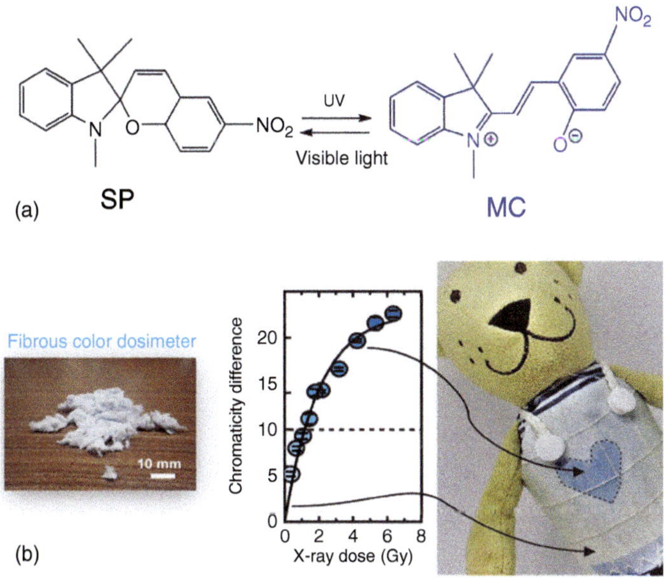

Figure 3.5 (a) Reversible photochromic reaction between the SP- and MC-forms and (b) photographic images of the composite resin dosimeter. Source: Reproduced with permission from Kinashi et al. [42]/American Chemical Society.

fluorescence emission peak at 380 nm due to the transition of the Eu^{2+} $4f^6 5d^1$ (^2e_g) state to the $4f^7$ ($^8S_{7/2}$) ground state. Interestingly, the emission peak excited the phase transition of SP to form the MC structure, thus, resulting in a color change. The LOD of the X-ray dose was 1.2 Gy. Finally, the authors developed clothes out of the PS-based composite microfibers and placed them on a bear doll. On exposure to X-rays, the rays were observed to pass through a heart-shaped hollow lead plate, and the color of the heart-shaped part changed from white to blue (Figure 3.5). This study provided a beneficial platform for developing wearable and flexible dosimeters, although the thermal stability and flexibility of the dosimeters need to be improved.

3.3.3 Fluorescent Polymer Dosimeters

On exposure to the ionizing radiation, water radiolysis yields H radicals, OH radicals, e_{aq}^-, and others. Thus, the fluorescent probes that can selectively and effectively respond to these free radicals can be used as ionizing radiation dosimeters. For instance, a variety of dosimeters have been prepared by simply mixing the reactive oxygen species (ROS)-responsive fluorescence probes with nanoclay, such as benzoin acid (BA), trimesic acid (TMA), coumarin-3-carboxylic acid (CCA), terephthalic acid (TPA), and pyromellitic acid (PMA, Figure 3.6) [43, 44]. Under X-ray irradiation, these probes are converted into fluorescent compounds via interaction with the OH radicals due to the water radiolysis. Consequently, the dose–response of the dosimeters can be evaluated using a fluorescent spectrometer.

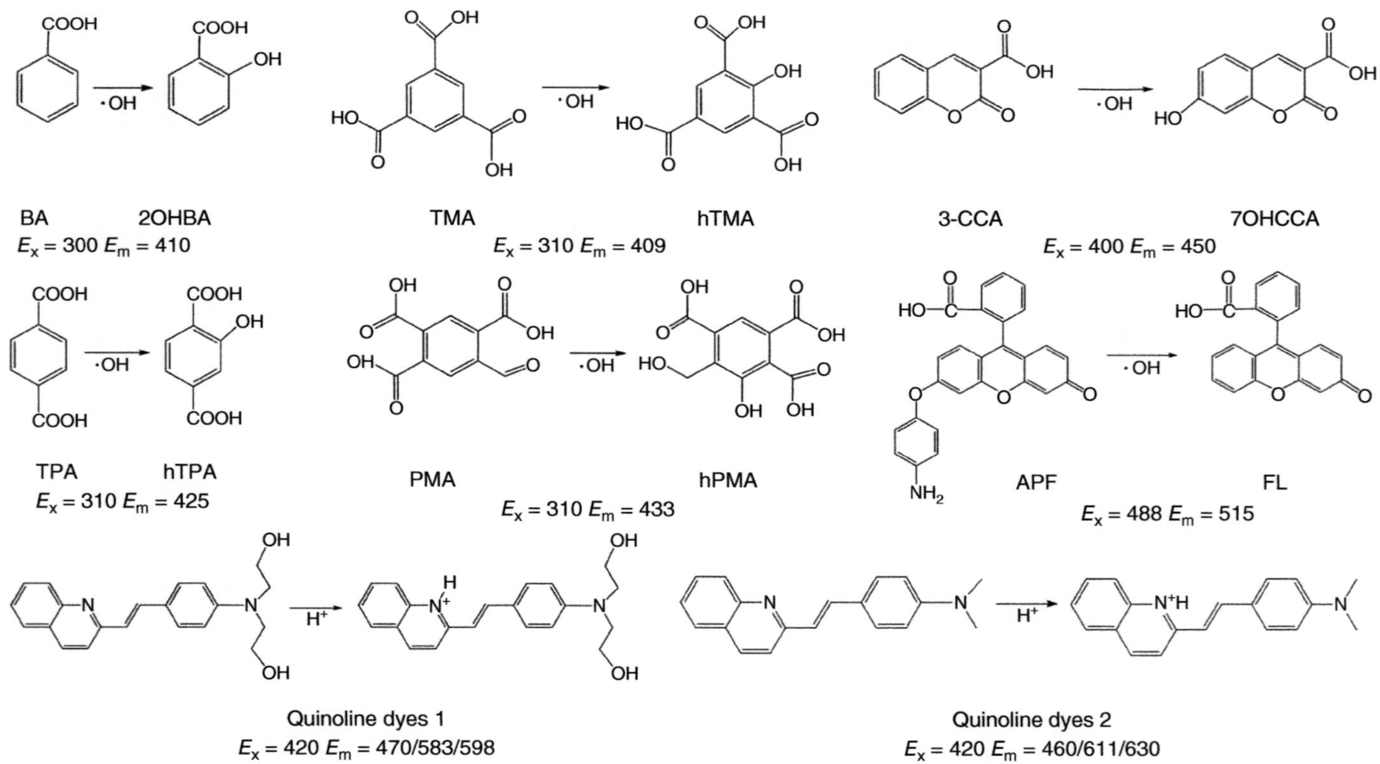

Figure 3.6 Representative fluorescent probes for measuring doses of ionizing radiation.

Recently, the Hu group developed a ratiometric fluorescence dosimeter based on polyacrylamide nanogels containing CCA and 5(6)-carboxytetramethylrhodamine (TAMRA). The CCA moiety reacted with the OH radicals due to the radiolysis of water, thus, forming a blue fluorescent substance 7-OHCCA (400/450 nm, $\lambda_{ex}/\lambda_{em}$). In contrast, the fluorescence character of TAMRA was stable under the ionizing radiation, thus, maintaining the red fluorescence signal (530/580 nm). Therefore, as the X-ray dose was increased from 0 to 20 Gy, the fluorescence ratio (I_{450}/I_{580}) was linearly increased. Accordingly, the nanogel solution changed from red (0 Gy) to purple (10 Gy), followed by blue (20 Gy). The authors determined the LOD of the nanogel sensor to be 0.1 Gy [45]. To enhance the emitted wavelength of the fluorescent probes, the Hu group further used aminophenyl fluorescein (APF) to modify the poly(acrylamide-*co*-*N*-(3-amino-propyl) methyl acrylamide) (PAAm-*co*-APMA) nanogels. Owing to the generation of the OH radicals, APF transitioned to fluorescein (FL), thereby emitting a green fluorescence signal (488/515 nm, Figure 3.7a,b) [46]. The fluorescence intensity of the nanogels at 515 nm exhibited an excellent linear relationship with the radiation dose in the range of 0–15 Gy, and the LOD value reached 0.5 Gy. The nano-dosimeters exhibited an optimal temperature and time stability (Figure 3.7c,d). The nanosize and positive Zeta potential imparted the nano-dosimeters a high potential for intracellular dosimetry. In addition, the gel dosimeter was also used to measure the X-ray dose in phantoms and tissues under clinical photon beams.

In another study, Pei et al. reported an acrylic resin film for detecting the X-ray dose. In brief, a pH-sensitive fluorescent quinoline derivative was embedded in the poly(methyl methacrylate) (PMMA) or polyvinyl chloride (PVC) films. The authors reported that PMMA and PVC were prone to yield acid moieties under γ-rays, which resulted in the protonation reaction of the quinoline substance. Thus, the emission wavelength of the quinoline derivative exhibited a redshift. The dosimeters responded to a wide range of the γ-ray dosage (80–4060 Gy), with a significant fluorescent color change [47].

3.3.4 Polymer/Metal Nanomaterials Dosimeters

The noble metal nanomaterials (e.g. gold and silver) possess a nanoscale size (1–100 nm), thus, exhibiting unique physical and chemical properties [48]. For instance, gold nanoparticles yield the local surface plasmon effect and different colors depending on the morphology and size [49]. Also, these metal nanoparticles often possess a large atomic number, thus, exhibiting the radiation sensitization effect [50]. The gold nanoparticles decompose gradually under ionizing radiation field, thus, resulting in a blue shift of UV absorbance [49]. Therefore, the absorption dose can be easily interpreted by observing the shift in the characteristic UV absorption peak of the nanoparticles. Besides, the R_2 value (from MRI) of the metal hydrogel dosimeter also increases with the dose [51].

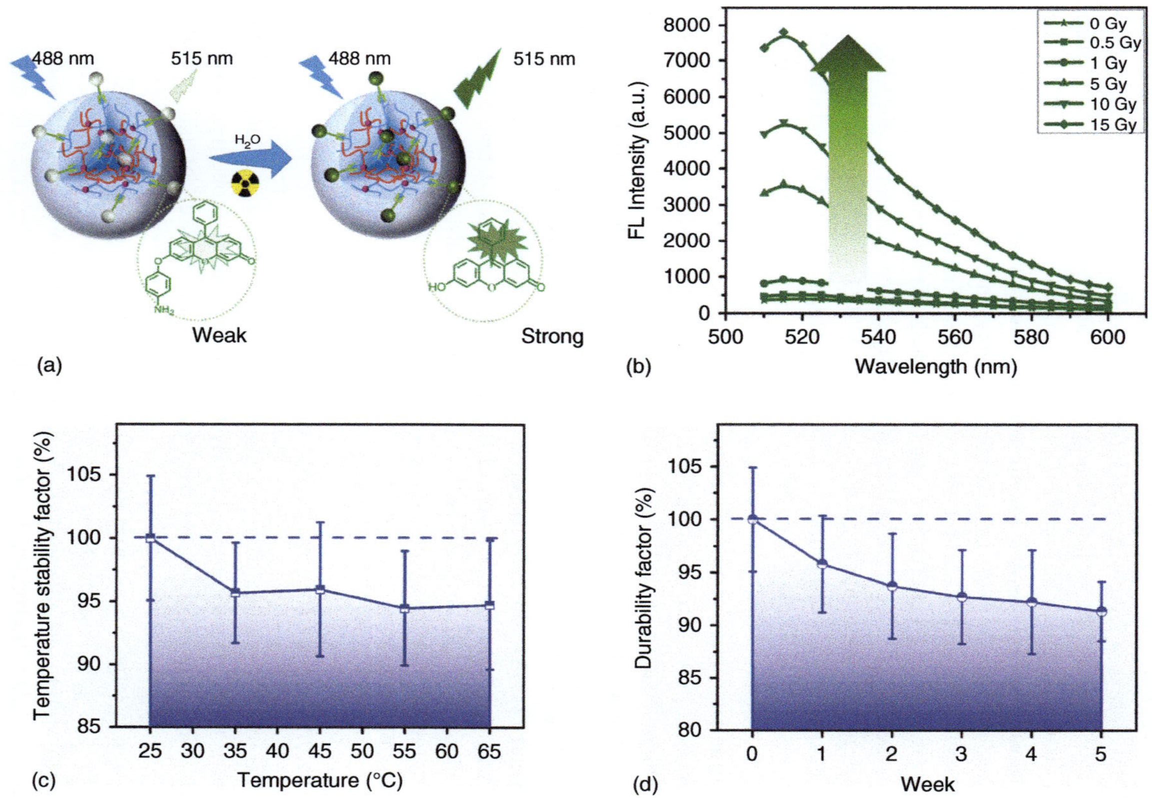

Figure 3.7 (a) The schematic illustration of the sensing mechanism, (b) the FL emission spectra, (c) temperature stability, and (d) durability of PAA–APF nanogel sensors when exposed to 1 Gy of X-rays. Source: Reproduced with permission from Jiang et al. [46]/American Chemical Society.

The metal nanoparticles can be prepared by radiation reduction of the metal ions under γ-rays [52]. Based on this phenomenon, Soliman et al. mixed the radiation-sensitive $AgNO_3$ with gelatin, and the mixture was subsequently exposed to γ-rays. The Ag^+ ions were reduced to generate the yellow Ag nanoparticles. On increasing the dose of γ-rays from 0 to 100 Gy, the gel color gradually changed to yellow brown. Thus, the dose amount could be readily obtained by measuring the absorbance of the UV characteristic peak of the Ag nanoparticles/gelatin at 450 nm. The signal stability was well retained on storing the irradiated Ag nanoparticles/gelatin at 6 °C in dark [53].

Recently, Rege et al. reported a novel hydrogel dosimeter composed of $HAuCl_4$, cetyl trimethylammonium bromide (CTAB, as a cationic emulsifier) and agarose [54]. First, a small amount of ascorbic acid was added to reduce Au(III) to obtain Au(I). After X-ray irradiation using a certain dose, the negative electrons and H radicals due to the radiolysis of water acted as the reducing agents. Thus, Au(I) in the agarose hydrogels was reduced to Au(0), thus, forming Au nanoparticles exhibiting a maroon color. As a result, the X-ray dose could be estimated by measuring the UV absorbance of Au-agarose. As shown in Figure 3.8, the characteristic absorbance

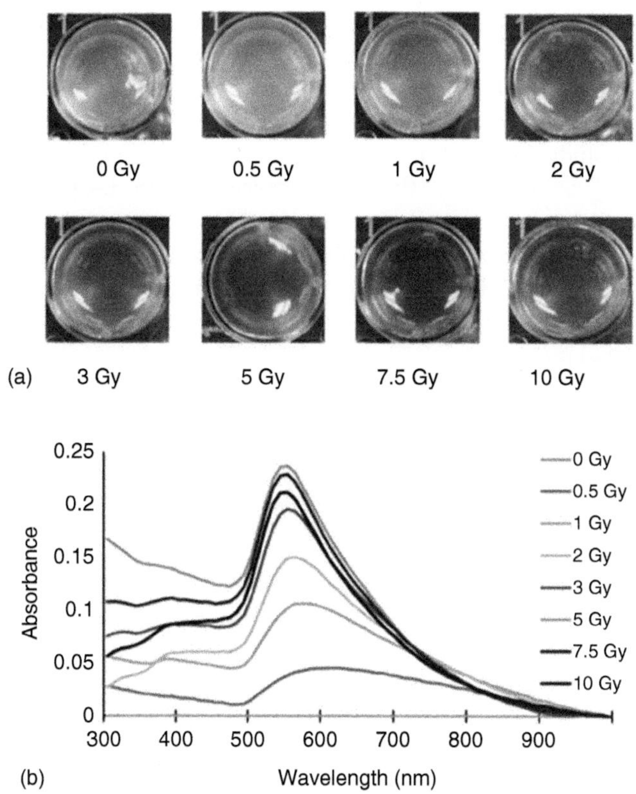

Figure 3.8 (a) Photographs and (b) UV–vis absorbance spectra of the $HAuCl_4$-containing agarose gel following exposure to different doses of X-rays. Source: Reproduced with permission from Pushpavanam et al. [54]/John Wiley & Sons, Inc.

peaks at ~550 nm could be attributed to the gold nanoparticles. The results showed that the minimum dose of the gel dosimeter was 0.5 Gy, whereas the linear dose range was 0–3 Gy [55]. The system was also suitable for detecting the high-energy electrons and proton beams [56], along with the applicability for the planar dose measurement [57]. These characteristics confirmed a high potential for clinical application of the developed system.

3.4 Ionizing Radiation-Responsive Polymer Systems for Therapy

In a related study, Tanabe et al. prepared the DNA amphiphiles composed of oligodeoxynucleotides as the hydrophilic part and alkyl chains as the hydrophobic part [13]. The hydrophilic and hydrophobic components were linked by using a radiation-responsive disulfide bond. In aqueous solutions, the DNA amphiphiles self-assembled to generate the nanosized aggregates, with the hydrophobic dyes and drugs in the core. Upon X-ray irradiation, the exchange reaction of the disulfide bond occurred due to the reduction in the hydrated e_{aq}^- and H radicals (Figure 3.9). This resulted in the dissociation of the aggregates, thus, releasing the encapsulated dyes and drugs. Using the living A549 cells, the *in vitro* experiments revealed that the aggregates penetrated in the cells, thus, loading them with drugs. The X-ray irradiation triggered the dissociation of the aggregates, thus, leading to the drug release. In another study, Wu et al. reported a γ-ray-sensitive drug-delivery system prepared by the conjugation of the G4.5 dendrimer with L-Cys via the *N*-ethyl-*N*′-(3-(dimethylamino) propyl) carbodiimide/*N*-hydroxysuccinimide (EDC/NHS) reaction. The resultant G4.5-L-Cys spherical nanoparticles contained a large amount of the S—S bonds. The G4.5-L-Cys nanoparticles could effectively release the preloaded doxorubicin (DOX) under γ irradiation *in vitro* and *in vivo* (using zebrafish as a model) [57].

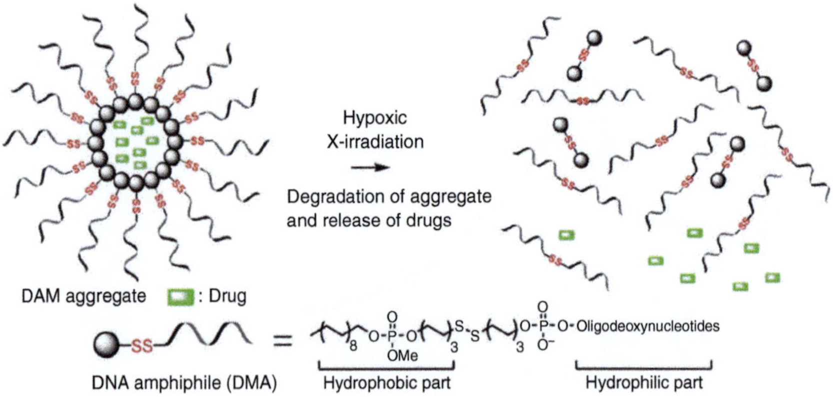

Figure 3.9 Schematic illustration of disulfide bond-containing DNA amphiphiles that can respond to X-rays. Source: Reproduced with permission from Tanabe et al. [13]/American Chemical Society.

As an essential trace element in humans, selenium exhibits excellent biological characteristics that enable its widespread use in treating the human diseases [58, 59]. In recent years, selenium-containing polymers have attained widespread research attention [60, 61]. The energy of the selenium–selenium bond is relatively low, therefore, the low energy ionizing radiation can rupture the Se—Se bonds.

Ma et al. reported a γ-ray-responsive diselenated amphiphilic block copolymer, PEG–PUSeSe–PEG (PEG refers to poly(ethylene glycol) and PU refers to polyurethane) [62]. The block copolymer contained the hydrophobic diselenide PU block and hydrophilic PEG block. The amphiphilic block copolymer could aggregate to form micelles in water, leading to the loading of the anticancer drugs in the core. As the γ-ray dose was increased to 50 Gy, the aggregates started to collapse to form irregular aggregates. The radiation sensitivity was attributed to the reaction of Se–Se with the OH radicals produced during water radiolysis. As such, the preloaded drugs were released gradually after γ-ray irradiation. The cytotoxicity of the polymers to the HepG2 cells at 1×10^{-4} mg/ml was negligible and significantly accelerated the cell growth. This study provided a platform for the combined radiotherapy and chemotherapy analyses. Based on this concept, Choi et al. synthesized the ebselen-loaded nanoparticles using a diselenated methoxy poly(ethylene glycol)-*b*-poly(DL-lactide-*co*-glycolide) copolymer [63]. The nanoparticles collapsed on exposure to the photon beams, to accelerate the release of ebselen and prevent the radiation damage to the cells.

In addition, ionizing-radiation-responsive hydrogels have attracted increasing attention in the biomedicine and nanotechnology fields [64–66]. The gel–sol transitions usually occur for the supramolecular hydrogels, which rely on the noncovalent interactions. These stimuli-responsive hydrogels provide a "smart" pathway for loading/releasing the drugs of interest in the laboratories/clinics. In a related study, Cao et al. reported a γ-ray supramolecular responsive hydrogel composed of a peptide amphiphile and a diselenide-containing polymer (Figure 3.10) [64]. The hydrogel was fabricated based on the electrostatic interactions and hydrogen bonding between the peptide and polymer matrix. Initially, the assembled supramolecular hydrogel exhibited a diameter of ~9 nm nanofibers. On exposure to the γ-ray irradiation (500 Gy dose), the hydrogel was disassembled, which was attributed to the cleavage of the Se—Se bonds due to the oxidative radicals (e.g. OH radicals).

In another study, Zhang et al. reported the X-ray-responsive diselenide polymer-based nanocarriers for effective delivery of the anticancer drugs [67]. The DOX-loaded nanocarriers were prepared by the (i) synthesis of di(1-hydroxylundecyl) diselenide, (ii) subsequent polymerization of the monomers with isophorone diisocyanate (IPDI) and PEG to form the Se-polymer, and (iii) assembly of the Se-polymer in the presence of the chemotherapeutic DOX (Figure 3.11). The transmission electron microscopy (TEM) analysis revealed that the assembly had an average diameter of ~83 nm. In the presence of 100 μM H_2O_2 (matching the biological ROS level in cancers), 2 Gy X-ray rapidly triggered the disassembly of the Se-nanoparticles, thus, releasing their cargo. Based on the density functional theory calculations, the authors suggested that H_2O_2 reacted with the selenyl

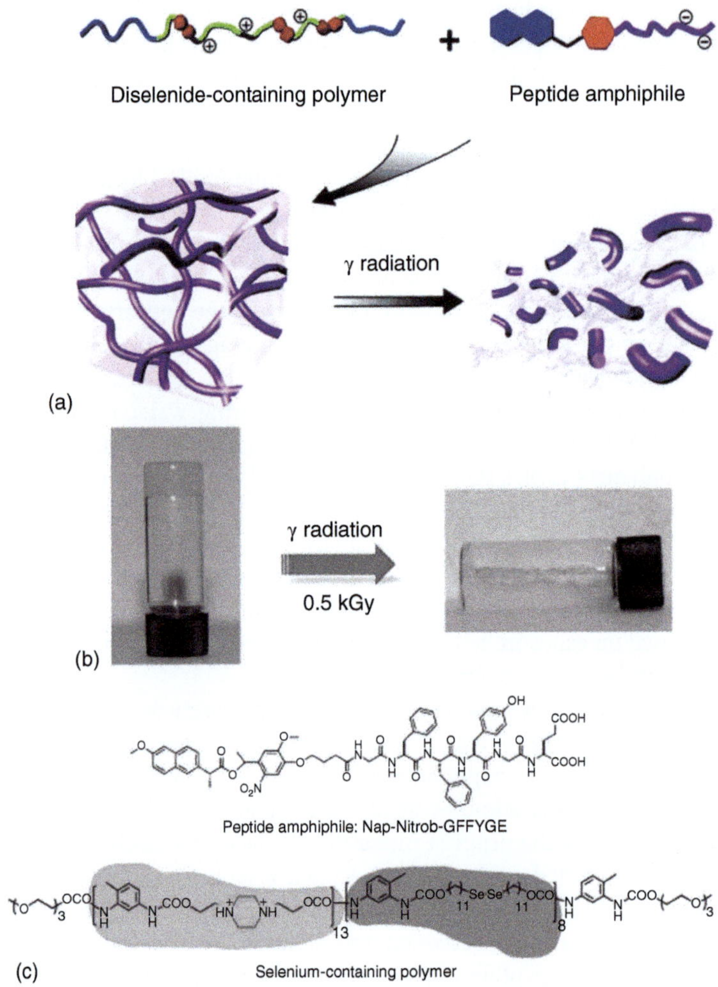

Figure 3.10 (a) Schematic illustration, (b) photographs of gel–sol transitions for the supramolecular hydrogels. (c) Chemical structure of diselenide-containing polymer and peptide amphiphile. Source: Reproduced with permission from Cao et al. [64]/John Wiley & Sons, Inc.

moieties owing to X-ray radiolysis, thus, forming SeOOH. This process effectively prevented the selenyl moieties from reforming the covalent Se—Se bonds, thus, the absorbed X-ray dose was lower than the dose reported in other literature reports. Furthermore, the *in vivo* experiments revealed that DOX was effectively delivered to the tumor site due to the disassembly of the nanocarriers under X-ray irradiation, thus, presenting new insights into cancer chemoradiotherapy.

Te is another bioactive element, though it is not a vital trace element *in vivo*. Cao et al. synthesized a novel Te-containing polymer micellar structure with an original hydrodynamic diameter of ~30 nm [68]. The heavy Te atoms were located in the core of the micellar structure. Under 2 Gy γ-ray radiation, the hydrodynamic diameter

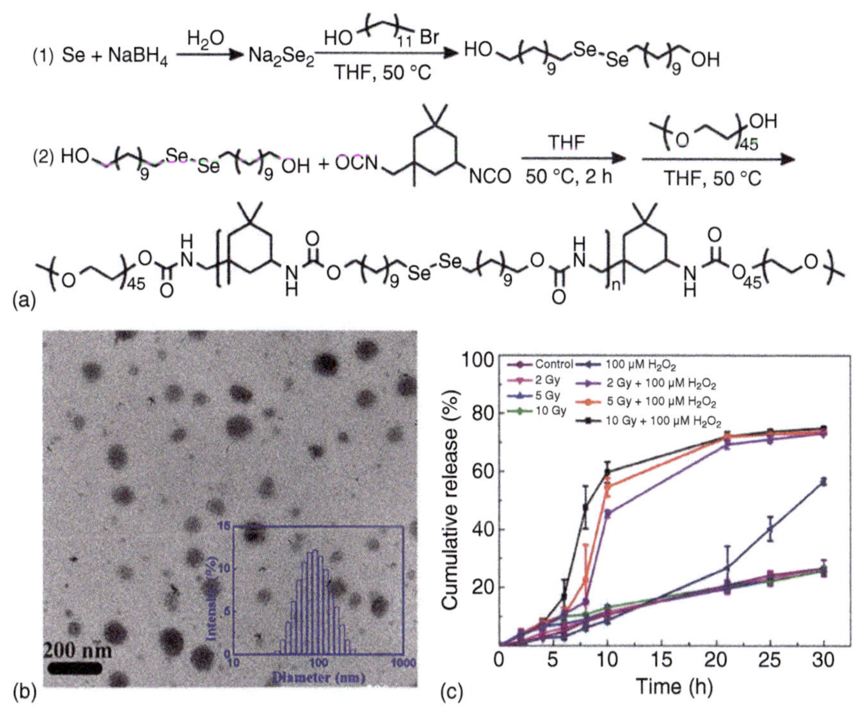

Figure 3.11 (a) The synthesis routes of the Se-polymer. (b) TEM image of the D-nanoparticles in PBS solution and particle size distribution histograms of the sample. (c) *In vitro* release of NR from NR-loaded nanoparticles under different conditions. Source: Reproduced with permission from Zhang et al. [67]/American Chemical Society.

of the micelle was observed to increase to ~200 nm due to the oxidation of Te on the polymer backbone (Figure 3.12). The Te-containing polymer provided a good platform for radiotherapy due to the sensitivity of Te [69]. A low ionizing radiation dose could trigger the disassembly of the Te-containing polymer-based aggregates, thus, releasing the preloaded drugs. However, Te is sensitive to air and light, which limit its use in clinical applications.

3.5 Conclusion

Since the discovery of radioactive rays in the nineteenth century, the radiation (i.e. α-particles, electrons, and photons)–matter interactions have been the focus of intense research. Based on the underlying mechanism, ionizing radiation affects many aspects, such as nuclear power, medical diagnosis, treatment, and food processing. In these fields, measuring the radiation dose is of vital importance. Therefore, a wide variety of dosimeters have been documented, including polymer/dye, polymer fluorescent, and polymer/noble metal nanoparticle dosimeters. The absorbed radiation dose can be readily read through the optical signals of these dosimeters, thus, avoiding the cumbersome processes associated with the

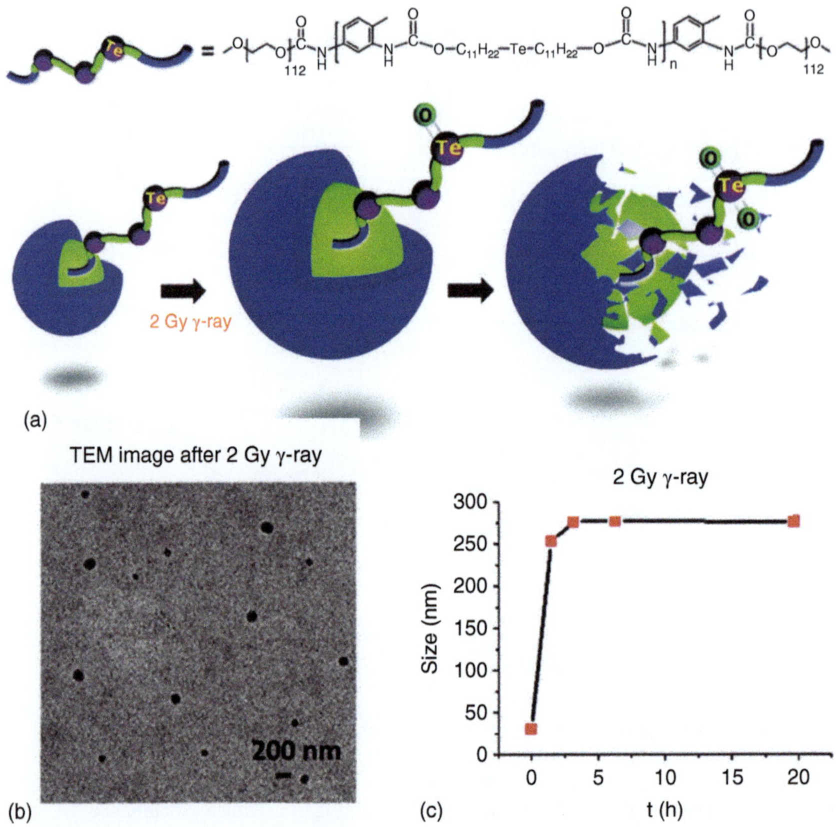

Figure 3.12 (a) Schematic illustration of chemical structure and γ-ray responsive mechanism of Te-containing polymer and (b) TEM and dynamic light cattering (DLS) data of the Te-containing polymer after 2 Gy γ-ray radiation. Source: Reproduced with permission from Cao et al. [68]/Royal Society of Chemistry.

traditional dosimeters. It is believed that the novel dosimeters can replace the traditional personal dose-monitoring devices, thus, imparting long-term developments in microenvironmental dosimetry. The ionizing radiation-triggered drug release is another vital field, especially the combined radiotherapy and chemotherapy analyses. For this purpose, a number of studies have focused on thoil-/selenium-/tellurium-containing polymers, which can lead to redox reactions as a result of the radicals generated during the radiolysis of the water molecules. Under ionizing radiation, these polymer-based systems are observed to be broken, thus, releasing the preloaded chemotherapeutic drugs (e.g. DOX). As the dose of the ionizing radiation is at the clinical level, thus, this strategy presents a high potential for application in the radiotherapy and chemotherapy fields.

Acknowledgments

We are thankful for funding from the National Natural Science Foundation of China (51873137) and the Priority Academic Program Development of Jiangsu Higher Education Institutions (PAPD).

References

1 Aziz, A., Pane, S., Iacovacci, V. et al. (2020). Medical imaging of microrobots: toward in vivo applications. *ACS Nano* 14 (9): 10865–10893.
2 Ehlermann, D.A.E. (2016). The early history of food irradiation. *Radiat. Phys. Chem.* 129: 10–12.
3 Nambiar, S. and Yeow, J.T.W. (2012). Polymer-composite materials for radiation protection. *ACS Appl. Mater. Interfaces* 4 (11): 5717–5726.
4 Beyea, J., Franke, B., and Siedband, M. (2014). X-ray doses from radar received by Belgian professional military radar operators. *Int. J. Cancer* 134 (12): 2995–2996.
5 Nasef, M.M. (2014). Radiation grafted membranes for polymer electrolyte fuel cells: current trends and future directions. *Chem. Rev.* 114 (24): 12278–12329.
6 Hickling, S., Xiang, L., Jones, K.C. et al. (2018). Ionizing radiation-induced acoustics for radiotherapy and diagnostic radiology applications. *Med. Phys.* 45 (7): e707–e721.
7 Shore, R.E. (2014). Radiation impacts on human health: certain, fuzzy, and unknown. *Health Phys.* 106 (2): 196–205.
8 Hu, L., Zhang, Q., Li, X., and Serpe, M.J. (2019). Stimuli-responsive polymers for sensing and actuation. *Mater. Horiz.* 6 (9): 1774–1793.
9 Cao, W., Gu, Y.W., Meineck, M., and Xu, H.P. (2014). The combination of chemotherapy and radiotherapy towards more efficient drug delivery. *Chem. Asian J.* 9 (1): 48–57.
10 Fan, F.Q., Gao, S.Q., Ji, S.B. et al. (2018). Gamma radiation-responsive side-chain tellurium-containing polymer for cancer therapy. *Mater. Chem. Front.* 2 (11): 2109–2115.
11 Bouché, M., Dong, Y.C., Sheikh, S. et al. (2021). Novel treatment for glioblastoma delivered by a radiation responsive and radiopaque hydrogel. *ACS Biomater. Sci. Eng.* 7 (7): 3209–3220.
12 Tanabe, K., Asada, T., and Nishimoto, S. (2012). Aggregate formation and radiolytic degradation of amphiphilic DNA block copolymer possessing disulfide bond. *Bioorg. Med. Chem. Lett.* 22 (23): 7045–7047.
13 Tanabe, K., Asada, T., Ito, T., and Nishimoto, S. (2012). Radiolytic reduction characteristics of drug-encapsulating DNA aggregates possessing disulfide bond. *Bioconjugate Chem.* 23 (9): 1909–1914.

14 Li, T., Pan, S., Zhuang, H. et al. (2020). Selenium-containing carrier-free assemblies with aggregation-induced emission property combine cancer radiotherapy with chemotherapy. *ACS Appl. Bio Mater.* 3 (2): 1283–1292.

15 Liu, J., Ma, X.N., and Lang, M.D. (2018). Synthesis and characterization of dynamic covalent polymers based on ditelluride bonds. *Acta Polymer. Sin.* 12: 1514–1523.

16 Wegst, A.V. (1980). Radiation detection and measurement, by G. F. Knoll. *Med. Phys.* 7 (4): 397–398.

17 Rosenfeld, A.B., Biasi, G., Petasecca, M. et al. (2020). Semiconductor dosimetry in modern external-beam radiation therapy. *Phys. Med. Biol.* 65 (16): 16TR01.

18 González, P.R., Ávila, O., Mendoza-Anaya, D. et al. (2020). Synthesis, luminescent and dosimetric properties of $Li_2B_4O_7$:Cu,Ag,Tm. *J. Mater. Sci. – Mater. Electron.* 31 (15): 12191–12205.

19 Devic, S. (2011). Radiochromic film dosimetry: past, present, and future. *Phys. Med.Eur. J. Med. Phys.* 27 (3): 122–134.

20 Farhood, B., Geraily, G., and Abtahi, S.M.M. (2019). A systematic review of clinical applications of polymer gel dosimeters in radiotherapy. *Appl. Radiat. Isot.* 143: 47–59.

21 Saini, A.S. and Zhu, T.C. (2002). Temperature dependence of commercially available diode detectors. *Med. Phys.* 29 (4): 622–630.

22 Dong, S.L., Chu, T.C., Lan, G.Y. et al. (2002). Characterization of high-sensitivity metal oxide semiconductor field effect transistor dosimeters system and LiF:Mg,Cu,P thermoluminescence dosimeters for use in diagnostic radiology. *Appl. Radiat. Isot.* 57 (6): 883–891.

23 Ningnoi, T. and Ehlermann, D.A.E. (1994). Effects of temperature and humidity during irradiation on the response of radiachromic film dosimeters. *Radiat. Phys. Chem.* 43 (6): 569–572.

24 Yegingil, Z., Nur, N., Dogan, T. et al. (2012). Effects of annealing and high radiation dose on the thermoluminescence characteristics of natural fluorite. *Radiat. Meas.* 47 (10): 981–987.

25 Kirby, T.H., Hanson, W.F., and Johnston, D.A. (1992). Uncertainty analysis of absorbed dose calculations from thermoluminescence dosimeters. *Med. Phys.* 19 (6): 1427–1433.

26 Frahn, M.S., Abellon, R.D., Jager, W.F. et al. (2001). Synthesis and characterization of a new fluorogenic probe molecule *N*-(1-pyrene)methacrylamide for monitoring radiation-induced polymerization. *Nucl. Instrum. Methods Phys. Res., Sect. B* 185 (1–4): 241–247.

27 Warman, J.M., de Haas, M.P., and Luthjens, L.H. (2009). High-energy radiation monitoring based on radio-fluorogenic co-polymerization. I: Small volume in situ probe. *Phys. Med. Biol.* 54 (10): 3185–3200.

28 Sun, W.H., Chen, L.J., Tian, J.P. et al. (2013). Degradation of a monoazo dye Alizarin Yellow GG in aqueous solutions by gamma irradiation: decolorization and biodegradability enhancement. *Radiat. Phys. Chem.* 83: 86–89.

29 LaVerne, J.A., Tandon, L., Knippel, B.C., and Montoya, V.M. (2005). Heavy ion radiolysis of methylene blue. *Radiat. Phys. Chem.* 72 (2, 3): 143–147.

30 Abdel-Fattah, A.A., Beshir, W.B., Hassan, H.M., and Soliman, Y.S. (2017). Radiation-induced coloration of nitro blue tetrazolium gel dosimeter for low dose applications. *Radiat. Meas.* 100: 18–26.

31 Alqathami, M., Adamovics, J., Benning, R. et al. (2013). Evaluation of ultra-sensitive leucomalachite dye derivatives for use in the PRESAGE® dosimeter. *Radiat. Phys. Chem.* 85: 204–209.

32 Chung, W.H. and Miller, A. (1994). Film dosimeters based on methylene blue and methyl orange in polyvinyl alcohol. *Nucl. Technol.* 106 (2): 261–264.

33 Dajka, K., Takacs, E., Solpan, D. et al. (2003). High-energy irradiation treatment of aqueous solutions of C.I. Reactive Black 5 azo dye: pulse radiolysis experiments. *Radiat. Phys. Chem.* 67 (3, 4): 535–538.

34 Van Vlierberghe, S., Dubruel, P., and Schacht, E. (2011). Biopolymer-based hydrogels as scaffolds for tissue engineering applications: a review. *Biomacromolecules* 12 (5): 1387–1408.

35 Li, W.X., Feng, R.Y., Wang, R.S. et al. (2018). Polyelectrolyte-based physical adhesive hydrogels with excellent mechanical properties for biomedical applications. *J. Mater. Chem. B* 6 (29): 4799–4807.

36 Baldock, C. (2017). Review of gel dosimetry: a personal reflection. *J. Phys. Conf. Ser.* 777: 012029.

37 Titus, D., Samuel, E.J.J., and Mohana Roopan, S. (2016). Current scenario of biomedical aspect of metal-based nanoparticles on gel dosimetry. *Appl. Microbiol. Biotechnol.* 100 (11): 4803–4816.

38 Adliene, D., Urbonavicius, B.G., Laurikaitiene, J., and Puiso, J. (2020). New application of polymer gels in medical radiation dosimetry: plasmonic sensors. *Radiat. Phys. Chem.* 168: 108609.

39 Abtahi, S.M. (2016). Characteristics of a novel polymer gel dosimeter formula for MRI scanning: dosimetry, toxicity and temporal stability of response. *Phys. Med. Eur. J. Med. Phys.* 32 (9): 1156–1161.

40 Fan, J., Bao, B.W., Wang, Z.H. et al. (2020). High tri-stimulus response photochromic cotton fabrics based on spiropyran dye by thiol-ene click chemistry. *Cellulose* 27 (1): 493–510.

41 Asai, K., Koshimizu, M., Fujimoto, Y., and Asai, K. (2017). Isomerization behavior of spiropyran-based compounds upon X-ray irradiation. *Radiat. Meas.* 106: 166–169.

42 Kinashi, K., Iwata, T., Tsuchida, H. et al. (2018). Composite resin dosimeters: a new concept and design for a fibrous color dosimeter. *ACS Appl. Mater. Interfaces* 10 (14): 11926–11932.

43 Maeyama, T. and Hase, S. (2018). Nanoclay gel-based radio-fluorogenic gel dosimeters using various fluorescence probes. *Radiat. Phys. Chem.* 151: 42–46.

44 Watanabe, Y., Maeyama, T., Mochizuki, A. et al. (2020). Verification of dose distribution in high-dose-rate brachytherapy using a nanoclay-based radio-fluorogenic gel dosimeter. *Phys. Med. Biol.* 65 (17): 175008.

45 Li, W.X., Nie, J., Hu, R. et al. (2019). A nanogel sensor for colorimetric fluorescence measurement of ionizing radiation doses. *Chem. Commun.* 55 (65): 9614–9617.

46 Jiang, L., Li, W.X., Nie, J. et al. (2021). Fluorescent nanogel sensors for X-ray dosimetry. *ACS Sensors* 6 (4): 1643–1648.

47 Pei, B., Su, H., Chen, B. et al. (2020). Quantifiable polymeric fluorescent ratiometric γ-ray chemosensor. *ACS Appl. Mater. Interfaces* 12 (37): 42210–42216.

48 Wolfbeis, O.S. (2015). An overview of nanoparticles commonly used in fluorescent bioimaging. *Chem. Soc. Rev.* 44 (14): 4743–4768.

49 Tao, Y., Li, M.Q., Liu, X.Y. et al. (2020). Dual-color plasmonic nanosensor for radiation dosimetry. *ACS Appl. Mater. Interfaces* 12 (20): 22499–22506.

50 Cui, L., Her, S.Y., Borst, G.R. et al. (2017). Radiosensitization by gold nanoparticles: will they ever make it to the clinic? *Radiother. Oncol.* 124 (3): 344–356.

51 Marques, T., Schwarcke, M., Garrido, C. et al. (2010). Gel dosimetry analysis of gold nanoparticle application in kilovoltage radiation therapy. *J. Phys. Conf. Ser.* 250: 012084.

52 Li, T., Park, H.G., and Choi, S.-H. (2007). γ-Irradiation-induced preparation of Ag and Au nanoparticles and their characterizations. *Mater. Chem. Phys.* 105 (2, 3): 325–330.

53 Soliman, Y.S. (2014). Gamma radiation induced synthesis of silver nanoparticles in gelatin and its application for radiotherapy dose measurements. *Radiat. Phys. Chem.* 102: 60–67.

54 Pushpavanam, K., Inamdar, S., Chang, J. et al. (2017). Detection of therapeutic levels of ionizing radiation using plasmonic nanosensor gels. *Adv. Funct. Mater.* 27 (21): 1606724.

55 Pushpavanam, K., Inamdar, S., Dutta, S. et al. (2020). Plasmonic gel nanocomposites for detection of high energy electrons. *J. Mater. Chem. B* 8 (22): 4930–4939.

56 Inamdar, S., Pushpavanam, K., Lentz, J.M. et al. (2018). Hydrogel nanosensors for colorimetric detection and dosimetry in proton beam radiotherapy. *ACS Appl. Mater. Interfaces* 10 (4): 3274–3281.

57 Wu, S.-Y., Chou, H.-Y., Yuh, C.-H. et al. (2018). Radiation-sensitive dendrimer-based drug delivery system. *Adv. Sci.* 5 (2): 1700339.

58 Rotruck, J.T., Pope, A.L., Ganther, H.E. et al. (1973). Selenium: biochemical role as a component of glutathione peroxidase. *Science* 179 (4073): 588–590.

59 Boyd, R. (2011). Selenium stories. *Nat. Chem.* 3 (7): 570–570.

60 Xia, J.H., Li, T.Y., Lu, C.J., and Xu, H.P. (2018). Selenium-containing polymers: perspectives toward diverse applications in both adaptive and biomedical materials. *Macromolecules* 51 (19): 7435–7455.

61 Cao, W. and Xu, H.P. (2019). Selenium-containing supra-amphiphiles. *Mater. Chem. Front.* 3 (10): 2010–2017.

62 Ma, N., Xu, H.P., An, L.P. et al. (2011). Radiation-sensitive diselenide block co-polymer micellar aggregates: toward the combination of radiotherapy and chemotherapy. *Langmuir* 27 (10): 5874–5878.

63 Choi, C., Lee, H.Y., Jeong, Y.I., and Nah, J.W. (2017). Synthesis of methoxy poly(ethylene glycol)-*b*-poly(DL-lactide-*co*-glycolide) copolymer via diselenide linkage and fabrication of ebselen-incorporated nanoparticles for radio-responsive drug delivery. *J. Ind. Eng. Chem.* 47 (25): 112–120.

64 Cao, W., Zhang, X.L., Miao, X.M. et al. (2013). γ-Ray-responsive supramolecular hydrogel based on a diselenide-containing polymer and a peptide. *Angew. Chem. Int. Ed.* 52 (24): 6233–6237.

65 Hailemeskel, B.Z., Addisu, K.D., Prasannan, A. et al. (2018). Synthesis and characterization of diselenide linked poly(ethylene glycol) nanogel as multi-responsive drug carrier. *Appl. Surf. Sci.* 449: 15–22.

66 Lu, W.H., Xu, X.X., Imbernon, L. et al. (2020). On-demand dissoluble diselenide-containing hydrogel. *Biomacromolecules* 21 (8): 3308–3317.

67 Zhang, L.X., Zhang, S.T., Xu, J.Y. et al. (2020). Low-dose X-ray-responsive diselenide nanocarriers for effective delivery of anticancer agents. *ACS Appl. Mater. Interfaces* 12 (39): 43398–43407.

68 Cao, W., Gu, Y.W., Li, T.Y., and Xu, H.P. (2015). Ultra-sensitive ROS-responsive tellurium-containing polymers. *Chem. Commun.* 51 (32): 7069–7071.

69 Pan, S.J. and Xu, H.P. (2021). ROS-responsive tellurium-containing polymers. *Acta Polymer. Sin.* 52 (8): 857–866.

4

Shrink and Wrinkle – Thermally Responsive Substrates for Thin-Film Structuring

Eduardo González-Martínez[1] and Jose Moran-Mirabal[1,2]

[1] *McMaster University, Department of Chemistry and Chemical Biology, 1280 Main Street West, Hamilton, ON L8S 4M1, Canada*
[2] *McMaster University, Brockhouse Institute for Materials Research, 1280 Main Street West, Hamilton, ON L8S 4M1, Canada*

4.1 Structured Thin Films

According to Greene [1], the use of the adjective "thin" in the term "thin film," which is used to describe layers with thicknesses from less than a single atomic layer to hundreds of microns, is ambiguous and poorly defined. While the definition of what constitutes a thin film may be a matter of debate, what is unquestionable is the fact that these materials have become key components of the modern world. The first applications of thin films date back to more than 5000 years ago, when the ancient Egyptians used gold thin films of less than 300 nm to decorate tombs [2]. Nowadays, thin films have countless applications in technologies that have been and are essential for human progress, such as electronic devices, data storage devices, medical devices, or food industry.

Thin films can be deposited using either top-down or bottom-up methods. In the former, the starting material is in bulk form and is usually exfoliated or etched to produce films, whereas in the latter approach, the thin films are produced by building atomic or molecular layers on top of the substrate. The most commonly used procedures are bottom-up, which include vapor phase deposition, either through physical or chemical methods. Physical vapor deposition (PVD) includes methods such as molecular beam epitaxy, laser pulse deposition, thermal evaporation, or sputtering. The materials fabricated using these procedures possess better mechanical, optical, and anticorrosive properties than the substrate material onto which they are deposited. Through PVD, almost all inorganic materials and some types of organic compounds can be deposited, the process is environmentally friendly, and more than one method can be combined to obtain a film with a targeted composition. Some disadvantages of PVD methods are the difficulty in coating samples with complex shapes, the processes are expensive and use high vacuum and temperature conditions. PVD techniques are mostly used in the aerospace industry,

Smart Stimuli-Responsive Polymers, Films, and Gels, First Edition.
Edited by Liang Hu, Yongfeng Gao, and Michael J. Serpe.

in microelectronic components, to coat cutting tools and for optical device fabrication [3–5]. On the other hand, chemical vapor deposition (CVD) encompasses methods such as plasma-enhanced CVD, laser CVD, or atomic layer deposition. Some advantages of CVD are that the deposition is not restricted to line-of-sight deposition, which allows coating of more complex shapes. It generally does not need ultrahigh vacuum, the deposition rate is high, and it can be more cost-effective than PVD. However, for this process a precursor with high vapor pressure is always required, it usually needs to be carried out at high temperatures, and a gas phase reaction may occur. Materials produced through CVD have applications in the production of refractory materials, anticorrosive coatings, and the fabrication of optoelectronic devices [4, 6].

Technologies related to thin-film fabrication are well established; however, the nano and microstructuration of the films after deposition remain expensive and are time-consuming processes. There have been several approaches to overcome this challenge, one of them being the formation of wrinkled patterns on film surfaces. The generation of wrinkles involves two structural components: a soft substrate and a rigid thin film, also known as the "skin." The most common materials used as thin films in processes that use wrinkling as a structuring approach are semiconductors, polymers, or metals. The evolution of wrinkles in thin films made from such materials arises from a mismatch in stiffness and/or the coefficient of thermal expansion (CTE) of the substrate and the skin [7, 8].

Bowden's 1998 pioneering work described the generation of wrinkles on thin films through the manipulation of film–substrate CTE mismatch [9]. In this work, the authors deposited 50-nm-thick gold films on a thermally expanded poly (dimethylsiloxane) (PDMS) substrate. Upon cooling, the PDMS shrunk and the mismatch in the mechanical properties between the polymer and the metal films generated a buckling pattern on the gold surface with wavelengths between 20 and 50 μm. Following this work, several works reported methodologies to produce wrinkles on thin films, but these presented challenges, such as the complexity of the protocols used, the fabrication cost and time, the use of molds, or the formation of wrinkles with very large wavelengths. To overcome such limitations, Fu et al. [10] developed a fast and inexpensive methodology for the fabrication of wrinkled thin metal film surfaces. In this process, bi-axially prestressed polystyrene (PS) sheets were coated with thin gold films of a range of thicknesses, and the resulting samples were shrunk in an oven at 160 °C. Because the glass transition temperature of PS is ~100 °C, heating the samples above this temperature caused the substrates to relax and contract by ~60% in each of the transverse dimensions. The stress induced by the relaxation process, the viscous flow of the PS, and the mismatch in the CTE between the thin gold film and the PS produced anisotropic wrinkle patterns on the gold surface (Figure 4.1a). The wavelengths of the wrinkled films were measured to be <1 μm and a linear correlation between film thickness and wavelength was observed (Figure 4.1b). The authors also demonstrated the ability to generate oriented wrinkles by clamping the PS substrate along one axis during the heating step, shrinking the sample uniaxially (Figure 4.1c). The wavelength of the uniaxially wrinkled surfaces also correlated linearly with the thickness of the film (Figure 4.1d). This work

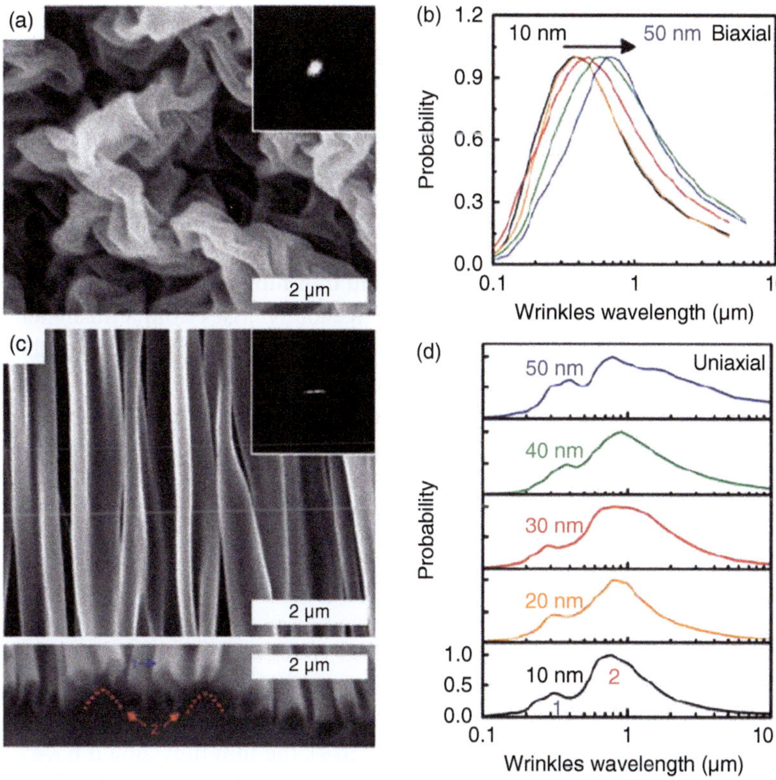

Figure 4.1 Characterization of nanowrinkles formed through thermal shrinking of prestressed polystyrene. (a) Scanning electron microscopy (SEM) image of biaxial wrinkles formed on a 10-nm-thick gold film. Inset: 2D-fast Fourier transform (FFT) spectrum. (b) Wavelength distributions of biaxial wrinkles made from gold films of various thicknesses. (c) SEM image of uniaxial wrinkles formed on a 10-nm-thick gold film. Inset: 2D-FFT spectrum. Bottom panel: Cross-sectional SEM view of the buckled surface. (d) Wavelength distributions of uniaxial wrinkles made from gold films of various thicknesses. Source: Reproduced with permission from Fu et al. [10]/John Wiley & Sons, Inc.

was a breakthrough because it introduced a simple and inexpensive method for the generation of microstructured thin-film surfaces that have found applicability in a broad range of fields, as detailed below.

4.2 Measuring the Mechanical Properties of Thin Films Using Thermal Wrinkling

The application of polymer and nanocomposite thin films requires precise knowledge of their mechanical properties. The use of conventional methods, such as nanoindentation or scanning probe microscopy, to measure the mechanical properties of polymeric thin films with sub-micron thicknesses has proved challenging and not feasible. The small forces involved in the deformation of films with thicknesses in the tens to hundreds of nanometers mean that classical methodologies are not sensitive enough to provide an accurate measurement. To circumvent these limitations, Stafford et al. [11] introduced strain-induced elastic buckling instability for mechanical measurements (SIEBIMMs), a simple methodology to determine the elastic modulus that takes advantage of the buckling pattern produced when a thin film is subjected to small compressive strains (<1%). This method was built on the methodology by Bowden et al. [9], by coating silicon wafers with thin PS films that were then transferred to an elastic PDMS sheet. The PDMS/film laminate was mounted onto a custom-built strain stage, and thin-film buckling was induced by applying a compressive strain. The buckling instability arises from a mismatch in the elastic properties (described by the modulus and Poisson's ratio) and/or the CTE between the thin film and the underlying substrate.

The buckling theory, described by Groenewold [12], identifies a critical wavelength (λ) that minimizes the total strain energy in the system and is dependent on the material properties of both the film and substrate. Assuming a sinusoidal waveform of the buckling instability, λ is defined as:

$$\lambda = 2\pi h \left[\frac{\left(1 - \nu_S^2\right) E_F}{3\left(1 - \nu_F^2\right) E_S} \right]^{1/3} \tag{4.1}$$

where h is the thickness of the thin film, ν is Poisson's ratio, and E is elastic modulus (F and S subscripts refer to the polymeric film and the substrate, respectively). For this relationship to hold, Stafford et al. described the following requirements: the thin film should be subjected to low strains ($\varepsilon \ll 10\%$), the thin film should have a much larger modulus than the substrate ($E_F/E_S \gg 1$), the substrate must be much thicker than the film, and the amplitude of the buckle should be much smaller than λ. If the properties of the substrate, and the thickness and the Poisson's ratio of the thin film are known, the modulus can be calculated using this simple relationship by measuring the wavelength of the resulting buckles. The wavelength of the buckles can be determined using laser diffraction, reflectance optical microscopy, atomic force microscopy, and digital image processing [11, 13]. Since the introduction of SIEBIMM in 2004, this method has been extensively used for the determination of the mechanical properties of a variety of materials, including liquid crystal elastomers [14], polyelectrolytes [15], magnetic nanofilms [16], and cellulose films [17], among many others. Despite its simplicity, SIEBIMM presents two major drawbacks: (i) crack formation across the film is common, and can interfere with modulus measurement by locally releasing stress and changing the wrinkle wavelength [11, 17]; and (ii) the characterization of highly hygroscopic materials needs to be performed under carefully controlled temperature and relative humidity (RH) conditions to minimize the impact of water absorption on the film's thickness and modulus [18].

As an alternative to SIEBIMM, Gill et al. [19] introduced the use of thermal shrinking as a method to produce wrinkles on nanobiocomposite thin films and measure their mechanical properties. As mentioned above, structuring by thermal shrinking relies on the compressive stress exerted by polymeric substrates when heated above their glass transition temperature. In the case of Gill, the substrates used were prestressed PS sheets. One advantage that this study demonstrated was that the elastic modulus of hygroscopic nanobiocomposite films could be determined in a humidity-independent manner. A layer-by-layer (LbL) procedure was used to deposit films containing cellulose nanocrystals (CNCs) and water-soluble biopolymers (xyloglucan [XG], or polyethylenimine [PEI]) on PS substrates. The films were subsequently shrunk by heating the laminated substrates in an oven at 135 °C. The thickness of the films was tuned by adding more LbL cycles and three different film compositions were studied: $CNCXG_{0.1\%}$, $CNC–PEI_{0.1\%}$, and $CNC–PEI_{1\%}$.

The elastic moduli were calculated by defining the parameter η:

$$\eta = \frac{E_F\left(1 - v_s^2\right)}{3E_S\left(1 - v_f^2\right)} \tag{4.2}$$

To determine η, the authors took advantage of the fact that a characteristic wrinkle length (ζ), akin to the one described by Groenewold [12], could be measured from the two-dimensional (2D) fast Fourier transform (FFT) power spectra (Figure 4.2a) obtained from scanning electron microscopy (SEM) images of the structured films. They observed that values for ζ varied linearly with thickness for the different film

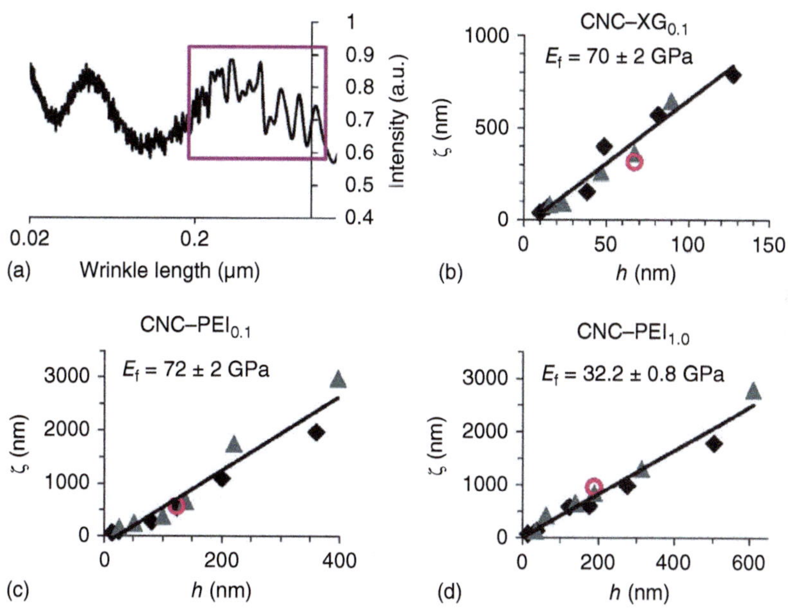

Figure 4.2 Calculation of the elastic modulus from the morphological analysis of the wrinkled nanobiocomposite films. (a) Intensity versus characteristic wrinkle length (ζ) plots were generated via 2D FFT analysis for the corresponding wrinkled structures. The plots were averaged and the wrinkle length values at the highest intensities were determined (peaks in boxed area) and averaged to calculate ζ. Plots of ζ versus film thickness (h) were generated for (b) $CNC–XG_{0.1\%}$, (c) $CNC–PEI_{0.1\%}$, and (d) $CNC–PEI_{1.0\%}$ films, where the slopes were used with Eq. (4.2) to calculate the elastic moduli. Full (◆) and half (▲) bilayers are indicated, $n \geq 3$, all $R^2 > 0.94–0.97$. Open circles indicate films that were hydrated by immersion in water for 30 minutes prior to structuring. Source: Reproduced from Gill et al. [19] with permission.

compositions studied (Figure 4.2b–d). Then, the parameter η was calculated from a linear regression of the characteristic wavelength (ζ) versus thickness (h), according to the semi-empirical relationship:

$$a\zeta = 2\pi\eta^{2/3}h \tag{4.3}$$

where the parameter a is a constant that takes into account the effects of the high strain applied to the substrate and the fact that the substrate becomes viscoelastic at the shrinking temperature; a was calibrated using wrinkled thin gold films as reference materials.

The elastic moduli determined using the thermal shrinking method (32–72 GPa, Figure 4.2) were found to be within a reasonable range, given the relative composition of the films and the moduli previously reported for CNCs (57–143 GPa) [20, 21], PEI (0.3 GPa) [17], and XG (5.95 GPa) [22]. The authors also observed that the moduli measured were independent of the degree of sample hydration at the time of shrinking, given that the values of dry and hydrated films were similar (Figure 4.2, open magenta circles). This observation led to the conclusion that the thermal structuring method is unaffected by the degree of hydration in the films and yields a value for the modulus that depends only on the intrinsic film composition. The method developed by Gill has been further used for the determination of the mechanical properties of other materials, like different molecular weight XG–CNC composites [13], or nanocomposite hydrogels [23]. Thus, the use of thermal shrinking for the determination of mechanical properties of thin polymer thin films is a new approach that can address challenges in the mechanical measurements of highly hygroscopic films and opens the possibility for its application to a wide range of biologically relevant materials. The simplicity and elegance of the method should accelerate its acceptance and use in the scientific community.

Recently, Stimpson et al. [24] developed a modified version of this method by thermally shrinking films deposited on the prestressed PS substrate in a uniaxial fashion. To produce uniaxially oriented wrinkled films, the thermoresponsive substrate was constrained along one axis via a clamping device. The elastic moduli of CNC–PEI films were determined through three different buckling-based methods (biaxial thermal shrinking, uniaxial thermal shrinking, and SIEBIMM). The authors concluded that the methods yielded similar trends (Table 4.1) and comparable values for the elastic moduli of each CNC–PEI film composition.

4.2.1 Thermally Structured Thin Films for Cell Culture

Most of the information about cellular behavior *in vitro* has been traditionally obtained through 2D cell culture despite the fact that tissues, organs, and blood vessels are three-dimensional (3D) constructs. Yet, 2D culture does not accurately mimic the 3D structure of the natural microenvironment that surrounds the cell [25, 26], and the extracellular matrix (ECM) produced by cells during *in vitro* culture is usually different from the natural chemical composition, as are the mechanical properties of the culture matrix. In recent years, 3D structures have been used as an alternative to 2D culture to better mimic the cell's natural microenvironment.

Table 4.1 Slopes from linear regression analysis of wavelength versus film thickness data, and elastic moduli calculated for CNC–PEI films, buckled via all three methods. Errors reported are propagated from error in the fits. SIEBIMM wavelength data could not be collected for $CNC_{0.03}$–$PEI_{0.1}$ films due to the lack of an available objective with sufficient magnification and a working distance that could fit between the clamps of the strain stage.

	Slope (nm wavelength/nm thickness)			Elastic modulus (GPa)		
Film composition	**Biaxial thermal**	**Uniaxial thermal**	**SIEBIMM**	**Biaxial thermal**	**Uniaxial thermal**	**SIEBIMM**
$CNC_{0.03}$–$PEI_{0.1}$	3.0 ± 0.2	4.6 ± 0.6	—	21 ± 2	15 ± 3	—
$CNC_{3.0}$–$PEI_{0.1}$	4.2 ± 0.2	7.7 ± 0.7	110 ± 10	35 ± 3	$33 = 5$	40 ± 10
$CNC_{3.0}$–$PEI_{1.0}$	4.3 ± 0.3	9.4 ± 0.7	90 ± 20	37 ± 5	44 ± 5	20 ± 20

Source: Reproduced from Stimpson et al. [24] with permission.

This approach has important advantages over 2D cultures, such as the more physiologically relevant contact between cells. The fact that the cells are in contact with a 3D ECM, that co-cultures of different cells can be obtained in different spatial arrangements, and in the case of cancer cell cultures, that the resistance to anticancer drugs can be recapitulated [27–29]. A variety of methods to produce 3D scaffolds for cell culture have been reported, with the most common being bulk hydrogels, electrospinning, and 3D printing [30–33]. In addition to these, thermally structured thin films have emerged as simple and versatile alternatives to evaluate the effect of micro and nanostructured surfaces on cell behavior.

One of the earliest applications of thermally structured substrates to cell work focused on studying the behavior of cardiomyocytes (CMs). The heart is one of the most complex human organs, as it possesses a complex architecture that encompasses structures from the nanoscale to the microscale. The 2D substrates used traditionally for CM culture do not produce cells with the morphological, mechanical, or physiological characteristics that they present in native tissue. To overcome this issue, Luna et al. used thermally structured thin films as templates to create PDMS-based biomimetic cell culture substrates and studied CM alignment on them [34]. To create the substrates, uniaxially aligned wrinkles were first generated by depositing metal films onto prestressed PS films, clamping the substrates along one axis, and shrinking them at 160 °C. Then, an elastomer-crosslinker PDMS mix was poured onto the wrinkled metal mold and cured to generate the complementary wrinkle pattern on the PDMS, which was then peeled and coated with fibronectin and laminin. Using the wrinkled PDMS surfaces, the authors observed and quantified the alignment and cell-tissue morphology of murine neonatal CM and human embryonic stem cell (hESC)-derived CM after two to seven days of *in vitro* culture (Figure 4.3). They also demonstrated that the cells cultured on the microstructured substrates had better adhesion to the surface and more coherent beating than that of cells cultured on flat surfaces. Moreover, CM derived from hESCs were also cultured on the wrinkled substrate showing a similar behavior to that of the murine neonatal CM. The results presented in this work open the

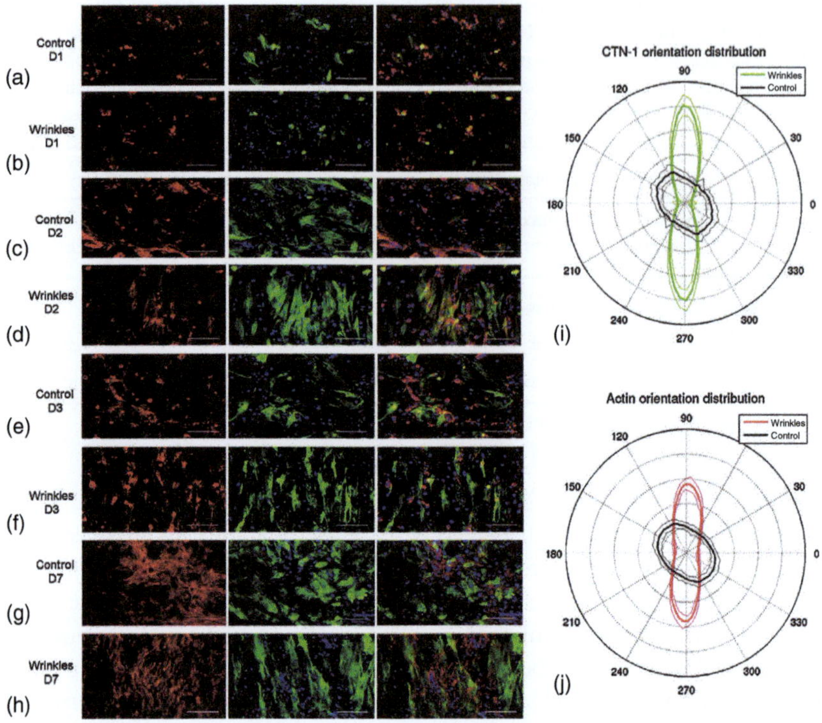

Figure 4.3 Alignment of neonatal cardiac cells on wrinkles. (a–h) Fluorescent micrographs of the cardiac cells unaligned on the flat substrates (controls) and aligned on the wrinkle substrates and on days 1, 2, 3, and 7. First column: actin (red). Second column: cardiac troponin I (CTN-I; green). Third column: merged images of CTN-I and actin staining. Blue is nuclear staining 4′,6-diamidino-2-phenylindole (DAPI). Scale bars = 100 mm.
(i, j) Anisotropy analysis of control (black) versus red or green (on wrinkles) by computing a histogram of gradient orientations where the contribution of each pixel was weighed by the gradient magnitude averaged over all four days. The direction of the wrinkles is defined at 90°. Thinner lines indicate standard deviation. Source: Reproduced with permission Luna et al. [34]/Mary Ann Liebert, Inc.

possibility to use thermally shrunk thin films as substrates for the contact guided alignment of CM, and therefore, their application for pharmacological studies and heart injury repair.

Using a similar platform, the effect of two drugs, E-4031 and isoprenaline, on hESC-CM cultures was also studied [35]. As in the example above, it was observed that the substrates induced contact alignment of the hESC-CM cells, and their behavior was analyzed through the correlation of bright field microscopy images. The authors found that cells treated with E-4031 showed an increase in contractile duration, exclusively due to prolonged relaxation peak, whereas cells supplemented with isoprenaline exhibited increased contractile frequency and acceleration. Moreover, they suggested that topography may help induce more mature ion channel development. Overall, by using a simple and inexpensive structured platform, they were able to detect bradycardia and tachycardia in cells grown *in vitro*.

According to the authors, around 30% of the withdrawals of drugs from trials are due to cardiotoxicity-related issues. In this sense, the wrinkled topography platform could be used as a drug screening method for early detection of cardiotoxicity during the first step in the development of new drugs.

The group of Khine and coworkers has further reported the use of wrinkled surfaces to study the effect of periodic topography on stem cell differentiation and function. Chen et al. [36] reported the subcellular time-lapsed response of feeder-free pluripotent hESCs without soluble differentiation factors to wrinkled topographies. They observed that more than 40% of hESC cells were aligned within 15° of the direction of the wrinkles within the first four hours of seeding. When comparing these results with those obtained by exposing the cells to shear stress [37], they noted that to obtain a similar degree of cell alignment 10 hours of exposure to 1.5 Pa shear stress was required. This work also measured the nuclear area and circularity of aortic smooth muscle cells (AoSMCs), mouse embryonic fibroblasts (MEFs), and hESCs, and found that only the latter exhibited highly compliant nuclei that deformed due to the topography. Using similar substrates, Wang et al. [38] studied the susceptibility of human pluripotent stem cell (PSC)-derived ventricular CMs to multicellular arrhythmogenicity. They cultured hESCs and ventricular CMs on Matrigel-coated wrinkled PS films and studied the electrophysiology of the cultures. One of the most interesting conclusions of their study was that the dynamic stability of PSC-derived grafts needs to be assessed before transplantation to patients with prominent preexisting heterogeneity, which likely further increases dispersion and therefore susceptibility to arrhythmias.

In a direct application of wrinkled substrates, without pattern transfer through molding, Chen et al. used oxygen plasma treatment to generate *in situ* an oxidized thin film on a prestressed PS sheet substrate [36]. Then, they produced a pattern of uniaxially oriented wrinkles via thermal shrinking of the clamped substrate. The topography of the buckles could be tuned by changing the plasma treatment time. Surprisingly, it was found that this process produced hierarchical wrinkles with sizes at two different scales. Under low magnification, a microscale pattern was observed, with wrinkles of wavelengths ranging from 1 to 7 μm depending on the plasma treatment time, while at higher magnifications, the wavelengths of the second wrinkling pattern ranged from 100 to 400 nm. Given that type I collagen, the main component in the ECM, forms fibers with characteristic diameters of 20 to 100 nm, and these fibers can combine to produce bundles with sizes several order of magnitude higher [39], the hierarchically wrinkled surfaces were explored as a biomimetic substrate for cell culture. The structured samples were coated with a thin layer of Matrigel and MEFs, AoSMCs, and hESCs were cultured on them to evaluate their ability to provide contact guidance to the cells. After 24 hours of culture on the structured surfaces, MEF cells showed signs of alignment, while the control showed no indication of contact guidance. The samples that exhibited the highest number of cells aligned to the wrinkles were those with wavelengths closer to that of collagen fibrils, showcasing the potential of wrinkled substrates as biomimetic surfaces.

Polymeric thin films have also been structured on prestressed PS sheets and used to directly study the effect of topography and surface chemistry on cellular behavior.

De France et al. [23] spin coated onto pre-stressed PS sheets alternate layers of poly(oligoethylene glycol methacrylate) (POEGMA)modified as aldehyde and hydrazide-POEGMA hydrogels reinforced with CNCs. The films were subsequently shrunken unconstrained and uniaxially constrained. The effect of deposition cycles and crosslink density on the wrinkle size and mechanical properties was evaluated. The authors found that the wrinkle size increased linearly with the number of deposition cycles, whereas the crosslink density had a stronger impact. The modulus values obtained were also similar to those of rigid CNC–polymer nanocomposite thin films, suggesting that the hydrogel films generated have both a high crosslink density and dry-networked structure. As a continuation of this work, the effect of electrospun mats on wrinkled hydrogel thin films on the culture of myoblast cells was studied [40]. In this case, POEGMA–CNC fibers were deposited by electrospinning on top of the LbL deposited POEGMA–CNC hydrogel, and the substrates were biaxially or uniaxially shrunk in an oven at 130 °C. The electrospun fibers were deposited in three different manners, randomly oriented on the biaxially shrunk films, parallel-oriented in uniaxially wrinkled films, and perpendicular-oriented in uniaxially wrinkled films. The authors found that myoblasts grow randomly on biaxially wrinkled hydrogels, while in uniaxial ones, there is a preferential orientation along the direction of the wrinkles. Furthermore, the orientation of the nanofiber played an important role in cell orientation, with parallel-oriented fibers enhancing the alignment of cells and perpendicular-oriented slightly hindering alignment to them. The developed methodology is simple and allows control on fiber and wrinkle alignment, which is difficult to achieve with other technologies. This method could be used for the encapsulation of cells for drug delivery, cell screening, and tissue implants.

Zhou et al. [41] also studied the effect of the topography of shrunk polymeric thin films on fibroblast morphology. The authors fabricated wrinkled thin films by spin coating a prestressed PS sheet with a polyampholyte, p(APM-*co*-AA) [poly(*N*-(3-aminopropyl) methacrylamide hydrochloride-*co*-acrylic acid)]. These films were then thermally shrunken at 135 °C for 15 minutes and heated at 160 °C for 45 minutes. It was found that during the second heating step, the polymer self-crosslinked, completely avoiding the use of small crosslinker agents that could be cytotoxic in cell or tissue culture applications. The authors used different concentrations of the polymer to produce films of various thicknesses, and therefore, different topography after the shrinking process. Then, the morphology of fibroblasts cultured on those films was studied. The area of the fibroblast decreased from the PS control to the flat polymeric thin film. This phenomenon was attributed to the increase in the hydrophilicity of the surface that hindered the adhesion process of the cells. After microstructuring, the authors noted that the area increased at a concentration of 4 wt% compared with 2 wt% or the flat polymeric films. Nevertheless, at a polymer concentration of 6 wt% and above the trend reversed. This behavior was attributed to fact that the wrinkle height and the depth of the microstructured films increase with an increase in the film thickness. At 4 wt%, the wrinkle separation is small enough, which allows the cells to span across them; however above 6 wt%, the grooves are large enough to allow the cells to adhere within the valleys between wrinkles. In this work, the influence of the hydrophobic properties of the surface on the cell morphology was also studied.

Finally, recent work has structured rigid glassy thin films through thermal wrinkling and used them to evaluate the impact of topography on murine bone marrow-derived macrophage (BMDM) morphology and function. Macrophages are the first line of defences in the immune system of vertebrates, and are responsible for the phagocytosis of foreign materials entering the organism, initiation of the inflammatory response, and wound healing, among other important functions. It has been suggested that the topographical features of a surface can determine whether macrophages display wound healing or pro-inflammatory phenotypes [42, 43]. Using wrinkled glassy surfaces, Makaremi et al. studied the effect of surface topography on the phagocytic capacity, pro-inflammatory cytokine IL-6 production, and lateral diffusion of the membrane-associated toll-like receptor 2 (TLR2) [44]. In this study, BMDM cells were cultured on flat and thermally wrinkled SiO_2 films with thicknesses of 2 and 50 nm. SEM images (Figure 4.4) showed that the morphology of the macrophages grown on the flat and 2 nm wrinkled surface had "elongated" or "fried-egg" morphologies with numerous filopodia extensions, similar to the morphologies seen when macrophages were cultured on 2D plastic or glass surfaces. On the other hand, macrophages cultured on the 50 nm wrinkled substrates were more frequently "star-shaped" with anchor points on the ridges of the wrinkles and a reduced number of filopodia. More importantly, it was observed that the BMDM cells grown on microstructured substrates had significantly increased phagocytic abilities. Conversely, these morphological changes did not alter the secretion of pro-inflammatory IL-6 cytokine or the lateral diffusion of the transmembrane receptor TLR2. This study highlights the potential of structured surfaces to modulate immune function, which can be leveraged to develop immune function assays and rationally design implant materials that are more easily integrated into the host organism.

The use of thermal shrinking of thin films for the generation of substrates with topographical features on the nano to microscale as biomimetic surfaces for cell culture is an excellent alternative to current methodologies. The methods used for the fabrication of the substrate are simple, inexpensive, and offer tunability in the size and morphology of the topographical features. The fact that the topography and the chemical properties of the surface can be easily modified, offers an enormous advantage for culturing different types of cells and maintaining the 3D microenvironment. A variety of materials can be structured that offers a unique opportunity to study the behavior of cells in 3D materials, which previously could only be studied in 2D. In addition, the exceptional capabilities of these materials for cell alignment open a new spectrum of possibilities for drug screening for cardiotoxicity, drug response studies, and *in vivo* implants.

4.2.2 Wrinkled Conductive Thin Films for Wearable Electronics

The demand for precise point-of-need biometric measurements in the modern world has pushed toward the development of wearable electronic devices. A

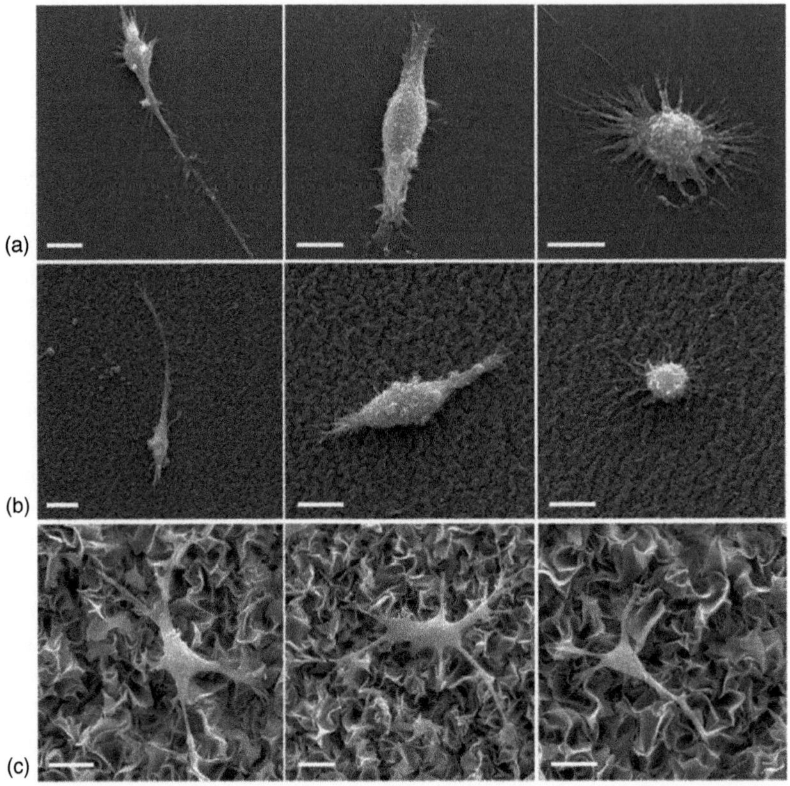

Figure 4.4 Representative SEM images of macrophages adhered to SiO_2 substrates, showing differences in cell morphology in response to rougher topography. (a) Flat 50 nm-thick control, (b) wrinkled 2 nm thick film, and (c) wrinkled 50 nm thick film. All scale bars represent 10 μm. Source: Reproduced with permission from Makaremi et al. [44]/John Wiley & Sons, Inc.

wearable device can be defined as: "an electronic device (that is) constantly worn by a person as unobstructively as clothing to provide intelligent assistance that augments memory, intellect, creativity, communication, and physical senses" [45]. These type of devices needs to conform to curved surfaces and perform under normal physical activities, which means that they must be flexible, bendable, and stretchable, all the while maintaining constant electronic properties. Wearable devices can be classified according to the property responsible for signal generation, with the most common being mechanical, chemical, electrical, and optical devices. Mechanical sensors have been extensively studied because multiple body-to-signal transduction methods can be used in them. Frequently, the output measured in a mechanical sensor is the change in the resistance of a film due to the application of mechanical deformation (piezoresistivity). The figure of merit that quantifies the piezoresistive behavior is the gauge factor (GF), defined as:

$$GF = \frac{\Delta R/R_0}{\varepsilon} \tag{4.4}$$

where $\Delta R/R_0$ is the relative change in the resistance of the material and ε is the mechanical strain. Therefore, to evaluate the suitability of new material as a piezoresistive sensor for wearable electronics, the response of the resistivity under strain must be determined.

One of the first works that explored the suitability of thermally structured thin films for wearable electronics was presented by Zhu and Moran-Mirabal [46]. The flexible devices were fabricated by spin coating a photoresist sacrificial layer on a prestressed PS sheet. Then a vinyl stencil mask was placed on the films with the electrode shape, a gold thin film was deposited, the mask was removed, and the film was structured through thermal shrinking. Afterwards, the sacrificial photoresist layer was dissolved, and the wrinkled film was transferred to an elastic substrate. The electrode response under strain is shown in Figure 4.5a. The authors measured the resistance of the films at 5% strain increments until the PDMS cohesively failed; at strains below the failure threshold, all the electrodes remained highly conductive. Depending on the desired application and the area of the human body where the sensors are going to be placed, they may have different shapes. Therefore, the authors evaluated the resistance under strain for electrodes of different form factors (defined as the length-to-width ratio, L/W). They found that at higher form factors, and the same strain, the normalized resistance increased (meaning they had a larger GF). The authors attributed this behavior to the decrease in the available conductive paths because of film cracking occurring perpendicular to the stretching axis. For electrodes with high L/W, the number of cracks necessary to span the width of the electrode, resulting in a total loss of conductivity, is smaller than for electrodes with low form factor. This behavior can be observed in Figure 4.5a, where the resistance for electrodes with $L/W = 3$, 5, and 7.5 at 100% strain is, on average, c. 10%, 30%, and 100% higher than their initial resistance. Another important parameter to study in the piezoresistive sensor is the resistance along the two axes (biaxial stretchability). The value of resistance should be stable in both axes, in this case, the values obtained between 0% and 50% remained constant for both axes in devices with a form factor of 1 (Figure 4.5b). Another important consideration for wearable electronics is that the device can be used for a long time i.e. the resistivity after each stretching cycle should remain as constant as possible. In this work, the authors evaluated the fabricated device for 100 strain cycles (Figure 4.5c). It was observed that in the first two cycles, the resistance increased slightly at 50% strain, which the authors attributed to the development of the necessary cracks to relieve tensile strain. However, after the first two cycles, the resistance remained constant for the stretched and relaxed states. Finally, the bendability of the sensor was evaluated at different values of a radius of curvature (Figure 4.5d). They wrapped the devices around a Teflon mold and then released them, and the resistance was measured before and after the bending process over 100 cycles. Figure 4.5d shows that the relative resistance after bending is close to 1 and remained unchanged over the 100 cycles for all the evaluated strains.

In a similar approach, Park et al. [47] developed a flexible wrinkled carbon nanotube (CNT)-based sensor to monitor human movement. A thin film made of CNTs was spray-gun deposited on a prestressed PS sheet and then shrunk at 150 °C to produce the characteristic wrinkle pattern. The CNT film was then transferred to an

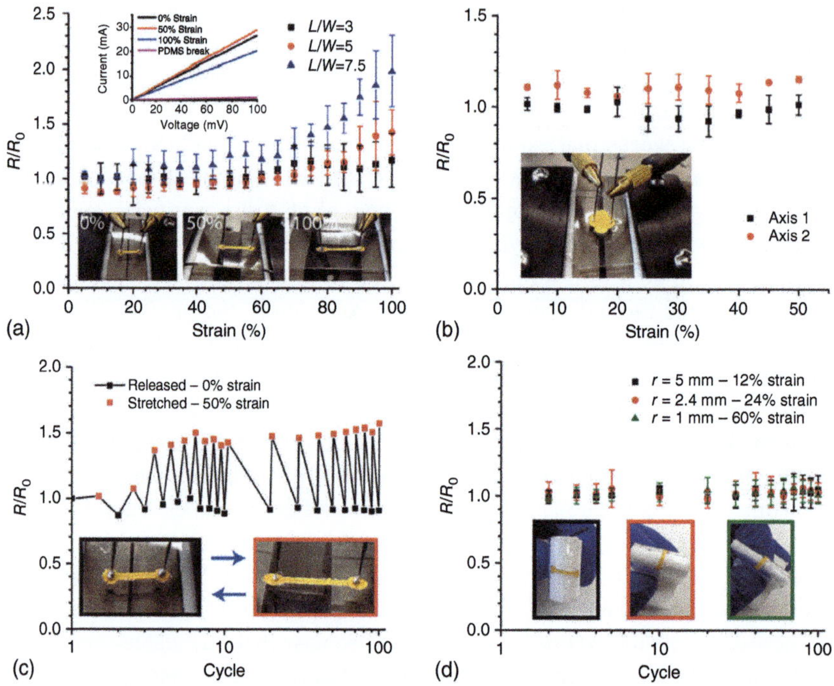

Figure 4.5 Stretchability of wrinkled electrodes. (a) Relative resistance change measurement under various strains for stretchable electrodes on PDMS substrates with $L/W = 3$ (black square, $W = 2$ mm), 5 (red circle, $W = 0.8$ mm), and 7.5 (blue triangle, $W = 0.8$ mm). Top inset graph shows typical $I-V$ curves where resistance was extracted for gold electrodes ($L/W = 5$) under 0%, 50%, and 100% strains and when PDMS broke. Bottom inset photos show the two-probe measurement for a gold electrode/PDMS device under 0%, 50%, and 100% tensile strain, respectively. (b) Biaxial stretchability. The relative resistance shows the same trend when Au/PDMS device was stretched along two orthogonal axes, indicating similar electric properties of both axes. Inset is a photo of flower-shaped Au film on PDMS at 0% strain for biaxial stretching measurement. (c) Relative resistance of an Au/PDMS electrode ($W = 0.8$ mm, $L/W = 7.5$) stretched from 0% to 50% strain over 100 stretching cycles. Inset photos show the device at 0% (black, left inset photo) and 50% strain (red, right inset photo). (d) Relative resistance of Au/PDMS electrodes ($W = 0.8$ mm, $L/W = 7.5$) released after being bent to 12% (black, left inset photo), 24% (red, middle inset photo), and 60% (green, right inset photo) bending strains. Inset photos show the electrode bent along holders with radii of 5 (black), 2.5 (red), and 1 mm (green). Each stretching and bending test were performed with three replicate electrodes. Source: Reproduced with permission from Zhu and Moran-Mirabal [46]/John Wiley & Sons, Inc.

elastic substrate, in this case, Ecoflex instead of PDMS. The former has an elastic modulus of ~125 kPa, similar to that of the human epidermis, while the latter has a modulus of ~3 MPa, significantly more rigid than human epidermis, which may cause additional stress on wearable devices. The fabricated sensor remained conductive up to ~750% strain. The sensor response to strain was evaluated using SEM images. They found that at the initial stage of the strain (at 100%), the wrinkles in the pattern started to align in the direction of the applied load, therefore, the change in the resistance of the film was minor. Then, at strains >400%, the wrinkle structures completely aligned uniaxially. Above this point, fracture nucleation and propagation on the film started; however, additional strain relief came from the percolating network of CNTs that allowed the film to remain conductive up to 750% strain, where the fractures propagated completely resulting in sensor failure. The authors evaluated the performance of the device to monitor motion in the elbow, knee, and finger with excellent results. They found that the maximum strain for the elbow was 180%, for the knee 230%, and for the finger 33% at a 90° bend. In a similar line of work, Pegan et al. [48] designed a sensor for real-time respiratory monitoring. In this case, they used a platinum thin film deposited on a prestressed PS substrate and posterior thermal shrinkage and transferred to an Ecoflex elastomeric substrate. The sensor was attached to an athletic tape and mounted in the intercostal region of a test subject. The authors measured the change in the chest wall displacement during respiration, and using spirometer readings, they were able to correlate approximate lung volumes to sensor data.

Another type of wearable electronic device is based on changes in capacitance. Such sensors can be described using the classic equation of capacitance:

$$C_0 = \frac{\varepsilon_0 \varepsilon_r l_0 \omega_0}{d_0} \tag{4.5}$$

where ε_0 is the permittivity constant, ε_r is the permittivity of the cavity that separates the plates, and l_0, ω_0, and d_0 are the initial length, width, and distance between plates of the capacitor. Upon application of a given uniaxial strain (ε), the distance between the parallel plated changes, resulting in an alteration in all parameters from Eq. (4.5), except ε_0. The response of the sensor under these conditions can be quantified according to Eq. (4.6) (which can be simplified by assuming the same Poisson's ratio (ν) for the electrode and dielectric layers) [49].

$$C = \varepsilon_0 \varepsilon_r \frac{(1+\varepsilon) l_0 \left(1 - \nu_{\text{electrode}} \varepsilon\right) \omega_0}{\left(1 - \nu_{\text{dielectric}} \varepsilon\right) d_0} = \varepsilon_0 \varepsilon_r \frac{(1+\varepsilon) l_0 \omega_0}{d_0} = (1+\varepsilon) C_0 \tag{4.6}$$

Eq. (4.6) shows that the capacitance after the sensor is subjected to a uniaxial strain is directly proportional to the initial capacitance, and the proportionality factor depends only on the strain. On the other hand, the GF (sensitivity) of these devices can be obtained according to:

$$\text{GF} = \frac{\Delta C / C_0}{\varepsilon} = \frac{\left(C - C_0\right) / C_0}{\varepsilon} = \frac{\left((1+\varepsilon) C_0 - C_0\right) / C_0}{\varepsilon} = 1 \tag{4.7}$$

In principle, the maximum GF that can be achieved in a capacitive sensor is 1. It can be concluded then that capacitive sensor possesses an excellent linear response, as opposed to piezoresistive-based devices, but has lower sensitivity.

Nur et al. were able to overcome this limitation by using wrinkled gold as the dielectric layer material [50]. The authors argued that when this material is used, an additional degree of freedom is introduced to the parallel plate capacitor structure through an out-of-plane deformation. This additional degree of freedom results in two additional factors that enhance the change in capacitance: the suppression of the width contraction and the free contraction of the dielectric layer. For this model, the response of the capacitance is determined according to Eq. (4.8) and the theoretical value of 1 for the GF can be overcome. In this work, a GF of 3.05 was obtained with a stretchability up to 140% strain, negligible hysteresis, and high linearity.

$$C = \frac{(1+\varepsilon)\,C_0}{\left(1 - \nu_{\text{dielectric}}\varepsilon\right)\,d_0} \tag{4.8}$$

Recently, Kim et al. applied this design to develop a capacitive sensor to monitor beat-to-beat blood pressure [51]. Monitoring arterial pulse, especially the systolic and diastolic blood pressures, can reveal important information about cardiovascular health. The authors used wrinkled gold thin films as electric conductors that were transferred to an elastomer substrate. The working principle of the device is shown in Figure 4.6a. When the blood flows through the radial artery, it compresses the silicone elastomer, resulting in a change in the distance between the capacitor plates that produces a concomitant change in the capacitance. Figure 4.6b shows the response of the sensors for one subject after the device is attached to the wrist (top)

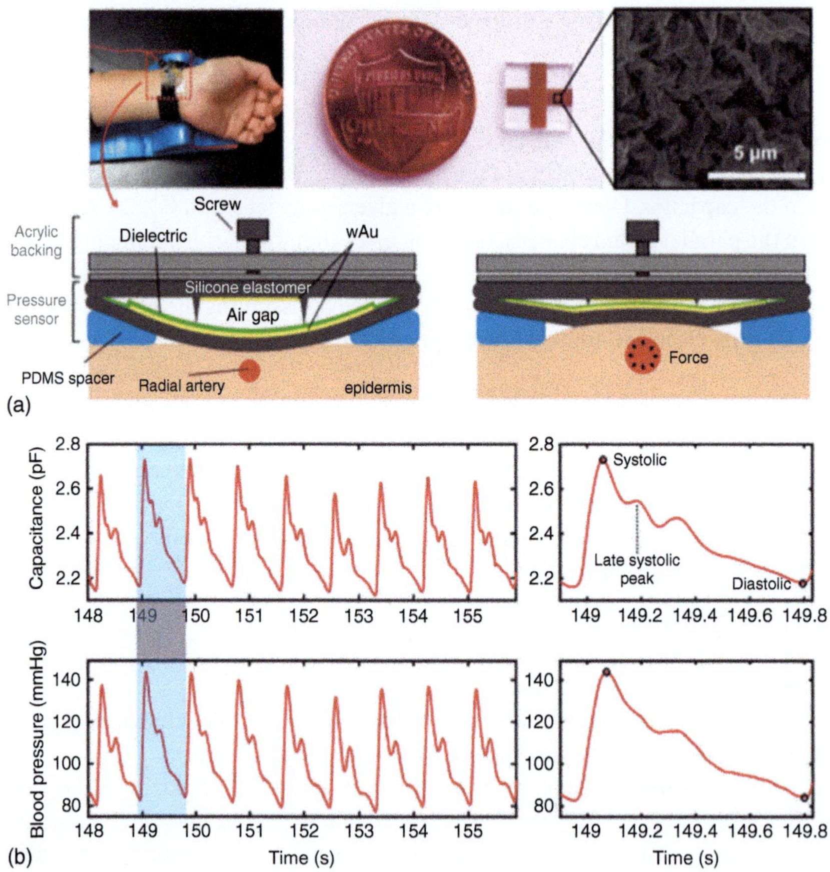

Figure 4.6 Schematic illustration of the pressure sensor when placed on the wrist above the radial artery. On the right, the pressure sensor is deformed as blood pulses through the radial artery. Screw is used to add incremental pressure to flatten the radial artery. Source: Reproduced with permission from Nguyen and Khine [52]/John Wiley & Sons, Inc.

compared with the results obtained from the same subject using an Food and Drug Administration (FDA)-approved noninvasive blood pressure (NIBP) monitoring device (bottom). The excellent pressure sensitivity and quick response time allowed for detection of the late systolic peak in the radial artery pulse waveform, which is not easily discernible in the NIBP device signal.

4.2.3 Wrinkled Electrochemical Sensors

The simplicity, low cost, and high sensitivity of electrochemical techniques make them one of the most used platforms in sensor fabrication. A main line of work in this field is the design of miniaturized electrochemical sensors for point-of-care applications. The challenge when designing a small and accurate sensor is that of reducing the size of the working electrode while maintaining or even increasing the electroactive surface area (ESA), and consequently the sensitivity. There have been multiple approaches to achieve this goal; one that has gained attention over the last decade is based on the use of wrinkled gold thin films. Gabardo et al. [7] fabricated high ESA miniaturized electrodes by sputtering gold on prestressed PS sheets that were previously coated with a self-adhesive vinyl stencil. After the sputtering process, the stencil was removed, and the electrodes were shrunk in an oven. As mentioned above, the mismatch in the mechanical properties of the thin film and the polymeric substrate creates a pattern of wrinkles on the gold surface. The vinyl mask was cut with a vinyl cutter, which allowed the rapid fabrication and prototyping of a wide variety of device designs (Figure 4.7).

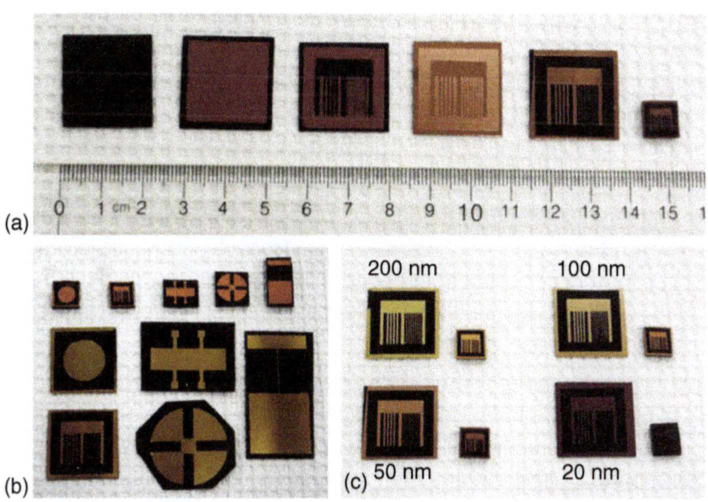

Figure 4.7 Bench-top fabrication techniques can be used to produce patterned electrodes with critical dimensions >100 μm. (a) Depiction of the fabrication of patterned electrodes through vinyl masking and prestressed PS substrate shrinking. (b) The bench-top approach developed allows the fabrication of patterned electrodes with a wide variety of geometrical shapes. (c) Patterned electrodes were fabricated from a variety of gold film thicknesses. Source: Reproduced with permission from Gabardo et al. [7]/John Wiley & Sons, Inc.

Through this simple and cost-effective methodology, the footprint of the electrode can be reduced by 94%. The ESA of the electrodes was evaluated in sulfuric acid using cyclic voltammetry. The authors found an increase of ~600% in the ESA of the structured electrodes compared with flat electrodes of the same geometrical footprint. After this research was published, a multitude of other researches has been published using similar methodologies for applications, such as DNA-based technologies, lipidic membrane sensors, or microbial fuel cells, among others [53–57]. Some studies have even integrated the high-ESA electrodes fabricated using this method within microfluidic devices [58, 59], and used them to sense the real-time adhesion of murine 3T3 fibroblast cells [60].

The glucose sensors available in the market typically rely on the oxidation of glucose by the enzyme glucose oxidase, in a process that generates hydrogen peroxide that is electrochemically sensed. Although these devices are highly specific, the use of an enzyme makes them sensitive to subtle changes in temperature, pH, or humidity. There has been increased interest in the development of enzyme-free glucose sensors and structured gold has arisen as an excellent alternative. Thermally wrinkled gold thin films have, therefore, been used for this purpose [61–63]. Recent work on this topic by Imamura et al. [64] fabricated an elastomer gold electrode able to sense nanomolar glucose concentrations. The authors deposited a sacrificial layer of poly(methyl methacrylate) (PMMA) onto a polyolefin (PO) substrate. Afterwards, they covered the substrate with an adhesive mask that left the desired shape exposed and sputtered an 80 nm-thick film of gold. In the next step, the electrode was shrunk at 140 °C. Like prestressed PS, the PO shrank at temperatures above its glass transition and the mismatch in the CTE created wrinkled patterns in the gold film. Then, they transferred the electrode onto an elastomer substrate by dissolving the sacrificial layer. Interestingly, the geometrical area of the electrodes was further decreased after the transfer step. As mentioned above, the ESA of the electrode fabricated using this methodology, is several times higher than that of the planar electrodes of the same geometrical footprint. In this study, they were able to further increase the ESA of the electrodes by subjecting them to stretching cycles. The limit of detection of glucose in physiologically relevant conditions was 268 nM for electrodes before stretching and 22.2 nM after the electrodes were subjected to stretching cycles. The limit of detection obtained in this work is currently the lowest registered for enzyme-free glucose detection and is in the range of applicability for glucose sensing in sweat samples.

Commonly used methods for depositing gold are sputtering and e-beam deposition. During sputtering deposition, the substrate is placed at the bottom of a chamber at high vacuum (anode). Simultaneously, a high purity gold target is arranged at the top of the chamber (cathode). The target is then bombarded with a high-energy gas plasma (commonly argon) and, upon collision with the surface, atoms are expelled from the target. These atoms travel to the substrate and condense on the surface producing a thin film. E-beam deposition is a similar technique, but in this case, the target is bombarded with an electron beam from a charged tungsten filament. Both methods have in common that they are costly, time-consuming, and require specialized and complex equipment. Replacing these technologies with an easily accessible, simple, and low-cost method is another challenge in the design of thermally wrinkled electrochemical sensors. Electroless deposition is an attractive route to overcome such a challenge. This method is simple, low cost, and allows the fabrication of thin films on a variety of substrates. In recent work, Gonzalez-Martinez et al. [65] used drop cast dopamine films on prestressed PS sheets to fabricate high-quality and low-cost noble metal thin films. They used a four-step process for the fabrication of gold electrodes. Prestressed PS templates masked with vinyl were obtained using the method previously developed by Gabardo et al. [7] then, these templates were drop-cast with a dopamine solution to create a polydopamine (PDA) adhesive layer within defined patterns. Afterward, gold nanoparticles were deposited onto the PDA layer as nucleation sites, and finally the nanoparticle coated substrates were submerged in an electroless bath containing tetrachloroauric acid and hydrogen peroxide. Using this approach, the cost of the electrodes was reduced from \$3.80 to \$0.65 USD and the ESA was almost twofold higher than that of wrinkled electrodes fabricated using sputtering. The increase in ESA arises from an increase in roughness when the electrodes are fabricated through the electroless process. This methodology was further used for the fabrication of solution-processed structured electrochemical cells with better analytical performance and higher ESA than that of the screen-printed electrodes available in the market.

The world is currently facing one of the biggest challenges of our time, the COVID-19 pandemic, and science is one of the cornerstones in the fight against this disease. In recently published work, Zakashansky et al. [66] fabricated a sensor to detect the SARS-CoV-2 spike protein in saliva. They used wrinkled gold thin films deposited on prestressed PS modified with a DNA aptamer specific to the protein. The DNA aptamer was modified at the 3′ end with a thiol group and the 5′ with the redox probe methylene blue (MB). In this system, the thiolated end of the aptamer interacts with the gold surface and the MB is responsible for the electrochemical signal at −0.28 V. Saliva samples diluted in phosphate buffer were used as blanks, and the test samples contained the SARS-CoV-2 spike protein. Once the aptamer interacted specifically with the protein the signal of the MB decreased. Using this methodology, the authors were able to sense concentrations of the spike protein in the range of femtograms per milliliter.

4.2.4 Current Challenges and Future Perspectives for the Use of Wrinkled Thin Films

In this section, we outline some of the current challenges and opportunities facing the relatively new technology of thin-film structuring through thermal shrinking. To date, only uniaxial and biaxial shrinking methods have been reported; a future goal should be to develop methodologies that allow total spatial control of the features on the wrinkled surface. This could lead to patterned surfaces where wrinkling patterns with a variety of symmetries and sizes are produced on a single surface.

As discussed in this chapter, the simplicity of thermal shrinkage makes an excellent alternative for the study of the mechanical properties of soft, thin films, but this approach has mostly been applied to CNC-based films. There is a large opportunity to measure the elastic moduli of thin films made of different polymers and composites to expand the applicability of the technique. Another important goal is to standardize this metrology method against other established methods, such as nanoindentation.

In another area discussed in this chapter, the use of wrinkled films to study cells has focused mainly on contact alignment of muscle cells. However, the multiscale topography and diversity of materials that can be structured through this method offer enormous opportunities for mimicking natural microenvironments for *in vitro* cell culture. So far there are only a few works that study the combination of other 3D scaffold fabrication methods with wrinkled films. The use of this fabrication method combined with 3D bioprinting for the study of different cells, such as cancer cell culture, remains an unexplored field.

A particularly interesting area of further application is that of flexible electronics, where almost all wearable electronic devices fabricated using the shrink and wrinkle method are based on piezoresistive or capacitive effects. In this sense, there is an opportunity to develop chemosensors that are less invasive than those currently in the market. The fabrication of flexible devices for noninvasive chemical-based wearable sensors is the next logical step; in fact, the excellent limit of detection for glucose sensing obtained by Imamura et al. [64] in a flexible device opens the possibility for their use as noninvasive glucose sensors. On the other hand, the use of this methodology for the development of optical wearable sensors is also an interesting alternative that should be explored. Sensors developed in research usually have short lifetimes and the integration into human skin is nontrivial; this area offers many future opportunities. To develop fully functional electrochemical devices for detection of biofluids, the antifouling capacities of existing electrodes also need to be enhanced. This would allow the use of these devices in a broader range of applications, such as environmental sensing and the food industry. Finally, the integration of platforms for multi-analyte detection could be the next challenge to overcome in the near future for sensing applications.

References

1 Greene, J.E. (2017). Review article: tracing the recorded history of thin-film sputter deposition: from the 1800s to 2017. *J. Vac. Sci. Technol. A* 35 (5): 05C204.

2 Greene, J.E. (2014). Tracing the 5000-year recorded history of inorganic thin films from ~3000 BC to the early 1900s AD. *Appl. Phys. Rev.* 1 (4): 041302.

3 Makhlouf, A.S.H. (2011). Current and advanced coating technologies for industrial applications. In: *Nanocoatings and Ultra-Thin Films*, Chapter 1 (ed. A.S.H. Makhlouf and I. Tiginyanu), 3–23. Woodhead Publishing.

4 Singh, J. and Wolfe, D.E. (2005). Review nano and macro-structured component fabrication by electron beam-physical vapor deposition (EB-PVD). *J. Mater. Sci.* 40 (1): 1–26.

5 Benelmekki, M. and Erbe, A. (2019). Nanostructured thin films–background, preparation and relation to the technological revolution of the 21st century. In:

Frontiers of Nanoscience, Chapter 1 (ed. M. Benelmekki and A. Erbe), 1–34. Elsevier.

6 Pierson, H.O. (1992). Introduction and general considerations. In: *Handbook of Chemical Vapor Deposition*, Chapter 1 (ed. H.O. Pierson), 1–16. Oxford: William Andrew Publishing.

7 Gabardo, C.M., Zhu, Y., Soleymani, L., and Moran-Mirabal, J.M. (2013). Bench-top fabrication of hierarchically structured high-surface-area electrodes. *Adv. Funct. Mater.* 23 (24): 3030–3039.

8 Mei, Y., Kiravittaya, S., Harazim, S., and Schmidt, O.G. (2010). Principles and applications of micro and nanoscale wrinkles. *Mater. Sci. Eng., R* 70 (3): 209–224.

9 Bowden, N., Brittain, S., Evans, A.G. et al. (1998). Spontaneous formation of ordered structures in thin films of metals supported on an elastomeric polymer. *Nature* 393 (6681): 146–149.

10 Fu, C.-C., Grimes, A., Long, M. et al. (2009). Tunable nanowrinkles on shape memory polymer sheets. *Adv. Mater.* 21 (44): 4472–4476.

11 Stafford, C.M., Harrison, C., Beers, K.L. et al. (2004). A buckling-based metrology for measuring the elastic moduli of polymeric thin films. *Nat. Mater.* 3 (8): 545–550.

12 Groenewold, J. (2001). Wrinkling of plates coupled with soft elastic media. *Physica A* 298 (1): 32–45.

13 Stimpson, T.C., Cathala, B., Moreau, C. et al. (2020). Xyloglucan structure impacts the mechanical properties of xyloglucan–cellulose nanocrystal layered films—a buckling-based study. *Biomacromolecules* 21 (9): 3898–3908.

14 Agrawal, A., Luchette, P., Palffy-Muhoray, P. et al. (2012). Surface wrinkling in liquid crystal elastomers. *Soft Matter* 8 (27): 7138–7142.

15 Nolte, A.J., Rubner, M.F., and Cohen, R.E. (2005). Determining the Young's modulus of polyelectrolyte multilayer films via stress-induced mechanical buckling instabilities. *Macromolecules* 38 (13): 5367–5370.

16 Taccola, S., Pensabene, V., Fujie, T. et al. (2017). On the injectability of free-standing magnetic nanofilms. *Biomed. Microdevices* 19 (3): 51.

17 Cranston, E.D., Eita, M., Johansson, E. et al. (2011). Determination of Young's modulus for nanofibrillated cellulose multilayer thin films using buckling mechanics. *Biomacromolecules* 12 (4): 961–969.

18 Kan, K. and Cranston, E. (2013). Mechanical testing of thin film nanocellulose composites using buckling mechanics. *Tappi J.* 12: 9–17.

19 Gill, U., Sutherland, T., Himbert, S. et al. (2017). Beyond buckling: humidity-independent measurement of the mechanical properties of green nanobiocomposite films. *Nanoscale* 9 (23): 7781–7790.

20 Mariano, M., El Kissi, N., and Dufresne, A. (2014). Cellulose nanocrystals and related nanocomposites: review of some properties and challenges. *J. Polym. Sci., Part B: Polym. Phys.* 52 (12): 791–806.

21 Moon, R.J., Martini, A., Nairn, J. et al. (2011). Cellulose nanomaterials review: structure, properties and nanocomposites. *Chem. Soc. Rev.* 40 (7): 3941–3994.

22 Kochumalayil, J., Sehaqui, H., Zhou, Q., and Berglund, L.A. (2010). Tamarind seed xyloglucan – a thermostable high-performance biopolymer from non-food feedstock. *J. Mater. Chem.* 20 (21): 4321–4327.

23 De France, K.J., Babi, M., Vapaavuori, J. et al. (2019). 2.5D Hierarchical structuring of nanocomposite hydrogel films containing cellulose nanocrystals. *ACS Appl. Mater. Interfaces* 11 (6): 6325–6335.

24 Stimpson, T.C., Osorio, D.A., Cranston, E.D., and Moran-Mirabal, J.M. (2021). Direct comparison of three buckling-based methods to measure the elastic modulus of nanobiocomposite thin films. *ACS Appl. Mater. Interfaces* 13 (24): 29187–29198.

25 Cecilia, A., Baecker, A., Hamann, E. et al. (2017). Optimizing structural and mechanical properties of cryogel scaffolds for use in prostate cancer cell culturing. *Mater. Sci. Eng., C* 71: 465–472.

26 Nilghaz, A., Hoo, S., Shen, W. et al. (2018). Multilayer cell culture system supported by thread. *Sens. Actuators, B* 257: 650–657.

27 Arhoma, A., Chantry, A.D., Haywood-Small, S.L., and Cross, N.A. (2017). SAHA-induced TRAIL-sensitisation of multiple myeloma cells is enhanced in 3D cell culture. *Exp. Cell. Res.* 360 (2): 226–235.

28 Lelièvre, S.A., Kwok, T., and Chittiboyina, S. (2017). Architecture in 3D cell culture: an essential feature for in vitro toxicology. *Toxicol. in Vitro* 45: 287–295.

29 Ravi, M., Paramesh, V., Kaviya, S.R. et al. (2015). 3D cell culture systems: advantages and applications. *J. Cell. Physiol.* 230 (1): 16–26.

30 De France, K.J., Yager, K.G., Chan, K.J.W. et al. (2017). Injectable anisotropic nanocomposite hydrogels direct in situ growth and alignment of myotubes. *Nano Lett.* 17 (10): 6487–6495.

31 González-Martínez, E., Díaz-Águila, C.R., Rodríguez Cruz, M. et al. (2020). Growing spheroids of lung adenosquamous carcinoma on electrospun poly(ε-caprolactone). *Bioinspired, Biomim. Nanobiomaterials* 9 (4): 252–256.

32 Herreros-Pomares, A., Zhou, X., Calabuig-Fariñas, S. et al. (2021). 3D printing novel in vitro cancer cell culture model systems for lung cancer stem cell study. *Mater. Sci. Eng., C* 122: 111914.

33 Malakpour-Permlid, A., Buzzi, I., Hegardt, C. et al. (2021). Identification of extracellular matrix proteins secreted by human dermal fibroblasts cultured in 3D electrospun scaffolds. *Sci. Rep.* 11 (1): 6655.

34 Luna, J.I., Ciriza, J., Garcia-Ojeda, M.E. et al. (2011). Multiscale biomimetic topography for the alignment of neonatal and embryonic stem cell-derived heart cells. *Tissue Eng. Part C* 17 (5): 579–588.

35 Chen, A., Lee, E., Tu, R. et al. (2014). Integrated platform for functional monitoring of biomimetic heart sheets derived from human pluripotent stem cells. *Biomaterials* 35 (2): 675–683.

36 Chen, A., Lieu, D.K., Freschauf, L. et al. (2011). Shrink-film configurable multiscale wrinkles for functional alignment of human embryonic stem cells and their cardiac derivatives. *Adv. Mater.* 23 (48): 5785–5791.

37 van der Meer, A.D., Poot, A.A., Feijen, J., and Vermes, I. (2010). Analyzing shear stress-induced alignment of actin filaments in endothelial cells with a microfluidic assay. *Biomicrofluidics* 4 (1): 11103.

38 Wang, J., Chen, A., Lieu, D.K. et al. (2013). Effect of engineered anisotropy on the susceptibility of human pluripotent stem cell-derived ventricular cardiomyocytes to arrhythmias. *Biomaterials* 34 (35): 8878–8886.

39 Birk, D.E., Silver, F.H., and Trelstad, R.L. (1991). Matrix assembly. In: *The Cell Biology of the Extracellular Matrix* (ed. E.D. Hay). New York: Academic Press.

40 De France, K.J., Xu, F., Toufanian, S. et al. (2021). Multi-scale structuring of cell-instructive cellulose nanocrystal composite hydrogel sheets via sequential electrospinning and thermal wrinkling. *Acta Biomater.* 128: 250–261.

41 Zhou, C., Zhao, J., Saem, S. et al. (2018). Self-cross-linking p(APM-*co*-AA) microstructured thin films as biomimetic scaffolds. *ACS Appl. Bio Mater.* 1 (5): 1512–1522.

42 Ogle, M.E., Segar, C.E., Sridhar, S., and Botchwey, E.A. (2016). Monocytes and macrophages in tissue repair: implications for immunoregenerative biomaterial design. *Exp. Biol. Med.* 241 (10): 1084–1097.

43 Champion, J.A. and Mitragotri, S. (2006). Role of target geometry in phagocytosis. *Proc. Natl. Acad. Sci. U.S.A.* 103 (13): 4930.

44 Makaremi, S., Luu, H., Boyle, J.P. et al. (2019). The topography of silica films modulates primary macrophage morphology and function. *Adv. Mater. Interfaces* 6 (21): 1900677.

45 Ko, F.K., El-Aufy, A., Lam, H., and Macdiarmid, A.G. (2005). Electrostatically generated nanofibres for wearable electronics. In: *Wearable Electronics and Photonics*, Chapter 2 (ed. X. Tao), 13–40. Woodhead Publishing.

46 Zhu, Y. and Moran-Mirabal, J. (2016). Highly bendable and stretchable electrodes based on micro/nanostructured gold films for flexible sensors and electronics. *Adv. Electron. Mater.* 2 (3): 1500345.

47 Park, S.-J., Kim, J., Chu, M., and Khine, M. (2016). Highly flexible wrinkled carbon nanotube thin film strain sensor to monitor human movement. *Adv. Mater. Technol.* 1 (5): 1600053.

48 Pegan, J.D., Zhang, J., Chu, M. et al. (2016). Skin-mountable stretch sensor for wearable health monitoring. *Nanoscale* 8 (39): 17295–17303.

49 Amjadi, M., Kyung, K.-U., Park, I., and Sitti, M. (2016). Stretchable, skin-mountable, and wearable strain sensors and their potential applications: a review. *Adv. Funct. Mater.* 26 (11): 1678–1698.

50 Nur, R., Matsuhisa, N., Jiang, Z. et al. (2018). A highly sensitive capacitive-type strain sensor using wrinkled ultrathin gold films. *Nano Lett.* 18 (9): 5610–5617.

51 Kim, J., Chou, E.-F., Le, J. et al. (2019). Soft wearable pressure sensors for beat-to-beat blood pressure monitoring. *Adv. Healthc. Mater.* 8 (13): 1900109.

52 Nguyen, T. and Khine, M. (2020). Advances in materials for soft stretchable conductors and their behavior under mechanical deformation. *Polymers* 12 (7): 1454.

53 Abbaszadeh Amirdehi, M., Saem, S., Zarabadi, M.P. et al. (2018). Microstructured anodes by surface wrinkling for studies of direct electron transfer biofilms in microbial fuel cells, advanced materials. *Interfaces* 5 (13): 1800290.

54 Bockaj, M., Fung, B., Tsoulis, M. et al. (2018). Method for electrochemical detection of brain derived neurotrophic factor (BDNF) in plasma. *Anal. Chem.* 90 (14): 8561–8566.

55 Saha, S., Chan, Y., and Soleymani, L. (2018). Enhancing the photoelectrochemical response of DNA biosensors using wrinkled interfaces. *ACS Appl. Mater. Interfaces* 10 (37): 31178–31185.

56 Sokunthearath, S., Shahid, O., Khondker, A. et al. (2020). Benchtop-fabricated lipid-based electrochemical sensing platform for the detection of membrane disrupting agents. *Sci. Rep. (Nature Publisher Group)* 10 (1): 1–12.

57 Li, S., Lin, L., Chang, X. et al. (2021). A wrinkled structure of gold film greatly improves the signaling of electrochemical aptamer-based biosensors. *RSC Adv.* 11 (2): 671–677.

58 Sonney, S., Shek, N., and Moran-Mirabal, J.M. (2015). Rapid bench-top fabrication of poly(dimethylsiloxane)/polystyrene microfluidic devices incorporating high-surface-area sensing electrodes. *Biomicrofluidics* 9 (2): 026501.

59 Gabardo, C.M., Adams-McGavin, R.C., Vanderfleet, O.M., and Soleymani, L. (2015). Rapid prototyping of microfluidic devices with integrated wrinkled gold micro–/nano textured electrodes for electrochemical analysis. *Analyst* 140 (16): 5781–5788.

60 Saem, S., Zhu, Y., Luu, H., and Moran-Mirabal, J. (2017). Bench-top fabrication of an all-PDMS microfluidic electrochemical cell sensor integrating micro/nanostructured electrodes. *Sensors* 17 (4): 732.

61 Adams-McGavin, R.C., Chan, Y., Gabardo, C.M. et al. (2017). Nanoporous and wrinkled electrodes enhance the sensitivity of glucose biosensors. *Electrochim. Acta* 242: 1–9.

62 Chan, Y., Skreta, M., McPhee, H. et al. (2019). Solution-processed wrinkled electrodes enable the development of stretchable electrochemical biosensors. *Analyst* 144 (1): 172–179.

63 González-Martínez, E., Saem, S., Beganovic, N.E., and Moran-Mirabal, J. (2021). Electrochemical nano-roughening of gold microstructured electrodes for the enhanced detection of copper and glucose. *ChemRxiv*.

64 Imamura, A., Zakashansky, J., Cho, K. et al. (2020). Stretchable sensors for nanomolar glucose detection. *Adv. Mater. Technol.* 5 (4): 1900843.

65 Gonzalez-Martinez, E., Saem, S., Beganovic, N.E., and Moran-Mirabal, J. (2021). Fabrication of microstructured electrodes via electroless metal deposition onto polydopamine-coated polystyrene substrates and thermal shrinking. *Nano Select* 2 (10): 1926–1940.

66 Zakashansky, J.A., Imamura, A.H., Salgado, D.F. et al. (2021). Detection of the SARS-CoV-2 spike protein in saliva with Shrinky-Dink© electrodes. *Anal. Methods* 13 (7): 874–883.

5

Design of Nanocomposite Microgels Prepared by Seeded Emulsion Polymerization in the Presence of Microgels

Takumi Watanabe[1] *and Daisuke Suzuki*[1,2]

[1] *Shinshu University, Graduate School of Textile Science and Technology, 3-15-1 Tokida Ueda, Nagano 386-8567, Japan*
[2] *Shinshu University, Research Initiative for Supra-Materials, Interdisciplinary Cluster for Cutting Edge Research, 3-15-1 Tokida Ueda, Nagano 386-8567, Japan*

5.1 Background on Composite Hydrogels

The properties and applications of polymer materials can be improved by combining polymers with different properties. Such polymer complexes prepared by mixing multiple components are often called nanocomposite materials [1]. In particular, combining a hydrophobic and a hydrophilic polymer is a powerful strategy to obtain excellent properties that cannot be achieved using a single component. For example, the construction of nanocomposite structures has expanded the functionality and applications of hydrogel materials, which have three-dimensional (3D) networks of hydrophilic (or amphiphilic) polymers [2–6]. Although the weak mechanical strength of typical hydrogel materials is problematic for some applications, nanocomposite microgels that contain hydrophobic materials such as clay [7], carbon-based nanomaterials [8, 9], and nanoparticles [10, 11] show excellent mechanical properties. Nanocomposite hydrogels obtained from *in situ* radical solution polymerization in the presence of clay exhibit attractive mechanical properties, given that the clay nanosheets serve as cross-linkers in the polymer networks [7]. Furthermore, the incorporated materials, such as magnetic nanoparticles and carbon nanosheets, can endow the hydrogels with mechanical strength and other functionalities, such as magnetic properties [12, 13]. The hydrogels can be downsized to the colloidal scale (50 nm to several μm) for dispersal in water as colloid particles, which are expected to exhibit the same behavior as the bulk hydrogels. Thus, nanocomposite or composite hydrogels promise great potential as next-generation hydrogel materials not only at the bulk scale, but also as hydrogel microspheres [14–16]. To broaden the range of applications for microgels, the incorporation of various functional properties is desirable and has been achieved using a combination of synthetic techniques. In this chapter, we will discuss the synthesis, characterization, and application of nanocomposite microgels, especially nanocomposite microgels prepared by seeded emulsion polymerization (SEP),

Smart Stimuli-Responsive Polymers, Films, and Gels, First Edition.
Edited by Liang Hu, Yongfeng Gao, and Michael J. Serpe.

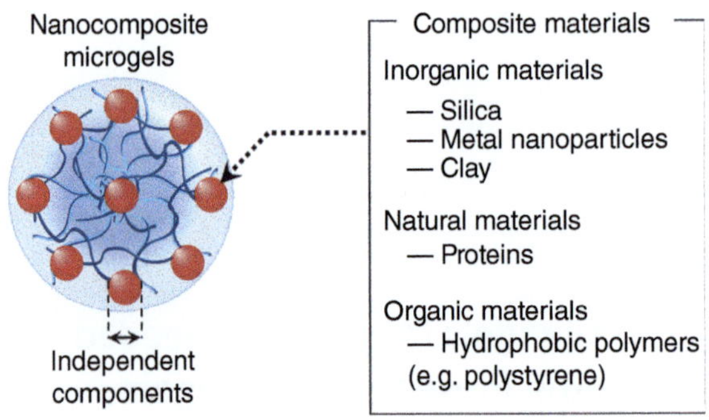

Figure 5.1 Schematic illustration of nanocomposite microgels.

which is emulsion polymerization conducted in the presence of microgels [17] (Organic materials in Figure 5.1).

5.2 Background on Composite Microgels

To design nanocomposite microgels, the properties of microgels, such as the cross-linking and functional-group distribution in microgels, should be well understood. To date, various types of microgels have been synthesized in many ways. Initially, radical copolymerization was the standard strategy for the functionalization of microgels (Figure 5.2a). In this polymerization method, water-soluble monomers and water-soluble initiators were used, and the resulting polymers showed thermo-responsivity [17, 18]. For example, poly(*N*-isopropyl acrylamide) (pNIPAm) microgels, which represent one of the most widely studied types of microgels, have been prepared by aqueous free-radical precipitation polymerization [19, 20]. The properties of these microgels, including size [21–24], cross-linking density and/or distribution [25, 26], as well as charge density and distribution [27–30], can be controlled via the polymerization conditions and the concentrations of the monomers. For example, the precipitation polymerization of a variety of vinyl monomers results in the introduction of various functional groups into the microgels. Since Kawaguchi et al. first reported the preparation of poly(acrylamide-*co*-methacrylic acid (MAc)) microgels by precipitation polymerization [31], various functional monomers, such as other carboxy-group-containing monomers [27–31], amino-group-containing monomers [32, 33], metal-catalyst-containing monomers [34–36], and hydrophobic monomers [37], have been copolymerized in microgels using precipitation polymerization. Due to the mechanism of precipitation polymerization [20], comonomers that react more slowly than NIPAm tend to be more localized on the surface of the microgels (and conversely, those that react more rapidly are more localized at the center of the microgels) [30]. Such core–shell microgels typically are categorized as composite microgels, but not nanocomposite microgels.

The construction of core–shell structures via seeded precipitation polymerization is another essential method for improving microgel properties (Figure 5.2b). Since the introduction of this strategy by Jones and Lyon in 2000 [38, 39], a variety of

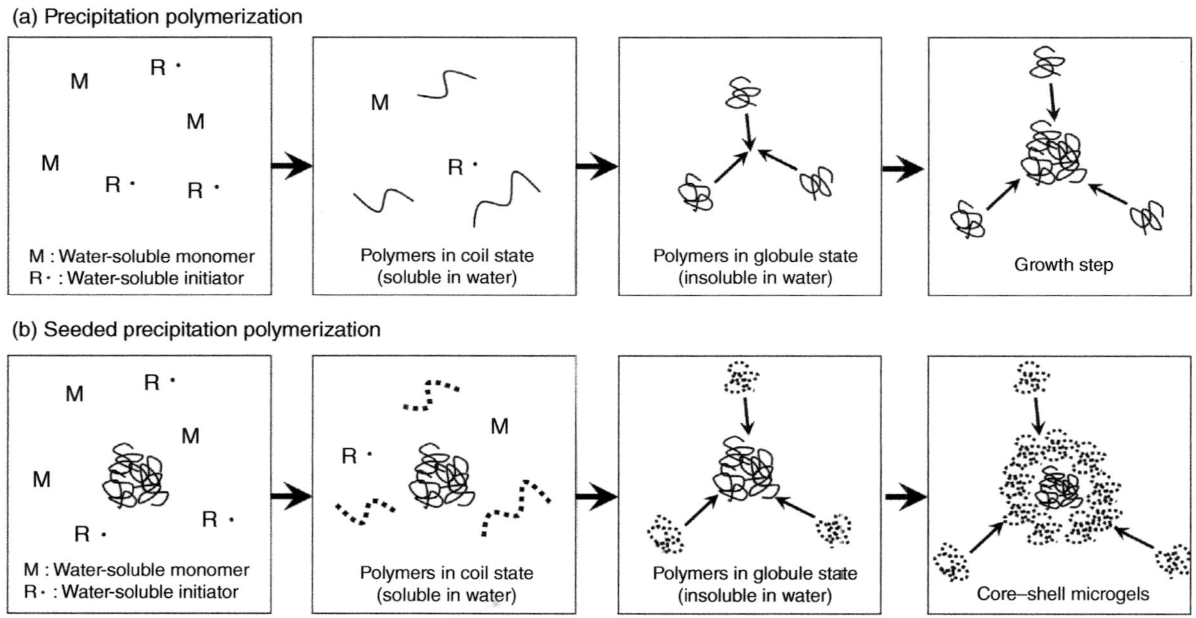

Figure 5.2 Schematic illustration of the typical polymerization process for (a) precipitation polymerization and (b) seeded precipitation polymerization.

functional microgels with a core–shell structure have been synthesized [40–48]. Seeded precipitation polymerization refers to the precipitation polymerization conducted in the presence of pre-synthesized microgels and has several advantages over the radical copolymerization in precipitation polymerization in terms of functionalization of the microgels. For example, when an excess of comonomer MAc is used in the precipitation polymerization of NIPAm, MAc is incorporated at the center of the microgels and on the surface of the microgels, despite the fact that MAc comonomers are usually incorporated in the center of microgels on account of their relatively high reactivity ratio [27–32]. On the other hand, the construction of a core–shell structure in terms of charge density [40], cross-linking density, volume-phase-transition temperature, and other properties can be achieved by seeded precipitation polymerization [46]. This is a practical approach for certain microgel applications, because one of the most significant advantages of microgels over conventional solid particles is the effective use of their 3D internal space. As such core–shell microgels, in contrast to bulk hydrogel materials, have so far typically been prepared using water-soluble vinyl monomers [17, 18], it is not surprising that the fabrication of nanocomposite structures with hydrophobic materials has not been reported yet.

5.3 Conventional Emulsion Polymerization and SEP

The radical polymerization of water-insoluble monomers typically affords solid-like microspheres. The most typical synthetic route to polymer colloids is emulsion polymerization, which has been used as a method to synthesize latex since the 1930s, and a kinetic model for the process was proposed in the 1940s [49, 50]. Conventional emulsion polymerization is based on a water-insoluble monomer such as styrene, water, a water-soluble surfactant, and a water-soluble initiator (Figure 5.3a). As hydrophobic monomers such as styrene do not easily dissolve in water, the polymerization of such monomers in the absence of surfactant usually requires relatively long polymerization times, while the monomer-conversion rate is often low. In contrast, when an appropriate surfactant is present, the hydrophobic monomer is dissolved in the hydrophobic regions of the surfactant micelles, commensurate with an increased bulk concentration of styrene in the aqueous solution [50]. Consequently, the monomer-conversion and polymerization kinetics increase with increasing surfactant concentration. Thus, emulsion polymerization represents an efficient approach for the synthesis of hydrophobic particles that consist of hydrophobic monomers.

Subsequently, several SEP strategies have been proposed for the synthesis of nanocomposite particles, whereby the most important advantage of this approach is the possibility to control the particle morphology (Figure 5.3b). Initially, SEP was used to examine the polymerization kinetics and the mechanism of the emulsion polymerization of styrene [50–52]. In contrast to conventional emulsion polymerizations, where the surfactant micelles are the polymerization sites, the number of polymerization sites for hydrophobic monomers can be accurately

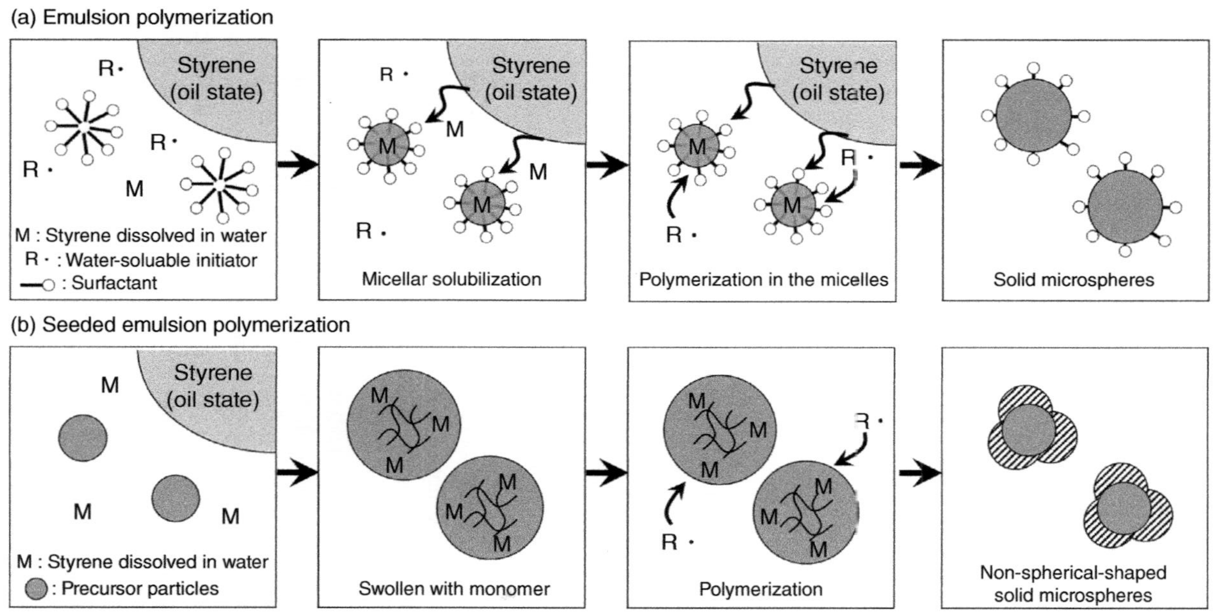

Figure 5.3 Schematic illustration of the typical polymerization processes of (a) emulsion polymerization and (b) SEP.

controlled using solid-microsphere seeds. Then, SEP was widely used to control the particle size and the morphology of particles via the phase separation between the seed microspheres and the hydrophobic monomers [53–60]. When SEP is carried out using the same hydrophobic monomer as in the seed particles, micron-sized spherical particles can be generated. When the hydrophobic monomer is different to the chemical constituents of the seed particles, non-spherical composite microspheres are formed via phase separation. The particle morphology is determined by the thermal dynamics and polymerization kinetics of the SEP, and thus, the polymerization conditions (e.g. temperature, solvent polarity, surfactant addition, and stirring speed) and the particle properties (e.g. size and cross-linking density) have an influence on the morphology of the nanocomposite microspheres. SEP is an efficient synthetic approach to preparing non-spherical hydrophobic particles such as snowman-like [53, 54], octopus-ocellatus-like [57], hollow [58, 59], and raspberry-like microspheres [60]. However, in most SEP strategies used thus far, solid microspheres (e.g. polystyrene or poly(methyl methacrylate) microspheres) have been used as seeds due to their compatibility with lipophilic or hydrophobic monomers with a few exceptions [61–67].

5.4 Nanocomposite Microgels Prepared by SEP in the Presence of Microgels

As mentioned above, SEP typically affords solid-like nanocomposite microspheres. However, by using microgels as the seed particles, SEP can also provide a novel type of nanocomposite microgels (Figure 5.4a). Kraft et al. have conducted the SEP of styrene using pNIPAm microgels as seeds [61]. A series of non-spherical nanocomposite microgels was obtained via the phase separation of styrene inside the pNIPAm microgels. Although this method provides rough control of the surface morphology and anisotropy of the nanocomposite microgels, a pre-swelling period (equilibrium swelling) is required. Moreover, the aggregation of the individual microgels during the polymerization is difficult to completely prevent. Suzuki et al. have obtained nanocomposite microgels that contain solid nanoparticles in a microgel network from the SEP of glycidyl methacrylate (GMA) or styrene using water-swellable pNIPAm microgels [62–64, 66, 67]. Even though there was no pre-swelling period before the polymerization in this approach, the polymerization of GMA and styrene can occur in the aqueous phase, i.e. not only within the microgels. However, hydrophobic nanoparticles were surprisingly formed only within the microgels. For example, raspberry-shaped composite microgels were formed when styrene was used as the monomer for the SEP (Figure 5.4b). Unlike conventional emulsion polymerization, this polymerization strategy did not require an equilibrium period for the swelling of the microgels by styrene. The amount and size of the attached polystyrene nanoparticles increased when the styrene concentration used during the SEP was increased from 50 to 100 mM. The nanocomposite microgels precipitated during polymerization at high styrene concentrations (~200 mM). Conversely, aggregation of the polystyrene nanoparticles was not observed when the

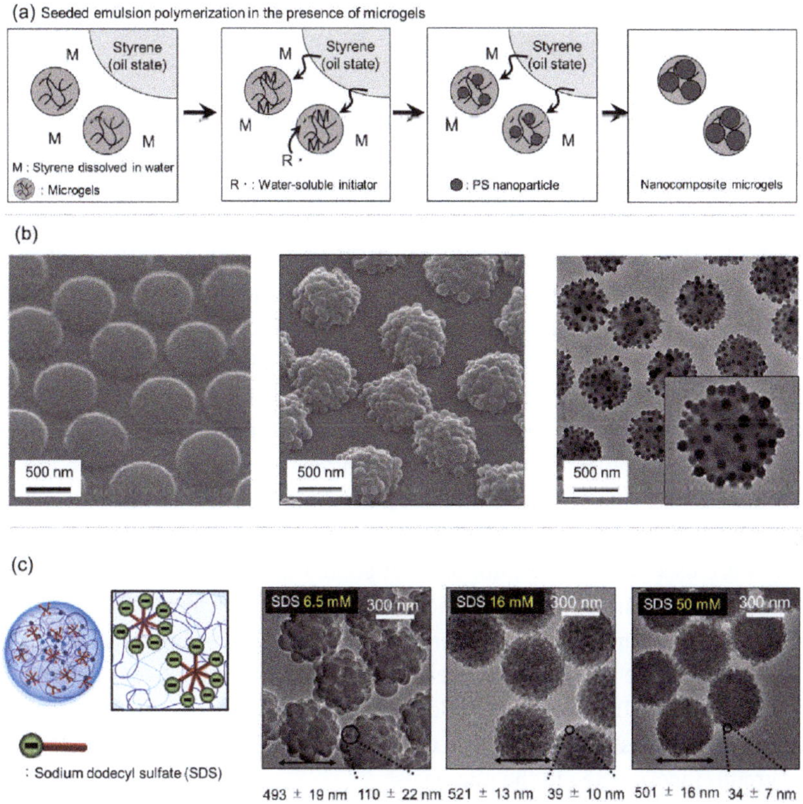

Figure 5.4 (a) Schematic of the seeded emulsion polymerization of styrene in the presence of microgels. (b) Representative FE-SEM and TEM images of the seed microgels (left) and nanocomposite microgels (center, right) in the dry state. (c) Representative results for the nanocomposite microgels prepared by the SEP of styrene in the presence of the surfactant SDS. Source: Reproduced with permission from Suzuki and Kobayashi [63]/American Chemical Society.

SEP was conducted in the presence of sodium dodecyl sulfate (SDS) [63]. The size of the polystyrene nanoparticles immobilized on the parent microgels decreased with increasing amount of SDS during the SEP (Figure 5.4c). The thus-obtained raspberry-shaped composite microgels show thermo-responsive properties and differ from pNIPAm microgels with respect to volume and surface-charge density. Moreover, SDS surfactants usually adsorb onto pNIPAm microgels in an aqueous solution and form aggregates inside the microgels, which was confirmed by scattering techniques [65]. It is therefore feasible to expect that the styrene monomers were most likely solubilized and polymerized in the SDS aggregates inside the microgels, resulting in an increased number of polystyrene particles incorporated into the parent microgels, thus endowing the resulting nanocomposites with high colloidal stability [63].

5.5 Design of the Internal Structure of the Nanocomposite Microgels

The incorporation of charged groups, such as the carboxyl group, into the microgel seeds affects the morphology of the resulting nanocomposite microgels prepared by SEP [64, 66, 67]. Due to the different reactivity ratios of methacrylic acid and NIPAm, the MAc fed in during the precipitation polymerization of NIPAm was localized at the center of the resulting pNIPAm microgels [30]. When these poly(NIPAm-*co*-MAc) microgels were used as seeds for the SEP of styrene, raspberry-shaped nanocomposite microgels were obtained, similar to the case of pure pNIPAm microgels [64]. However, the internal structure of the resulting nanocomposite microgels was different. According to the TEM images of ultra-thin cross-sections of the nanocomposite microgels, the polystyrene nanoparticles were localized throughout the entire microgel when the pNIPAm microgels were used as seeds for the SEP of styrene [66]. Conversely, the polystyrene nanoparticles were localized only on the surface of the microgels when the poly(NIPAm-*co*-MAc) microgels were used as seeds (Figure 5.5a). It should also be noted here that the SEPs were conducted under basic conditions, e.g. at pH = 10, which is higher than the pK_a of MAc. When the SEP was performed under acidic conditions such as at pH = 3, polystyrene nanoparticles were formed at the center of the microgels even when the poly(NIPAm-*co*-MAc) microgels were used as seeds. These results indicated that the polystyrene formed in the pNIPAm-based microgels avoided the charged groups incorporated in the microgels.

Therefore, the morphology of the nanocomposite microgels can be controlled by changing the charge distribution inside the microgels. When fumaric acid (FAc) was used as the comonomer during the precipitation polymerization of NIPAm, FAc was localized on the surface of the microgels due to its lower reactivity ratio compared to NIPAm [66, 67]. The SEP of styrene in the presence of poly(NIPAm-*co*-FAc) microgels also provided raspberry-shaped nanocomposite microgels. However, unlike the nanocomposites produced by the SEP of styrene using pNIPAm or poly(NIPAm-*co*-MAc) microgels as seeds, the polystyrene-poly(NIPAm-*co*-FAc) nanocomposite microgels maintained their colloidal stability after the SEP, even at high styrene monomer concentrations (e.g. 300 mM). The electrophoretic mobility of the resulting nanocomposite microgels changed in response to temperature and pH, indicating that the stimulus-responsive pNIPAm-based hydrogels were localized on the surface of the nanocomposite microgels. Importantly, the interior of the poly(NIPAm-*co*-FAc) microgels was completely occupied by polystyrene (Figure 5.4b). Therefore, the carboxyl groups incorporated on the surface of the microgels afforded the synthesis of nanocomposite microgels with a large amount of polystyrene located exclusively in the interior of the parent microgels. Since the polystyrene was formed inside the pNIPAm microgels, avoiding the polyelectrolytes, the polyelectrolytes remain on the surface of the nanocomposite microgels after the SEP, resulting in the observed high colloidal stability.

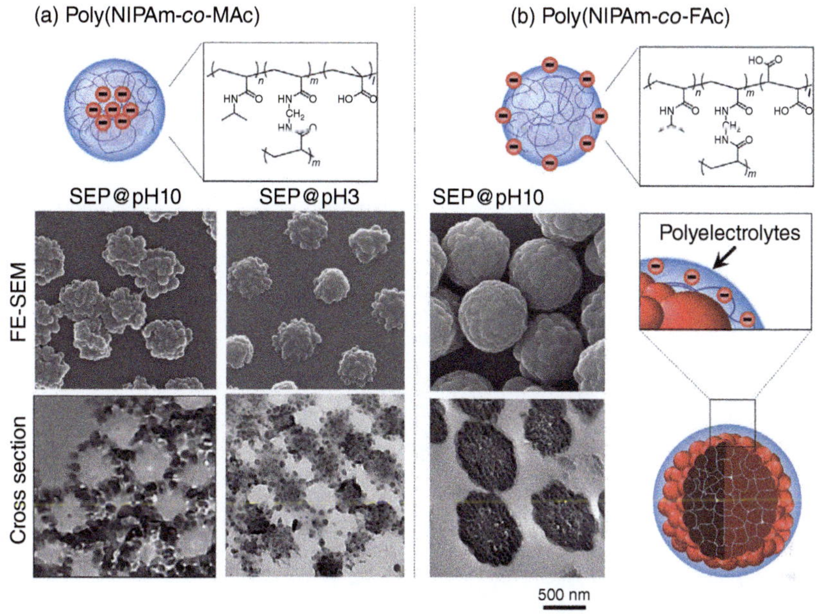

Figure 5.5 Representative results for nanocomposite microgels prepared by the SEP of styrene in the presence of (a) poly(NIPAm-*co*-MAc) microgels and (b) poly(NIPAM-*co*-FAc) microgels. Source: Reproduced with permission from Watanabe et al. [66, 67]/American Chemical Society.

5.6 Synthesis of Multi-layered Nanocomposite Microgels

The use of core–shell microgels as seeds for the SEP of styrene affords multi-layered nanocomposite microgels [68]. As shown in Figures 5.4 and 5.5, the styrene monomer can recognize the microenvironment inside the microgels, especially the polarity, during SEP. Therefore, by constructing multi-layered charge regions inside the seed microgels, the morphology of the nanocomposite microgels can be controlled in detail. For instance, when two-layered core–shell microgels were used as seeds for the SEP of styrene, three-layered nanocomposite microgels were obtained (Figure 5.6) [68]. The core–shell microgels were first synthesized by seeded precipitation polymerization in the presence of pNIPAm microgels; the shell layer was composed of poly(NIPAm-*co*-MAc) hydrogel. Due to the MAc:NIPAm reactivity ratio during the free-radical precipitation polymerization, the carboxyl groups of MAc were localized between the pNIPAm hydrogel layers (Figure 5.6a). When these core–shell microgels were used as seeds for the SEP of styrene, multi-layered nanocomposite microgels were obtained (Figure 5.6b–e). Ultra-thin cross-sectional images showed a hydrogel layer between the polystyrene shell (third layer) and the composite core (first layer). The thickness and number of layers in the multi-layered nanocomposite microgels can be controlled by the spatial distribution of the carboxy groups in the seed microgels used for SEP [68]. Interestingly, the obtained results

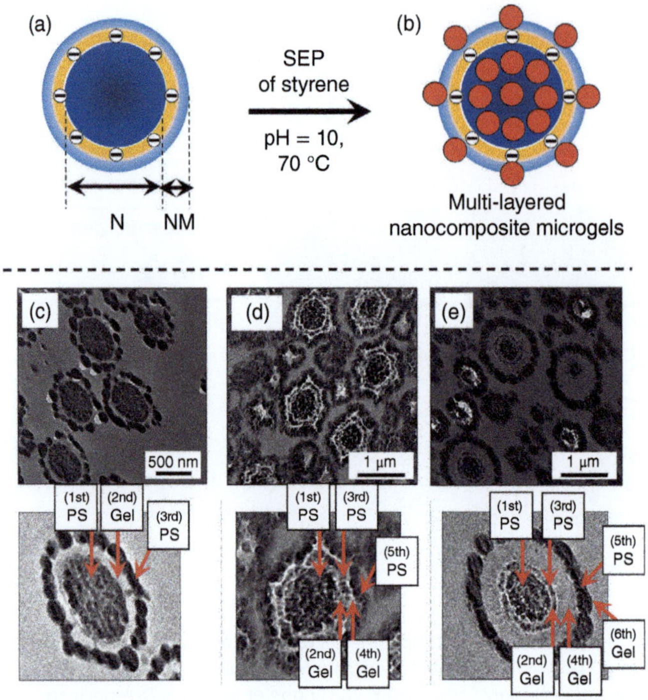

Figure 5.6 (a) Schematic illustration and (b–d) representative results of the multilayered nanocomposite microgels prepared by the SEP of styrene in the presence of core–shell microgels. N, polyNIPAm hydrogels; NM, poly(NIPAm-*co*-MAc) hydrogels; PS, polystyrene. Source: Reproduced with permission from Watanabe et al. [68]/John Wiley & Sons, Inc.

suggest that novel multi-layered nanocomposite microgels can be synthesized via the SEP of styrene, given that the polymerization of the styrene monomer occurs only in the uncharged (neutral) region of the microgels. In addition, the 3D distribution of charge moieties within the microgels was revealed based on the special distribution of the polystyrene nanoparticles that were used as probes for the microscopic characterization. The observed multi-layered nanocomposite structure suggests that the multi-layer that consists of charged groups is generated within the microgels by seeded precipitation polymerization. This study has thus not only led to the formation of novel multi-layered nanocomposite microgels, but also to the characterization of pH-responsive microgels, which promise potential in other colloidal applications.

5.7 Characterization of Nanocomposite Microgels

As the properties of microgels such as interfacial activity are correlated with their nanostructures, the determination of the 3D nanocomposite structures, including the inner structure of the microspheres, is needed. Like microgels, solid or metal nanocomposite microgels are deformed in the dried state and by attachment to air/water and solid/water interfaces [69, 70]. Conventional electron microscopy methods such as scanning electron microscopy (SEM) and transmission electron microscopy (TEM) typically require a drying process during observation (or sample preparation), which causes deformation of the nanocomposite microgels [68]. Even though some characterization methods, including scattering approaches [71–73], cryo-SEM (TEM) [74–76], *in situ* AFM [70, 77, 78], and super-resolution microscopy [79], can provide structural information on swollen microgels, the thus-obtained structural information is limited to the 3D structure of the interior of these nanocomposite microgels. More recently, some advanced characterization methods have revealed the detailed 3D nanostructure of microgels in the swollen state; however, they require pretreatment processes such as the immobilization of fluorescent molecules. Therefore, to develop nanocomposite microgels with enhanced properties, both functionalized microgels with well-defined structures and high-quality methods to characterize their structure are needed.

In this context, cryo-electron tomography (Cryo-ET) techniques [80, 81] could become a strong characterization method to accurately determine the 3D structure of solid–hydrogel nanocomposite microgels in the swollen state [67, 68]. In contrast to the aforementioned conventional characterization techniques, cryo-ET permits the determination of the morphology of nanocomposite microgels in the swollen (iced) state [67], which provides raw structural information on nanocomposite microgels in a hydrated state, while pre-staining the microgels is not required. Moreover, using this technique, the tomographic reconstruction of 61 projections collected over a tilt-angle range from −60° to +60° can provide detailed 3D structural information on nanocomposite microgels [67, 68, 80, 81]. Ultra-thin cross-sectional images and cryo-ET images of nanocomposite microgels prepared by the SEP of styrene in the presence of pNIPAm microgels and the surfactant SDS are shown in Figure 5.7. As shown in Figure 5.7a, the detailed structure of the nanocomposites could not be determined from the ultra-thin cross-sectional images given that the nanocomposite microgels were dehydrated during sample preparation, resulting in high deformation in the resin. However, cryo-ET measurements showed the morphology of the nanocomposite microgels and provided 3D segmentation images of the nanocomposite microgels, which revealed the existence of free space inside the nanocomposite microgels. Moreover, based on the cryo-ET analysis, the number of polystyrene particles inside the nanocomposite microgels could be determined [68].

Taking advantage of the high resolution of cryo-ET, the obtained segmentation images were used to provide new insight into the polymerization mechanism of the SEP of styrene in the presence of microgels. When the SEP of styrene was conducted using a styrene concentration of 300 mM in the presence of poly(NIPAm-*co*-FAc) microgels, the interior of the parent microgels was occupied by polystyrene (Figure 5.5b). The segmentation images also revealed that there was no free space inside the nanocomposite microgels (Figure 5.7b). Conversely, the space was observed inside the nanocomposite microgels prepared by SEP with SDS (Figure 5.7c), even though a large amount of polystyrene was formed within the microgels. The polystyrene nanoparticles were immobilized individually inside

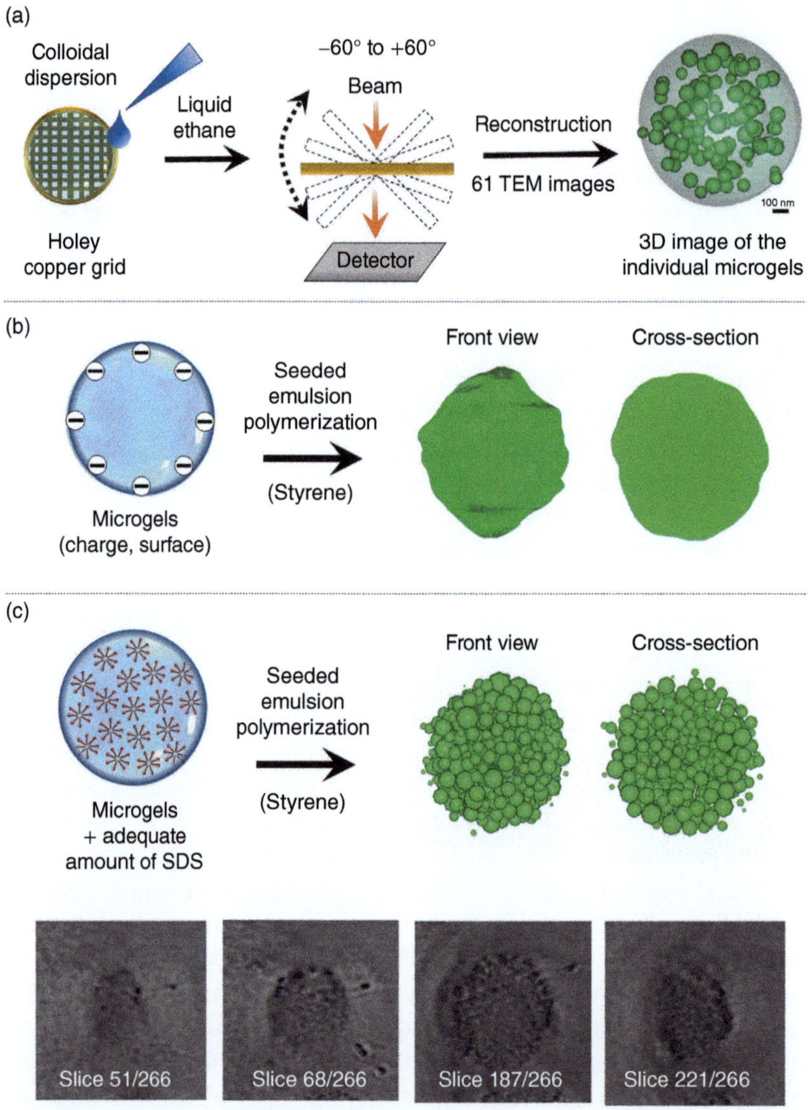

Figure 5.7 (a) Schematic illustration of the cryo-ET technique. Selected cryo-TEM tilt images and segmentation images of the nanocomposite microgels prepared by the SEP of styrene (b) without SDS and (c) with SDS. Source: Watanabe et al. [68]/with permission of John Wiley & Sons.

microgels and did not undergo fusion with one another. These results, in combination with the fact that the polystyrene particles generated within the pNIPAm microgels avoid the charged regions of the microgels, suggest that the internal space within the nanocomposite microgels determined by cryo-ET most likely corresponds to the presence of the charged moieties of SDS. It seems thus feasible to assume that the hydrophobic monomers are able to recognize microenvironments in the hydrogel microspheres at the molecular scale during the free-radical SEP [67].

5.8 Applications of Nanocomposite Microgels

Nanocomposite microgels may potentially find colloidal applications based on their non spherical shapes and hydrophobicity. In the present chapter, the emulsion properties of nanocomposite microgels will be discussed based on an examination of an emulsion stabilized by nanocomposite microgels prepared by the SEP of styrene [66–68].

Emulsions stabilized by colloidal particles [82, 83] have attracted substantial attention due to a variety of applications based on their fascinating properties, which include high adsorption energies of colloidal particles at oil/water interfaces ($\sim 10^2$–10^6 k_BT) that increase the long-term stability of such emulsions [84–86]. Especially microgel-stabilized emulsions (MS-Es) exhibit several advantages compared to conventional Pickering emulsions that are stabilized by solid microspheres (e.g. polystyrene or silica particles) [87–90]. For example, compared to conventional solid-particle-stabilized emulsions, the synthesis of MS-Es requires a lower energy input given that the microgels have a higher interfacial activity than solid microspheres [94–96]. Moreover, the stimuli responsivity of the microgels allows demulsifying remarkably stable MS-Es by changing the temperature and/or the pH value [91–93]. Furthermore, MS-Es can be employed as scaffolds for porous bulk materials given that MS-Es maintain, similar to microgels, high colloidal stability under high filling rates [87, 97]. However, MS-Es typically provide oil-in-water (O/W) emulsions, which are not suitable for some applications. In general, the emulsion type (W/O or O/W) is determined by particle wettability [98, 99]. When the particles display better affinity with the oil phase than the water phase, W/O emulsions will be formed in theory [99]. Since microgels typically show hydrophilic properties, it is difficult to realize O/W emulsions stabilized by microgels.

In this context, nanocomposite microgels, such as the composites shown in Figure 5.5, allow the preparation of W/O emulsions with MS-E properties due to the individual hydrophobic regions of the nanocomposite microgels [100]. Specifically, hydrophobized nanocomposite microgels generate W/O emulsions with different types of oils (including non-polar oils), which usually do not engage in the formation of W/O emulsions, unless the parent microgels have been subject to a hydrophobic treatment (Figure 5.8a). For instance, when an aqueous solution of polystyrene-poly(NIPAm-*co*-FAc) nanocomposite microgels and cyclohexane was prepared using a homogenizer, a stable particle-stabilized emulsion was obtained (Figure 5.8c), even though the parent microgels afford O/W emulsions (Figure 5.8b). Optical microscopy images clearly showed that the emulsion droplets were covered with nanocomposite microgels. The fluorescence microscopy image shown in Figure 5.8b,c revealed that the obtained particle-stabilized emulsion was a W/O emulsion, unlike typical MS-Es. However, O/W emulsions were obtained when 1-pentanol was used as the oil phase, indicating that the polarity of the oil phase influenced the obtained emulsion type even when the nanocomposite microgels were used as the stabilizer for the emulsion. Moreover, the volume fraction of oil used during emulsification influences the resulting emulsion type (Figure 5.8d). The type of emulsion can be controlled by changing the volume fraction of water

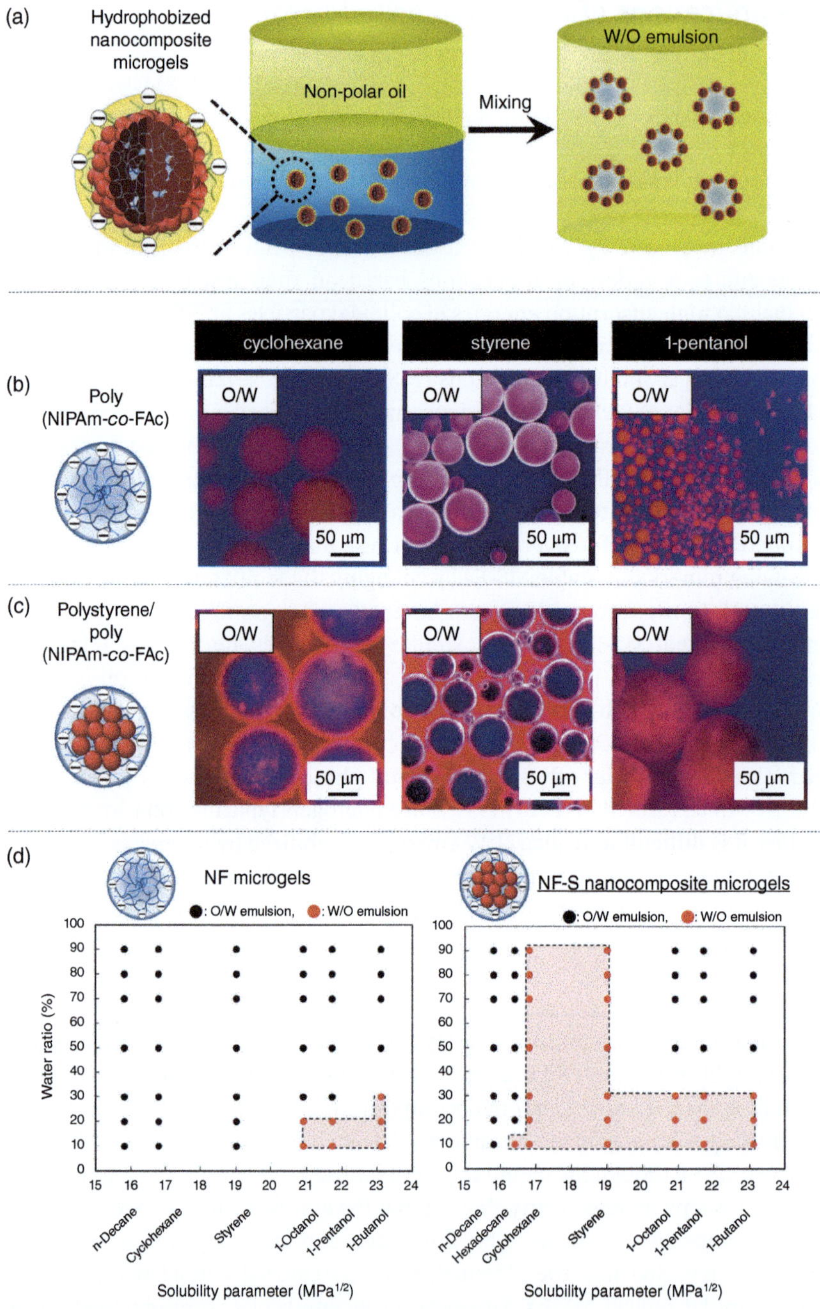

Figure 5.8 (a) Schematic illustration of emulsions stabilized by nanocomposite microgels. Typical fluorescence microscopy images of (b) microgel- and (c) nanocomposite-microgel-stabilized emulsions. (d) Phase diagram of the emulsion type (W/O or O/W) and water mixing ratio during emulsification using microgels and nanocomposite microgels as emulsifiers. Source: Watanabe et al. [100]/with permission of Royal Society of Chemistry.

relative to that of the polar oil (e.g. octanol). By constructing composite structures that comprise polystyrene and microgels, the range of preparation conditions under which W/O emulsions can be obtained has been expanded. Most importantly, nanocomposite microgels obtained from the SEP of styrene in the presence of microgels afford W/O emulsions with a wide range of oils (polar and non-polar oils) [100].

In addition to their hydrophobicity, the surface roughness of nanocomposite microgels also influences the emulsion type. When spherical polystyrene-core/poly (NIPAm-*co*-FAc)-hydrogel-shell core–shell microgels were used as a stabilizer for cyclohexane, O/W emulsions were formed. In this case, the shell thickness of the core–shell microgels was very thin (several nm as determined by dynamic light scattering). Conversely, polystyrene nanocomposite microgels provided W/O emulsions regardless of the preparation conditions (e.g. stirring speed or oil fraction). These results indicate that the surface roughness of the nanocomposite microgels should play an important role for achieving efficient contact between the microgel stabilizer and the water/oil interface [100]. The pinning effect, which was observed with non-spherical-shaped silica microspheres [101], may not be effective in this study due to the existence of the hydrogel layers on the surface of the nanocomposites.

5.9 Summary and Perspective

In this chapter, the effect of various microgel properties on the morphology of the nanocomposite structure was examined based on systematic emulsion polymerizations using a variety of microgels. The fabricated nanocomposites offer a variety of benefits relative to hydrated water-swollen microgels, such as the possibility to copolymerize functionalized vinyl monomers and to construct core–shell structures. Apart from the carboxyl group, amino group, and cross-linkers, hydrophobic components such as polystyrene can be incorporated into well-defined domains in the microgels by using SEP in the presence of microgels. This synthetic versatility offers a range of options to control the morphology of microgels. For example, the SEP of styrene in the presence of pNIPAm microgels furnishes raspberry-shaped nanocomposite microgels. The internal structure of the nanocomposite microgels can be controlled by changing the spatial distribution of charged groups within the microgels, which results in the formation of multi-layered nanocomposite microgels. The size of the polystyrene nanoparticles incorporated within the microgels can be controlled by the addition of surfactant or by changing the monomer concentration during the SEP.

But even though structural control of the nanocomposite microgels has been achieved, some challenges remain in this field. Although the discussed results clearly indicate that pNIPAm-based microgels can serve as polymerization sites for hydrophobic monomers such as styrene, it is not clear to what extent the gel microparticles provide a polymerization site for other hydrophobic monomers. However, as microgels provide a polymerization site for relatively hydrophilic monomers such as GMA and methyl methacrylate, the range of monomers applicable to SEP in the presence of microgels should be wide. For instance, if a functional polymer such as poly(2-methoxyethyl acrylate) could be immobilized in microgels [102], the resulting nanocomposite microgels could be used as high-quality separate carriers for dyes and bio-coatings.

Moreover, emulsions stabilized by multi-layered nanocomposite microgels can be expected to exhibit fascinating properties given that the microgel softness can be controlled not only by changing the cross-linking density, but also by varying the number of layers of the nanocomposite microgels. An optimization of the surface roughness and softness of the entire microgels by synthesizing nanocomposite structures with hydrophobic polymers can be expected to improve the properties (e.g. long-term stability and stimuli-responsive kinetics) of particle-stabilized emulsions. Nanocomposite microgels may thus be expected to lead to the development of versatile functional microgels and polymeric materials related to microgels.

References

1. Alemán, J.V., Chadwick, A.V., He, J. et al. (2007). Definitions of terms relating to the structure and processing of sols, gels, networks, and inorganic-organic hybrid materials (IUPAC Reccomendations 2007). *Pure Appl. Chem.* 79: 1801–1829.
2. Ko, H. and Javey, A. (2017). Smart actuators and adhesives for reconfigurable matter. *Acc. Chem. Res.* 50: 691–702.
3. Tanaka, T. and Fillmore, D.J. (1979). Kinetics of swelling of gels. *J. Chem. Phys.* 70: 1214–1218.
4. Matsuda, T., Kawakami, R., Namda, R. et al. (2019). Mechanoresponsive self-growing hydrogels inspired by muscle training. *Science* 363: 504–508.
5. Kamata, H., Akagi, Y., Kariya, Y.K. et al. (2014). "Nonswellable" hydrogel without mechanical hysteresis. *Science* 343: 873–875.
6. Hirai, T., Maruyama, H., Suzuki, T., and Hayashi, S. (1992). Shape memorizing properties of a hydrogel of poly(vinyl alcohol). *Appl. Polym. Sci.* 45: 1849–1855.
7. Haraguchi, K. and Takehisa, T. (2002). Nanocomposite hydrogels: a unique organic–inorganic network structure with extraordicary mechanical, optical, and swelling/de-swelling properties. *Adv. Mater.* 14: 1120–1124.
8. Cha, C., Shin, S.R., Annabi, N. et al. (2013). Carbon-based nanomaterials: multifunctional materials for biomedical engineering. *ACS Nano* 7: 2891–2897.
9. Shin, S.R., Jung, S.M., Zalabany, M. et al. (2013). Carbon-nanotube-embedded hydrogel sheets for engineering cardiac constructs and bioactuators. *ACS Nano* 7: 2369–2380.
10. Liu, K., Pan, X., Chen, L. et al. (2018). Ultrasoft self-healing nanoparticle-hydrogel composites with conductive magnetic properties. *ACS Sustainable Chem. Eng.* 6: 6395–6403.

11 Chen, S., Tang, F., Tang, L., and Li, L. (2017). Synthesis of Cu-nanoparticle hydrogel with self-healing and photothermal properties. *ACS Appl. Mater. Interfaces* 9: 20895–20903.

12 Liao, J. and Huang, H. (2020). Review on magnetic natural polymer constructed hydrogels as vehicles for drug delivery. *Biomacromolecules* 21: 2574–2594.

13 Liu, Y., Xu, K., Chang, Q. et al. (2018). Highly flexible and resilient elastin hybrid cryogels with shape memory, injectability, conductivity, and magnetic responsive properties. *Adv. Mater.* 28: 7758–7767.

14 Zhang, J., Xu, S., and Kumacheva, E. (2004). Polymer microgels: reactors for semiconductor, metal, and magnetic nanoparticles. *J. Am. Chem. Soc.* 126: 7908–7914.

15 Liu, K., Pan, X., Chen, L. et al. (2018). Ultrasoft self-healing nanoparticle-hydrogel composites with conductive and magnetic properties. *ACS Sustainable Chem. Eng.* 6: 6395–6403.

16 Honda, K., Sazuka, Y., Iizuka, K. et al. (2019). Hydrogel microellipsoids that form robust string-like assemblies at the air/water interface. *Angew. Chem. Int. Ed.* 58: 7294–7298.

17 Plamper, F.A. and Richtering, W. (2017). Functional microgels and microgel systems. *Acc. Chem. Res.* 50: 131–140.

18 Karg, M., Pich, A., Hellweg, T. et al. (2019). Nanogels and microgels: from model colloids to applications, recent developments, and future trends. *Langmuir* 35: 6231–6255.

19 Pelton, R.H. and Chibante, P. (1986). Preparation of aqueous latices with *N*-isopropylacrylamide. *Colloids Surf.* 20: 247–256.

20 Pelton, R. (2000). Temperature-sensitive aqueous microgels. *Adv. Colloid Interface Sci.* 85: 1–33.

21 Li, Z., Kwok, M.H., and Ngai, T. (2012). Preparation of responsive micrometer-sized microgel particles with a highly functionalized shell. *Macromol. Rapid Commun.* 33: 419–425.

22 Kwok, M., Li, Z., and Ngai, T. (2013). Controlling the synthesis and characterization of micrometer-sized PNIPAM microgels with tailored morphologies. *Langmuir* 29: 9581–9591.

23 Meng, Z., Smith, M.H., and Lyon, L.A. (2009). Temperature-programmed synthesis of micron-sized multi-responsive microgels. *Colloid Polym. Sci.* 287: 277–285.

24 Andersson, M. and Maunu, S.L. (2006). Structural studies of poly(*N*-isopropylacrylamide) microgels: effect of SDS surfactant concentration in the microgel synthesis. *J. Polym. Sci., Part B: Polym. Phys.* 44: 3305–3314.

25 Contreras-Cáceres, R., Schellkopf, L., Fernández-López, C. et al. (2015). Effect of the cross-linking density on the thermoresponsive behavior of hollow pNIPAM microgels. *Langmuir* 31: 1142–1149.

26 Acciaro, R., Gilányi, T., and Varga, I. (2011). Preparation of monodisperse poly(*N*-isopropylacrylamide) microgel particles with homogenous cross-link density distribution. *Langmuir* 27: 7917–7925.

27 Hoare, T. and McLean, D. (2006). Kinetic prediction of functional group distributions in thermosensitive microgels. *J. Phys. Chem. B* 110: 20327–20336.

28 Hoare, T. and Pelton, R. (2004). Highly pH and temperature responsive microgels functionalized with vinylacetic acid. *Macromolecules* 37: 2544–2550.

29 Hoare, T. and Pelton, R. (2006). Titrametric characterization of pH-induced phase transitions in functionalized microgels. *Langmuir* 22: 7342–7350.

30 Hoare, T. and Pelton, R. (2008). Characterizing charge and crosslinker distributions in polyelectrolyte microgels. *Curr. Opin. Colloid Interface Sci.* 13: 413–428.

31 Kawaguchi, H., Fujimoto, K., Saito, M. et al. (1993). Preparation and modification of monodisperse hydrogel microspheres. *Polym. Int.* 30: 225–231.

32 Guo, Z., Chen, Q., Gu, H. et al. (2018). Giant microgels with CO_2-induced on–off, selective, and recyclable adsorption for anionic dyes. *ACS Appl. Mater. Interfaces* 10: 38073–38083.

33 Duracher, D., Sauzedde, F., Elaissari, A. et al. (1998). Cationic amino-containing *N*-isopropyl-acrylamide-styrene copolymer latex particles: 1-particle size and morphology vs. polymerization process. *Colloid Polym. Sci.* 276: 219–231.

34 Suzuki, D., Sakai, T., and Yoshida, R. (2008). Self-flocculating/self-dispersing oscillation of microgels. *Angew. Chem. Int. Ed.* 47: 917–920.

35 Suzuki, D., Taniguchi, H., and Yoshida, R. (2009). Autonomously oscillating viscosity in microgel dispersions. *J. Am. Chem. Soc.* 131: 12058–12059.

36 Suzuki, D., Kobayashi, T., Yoshida, R., and Hirai, T. (2012). Soft actuators of organized self-oscillating microgels. *Soft Matter* 8: 11447–11449.

37 Zhang, Q., Zha, L., Ma, J., and Liang, B. (2007). Synthesis and characterization of novel, temperature-sensitive microgels based on *N*-isopropylacrylamide and *tert*-butyl acrylate. *J. Appl. Polym. Sci.* 103: 2962–2967.

38 Jones, C.D. and Lyon, L.A. (2000). Synthesis and characterization of multiresponsive core–shell microgels. *Macromolecules* 33: 8301–8306.

39 Jones, C.D. and Lyon, L.A. (2000). Synthesis and characterization of multiresponsive core–shell microgels. *Macromolecules* 33: 8301–8306.

40 Jones, C.D. and Lyon, L.A. (2003). Dependence of shell thickness on core compression in acrylic acid modified poly(*N*-isopropylacrylamide) core/shell microgels. *Langmuir* 19: 4544–4547.

41 Keerl, M., Pedersen, J.S., and Richtering, W. (2009). Temperature sensitive copolymer microgels with nanophase separated structure. *J. Am. Chem. Soc.* 131: 3093–3097.

42 Bradley, M., Vincent, B., and Burnett, G. (2007). Uptake and release of anionic surfactant into and from cationic core–shell microgel particles. *Langmuir* 23: 9237–9241.

43 Hu, X., Tong, Z., and Lyon, L.A. (2011). One-pot synthesis of microcapsules with nanoscale inclusions. *Macromol. Rapid Commun.* 32: 1461–1466.

44 Smith, M.H. and Lyon, L.A. (2011). Multifunctional nanogels for siRNA delivery. *Acc. Chem. Res.* 45: 985–993.

45 Hellweg, T. (2013). Responsive core–shell microgels: synthesis, characterization, and possible applications. *J. Polym. Sci., Part B: Polym. Phys.* 51: 1073–1083.

46 Saxena, S., Hansen, C.E., and Lyon, L.A. (2014). Microgel mechanics in biomaterial design. *Acc. Chem. Res.* 47: 2426–2434.

47 Dubbert, J., Nothdurft, K., Karg, M., and Richtering, W. (2015). Core–shell–shell and hollow double-shell microgels with advanced temperature responsiveness. *Macromol. Rapid Commun.* 36: 159–164.

48 Clarke, K.C., Dunham, S.N., and Lyon, L.A. (2015). Core/shell microgels decouple the pH and temperature responsivities of microgel films. *Chem. Mater.* 27: 1391–1396.

49 Harkins, W.D. (1947). A general theory of the mechanism of emulsion polymerization. *J. Am. Chem. Soc.* 69: 1428–1444.

50 Smith, W.V. and Ewart, R.H. (1948). Kinetics of emulsion polymerization. *J. Chem. Phys.* 16: 592–599.

51 Chern, C.S. (2006). Emulsion polymerization mechanisms and kinetics. *Prog. Polym. Sci.* 31: 443–486.

52 Thickett, S.C. and Gilbert, R.G. (2007). Emulsion polymerization: state of the art in kinetics and mechanisms. *Polymer* 48: 6965–6991.

53 Mock, E.B., De Bruyn, H., Hawkett, B.S. et al. (2006). Synthesis of anisotropic nanoparticles by seeded emulsion polymerization. *Langmuir* 22: 4037–4043.

54 Liu, Y., Ma, Y., Liu, L., and Yang, W. (2015). Facile synthesis of core–shell/hollow anisotropic particles via control of cross-linking during one-pot dispersion polymerization. *J. Colloid Interface Sci.* 445: 268–276.

55 Minami, H., Wang, Z., Yamashita, T., and Okubo, M. (2003). Thermodynamic analysis of the morphology of monomer-adsorbed, cross-linked polymer particles prepared by the dynamic swelling method and seeded polymerization. *Colloid Polym. Sci.* 281: 246–252.

56 Tu, F. and Lee, D. (2014). Shape-charging and amphiphilicity-reversing Janus particles with pH-responsive surfactant properties. *J. Am. Chem. Soc.* 136: 9999–10006.

57 Okubo, M., Kanaida, K., and Matsumoto, T. (1987). Production of anomalously shaped carboxylated polymer particles by seeded emulsion polymerization. *Colloid Polym. Sci.* 265: 876–881.

58 Okubo, M. and Minami, H. (1996). Control of hollow size of micron-sized monodispersed polymer particles having a hollow structure. *Colloid Polym. Sci.* 274: 433–438.

59 Li, B., Xu, Y., Wang, M., and Ge, X. (2013). Morphological control of multihollow polymer latex particles through controlled phase separation in the seeded emulsion polymerization. *Langmuir* 29: 14787–14794.

60 Duracher, D., Sauzedde, F., Elaissari, A. et al. (1998). Cationic amino-containing *N*-isopropyl-acrylamide-styrene copolymer latex particles: 1-particle size and morphology vs. polymerization process. *Colloid Polym. Sci.* 276: 219–231.

61 Kraft, D.J., Hilhorst, J., Heinen, M.A.P. et al. (2011). Patchy polymer colloids with tunable anisotropy dimensions. *J. Phys. Chem. B* 115: 7175–7181.

62 Suzuki, D., Yamagata, T., and Murai, M. (2013). Multilayered composite microgels synthesized by surfactant-free seeded polymerization. *Langmuir* 29: 10579–10585.

63 Suzuki, D. and Kobayashi, C. (2014). Raspberry-shaped composite microgel synthesis by seeded emulsion polymerization with hydrogel particles. *Langmuir* 30: 7085–7092.

64 Kobayashi, C., Watanabe, T., Murata, K. et al. (2016). Localization of polystyrene particles on the surface of poly(*N*-isopropylacrylamide-*co*-methacrylic acid) microgels prepared by seeded emulsion polymerization of styrene. *Langmuir* 32: 1429–1439.

65 Tam, K.C., Ragaram, S., and Pelton, R.H. (1994). Interaction of surfactants with poly(*N*-isopropylacrylamide) microgel latexes. *Langmuir* 10: 418–422.

66 Watanabe, T., Kobayashi, C., Song, C. et al. (2016). Impact of spatial distribution of charged groups in core poly(*N*-isopropylacrylamide)-based microgels on the resultant composite structures prepared by seeded emulsion polymerization of styrene. *Langmuir* 32: 12760–12773.

67 Watanabe, T., Song, C., Murata, K. et al. (2018). Seeded emulsion polymerization of styrene in the presence of water-swollen hydrogel microspheres. *Langmuir* 34: 8571–8580.

68 Watanabe, T., Nishizawa, Y., Minato, H. et al. (2021). Hydrophobic monomers recognize microenvironments in hydrogel microspheres during free-radical-seeded emulsion polymerization. *Angew. Chem. Int. Ed.* 132: 8934–8938.

69 Minato, H., Murai, M., Watanabe, T. et al. (2018). The deformation of hydrogel microspheres at the air/water interface. *Chem. Commun.* 54: 932–935.

70 Matsui, S., Kureha, T., Hiroshige, S. et al. (2017). Fast adsorption of soft hydrogel microspheres on solid surfaces in aqueous solution. *Angew. Chem. Int. Ed.* 129: 12314–12317.

71 Wong, J.E. and Richtering, W. (2008). Layer-by-layer assembly on stimuli-responsive microgels. *Curr. Opin. Colloid Interface Sci.* 13: 403–412.

72 Nickel, A.C., Scotti, A., Houston, J.E. et al. (2019). Anisotropic hollow microgels that can adapt their size, shape, and softness. *Nano Lett.* 19: 8161–8170.

73 Scotti, A., Bochenek, S., Brugnoni, M. et al. (2019). Exploring the colloid-to-polymer transition for ultra-low crosslinked microgels from three to two dimensions. *Nat. Commun.* 10: 1–8.

74 Crassous, J.J., Ballauff, M., Drechsler, M. et al. (2006). Imaging the volume transition in thermosensitive core–shell particles by cryo-transmission electron microscopy. *Langmuir* 22: 2403–2406.

75 Destribats, M., Lapeyre, V., Wolfs, M. et al. (2011). Soft microgels as Pickering emulsion stabilisers: role of particle deformability. *Soft Matter* 7: 7689.

76 Mihut, A.M., Stenqvist, B., Lund, M. et al. (2017). Assembling oppositely charged lock and key responsive colloids: a mesoscale analog of adaptive chemistry. *Sci. Adv.* 3: e1700321.

77 Nishizawa, Y., Matsui, S., Urayama, K. et al. (2019). Non-thermoresponsive decanano-sized domains in thermoresponsive hydrogel microspheres revealed by temperature-controlled high-speed atomic force microscopy. *Angew. Chem. Int. Ed.* 58: 8809–8813.

78 Nishizawa, Y., Minato, H., Inui, T. et al. (2021). Nanostructure and thermoresponsiveness of poly(*N*-isopropyl methacrylamide)-based hydrogel microspheres prepared via aqueous free radical precipitation polymerization. *RSC Adv.* 11: 13130–13137.

79 Bergmann, S., Wrede, O., Huser, T., and Hellweg, T. (2018). Super-resolution optical microscopy resolves network morphology of smart colloidal microgels. *Phys. Chem. Chem. Phys.* 20: 5074–5083.

80 Dubochet, J., Adrian, M., Chang, J.-J. et al. (1988). Cryo-electron microscopy of vitrified specimens. *Q. Rev. Biophys.* 21: 129–228.

81 Kremer, J.R., Mastronarde, D.N., and Mcintosh, J.R. (1996). Computer visualization of three-dimensional image data using IMOD. *J. Struct. Biol.* 116: 71–76.

82 Ramsden, W. (1903). Separation of solids in the surface-layers of solutions and 'suspensions' (observations on surface-membranes, bubbles, emulsions, and mechanical coagulation).—Preliminary account. *Proc. R. Soc. London* 72: 156–164.

83 Pickering, S.U. (1907). CXCVI.—Emulsions. *J. Chem. Soc. Trans.* 91: 2001–2021.

84 Iwasaki, Y., Takahata, Y., and Fujii, S. (2015). Self-setting particle-stabilized emulsion for hard-tissue engineering. *Colloids Surf., B* 126: 394–400.

85 Wu, J. and Ma, G.H. (2016). Recent studies of Pickering emulsions: particles make the difference. *Small* 12: 4633–4648.

86 Binks, B.P. and Olusanya, S.O. (2017). Pickering emulsions stabilized by coloured organic pigment particles. *Chem. Sci.* 8: 708–723.

87 Richtering, W. (2012). Responsive emulsions stabilized by stimuli-sensitive microgels: emulsions with special non-Pickering properties. *Langmuir* 28: 17218–17229.

88 Li, Z. and Ngai, T. (2013). Microgel particles at the fluid–fluid interfaces. *Nanoscale* 5: 1399–1410.

89 Schmitt, V. and Ravaine, V. (2013). Surface compaction versus stretching in Pickering emulsions stabilised by microgels. *Curr. Opin. Colloid Interface Sci.* 18: 532–541.

90 French, D.J., Brown, A.T., Schofield, A.B. et al. (2016). The secret life of Pickering emulsions: particle exchange revealed using two colours of particle. *Sci. Rep.* 6: 31401.

91 Ngai, T., Behrens, S.H., and Auweter, H. (2005). Novel emulsions stabilized by pH and temperature sensitive microgels. *Chem. Commun.* 331–333.

92 Fujii, S., Read, E.S., Binks, B.P., and Armes, S.P. (2005). Stimulus-responsive emulsifiers based on nanocomposite microgel particles. *Adv. Mater.* 17: 1014–1018.

93 Tsuji, S. and Kawaguchi, H. (2008). Thermosensitive Pickering emulsion stabilized by poly(*N*-isopropylacrylamide)-carrying particles. *Langmuir* 24: 3300–3305.

94 Destribats, M., Lapeyre, V., Sellier, E. et al. (2011). Water-in-oil emulsions stabilized by water-dispersible poly(*N*-isopropylacrylamide) microgels. *Langmuir* 27: 14096–14107.

95 Buchcic, C., Tromp, R.H., Meinders, M.B.J., and Stuart, M.A.C. (2017). Harnessing the advantages of hard and soft colloids by the use of core–shell particles as interfacial stabilizers. *Soft Matter* 13: 1326–1334.

96 Wiese, S., Tsvetkova, Y., Daleiden, N.J.E. et al. (2016). Microgel stabilized emulsions: breaking on demand. *Colloids Surf., A* 495: 193–199.

97 Li, Z., Ming, T., Wang, J., and Ngai, T. (2009). High internal phase emulsions stabilized solely by microgel particles. *Angew. Chem. Int. Ed.* 45: 8642–8645.

98 Finkle, P., Draper, H.D., and Hildebrand, J.H. (1923). The theory of emulsification 1. *J. Am. Chem. Soc.* 45: 2780–2788.

99 Destribats, M., Gineste, S., Laurichesse, E. et al. (2014). Pickering emulsions: what are the main parameters determining the emulsion type and interfacial properties? *Langmuir* 30: 9313–9326.

100 Watanabe, T., Takizawa, M., Jiang, H. et al. (2019). Hydrophobized nanocomposite hydrogel microspheres as particulate stabilizers for water-in-oil emulsions. *Chem. Commun.* 55: 5990–5993.

101 Zanini, M., Marschelke, C., Anachkov, S.E. et al. (2017). Universal emulsion stabilization from the arrested adsorption of rough particles at liquid–liquid interfaces. *Nat. Commun.* 8: 1–9.

102 Kureha, T., Nishizawa, Y., and Suzuki, D. (2017). Controlled separation and release of organoiodine compounds using poly(2-methoxyethyl acrylate)-analogue microspheres. *ACS Omega* 2: 7686–7694.

6

Compressible Microgels in Concentrated Suspensions: Phase Behavior, Flow Properties, and Scattering Techniques to Probe Their Structure and Dynamics

A. Scotti[1], U. Gasser[2], B. Zhou[2], A. Arenas-Gullo[3,5], A. de la Cotte[3], J. Rojo González[3,5], and A. Fernandez-Nieves[3,4,5]

[1]*RWTH Aachen University, Institute of Physical Chemistry, 52056 Aachen, Germany*
[2]*Paul Scherrer Institut, Laboratory for Neutron Scattering and Imaging, Forschungsstrasse 111, 5232 Villigen, Switzerland*
[3]*University of Barcelona, Department of Condensed Matter Physics, 08028 Barcelona, Spain*
[4]*ICREA-Institucio Catalana de Recerca i Estudis Avancats, 08010 Barcelona, Spain*
[5]*University of Barcelona, Institute of Complex Systems (UBICS), Martí i Franquès, 1-1. 08028 Barcelona, Spain*

6.1 Introduction

Microgels and macrogels are crosslinked polymer networks that, depending on the solvent quality, can be in a swollen or deswollen state. They both continue to have as much significant interest as they did since their creation [1–7]. A significant figure in the field is Toyoichi Tanaka. His work led to a much better understanding of the physics of gels [8]. Both microgels and macrogels have a lot in common. They are stimuli-sensitive in nature and exhibit fascinating behavior in connection with their volume-phase transition. However, they also exhibit significant differences. For example, while macrogels can take hours or days to equilibrate with their surroundings [9], for microgels, this can happen within a fraction of a second [10, 11]; this is an advantage in many cases.

Microgels are in the colloidal-size domain [12]. Their degree of swelling, which is related to the amount of solvent in the polymer network, is influenced by changes in external stimuli, such as temperature [13–15], pH [16, 17], and hydrostatic pressure [18–20]. Typically, a change in external stimuli modifies the solvent–polymer interaction leading to a coil-to-globule transition of the polymer composing the particles. Consequently, microgels exhibit a significant change in volume. The associated volume-phase transition is described by the Flory–Rehner theory of swelling [21–23]. Even in bad-solvent conditions, microgels still contain a significant amount of solvent, $\approx 40\%$ of their volume [24]; this microgel state is commonly

(Dated: May 23, 2022)

Smart Stimuli-Responsive Polymers, Films, and Gels, First Edition.
Edited by Liang Hu, Yongfeng Gao, and Michael J. Serpe.

referred to as deswollen. In contrast, the collapsed state corresponds to the state of the microgel where no solvent is present. The dimensions of microgels in the collapsed state can be determined from viscosity data [25, 26].

The most common method to obtain microgels with radii smaller than 400 nm and with a narrow size distribution is precipitation polymerization [27]. In this type of synthesis, the temperature is kept well above the lower critical solution temperature of the polymer. Therefore, the particles are synthesized under bad-solvent conditions, which imply that the equilibrium state of the polymer network is the deswollen state. Once synthesized, the balance between the solvent–polymer mixing, which tends to swell the microgels, the network elasticity which tends to deswell the microgels, and in the case of charged microgels, the counter-ion distribution in and out of the particle, determines the size of a microgel in suspension.

The swelling ratio, defined as the microgel radius in a good solvent divided by the radius of the same microgel in its deswollen state, provides a qualitative measure of the network elasticity. For thermoresponsive microgels – e.g. poly *N*-isopropylacrylamide (pNIPAM) with a lower-critical-solution temperature (LCST) of 32 °C in water [12, 14] – this can simply be obtained from the hydrodynamic radii measured at 20 and 40 °C using dynamic light scattering (DLS) in dilute conditions. Microgels synthesized with a low amount of crosslinker have a large swelling ratio, which implies that the polymer network has small resistance to swelling in a good solvent, i.e. they are softer than microgels synthesized with a higher amount of crosslinker [24]. Therefore, the microgel stiffness can be controlled both with external stimuli causing swelling or deswelling of the polymer network, or during the particle synthesis, by controlling the amount of crosslinker.

The compressibility of microgels in suspension has important consequences, particularly at high concentrations. It has been shown by both scattering [28–33] and microscopy techniques [34–37] that microgels can either change their shape maintaining their volume, e.g. by forming facets due to contacts with neighboring microgels, or isotropically deswell by reducing their volume while apparently maintaining their spherical shape. The compressibility of microgels is quantified by their bulk modulus, $K_p = -V\frac{\partial \pi}{\partial V}$. Once the suspension osmotic pressure $\pi > K_p$, microgels isotropically deswell. However, if this condition is reached at a concentration where microgels are already in direct contact, the particles will have deformed and faceted before deswelling [35].

The behavior discussed above highlights that in contrast to hard, incompressible colloids, the volume of a microgel in suspension changes with particle concentration. Their volume fraction $\phi = \frac{Nv}{V}$, with N the number of particles, v the volume of a single particle, and V the total suspension volume, depends on concentration through changes in v. Therefore, a generalized volume fraction is used both in simulations and experiments when discussing soft particles such as microgels [29, 30, 32, 33, 36–44]:

$$\zeta = \frac{Nv_0}{V} \tag{6.1}$$

where v_0 is the volume of the microgel in dilute suspension. This definition is useful, as it agrees with the true microgel volume fraction, ϕ, in the dilute limit. However, $\zeta > \phi$ when deswelling takes place. Thus, at high concentration, ζ provides the degree of overpacking of the suspension relative to what could be achieved for incompressible colloids.

One approach to determine ϕ is to experimentally measure the microgel radius, and thus its volume, as a function of ζ: $v(\zeta)$. This can be done in real space using super-resolved fluorescence microscopy (SRFM) [36, 37, 45–47] or in reciprocal space with small-angle neutron scattering (SANS) with contrast variation [28, 30–32, 41]. SRFM typically requires microgels that are immobilized during image acquisition. This has been demonstrated at relatively high concentrations [36, 37, 48] or with the microgels adsorbed and immobilized on a surface, for dilute or semidilute samples. This, however, leads to microgel deformation [49] and a change in elastic properties compared to microgels in bulk [49–52]. Furthermore, the addition of salt, typically mercaptoethylamine [36, 37, 48], to induce the blinking of the dye, limits the experimental conditions at which SRFM can be performed. For more details on this technique, we refer to reference [53]. Here, we will present the principles of scattering techniques, with a particular focus on experiments, and discuss cross-correlation DLS [54–56] and contrast variation SANS [26, 28–30, 32, 33, 41]; both techniques allow probing concentrated colloidal suspensions and have not been extensively discussed in the literature.

Another route to obtain $v(\zeta)$ is to consider the suspension osmotic pressure π. The relation $\pi(\zeta)$ can be directly measured using a membrane osmometer. Then, if via appropriate modeling, we can write down the expected value of π, which depends on ϕ, we can obtain the ϕ–ζ relation. For the case of ionic microgels, the suspension osmotic pressure is set by the counter-ions that, for entropic reasons, can escape the microgel electrostatic attraction. Cloitre and coworkers [57, 58] developed an equation that relates the measured osmotic pressure to the free counter-ions and the volume fraction ϕ occupied by the microgels in solution. This allows linking the measured $\pi(\zeta)$ with the calculated $\pi(\phi)$, to obtain the $\phi(\zeta)$ relation from which $v(\zeta)$ can be further obtained [42, 59].

It is important to note that the osmotic pressure of the suspension is not determined by the microgels *per se* but by the counter-ions originating from charged groups on the particle. These charges can be incorporated into the polymer network using charged monomers that are co-polymerized during the synthesis, e.g. NIPAM can be co-polymerized with methyl methacrylate (MMA), ethyl acrylate (EA), or butyl methacrylate (BA), to obtain so-called ionic microgels [60]. Noteworthy, also microgels synthesized without charged monomers contain charges in the particle periphery [12]. These charges originate from the initiator used in the synthesis, e.g. potassium or ammonium persulfate. Some of the counter-ions associated with these charges can leave the periphery of the microgel for entropic reasons, as they are bound to the microgel with energies comparable to the thermal energy k_BT [29, 42, 57–59]. Another key point is that the microgel bulk modulus increases with compression or deformation, as polymeric chains are rearranged within the microgel [44]. De Aguiar and coworkers [35] have shown that, depending on the microgel concentration in solution, two regimes can be distinguished: (i) right above random close packing, the work needed to deform a microgel, W_d, is smaller than the work required to deswell the microgel, W_s, and, therefore, faceting is observed; (ii) once microgels start to deform and rearrange the polymer within their volume, there is a critical concentration for which $W_s < W_d$ and deswelling is more favorable. The deformation of microgels and the successive isotropic deswelling has been observed using super-resolved fluorescent microscopy [36] and used to explain the behavior of the storage and loss moduli, G' and G'', of concentrated microgel suspensions. It is suggested that the possibility to deform microgels at high concentration opens new pathways for energy dissipation that can explain the significant increase in the shear modulus G'' with sample concentration, while the elastic modulus G' remains almost unchanged [37]. Molecular dynamics [61] and dissipative particle dynamics simulations [62] can reproduce the observed faceting of microgels with increasing concentration.

However, since microgels deswell when the suspension osmotic pressure $\pi > K_p$, microgels can deswell before faceting and even before making contact with neighboring microgels. This phenomenon can have dramatic consequences when microgels with different bulk moduli are mixed together. For instance, Iyer and Lyon observed that when a few large microgels are mixed with a majority of smaller microgels, the former spontaneously deswell and fit into the crystalline lattice of the smaller microgels without giving rise to point defects in the crystal [34]. Even microgels consisting of a charge-neutral polymer network, such as pNIPAM, are surrounded by counter-ion clouds [12]; of all the counter-ions, those that are not bound to the particle are free to explore the suspension and contribute to its osmotic pressure. In the dilute state, this pressure is the same inside and outside the microgels and, therefore, does not cause microgel deswelling. However, as the suspension concentration is increased, the counter-ion clouds percolate in the space between the microgels and all the initially bound counter-ions become effectively free to explore the space between the particles without leaving the percolated ion clouds. This percolation, therefore, increases the osmotic pressure outside but not inside the microgels, leading to deswelling when the pressure outside exceeds the microgel bulk modulus. The combination of scattering techniques with measurements of the solution osmotic pressure has shown that, when the percolation threshold is reached, microgels with smaller bulk modulus deswell first. As the largest microgels are usually the softest, the polydispersity is reduced spontaneously [29]. With deswelling, the particle bulk modulus increases; the new equilibrium situation results from the balance between the osmotic pressure and the new bulk modulus [44]. In the literature, there are several studies that report microgel deswelling at concentrations well below random close packing, i.e. without direct contact of the microgels. This can be explained by this osmotic deswelling effect [15, 34, 63, 64]. If the microgels are in direct contact when this condition is reached, deswelling is accompanied by deformation [32]. A similar mechanism can be used to explain the observed presence of a jamming transition after the formation of a glassy state [58].

This isotropic deswelling is directly linked to the phase behavior of polydisperse microgel suspensions; the spontaneous reduction of polydispersity at high concentrations enables the suspension to crystallize in spite of the high polydispersity of the fully swollen microgels, which would suppress crystallization in hard colloids [38]. This mechanism affects solutions of microgels both with monodisperse or polydisperse (including bimodal) size distributions. In general, osmotic deswelling decreases the solution polydispersity as soon as microgels with different compressibility are mixed together in solution. For instance, hollow microgels, which are polymeric crosslinked networks with a solvent-filled cavity in their center, have shown pronounced isotropic deswelling before direct contact when they are embedded in a matrix of regular microgels [31, 65].

In analogy to hard colloids, suspensions of microgels show a first-order phase transition between liquid and crystal phases. As expected, a coexistence of the two phases is observed between the freezing point, ζ_f, and the melting point, ζ_m, above which the suspension is fully crystalline unless it is trapped in a glassy state. In hard spheres, the freezing and melting points are $\phi_f = 0.494$ and $\phi_m = 0.545$ [66, 67]. Due to the soft particle–particle interactions, the boundaries of the liquid-crystal phase transition are shifted to higher generalized volume fraction, $0.56 \lesssim \zeta_f \lesssim 0.59$ and $\zeta_m \approx 0.61$, respectively [38, 39, 64, 68]. This shift increases with microgel softness, with ζ_f and ζ_m taking values $\zeta \gtrsim 0.8$ [32, 42, 43, 69]. Such a shift of the phase boundaries cannot be explained only in terms of softer particle–particle interactions. It also depends on the fact that, at such high concentrations, microgels are compressed and/or deformed. The suspension, therefore, is at volume fractions smaller than ζ. In fact, the highest volume fractions accessible to hard spheres are random close packing, $\phi_{\text{rcp}} = 0.64$, in amorphous states, and hexagonal close packing, $\phi_{\text{hcp}} \approx 0.74$, in crystalline states. Therefore, the true volume fraction of microgels, $\phi(\zeta)$ is expected to be lower than ζ, when $\zeta \gtrsim \phi_{\text{rcp}}$. In this chapter, we will also discuss the effect of microgel compressibility on the phase behavior and flow properties of the suspension. We will then end by summarizing what we have presented and providing a brief outlook on what we see as future relevant work in the field of colloidal microgels.

6.2 Swelling Thermodynamics

Due to their stimuli-sensitive nature, microgels can swell and deswell in response to changes in temperature, solvent composition, *p*H, or ionic concentration. The first two factors alter the balance between repulsive and attractive interactions between the polymer network and the solvent [70, 71]. The presence of fixed charges in the microgel backbone and/or addition of salt introduces ions in the solvent, which depending on their distribution, can create a difference in osmotic pressure inside and outside the microgel, thereby affecting the particle size. The effects above are well described by the equation of state, $\Pi_{\text{in}} = \Pi_{\text{mix}} + \Pi_{\text{el}} + \Pi_{\text{ionic}} = \Pi_{\text{out}}$; each individual contribution to Π_{in} is described below. In equilibrium with pure solvent, $\Pi_{\text{out}} = 0$. In the next paragraphs, we discuss each term in the equation of state from a thermodynamic point of view. The goal is to provide a starting point to understand the intrinsic nature of microgels and what distinguishes them from other colloids.

6.2.1 Polymer/Solvent Mixing

Polymer/solvent mixing occurs by breaking solvent–solvent and polymer–polymer bonds which are replaced by solvent–polymer bonds. These bonds are characterized by the bond energies e_{ss}, e_{pp}, and e_{sp}, respectively. The creation of one solvent–polymer bond is, therefore, associated with an energetic change $\Delta e = e_{sp} - (e_{ss} + e_{pp})/2$. To quantify the energetic and entropic contributions to the mixing process, we follow Flory and consider a lattice with n_0 sites randomly filled with n_1 solvent molecules and n_2 polymer chains occupying x lattice sites each. With $n_0 = n_1 + xn_2$ [23], the number of possible microstates is $\Omega_{mix} = n_0!/(n_1!n_2!)$, each being equally probable for an isolated system. The entropic contribution to the solvent/polymer mixing then is [23, 72]:

$$\Delta S_{mix} = -k_B \ln \Omega_{mix} = -k_B \left(n_1 \ln \phi_1 + n_2 \ln \phi_2\right), \tag{6.2}$$

where the probabilities to find a site occupied by a solvent or a polymer molecule are given by the volume fractions $\phi_1 = n_1/n_0$ and $\phi_2 = xn_2/n_0$. Note that Stirling's approximation is used, $\ln x! \approx x(\ln x - 1)$. Note also this is similar to the mixing of solvents, except for the fact that here n_2 represents the number of polymer chains and not the number of molecules of the second solvent.

We now define the probability that a solvent molecule can form a bond with a polymer chain as $z\phi_2$, with z the coordination number of the lattice. The total associated energy change is:

$$\Delta U_{mix} = zn_1\phi_2\Delta e = k_B T \chi_1 n_1 \phi_2 \tag{6.3}$$

with the Flory solvency parameter defined as $\chi_1 = z\Delta e/k_B T$. The mixing normally happens at fixed temperature T and pressure p. Therefore, it is best described in terms of the Gibbs free energy. The free energy change is $\Delta G_{mix} = \Delta U_{mix} - T\Delta S_{mix} - p\Delta V$. The change in volume, ΔV, is usually negligible, and thus we can obtain the mixing contribution to the osmotic pressure as

$$\Pi_{mix} = -\frac{1}{v_s}\left(\frac{\partial \Delta G_{mix}}{\partial n_1}\right)_{n_2,p,T} \approx -\frac{k_B T}{v_s}\left[\phi_2 + \ln(1-\phi_2) + \chi_1\phi_2^2\right] \tag{6.4}$$

where we have used the approximation of long polymer chains and assumed that $x \approx x + 1$. In this expression, v_s is the volume of a solvent molecule.

6.2.2 Elasticity

Due to crosslinks in the gel network, microgels cannot dissolve and have a tendency to relax toward the state where the network was formed. This elastic behavior greatly contributes to the microgel free energy. We start by thinking of the microgel as a number of crosslinked polymer chains that are all distorted in the same statistical manner. If L is the deformation and f is the resulting restoring force, we can express the free energy as $dF = -SdT - pdV + fdL$, where $f = \left(\frac{\partial F}{\partial L}\right)_{T,V} = \left(\frac{\partial U}{\partial L}\right)_{T,V} - T\left(\frac{\partial S}{\partial L}\right)_{T,V}$. We see that the restoring force has an energetic and an entropic contribution. For microgels in suspension, the entropic contribution is much larger than the energetic contribution [73]. The change in entropy for a uniform deformation was calculated by Flory and Rehner as [21, 73]

$$\Delta S_{\text{el}} = S_{\text{el}}^{\text{deformed}} - S_{\text{el}}^{\text{undeformed}} = -\frac{3k_B N_c}{2}[\alpha^2 - 1 - \ln\alpha] \tag{6.5}$$

where N_c is the number of chains and $\alpha = (V/V_0)^{1/3}$ is the swelling degree with V and V_0 the deformed and relaxed volumes, respectively. Equation (6.5) is valid under the assumption that the polymer is random and that, as a result, there are no interactions between the monomers. With the entropy change, we can calculate the variation in the Gibbs free energy associated with microgel elasticity:

$$\Delta G_{\text{el}} = -T\Delta S_{\text{el}} = \frac{3k_B T N_c}{2}[\alpha^2 - 1 - \ln\alpha] \tag{6.6}$$

From ΔG_{el}, we obtain the osmotic pressure change due to deformation by taking the derivative with respect to the number of solvent molecules n_1,

$$\Pi_{\text{el}} = -\frac{1}{v_s}\left(\frac{\partial \Delta G_{\text{el}}}{\partial \alpha}\right)_{T,P,n_2}\left(\frac{\partial \alpha}{\partial n_1}\right)_{T,P,n_2} = \frac{k_B T N_c}{V_0}\left[\frac{\phi_2}{2\phi_{2,0}} - \left(\frac{\phi_2}{\phi_{2,0}}\right)^{1/3}\right] \tag{6.7}$$

where $\phi_{2,0}$ is the polymer volume fraction in the relaxed state, which usually is the deswollen state of the microgel, as they are most frequently synthesized under bad-solvent conditions.

6.2.3 Ionic Effects

Microgels can be classified as neutral and ionic. The counter-ions originating from the charged groups in ionic microgels form a gas with a pressure that needs to be considered when discussing the swelling behavior of the microgel. As a first approximation, we assume that the counter-ion density is sufficiently low so that the gas can be approximated as ideal. When we consider a microgel with N_c crosslinks and N monomers between crosslinks, the monomer density is $c \approx \frac{NN_c}{v}$, where v is the microgel volume. If f is then the fraction of charged monomers, the number of counterions in the particle is $fcv \approx fNN_c$ and the number of configurations of the counterion gas is $\Omega \propto v^{fNN_c}$. Up to a constant, the entropy is then given by [70]

$$S_{\text{ionic}} = k_B \ln\Omega = k_B f N N_c \ln v \tag{6.8}$$

The chemical potential of the solvent molecule in the solution with counterions is defined as,

$$\mu_1 = -T\left(\frac{\partial S_{\text{ionic}}}{\partial n_1}\right)_{U,V,n_2} = -\frac{k_B T f N N_c v_s}{v} \tag{6.9}$$

with $v = v_s(n_1 + n_2 + fNN_c)$. Here, we assume that the volumes of a monomer, a solvent molecule, and a counterion are the same. Taking the chemical potential of a pure solvent solution μ_1^{pure}, and assuming that no salt is added to the microgel suspension, we can write the osmotic pressure of the ions inside the microgel as:

$$\Pi_{\text{ionic}} = -\frac{\Delta\mu_1}{v_s} = k_B T f c \tag{6.10}$$

which is, of course, analogous to the pressure of an ideal gas of ions confined inside the microgel.

6.2.4 Equation of State

Equations (6.4), (6.7), and (6.10) are the three main contributions to the osmotic pressure. Thermodynamic equilibrium with pure solvent is achieved when the sum of all three terms equals 0,

$$\Pi_{\text{mix}} + \Pi_{\text{el}} + \Pi_{\text{ionic}} = \frac{k_B T N_A}{v_s} \left(\frac{N_c v_s}{V_0 N_A} \left[\left(fN + \frac{1}{2} \right) \left(\frac{\phi_2}{\phi_{2,0}} \right) - \left(\frac{\phi_2}{\phi_{2,0}} \right)^{1/3} \right] \right.$$
$$\left. - \phi_2 - \ln(1 - \phi_2) - \chi \phi_2^2 \right) = 0. \tag{6.11}$$

Equation (6.11) is the Flory–Rehner equation of state describing the swelling behavior of polymer gels. It relates the particle size to thermodynamic parameters; χ_1 reflects the solvent quality, the ratio $\phi_2/\phi_{2,0}$ determines the polymer network elasticity, and fN is the number of counterions inside the particles. Equilibrium is achieved when the swelling tendency from the mixing and ionic effects is balanced with the deswelling tendency resulting from entropic chain elasticity. Note the equation above does not consider how, in the case of microgel suspensions, the osmotic pressure of the suspension bath increases with particle number density due to the decreased volume available to the free ions in solution. This contribution can eventually result in deswelling.

An alternative to the Flory–Rehner equation of state that includes this effect was proposed in reference [44]. In this work, the osmotic pressure inside and outside the microgels are:

$$\Pi_{\text{in}} = k\phi_{p,0}^2 \left(\frac{a_0^6}{a^6} - 1 \right), \quad \Pi_{\text{out}} = \frac{3k_B T s_1 \zeta}{4\pi(a_0^3 - s_2 a^3 \zeta)} \tag{6.12}$$

Here Π_{in} accounts for mixing and elastic contributions, with k an effective stiffness of the particle, $\phi_{p,0}$ and a_0 the particle volume fraction and particle radius in the swollen state, and a the particle radius at generalized volume fraction ζ. The solvent bath osmotic pressure is Π_{out}, with s_1 and s_2 two scaling factors. From Eq. (6.12), we see that at constant T, Π_{out} increases when ζ increases. Equilibrium then requires the particle size a decreases, simultaneously increasing Π_{in} and decreasing Π_{out}, so that ultimately $\Pi_{\text{in}} = \Pi_{\text{out}}$. Note that the presence of the polymer volume fraction within the microgel, $\phi_{p,0}$, indicates that more compressed microgels become denser and, therefore, stiffer and harder to further compress. This is a simple model based on intuition that has been applied in studies of the glass transition in microgel suspensions [61, 74, 75].

6.3 Experimental Techniques

In this section, we describe scattering techniques that have proved to be powerful tools to study both the dynamics and the characteristic lengths of microgel suspensions. After a general description of the basic principles of dynamic light scattering

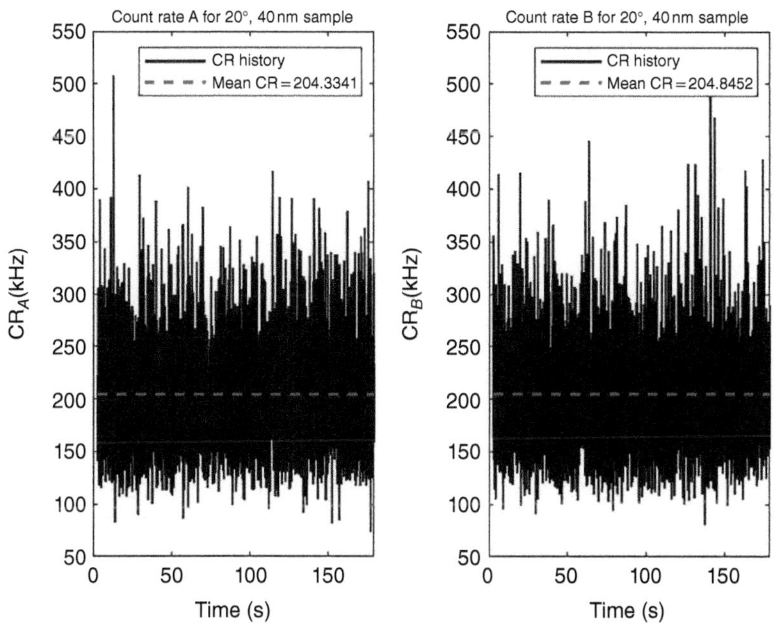

Figure 6.1 Fluctuations in the count rate for the two detectors used in a cross-correlation experiment. The scattering angle is $\theta = 20°$ and the particles have a diameter of 40 nm. Dashed lines correspond to the mean intensities.

and neutron scattering, we focus on how these techniques can be used to probe concentrated suspensions of microgels and how the problems related to multiple and selective scattering can be overcome by performing cross-correlation and contrast variation experiments in dynamic light scattering and SANS, respectively.

6.3.1 Dynamic Light Scattering

Light-scattering experiments allow studying statistical properties of systems of particles in the colloidal-size domain. In particular, by taking a measurement of the scattered intensity as a function of time, as shown in Figure 6.1, we can obtain information about the instantaneous microscopic density and configuration of the scatterers in the system.

To get an intuition on this, let's think of a set of scatterers fixed in space, as shown in Figure 6.2. If an incident plane wave of monochromatic light with wave vector $\vec{k}_0$ is scattered, with its wavelength and phase remaining constant, the scattered electric field corresponds to the sum of the scattered electric fields originating from all scatterers. The structural information of the system is accessible only when the distance between the scatterers and the wavelength of the scattered light is of the same order. This is the case for colloidal particles and visible light. If we now allow motion of the particles, the intensity measured at some instant t, is related to the electric field via [76]:

$$I(t) = \frac{1}{2}\sqrt{\frac{\epsilon_f}{\mu_0}}|E(t)|^2 \tag{6.13}$$

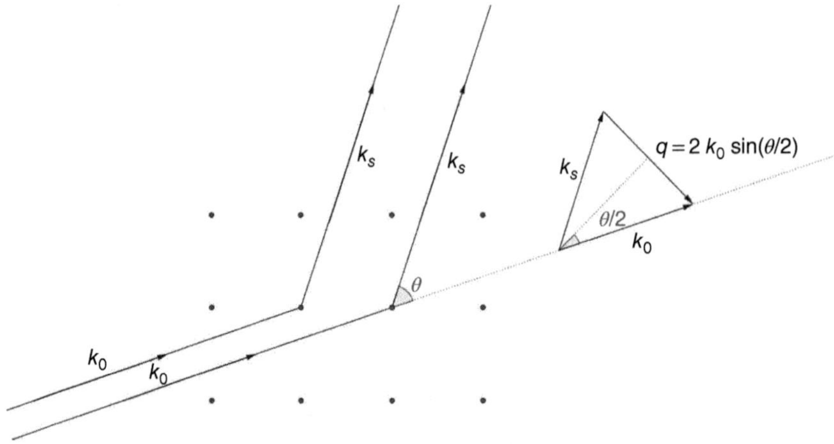

Figure 6.2 Light with an incoming wave vector $\vec{k}_0$ scattered by particles – blue points – and the corresponding scattering vector $\vec{q} = \vec{k}_0 - \vec{k}_s$, where $\vec{k}_s$ is the wave vector of the scattered wave. The scattering angle between $\vec{k}_0$ and $\vec{k}_s$ is θ.

with ϵ_f and μ_0 the dielectric constant of the fluid and the magnetic permeability of vacuum. For N identical particles, the instantaneous scattered intensity is:

$$I(\vec{q}, t) \sim \sum_{i,j=1}^{N} \exp\left[i\vec{q} \cdot (\vec{r}_i(t) - \vec{r}_j(t))\right] = |\rho(\vec{q}, t)|^2 \tag{6.14}$$

where $\vec{r}_i(t)$ and $\vec{r}_j(t)$ are the position vectors of the ith and jth particle, and $\vec{q}$ is the scattering vector, whose magnitude equals:

$$q = |\vec{q}| = \left|\vec{k}_0 - \vec{k}_s\right| = \frac{4\pi}{\lambda} \sin\left(\frac{\theta}{2}\right) \tag{6.15}$$

where $\vec{k}_0$ and $\vec{k}_s$ are the wave vectors of the incoming and scattered waves, respectively, λ is the wavelength in the medium, which for light, can be written as λ_0/n, with λ_0 the wavelength of the probe in vacuum and n the refractive index of the medium, and θ is the scattering angle. The proportionality in Eq. (14) can be understood after considering the scattered electric field,

$$\vec{E} = \sum_{j=1}^{N} \int_{V_j} d\vec{r} f(\vec{r}) \exp\left[i\vec{q} \cdot \vec{r}\right] \vec{E}_0 \tag{6.16}$$

where V_j is the volume of particle j located at location r_j, and $f(\vec{r})$ is the scattering strength, which depends on the dielectric constant of the particle relative to that of the solvent, with $\vec{r}$ the position of a point in the scattering volume. Note that V_j here is relative to a common lab frame. If we now change to a frame of reference centered at each of our colloidal particles, that is, we change variables from $\vec{r}$ to $\vec{r}' = \vec{r} - \vec{r}_j$, we then have:

$$\vec{E} = \sum_{j=1}^{N} \exp\left[i\vec{q} \cdot \vec{r}_j\right] \int_{V_j} d\vec{r}' f(\vec{r}') \exp\left[i\vec{q} \cdot \vec{r}'\right] \vec{E}_0 \tag{6.17}$$

Since the integral in this expression amounts to a multiplicative constant, multiplying $\vec{E}$ by its complex conjugate results in the proportionality expressed by Eq. (14). The subsequent equality follows after considering the microscopic density of an N-particle system,

$$\rho(\vec{r}, t) = \sum_{j=1}^{N} \delta(\vec{r} - \vec{r}_j(t)) \tag{6.18}$$

and the product of its Fourier transform,

$$\rho(\vec{q}, t) = \sum_{j=1}^{N} \exp\left[-i\vec{q} \cdot \vec{r}_j(t)\right] \tag{6.19}$$

with its complex conjugate.

6.3.1.1 Auto-correlation Experiments

We now distinguish between static light-scattering (SLS) experiments, where we focus on the mean value of the scattered intensity, and DLS experiments, where we measure the time dependence of the fluctuating intensity. In this section, we will not discuss SLS; see Ref. [76], for example, for details. The dynamic aspects of light scattering or the corresponding intensity fluctuations are best characterized through the intensity auto-correlation function:

$$G_I(\vec{q}, \tau) = \langle I(\vec{q}, t) I(\vec{q}, t + \tau) \rangle \tag{6.20}$$

where τ is the correlation time. In terms of the electric field, we have:

$$G_I(\vec{q}, \tau) = \frac{\epsilon_f}{4\mu_0} \langle E(\vec{q}, t) E^*(\vec{q}, t) E(\vec{q}, t + \tau) E^*(\vec{q}, t + \tau) \rangle \tag{6.21}$$

Using Wick's theorem, this equation can be rewritten as [76]:

$$\begin{aligned} G_I(\vec{q}, \tau) = \frac{\epsilon_f}{4\mu_0} & \left[\langle E(\vec{q}, t) E^*(\vec{q}, t) \rangle \langle E(\vec{q}, t + \tau) E^*(\vec{q}, t + \tau) \rangle \right. \\ & + \langle E(\vec{q}, t) E(\vec{q}, t + \tau) \rangle \langle E^*(\vec{q}, t) E^*(\vec{q}, t + \tau) \rangle \\ & \left. + \langle E(\vec{q}, t) E^*(\vec{q}, t + \tau) \rangle \langle E^*(\vec{q}, t) E(\vec{q}, t + \tau) \rangle \right]. \end{aligned} \tag{6.22}$$

The first term is nothing but I^2, where I is the mean scattered intensity. The second term is zero for nonzero wavevectors [76]. Finally, the third term is equal to $|G_E|^2$, where G_E is the electric field auto-correlation function, defined as

$$G_E(\vec{q}, \tau) = \frac{1}{2} \sqrt{\frac{\epsilon_f}{\mu_0}} \langle E(\vec{q}, t) E^*(\vec{q}, t + \tau) \rangle \tag{6.23}$$

Combining Eqs. (6.22) and (6.23), we obtain the Siegert relation:

$$G_I(\vec{q}, \tau) = I^2 + \left| G_E(\vec{q}, \tau) \right|^2 \tag{6.24}$$

If we now define the normalized correlation functions

$$\begin{aligned} g_I(\vec{q}, \tau) &= G_I(\vec{q}, \tau) / I^2, \\ g_E(\vec{q}, \tau) &= G_E(\vec{q}, \tau) / I, \end{aligned} \tag{6.25}$$

the Siegert relation becomes:

$$g_I(\vec{q},\tau) = 1 + |g_E(\vec{q},\tau)|^2 \tag{6.26}$$

This equation is true only for ideal experiments; experimental errors, such as a misalignment of the experimental setup, can affect it. We account for this by introducing an intercept, β, that quantifies these errors:

$$\begin{aligned}\beta &= g_I(\vec{q},0) - \lim_{\tau\to\infty} g_I(\vec{q},\tau)\\ &= \frac{\langle I(q,0)I(q,0)\rangle - \lim_{\tau\to\infty}\langle I(q,0)I(q,\tau)\rangle}{\lim_{\tau\to\infty}\langle I(q,0)I(q,\tau)\rangle} = \frac{\langle I^2(q)\rangle - I^2}{I^2}.\end{aligned} \tag{6.27}$$

The last relation is true because for long times, $\langle I(q,0)I(q,\tau)\rangle^2$ becomes uncorrelated. Using β, Eq. (6.26) then becomes:

$$g_I(\vec{q},\tau) = 1 + \beta|g_E(\vec{q},\tau)|^2 \tag{6.28}$$

Note that the upper limits of β, $g_I(\vec{q},\tau)$, and $|g_E(\vec{q},\tau)|^2$ are 1, 2, and 1, respectively. In a DLS experiment, one measures, oftentimes using a correlator, the normalized intensity time correlation function, $g_I(\vec{q},\tau)$, at a certain q as a function of the correlation time, τ.

6.3.1.2 Cross-correlation and 3D-DLS Experiments

In the simplest DLS experiment, a single beam passes through the sample and the autocorrelation function of the scattered intensity is measured using a correlator or calculated from the time dependence of the scattered intensity. When we only have single scattering, this is all good. However, multiple scattering can be non-negligible in many samples of interest, particularly at high particle number concentrations. In this case, successive scattering events prevent extracting the information that we are interested in. To deal with this, we can perform a cross-correlation experiment. There are several cross-correlation schemes, but we specifically focus on the so-called 3D cross-correlation [77].

In this scheme, two coherent beams cross inside the sample resulting in a finite scattering volume. There are also two photodetectors to measure the intensity at a fixed scattering angle θ; see Figure 6.3. Two incident beams and two detectors result in four simultaneous experiments, and an intensity cross-correlation function with four terms:

$$\begin{aligned}G_I(\vec{q},\tau) &= \langle I_1(\vec{q},t)I_2(\vec{q},t+\tau)\rangle\\ &= \langle [I_1^1(\vec{q},t) + I_1^2(\vec{q},t)]\,[I_2^1(\vec{q},t+\tau) + I_2^2(\vec{q},t+\tau)]\rangle\\ &= \langle I_1^1(\vec{q},t)I_2^1(\vec{q},t+\tau)\rangle + \langle I_1^2(\vec{q},t)I_2^1(\vec{q},t+\tau)\rangle\\ &\quad + \langle I_1^2(\vec{q},t)I_2^2(\vec{q},t+\tau)\rangle + \langle I_1^1(\vec{q},t)I_2^2(\vec{q},t+\tau)\rangle.\end{aligned} \tag{6.29}$$

The intensity terms in Eq. (6.30) have subscripts and superscripts corresponding, respectively, to the detector where the measurement takes place and to the incident beam, $I^{\text{beam}}_{\text{detector}}$. After clarifying the notation, it is easy to see that each of the four

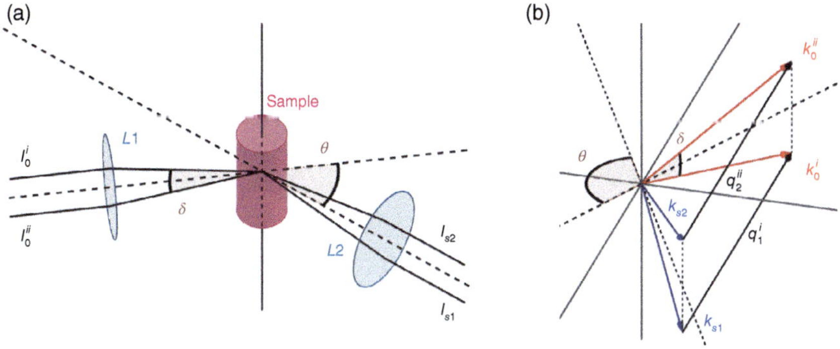

Figure 6.3 Experimental setup was designed by Pusey to perform 3D cross-correlation. Source: Pusey [77]/Elsevier/Public Domain. (a) Two parallel beams with intensities I_0^i and I_0^{ii} go through lens L1 that focuses the beams inside the samples. The angle between the beams is δ. The scattered beams, I_{s1} and I_{s2} are then going through a second lens, $L2$, that brings them toward the detectors. The scattering angle is defined as the angle between the plane of the triangle with vertex in the scattering volume and sides I_{s1} and I_{s2}, and the direction of the incoming beams. (b) Sketch of the scattering triangles for the two beams with incoming wave vectors k_0^i and k_0^{ii} and wave vector of the scattered wave equal to k_{s1} and k_{s2}. The angles δ and θ are defined in (a). The scattering vectors for the two beams are $\vec{q}_1^{\,i}$ and $\vec{q}_2^{\,ii}$; note that they are parallel and have the same magnitude.

terms in Eq. (6.30) corresponds to each of the four experiments referred to before. For example, the first one represents the correlation of the scattered intensity of beam 1 measured by detector 1 and the scattered intensity due to beam 1 measured by detector 2. The geometry of the experiment implies that only the fourth term corresponds to an experiment with the same scattering vector, $\vec{q}$. As a result, the first terms are uncorrelated and only contribute to the background. Thus,

$$G_I(\vec{q},\tau) = \frac{3}{4}I^2 + \langle I_1^1(\vec{q},t)I_2^2(\vec{q},t+\tau)\rangle \tag{6.30}$$

where we have assumed that our measurement is ideal and that each term has the same intensity contribution. We now write the correlation term as a function of the electric fields:

$$\begin{aligned}\langle I_1^1(\vec{q},t)I_2^2(\vec{q},t+\tau)\rangle &= \frac{\epsilon_f}{4\mu_0}\langle E_1^1(\vec{q},t)E_1^{1*}(\vec{q},t)E_2^2(\vec{q},t+\tau)E_2^{2*}(\vec{q},t+\tau)\rangle \\ &= \frac{\epsilon_f}{4\mu_0}\left[\langle E_1^1(\vec{q},t)E_1^{1*}(\vec{q},t)\rangle\langle E_2^2(\vec{q},t+\tau)E_2^{2*}(\vec{q},t+\tau)\rangle\right. \\ &\quad + \langle E_1^1(\vec{q},t)E_2^2(\vec{q},t+\tau)\rangle\langle E_1^{1*}(\vec{q},t)E_2^{2*}(\vec{q},t+\tau)\rangle \\ &\quad \left.+ \langle E_1^1(\vec{q},t)E_2^{2*}(\vec{q},t+\tau)\rangle\langle E_1^{1*}(\vec{q},t)E_2^2(\vec{q},t+\tau)\rangle\right].\end{aligned} \tag{6.31}$$

Proceeding as in the auto-correlation experiment, we obtain the Siegert relation for the cross-correlation situation:

$$G_I(\vec{q},\tau) = I^2 + \left|G_E(\vec{q},\tau)\right|^2 \tag{6.32}$$

where

$$\left|G_E(\vec{q},\tau)\right|^2 = \frac{\epsilon_f}{4\mu_0}\langle E_1^1(\vec{q},t)E_2^{2*}(\vec{q},t+\tau)\rangle\langle E_1^{1*}(\vec{q},t)E_2^2(\vec{q},t+\tau)\rangle \tag{6.33}$$

We now consider the following normalizations:

$$g_I(\vec{q},\tau) = G_I(\vec{q},\tau)/I^2, \tag{6.34}$$

$$\left|g_E(\vec{q},\tau)\right|^2 = \frac{\left|G_E(\vec{q},\tau)\right|^2}{(\epsilon_f/4\mu_0)\left|\langle E_1^1(\vec{q},t)E_2^{2*}(\vec{q},t)\rangle\right|^2} = \frac{\left|G_E(\vec{q},\tau)\right|^2}{\langle I^{(1)}(q)\rangle^2/4}. \tag{6.35}$$

Note $|g_E|^2$ is defined in terms of the averaged single-scattered intensity $\langle I^{(1)}(q)\rangle^2$. Putting it all together we have:

$$g_I(\vec{q},\tau) - 1 = \beta_0\frac{\langle I^{(1)}(q)\rangle^2}{4\langle I(q)\rangle^2}\left|g_E(\vec{q},\tau)\right|^2 \tag{6.36}$$

where, as in Eq. (6.28), we have included an empirical prefactor β_0 to account for experimental imperfections. The factors on the right-hand side of the equation can be grouped together to give:

$$\beta = \beta_0\frac{\langle I^{(1)}(q)\rangle^2}{4\langle I(q)\rangle^2} \tag{6.37}$$

If we perform measurements for very dilute systems, where there is no multiple scattering, we obtain $\beta^{\text{ref}} = \beta_0/4$, which we use to correct for the eventual multiple scattering. Explicitly, the single-scattered intensity is:

$$\langle I^{(1)}(q)\rangle = \langle I(q)\rangle\sqrt{\frac{\beta}{\beta^{\text{ref}}}} \tag{6.38}$$

The measured intensity depends on the scattering vector q and on the correlation time τ and offers the possibility to probe the dynamics of dense microgel suspensions on different length scales in the presence of multiple scattering. If the data are then averaged, and the experiment is performed at different angles, the obtained intensity, $I(q)$ will also give static structural information associated to the arrangement of the microgels in suspension. In Figure 6.4, we report examples of 3D-DLS measurements performed on concentrated suspensions of microgels taken from Refs [58, 78].

6.3.2 Small-angle Neutron-scattering

In general, scattering can be described with the formalism of quantum mechanics considering the incident and the scattered waves. The scattering event depends on the probe, light, X-rays, or neutron radiation, and how it interacts with the sample through interaction potential $U(r)$. Depending on our choice, the interaction with the scattering center happens *via* a different interaction potential. For instance, in light and X-ray scattering, the interaction between the photon and the scattering center – which is associated with the electron cloud surrounding the atoms composing the sample – it is mediated by the electromagnetic interaction.

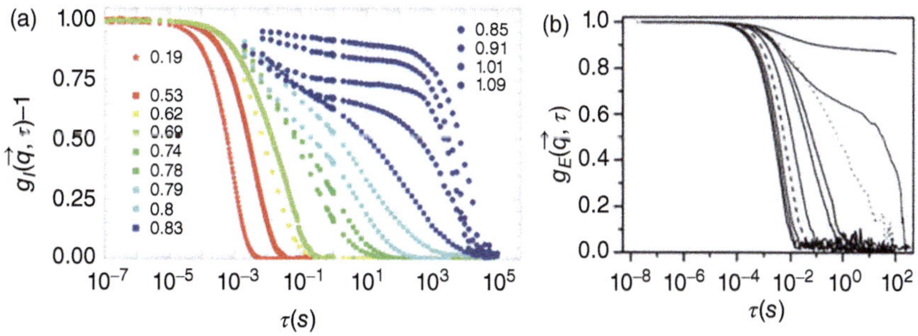

Figure 6.4 (a) Correlation functions of pNIPAM microgel suspensions with generalized volume fraction from 0.19 to 1.09, with corresponding data from left to right. Source: Image adapted from Philippe et al. [78]/with permission of American Physical Society. (b) Ensemble-averaged field autocorrelation functions for suspensions of microgels at concentrations between 0.1 and 12 mg/g, with corresponding data from left to right. Source: Image adapted from Pellet and Cloitre [58]/with permission of John Wiley and Sons. In both panels, the suspension transitions between a colloidal fluid phase, as characterized by a single decay in the correlation function, to a supercooled colloidal liquid phase, as reflected by the existence of an intermediate plateau region and to relaxation processes at short and long-correlation times.

In contrast, neutrons interact with the nuclei of the atoms composing the sample *via* the strong nuclear force. This interaction potential can be approximated with the so-called Fermi pseudopotential:

$$U(\vec{r}) = \frac{2\pi\hbar}{m_n}\sum_{i=1}^{N'} b_i\delta(\vec{r}-\vec{r}_i) \tag{6.39}$$

where m_n is the neutron mass, $\hbar$ is Planck's constant divided by 2π, N' is the number of atoms in the irradiated sample, and $\vec{r}$ and $\vec{r}_i$ are the positions of the neutron and the ith nucleus, respectively. The presence of the Dirac δ-function reflects the fact that the neutron wavelength is much larger than the nuclei and, therefore, that the nuclei can be considered as point-like scatterers. The other key aspect in Eq. (6.39) is the presence of the quantity b_i, the scattering-length of the ith nucleus. The probability that a neutron is scattered by nucleus-i is given by the scattering cross-section $\sigma_i = 4\pi b_i^2$.

In this section, we mainly focus on elastic neutron-scattering at small scattering angles. The term elastic refers to the fact that there is no exchange of energy between the neutrons and the sample. In addition, by focusing on small scattering angles, we can access length scales between 5 and 500 nm, which corresponds to the order of magnitude of the relevant characteristic length scales in microgel particles and suspensions. In fact, considering the scattering wave vector in Eq. (6.15) and that the typical wavelength of neutron beams used in elastic scattering experiments is between 0.1 and 1 nm, we see that to probe the distance of interest, l, mentioned above, a wavevector $q \approx 2\pi/l$ must be used, which corresponds to scattering angles θ between 0.1° and 10°. Since these angles are smaller than the angles used in normal scattering experiments, e.g. neutron diffraction, they are referred to as small angles.

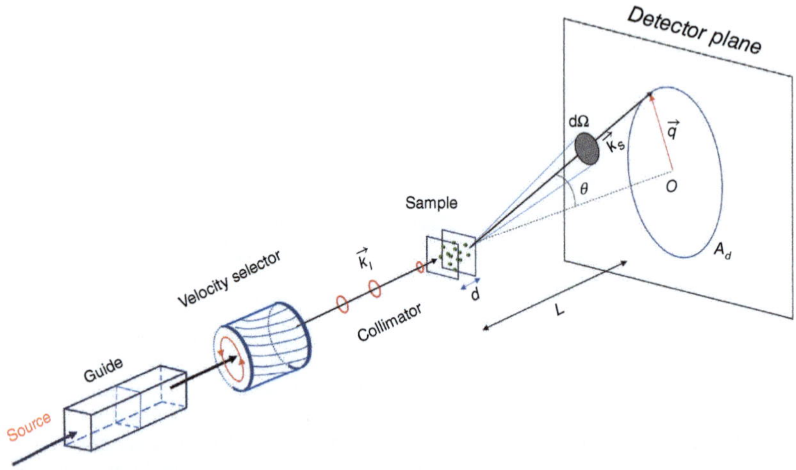

Figure 6.5 Schematic of a classic small-angle neutron-scattering instrument. The desired wavelength is chosen with a velocity selector. Neutron guides are used to bring the neutrons to the velocity selector, after which they encounter a collimation system. The resultant collimated beam impinges the sample, which is of thickness d. The neutrons are scattered and detected at a scattering angle θ along wave vector $\vec{k}_s$ within a solid angle $d\Omega$. They are counted with a position-sensitive area detector A_d at a distance L from the sample. To change the detected scattering vector $\vec{q}$ both the neutron wavelength and the sample-to-detector distance L can be changed.

6.3.2.1 SANS Setup

Figure 6.5 shows a schematic of a classic SANS instrument at a continuous neutron source. Neutrons originating from the source are transported using neutron guides toward a velocity selector that selects the neutron wavelength $\lambda \pm \Delta\lambda$. The neutron beam is then collimated with pin-holes before it is scattered by the sample. The scattered neutrons are counted with a position-sensitive area detector covering about a decade in q. The sample-detector distance can be varied to cover a q range from about 0.01 to 5 nm^{-1}.

6.3.2.2 Scattering Theory

As mentioned above, the neutrons interact with the nuclei in the irradiated sample, which act as scattering centers. Given the fact that we can consider the atoms as geometrical points without dimensions, the atomic form factors are isotropic and can be described with the scattering-lengths b_i. Every neutron scattered by a nucleus gives rise to a spherical wave and the final intensity is determined by the interference of these waves. The interaction between the incident neutron and a nucleus at position $\vec{r}$ in the sample is described by Fermi's pseudopotential, Eq. (6.39). The total wave function $\phi(\vec{R})$ in the detector at position $\vec{R}$ can be written as [79]:

$$\phi(\vec{R}) = \phi_I(\vec{R}) + \phi_S(\vec{R}) = \phi_I(\vec{R}) + \frac{2m_n}{\hbar^2} \int U(\vec{r})\phi_I(\vec{r}) \frac{e^{ik|\vec{R}-\vec{r}|}}{4\pi|\vec{R}-\vec{r}|} d\vec{r} \tag{6.40}$$

where $\phi_I(\vec{R})$ and $\phi_S(\vec{R})$ are the incident and scattered wave functions, respectively. The incident wave function can be approximated by a plane wave, $\phi_I(\vec{R}) = A\exp[i(k_I \cdot \vec{R})]$. To proceed, we make the following three assumptions—(i) there is no multiple scattering, and thus, once a neutron interacts with a nucleus it goes directly to the detector; (ii) the interaction potential is localized and can be approximated by Eq. (6.39), implying the interaction of the neutron with a scattering center does not depend on the location of other scattering centers (first Born approximation); (iii) $|\vec{r}| \ll |\vec{R}|$, i.e. we are in the far-field approximation. Under these assumptions, the scattered wave function $\phi_S(\vec{R})$ can be written as [80, 81]:

$$\phi(\vec{R}) = \phi_I(\vec{R}) + \frac{\exp[ikR]}{R}\sum_{j=1}^{N'} b_j \exp[i\vec{q}\cdot\vec{r}_j] \tag{6.41}$$

where $\vec{r}_j$ is the position of the jth nucleus in our particle, and b_j its scattering-length. The second part of Eq. (6.41) corresponds to a spherical wave, $\exp[ikR]/R$, of amplitude:

$$A(\vec{q}) = \sum_{j=1}^{N'} b_j \exp[i\vec{q}\cdot\vec{r}_j] \tag{6.42}$$

Considering now the experimental set-up sketched in Figure 6.5, the cross-section per unit of volume can be written as:

$$\frac{d\Sigma}{d\Omega}(\vec{q}) = \frac{1}{V}|A(\vec{q})|^2 \tag{6.43}$$

where V is the irradiated volume of the sample. Equations (6.42) and (6.43) describe the scattering event at the atomic level. We now notice that (i) the resolution of a SANS instrument cannot resolve atoms and that (ii) we are more interested in length scales associated with the particle size (70–200 nm), which contain thousands of nuclei. Therefore, we can consider the so-called scattering-length density (SLD), defined as:

$$\rho(\vec{r}) = \frac{1}{v}\sum_{i=1}^{N'} b_i^{\mathrm{coh}} \tag{6.44}$$

where v is the volume of the particle and b_i^{coh} is the coherent scattering length of atom i, which is the average value of b_i. Recall $\rho(\vec{r})$ has dimensions of an inverse length squared [82]. Due to the presence of isotopes and the possible nuclear spin incoherent, there is also an incoherent part of the scattering-length accounting for the fluctuations of b_i: $b_i^{\mathrm{inc}} = \langle b_i^2\rangle - \langle b_i\rangle^2$. In elastic scattering, this second part is isotropic, and therefore, it constitutes a background to the coherent scattering signal that can be easily accounted for.

The advantage of using the continuum representation is that we can substitute the discrete summations above with integrals and thus rewrite the cross section as:

$$\frac{\mathrm{d}\Sigma}{\mathrm{d}\Omega}(\vec{q}) = \frac{1}{V}\left|\int_V \mathrm{d}\vec{r}\rho(\vec{r})e^{i\vec{q}\cdot\vec{r}}\right|^2 \tag{6.45}$$

Equation (6.45) shows that the cross section is given by the squared absolute value of the Fourier transform of the distribution of scattering centers within the sample. This consideration is fundamental to developing models based on the radial distribution of scattering centers within the particles to interpret the results of any SANS experiment.

Colloidal suspensions consist of colloids dispersed in a medium. This offers the possibility to divide the SLD into two parts: the average SLD of the solvent, ρ_0, and the excess SLD related to the fluctuations of the SLD due to the presence of the colloids, $\delta\rho(\vec{r})$. Then:

$$\rho(\vec{r}) = \rho_0 + \delta\rho(\vec{r}) \tag{6.46}$$

Now note that since ρ_0 is constant, it contributes a Delta function at $q = 0$ to the scattering cross-section given by Eq. (6.45). If we exclude this case, which is only relevant in experiments without beamstop, where the direct beam dominates, we can therefore re-write Eq. (6.45) as:

$$\frac{d\Sigma}{d\Omega}(\vec{q}) = \frac{1}{V}\left\langle \left| \int_V d\vec{r}\ \delta\rho(\vec{r}) e^{i\vec{q}\cdot\vec{r}} \right|^2 \right\rangle \tag{6.47}$$

where $\langle \ldots \rangle$ indicates an average over all possible configurations. The meaning of Eq. (6.47) is that the cross section is nothing more than the Fourier transform of the distribution of scattering centers within the colloidal suspension.

Finally, we define the form factor of particle i, $F_i(\vec{q})$, in terms of the scattering amplitude at $q = 0$, f_i, as follows:

$$F_i(\vec{q}) = \frac{1}{f_i}\int_{v_i} d\vec{r}\ \delta\rho_i(\vec{r}) e^{i\vec{q}\cdot\vec{r}}\ , \tag{6.48}$$

$$f_i = \int_{v_i} d\vec{r}\ \delta\rho_i(\vec{r})\ , \tag{6.49}$$

where v_i is the volume of colloidal particle i. Using Eq. (6.46), $f_i = v_i <\Delta\rho_i>$.

6.3.2.3 Form Factor and Structure Factor

For the case of particles in suspension, the scattering signal can be split into the form factor, $P(q)$, and structure factor, $S(q)$. The form factor $P(q)$ is proportional to the scattering signal of a single particle and contains all the information on its shape and internal architecture, while the structure factor results from the arrangement of all particles in the suspension. Usually, size polydispersity needs to be taken into account for colloidal suspensions, and we often do this assuming that the particle size and orientation are not correlated with the particle positions, i.e. the colloidal structure is not affected by the polydispersity. This assumption is expected to be valid when the size polydispersity and solution concentration are not very high.

We consider a colloidal suspension with N particles and we write the position of a scattering center in the kth particle as $\vec{r}_{k,\alpha} = \vec{R}_k + \vec{u}_{k,\alpha}$, where the vector $\vec{R}_k$ gives the

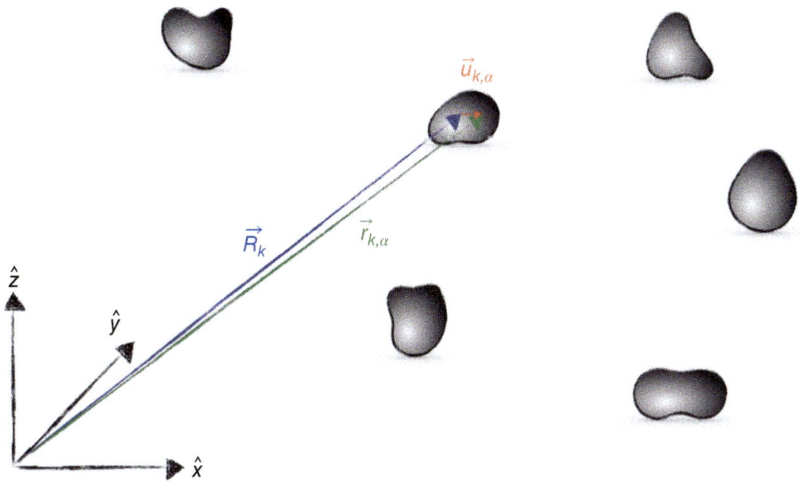

Figure 6.6 Lab frame and vectors $\vec{r}_{k,\alpha} = \vec{R}_k + \vec{u}_{k,\alpha}$ used to identify the kth colloid and the α scattering center inside the particle.

position of the kth particle and $\vec{u}_{k,\alpha}$ is the position of a scattering center within this particle with respect to $\vec{R}_k$, see Figure 6.6. With this, the scattering cross section can be expressed as [83]:

$$\frac{d\Sigma}{d\Omega}(\vec{q}) = \frac{1}{V}\left\langle \sum_{j,k=1}^{N} \int_{v_j}\int_{v_k} d\vec{u}_\alpha d\vec{u}_\beta \delta\rho_j(\vec{u}_\alpha)\delta\rho_k(\vec{u}_\beta) e^{i\vec{q}\cdot(\vec{R}_j-\vec{R}_k)} e^{i\vec{q}\cdot(\vec{u}_\alpha-\vec{u}_\beta)} \right\rangle$$
$$= \frac{1}{V}\sum_{j=1}^{N} \left\langle f_j^2\, |F_j(\vec{q})|^2 \right\rangle + \frac{1}{V}\sum_{j\neq k=1}^{N} \left\langle f_j f_k^* F_j(\vec{q}) F_k^*(\vec{q}) e^{i\vec{q}\cdot(\vec{R}_j-\vec{R}_k)} \right\rangle \qquad (6.50)$$

where $F_j(\vec{q})$ is the scattering amplitude of the jth particle defined in Eq. (6.48) and N is the number of particles. When all particles are identical with an SLD ρ_p, the scattering cross section is proportional to $|F(\vec{q})|^2$, which is usually written as $v^2\Delta\rho^2 P(\vec{q})$, where $\Delta\rho = \rho_p - \rho_0$ is the scattering contrast between the particles and the solvent. Therefore, the first term in Eq. (6.50) becomes $\frac{N}{V}v^2\Delta\rho^2\, P(\vec{q})$, and the second term can be rewritten as $\frac{v^2\Delta\rho^2 P(\vec{q})}{N}\sum_{j\neq k} < e^{i\vec{q}\cdot(\vec{R}_j-\vec{R}_k)} >$, considering that for identical spherical particles $\sum_{j\neq k=1}^{N} F_j(\vec{q})F_k^*(\vec{q}) = N|F(\vec{q})|^2$. Finally, grouping the common factors together, Eq. (6.50) becomes:

$$\frac{d\Sigma}{d\Omega}(\vec{q}) = \frac{N}{V}v^2\Delta\rho^2 P(\vec{q})S(\vec{q})$$
$$\text{with } S(\vec{q}) = 1 + \frac{1}{N}\sum_{j\neq k} < e^{i\vec{q}\cdot(\vec{R}_j-\vec{R}_k)} > \qquad (6.51)$$

When the particles are not all identical, e.g. due to polydispersity, the particle form factors vary and the information about the particle architecture and the suspension structure can no longer be fully separated as in Eq. (6.51). However, assuming that the particle size and structure do not correlate with the suspension structure, we can define the apparent structure factor,

$$S'(q) = 1 + \beta(q)(S(q) - 1), \tag{6.52}$$

$$\text{with } \beta(q) = |\langle F(q)\rangle|^2 / \langle |F(q)|^2 \rangle, \tag{6.53}$$

and the average form factor, $P(q) = \frac{1}{\mathcal{N}} \sum_j |F_j(\vec{q})|^2$, with $\mathcal{N} = \sum_j v_j^2 \Delta\rho_j^2$ [84, 85]. The effect of the correction factor $\beta(q)$ in Eq. (6.52) is to suppress the oscillations in $S(q)$. Microgels in suspension have a typical size polydispersity of $\sim$ 10%. Therefore, when describing the structure factor of microgel suspensions, Eq. (6.52) has to be considered. At high concentrations, the assumptions used to write the apparent structure factor may not hold, and partial structure factors must then be derived by solving the Ornstein–Zernike equation for the pair potential acting between the colloids [86]. If the suspension is dilute, $S(q) = 1$ and the scattering cross section is directly proportional to $P(q)$.

6.3.2.4 Contrast Variation

The scattering cross section of a particulate sample, such as a microgel suspension, Eq. (6.50), is proportional to the square of the scattering contrast $\Delta\rho$. Obviously, $\Delta\rho$ depends on the radiation used to probe the sample. As thermal neutrons interact by means of the strong nuclear force with the atomic nuclei composing the sample, the scattering strength for neutrons strongly varies with the nuclear structure and can show large variations for isotopes of a given chemical element [82]. It can even be positive or negative for isotopes of the same element. A negative scattering-length indicates that the Fermi pseudopotential in Eq. (39) is attractive between the given nucleus and the incoming neutron. Conversely, positive values of the scattering-length reflect a repulsion between the given nucleus and the incoming neutrons. For soft matter, the most important difference between the scattering-lengths pertains to those of hydrogen (H) and deuterium (D), $b_H = -3.7406 \cdot 10^{-15}$ m and $b_D = 6.671 \cdot 10^{-15}$ m, respectively. The substitution of hydrogen atoms with atoms of deuterium allows tuning the scattering contrast in many materials; this is true and important in all water-based suspensions. The mixture of water and heavy water allows to finely tune the contrast of the solvent without affecting the chemical composition of the sample in a significant way.

In Figure 6.7a, we illustrate the contrast matching of core–shell particles in an H_2O/D_2O solvent [87]. When the SLDs of the solvent, the particle core, and the particle shell all differ (solvent 1), both core and shell contribute to the scattering signal shown in Figure 6.7b. The solvent, however, can be adjusted to match the SLD of either the core (solvent 2) or the shell (solvent 3) to obtain the scattering signal and structural information of the unmatched part, the shell for solvent 2 and the core for solvent 3. Colloidal core–shell particles with a contrast between core and shell are naturally obtained when the core and shell consist of different materials, but this contrast can also result from adding a shell of, e.g. deuterated NIPAM, on a core of hydrogenated NIPAM.

SANS with contrast variation has proven to be a unique tool to measure the form factor of particles in concentrated solution. In this case, a mixture of deuterated

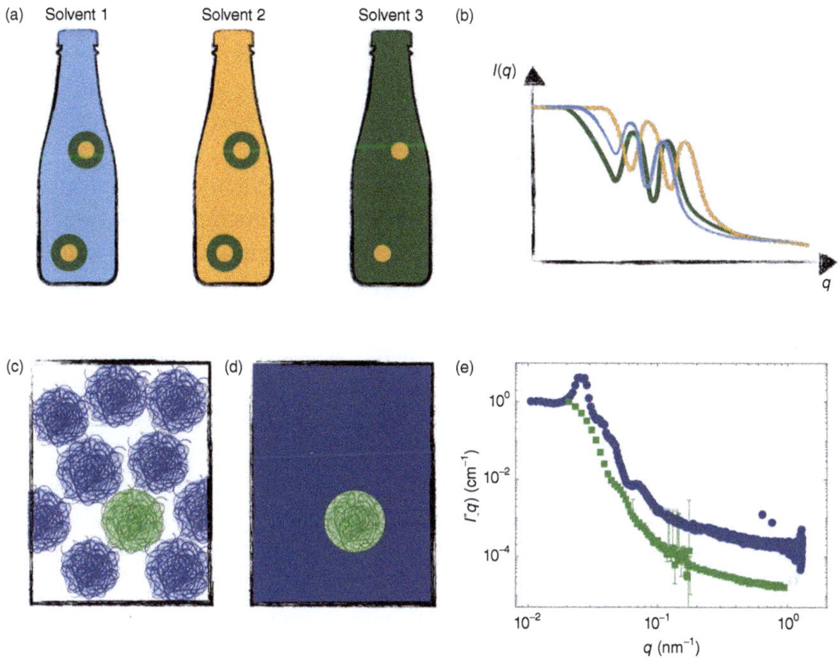

Figure 6.7 (a) Sketch of a sample composed of a core and a shell with different SLD (orange and green, respectively). The use of solvent with different contrasts (blue, green, and orange) matches different parts of the same object. (b) Example of the scattering profile of the core–shell particles where the (blue) particle, the (green) shell, and the (orange) core are visible. (c) Sketch of a sample composed of a majority of (blue) deuterated and (green) hydrogenated particles (e.g. microgels). (d) Sketch of a sample composed of a majority of (blue) deuterated and (green) hydrogenated particles (e.g. microgels) in a solvent that matches the scattering-length density of the deuterated particles. (e) Small-angle X-ray (blue circles) and neutron-scattering (green squares) intensities, $I(q)$, versus scattering vector, q.

and hydrogenated microgels is used. This is sketched in Figure 6.7c. Small angle X-ray scattering (SAXS) gives a scattering signal proportional to both $P(q)$ and $S(q)$, as both kinds of microgels have the same scattering contrast for X-rays. However, with SANS, the H_2O/D_2O solvent can be chosen to match the SLD of the deuterated microgels and this allows directly measuring the form factor of the small fraction of hydrogenated microgels in suspension, as their number is low and their structure factor is ≈ 1. SANS with contrast matching thus allows suppressing the signal from one species in the sample so that we can focus on the other species of interest. This approach (sketched in Figure 6.7c–e) is referred to as the tracing method. A second option to directly measure the form factor of the particles in a concentrated suspension is the zero-average contrast (ZAC) method, which uses a mixture of an equal number of deuterated and hydrogenated particles. Except for the substitution of hydrogen with deuterium, the two types of particles must be identical, that is, their size, internal structure, and also their elastic properties must be the same. The solvent (a H_2O/D_2O mixture) is chosen to have a SLD right between the SLD of the two-particle species: $\Delta\rho_{ZAC} = \rho_D - \rho_s = \rho_s - \rho_H$, where the subscripts refer to (D) deuterated and (H) hydrogenated pNIPAM, and to (s) the solvent. Under these conditions, the coherent interparticle scattering defining the structure factor $S(q)$ disappears, and only the average form factors of the deuterated and hydrogenated microgels contribute to the coherent scattering signal: $P_{ZAC}(q) = [P_H(q) + P_D(q)]/2$.

The intrinsic limitation of the ZAC method is the need of identical deuterated and hydrogenated particles. While a similar size and internal structure can be achieved by using surfactants and varying the ratio between initiator and monomers during particle synthesis, the incorporation of crosslinker and the suppression of self-crosslinking with the deuterated monomer are inevitable differences [88]. As a consequence, deuterated and hydrogenated microgels, synthesized with the same – nominal – concentration of crosslinker agents result in microgels with different compressibility. To avoid this problem, Schurtenberger and coworkers proposed to use a different amount of crosslinker for the synthesis of deuterated and hydrogenated microgels such that the two families of microgels have identical swelling ratios [30, 41]. However, even in this case, the synthesized microgels have slightly different sizes. As long as the D- and H-microgels show the same interaction and colloidal structure, the ZAC method can be used to obtain an average form factor $P_{ZAC}(q)$. The main advantage of ZAC with respect to the tracer method is that a 50%/50% mixture of H- and D-microgels is used and the scattering signal is, therefore, stronger than that from a small fraction of H-microgels measured in the tracer method. This last method, however, is more versatile, since it does not require the use of identical particles. This approach has been used to probe the deswelling of hydrogenated microgels embedded in a dense suspension of deuterated microgels with different sizes, both large hydrogenated in a majority of small protonated and vice versa [29]. Furthermore, it has been used to study the response of hollow microgels embedded in a "sea" of regular microgels synthesized with the same amount of crosslinker and total size [65] or again to study the different responses of hollow microgels, microgels with ultralow crosslinking, and regular microgels embedded in a "sea" of regular microgels [31]. In addition to these studies, the tracer method can be used to probe the response of microgels with the same structure and size. Recently, this method has been used to investigate the response of both hollow [33] and ultralow-crosslinked microgels [32] to crowding to rationalize their phase behavior.

For both the ZAC and the tracer method, it is fundamental to know the SLD of the microgels to properly choose the mixture of regular and heavy water for the suspension. The theoretical SLD of a material can be computed using Eq. (6.44) and the chemical formula and mass density of the material as input. For instance, for d7-NIPAM – $C_6D_7H_4NO$, we obtain:

$$\begin{aligned}\rho_{d7\text{-NIPAM}} &= \frac{d_{d7\text{-NIPAM}} N_A}{M_w}(6b_C^{coh} + 7b_D^{coh} + 4b_H^{coh} + b_N^{coh} + b_O^{coh}) \\ &= 4.784 \cdot 10^{-6}\,\text{Å}^{-2}, \end{aligned} \tag{6.54}$$

where $N_A = 6.022 \cdot 10^{23}\text{mol}^{-1}$ is Avogadro's constant, and $d_{d7\text{-NIPAM}} = 1.1$ g/cm^3 [29, 30, 89] and $M_w = 120$ g/mol are the density and molecular weight of d7-NIPAM.

Table 6.1 Coherent scattering-length for different atoms.

Element	b^{coh} (10^{-15} m)
^1_1H	−3.7406
^2_1H (D)	6.671
$^{12}_6C$	6.6511
$^{14}_7N$	9.37
$^{16}_8O$	5.803

Source: Values from Reference [82].

The values of b_i^{coh} are taken from Table 6.1. A particle with a SLD of $4.784 \cdot 10^{-6}$ Å^{-2} is contrast matched by a mixture of 80 wt% D_2O and 20 wt% H_2O, since in this case,

$$\rho_{20\%H_2O/80\%D_2O} = 0.2\left[\frac{d_{H_2O}N_A}{M_{H_2O}}(2b_H^{coh} + b_0^{coh})\right] + 0.8\left[\frac{d_{D_2O}N_A}{M_{D_2O}}(2b_D^{coh} + b_0^{coh})\right]$$
$$= 4.785 \cdot 10^{-6}\ \text{Å}^{-2}. \tag{6.55}$$

In the literature, values slightly different from the theoretical ones have been reported [26, 28–31, 33, 41, 65]. The main reason for this is that during the synthesis other components are incorporated in the polymer network – namely fragments of the initiator and the crosslinker agents. As a consequence of this, the match point of the particles slightly changes. Therefore, an experimental determination of the SLD of the microgels is always desirable.

To experimentally determine the SLD of microgels, the scattering profiles of dilute solutions in different mixtures of light and heavy water are measured with SANS at low q covering the Guinier regime of the microgels. Figure 6.8a shows an example of microgels synthesized with d7-NIPAM. The scattering intensity of these microgels is very low when the solvent is composed of (circles) 100 wt% and (upside triangles) 96 wt% D_2O, confirming that the match point is almost reached in pure D_2O. The most common method to obtain the solvent composition exactly matching the microgel is based on the Guinier approximation:

$$I(q) \propto \frac{d\Sigma}{d\Omega}(q) \overset{qR_g<1}{\approx} \frac{d\Sigma}{d\Omega}(0)\exp\left\{-\frac{q^2R_g^2}{3}\right\} \tag{6.56}$$

where R_g is the radius of gyration of the particle, and $d\Sigma/d\Omega(0)$ is the value of the scattering cross section at $q = 0$. In the q-range where the approximation $qR_g < 1$ holds, coherent scattering from the whole particle is probed. When this range of low q is reached, we can fit $\ln[I(q)]$ as a function of q^2 with a line. The slope is equal to $-\frac{R_g^2}{3}$ and the intercept equals $\ln\frac{d\Sigma}{d\Omega}(0)$. Fits of the data in the Guinier regime are shown in Figure 6.8b. In the case of dilute samples and for $q \to 0$, both $S(q)$ and $P(q)$ equal 1 and Eq. (6.56) directly probes $\Delta\rho^2$ and thus $\Delta\rho$ (black circles), see Figure 6.8c. The values of $\Delta\rho$ for the samples with solvent containing 100 wt% and

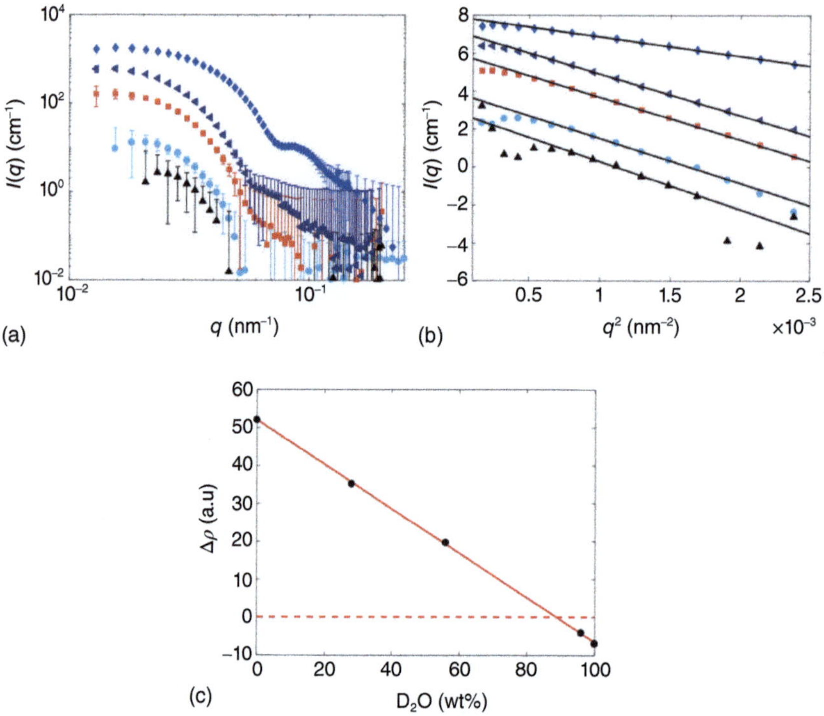

Figure 6.8 (a) Small-angle neutronscattering intensities, $I(q)$, versus scattering vector, q, of microgels synthesized with d7-NIPAM- $C_6D_7H_4NO$ – in solutions of water/heavy water with–100 (circles), 96 (upside triangles), 56 (squares), 28 (left-side triangles), and 0 (diamonds) wt% D_2O. All measurements have been conducted at (20.0 ± 0.1) °C. (b) Linear fits in the low-q region of the log[$I(q)$] versus q^2 for the data in (a). (c) Values of the contrast in scattering-length density, $\Delta\rho$, are obtained from the fits of the data in (b) versus the wt% of D_2O in solution. The solid red line is a linear fit to the data. The intercept of the solid line, with the dashed line indicating the zero, gives the wt% of D_2O needed in the H_2O/D_2O mixture to contrast match the deuterated particles. Source: Data taken from Ref. [31].

96 wt% of D_2O have been manually changed to negative. The D_2O content at which the solvent crosses the $\Delta\rho = 0$ line gives the match point of the microgels. As seen in Figure 6.8c, this is the case for 90 wt% D_2O [31]. In the literature, the match point of d7-NIPAM-based microgels is reported to be in the range from 83 wt% to 90 wt% D_2O [28–31, 33, 41].

An alternative method to access the SLD of microgels is based on the assumption that for samples using the same particles at a fixed concentration and only differing in the solvent composition, the scattering only varies due to the change in $\Delta\rho^2$. The example shown in Figure 6.9a is for d3-NIPAM-based microgels – ($C_6D_3H_8NO$). The contrast is found to significantly decrease for a D_2O concentration of the solvent approaching ≈ 55 wt%. The match point can be identified by finding the D_2O concentration where $\Delta\rho$ approaches zero. This is done in Figure 6.9b by comparing the $\sqrt{I(q)}$ values of six samples with various D_2O contents at the five q values highlighted by the vertical lines in Figure 6.9a. As before, the sign of

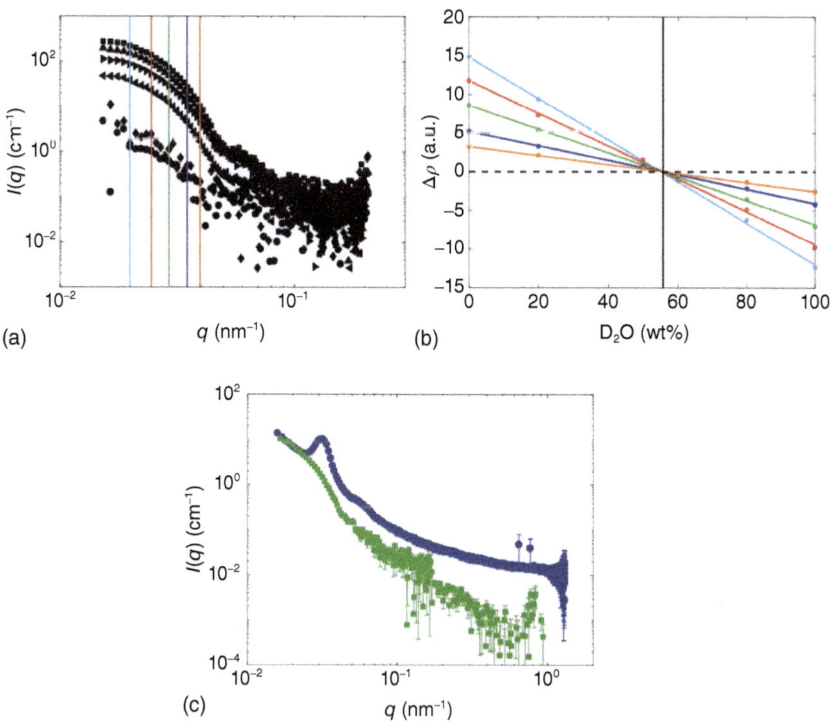

Figure 6.9 (a) Small-angle neutron-scattering intensities, $I(q)$, versus scattering vector, q, of microgels synthesized with d3-NIPAM – $C_6D_3H_8NO$ – in solutions of water/heavy water mixtures with: 100 (upside triangles), 80 (left-side triangles), 60 (circles), 50 (diamonds), 20 (right-side triangles), and 0 (squares) wt% D_2O. All measurements have been conducted at (20.0 ± 0.1) °C. The colored vertical solid lines represent fixed q values used to extract $\Delta\rho$ in panel (b). (b) Values of the contrast in scattering-length density, $\Delta\rho$, as a function of the wt% of D_2O in the solvent. Different colors correspond to the different q values chosen in panel (a) to extract $\Delta\rho$. The solid lines are linear fits to the data. The dashed black horizontal line indicates zero. The black solid vertical line represents the match point of the microgels synthesized with d3-pNIPAM. (c) Small-angle X-ray (blue circles) and neutron-scattering with contrast variation (green squares) intensities, $I(q)$, versus scattering vector, q. Source: Scotti et al. [32]/with permission of American Physical Society.

$\sqrt{I(q)}$ is chosen to be negative for samples with a D_2O content $\gtrsim 55$ wt% to allow fitting the data with a single straight line. All the linear fits shown in Figure 6.9b cross the $\Delta\rho = 0$ line (dashed black line) at the very same point, corresponding to (55.7 ± 0.3) wt% D_2O. This reveals that the match point of d3-pNIPAM-based microgels is 3.137×10^{-6} Å^{-2}. Again, the theoretical value of 2.949×10^{-6} Å^{-2} calculated with Eq. (6.44) is very close, yet not identical to the experimental value. Thus, deuterated microgels can be contrast matched to obtain the form factor of hydrogenated microgels in a concentrated sample. This is shown in Figure 6.9c, where SAXS and SANS data obtained with the sample are shown. While the SAXS data clearly show the structure peak characteristic of a concentrated suspension, in this case, in the q range from $3 \cdot 10^{-2}$ to $5 \cdot 10^{-2}$ nm^{-1}, the SANS data show no structure peak but only the form factor of the hydrogenated microgels in that q range.

6.4 Suspension Phase Behavior

The compressibility of microgels plays a key role in their equilibrium phase behavior and leads to a richer response to crowding when compared to hard, incompressible colloids. In addition to the deformability of an individual microgel, the interaction potential of microgels can be described as "soft." In contrast to hard colloids, the center-to-center distance of microgels can be smaller than the diameter in dilute solution and the interaction potential changes gradually. Therefore, in contrast to hard colloids for which the interaction can be approximated with a box-like potential that is zero for distances larger than the particle diameter and goes to infinity once particles come into contact, microgel interactions have been approximated with various "soft" potentials, including the Hertzian potential [45, 61, 90] and inverse-power-law potential of the type r^{-n} [39, 91, 92], with $n \in \mathbb{N}$ and r the microgel–microgel distance. Experimentally, more compressible microgels show a significantly smaller exponent n [32, 69, 93].

Hard colloids have a first-order phase transition from the liquid to the crystalline state. This was first observed in computer simulations [94–96]. However, the existence of this entropic freezing of purely repulsive hard spheres was only broadly accepted after Pusey and van Megan studied suspensions of hard-sphere-like poly(methyl methacrylate) (PMMA) particles, which confirmed the predicted transition with fluid-crystal coexistence between the freezing point $\phi_f = 0.494$ and the melting point $\phi_m = 0.536$, and with fully crystalline samples observed at higher volume fractions [66].

For microgel suspensions, we use the generalized volume fraction ζ as a proxy for the particle concentration. Experimental results show that, compared to hard spheres, the freezing and melting transitions in microgel suspensions occur at higher concentrations, $\zeta_f = 0.58 \pm 0.02$ and $\zeta_m = 0.62 \pm 0.03$, respectively [38–40, 68], see Figure 6.10a. Transitions at these high concentrations could be due to a soft interaction potential. Computer simulations using an inverse-power-law potential predicted the liquid-to-crystal phase transition for soft spheres to be located close to the ζ_f and ζ_m found experimentally [91, 92]. However, when this simple model is used to obtain measurable quantities, such as the radial distribution function $g(r)$ or the structure factor $S(q)$ of microgel solutions, it fails to reproduce experimental observations. For this reason, the Hertzian potential, which describes the interaction between elastic spheres, has been proposed as a theoretical representation of the microgel–microgel interaction [97], and it has been shown to correctly describe the fluid phase of microgels in concentrated suspensions [90]. This model, however, does not reproduce the presence of a re-entrant fluid phase at high generalized packing fractions observed with confocal laser microscopy in suspensions of pNIPAM [40]. Using computer simulations, Zaccarelli and coworkers examined the validity of the Hertzian interaction in crowded suspensions [61]. Based on these simulations and the observed presence of a more crosslinked core surrounded by a more compressible shell, it was proposed that an effective interaction, based on a cascade of Hertzian potentials corresponding to the different compressibilities within the single microgel, can be used [45]. This model represents a promising theoretical description of the microgel–microgel interaction that can be further tested against

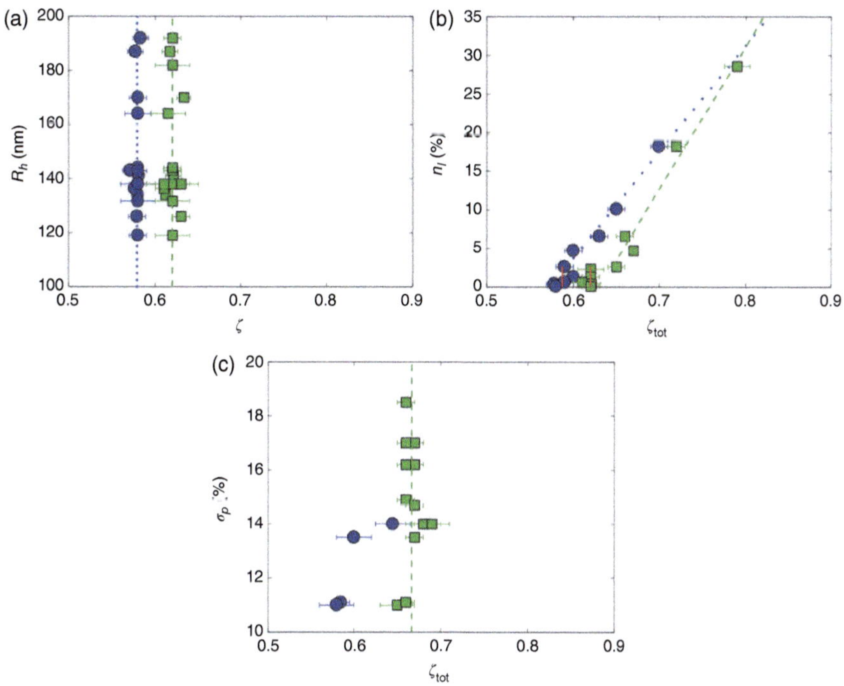

Figure 6.10 Values of ζ at the freezing (circles) and melting (squares) points were observed in (a) monodisperse samples – size polydispersity $\sigma_p \lesssim 10\%$ with microgel hydrodynamic radii, R_h, in the range between 110 and 200 nm. (b) Samples with bimodal size distribution with varying fractions of large particles, n_l. (c) Samples with size polydispersity σ_p between 11% and 19%. The behavior of the freezing and melting points is highlighted by the dotted and dashed lines, respectively. The solid lines in (b) represent the freezing and melting lines for $n_l \lesssim 2.5\%$. All phase behavior data are taken at $T = (18.3 \pm 0.5)$ °C. Source: Figure taken from Scotti et al. [38]/with permission of American Physical Society.

experimental data, e.g. the structure factors of crowded microgel suspensions well above the liquid phase.

The external fuzziness and, in general, the inhomogeneous internal architecture of microgels affect their interaction potential. In particular, microgels are able to adapt their shape and size depending on the suspension concentration. Iyer and Lyon observed the spontaneous deswelling of a few large microgels embedded in a "sea" of smaller microgels [34]. This "self-healing" mechanism is absent in other colloidal systems, such as hard colloids, soap bubbles, or multilamellar vesicles [98]. The combination of SANS and osmotic pressure measurements has shown that this isotropic deswelling of the larger microgels is triggered by the percolation of the counter-ion clouds surrounding peripherally charged microgels. This percolation is expected to occur when the effective particles, including their counter-ion clouds, effectively fill the available volume of the suspension. We, therefore, think of an effective particle radius $R_{eff} = R_p + \Delta R_i$, where ΔR_i is the radius increase due to the width of the counter-ion cloud, and we expect percolation when the volume fraction $\phi_{\text{eff}} = \frac{4\pi N R_{\text{eff}}^3}{3\ V_{\text{tot}}} \gtrsim 1$. At this point, the contribution of the counter-ions to the suspension osmotic pressure can be compared to the role of electrons in the conduction band of a metal; with percolated counter-ion clouds, the previously bound counter-ions are shared by all microgels and their translational degrees of freedom expand to all of the space filled by the clouds. Thus, they contribute to the suspension osmotic pressure [29]. Once the osmotic pressure becomes comparable to the microgel bulk modulus, the microgels isotropically deswell. This deswelling triggered by an increase in osmotic pressure set by the counter-ions causes the microgel deswelling observed in concentrated suspensions that can often occur before the microgels come into direct contact, i.e. at volume fractions $< \phi_{\text{rcp}}$.

This kind of self-healing capability also changes the role of size polydispersity for the liquid-to-crystal phase transition for suspensions with both monomodal and bimodal size distributions, as the suspension spontaneously reduces its polydispersity when the largest microgels are also the softest. Crystals indeed form in suspensions of microgels with a size polydispersity as high as 19% [38]. Similarly, once two families of microgels – large (l) and small (s) with a ratio between radii $R_l/R_s = 1.5$ and a number fraction of large microgels $n_l = 30\%$ – are mixed together, crystals form incorporating the large deswollen microgels in the lattice of the small microgels [38]. For these solutions, a shift of the freezing and melting points toward higher concentrations is observed with increasing n_l or size polydispersity, Figure 6.10b and c, respectively. As a result of the high polydispersity or a high n_l, the coexistence range of liquid and crystal coexistence gets narrower and both the freezing and the melting points are observed at $\zeta_m = 0.68$, when the polydispersity is 19%, or at $\zeta_m = 0.78$, when $n_l = 30\%$. The reason for this shift in ζ_f is that the concentration of the suspension has to rise to allow the counter-ion clouds to percolate within the available volume and set up an osmotic pressure high enough to overcome the bulk modulus of the larger microgels causing their deswelling, and thus minimizing their role as point defects.

If the microgels do not deswell before contact, i.e. if their bulk modulus is larger than the solution osmotic pressure, their fuzzy peripheries influence the solution equilibrium phase behavior. Face-centered cubic (or hexagonally close packed) crystals are the equilibrium structure formed by microgel suspensions at $\zeta > \zeta_m$ [17, 38]. Transient body-centered cubic (bcc) crystals have been observed for microgels made of poly(*N*-isopropylacrylamide-co-acrylic acid) (PNIPAm-co-AAc) [89]. The presence of this lattice can be rationalized by considering particles that interact via their soft, fuzzy coronae. On the one hand, the purely repulsive interaction favors the fcc lattice. On the other hand, the reduction of the contact area of the fuzzy coronae can maximize their entropy. This favors the bcc, which reduces the contact area compared to the fcc lattice [89]. Bcc crystals were also observed in other work [17]. Ultrasoft microgels can be obtained without the addition of crosslinker during the synthesis [88]. These so-called ultralow-crosslinked microgels show stable coexistence between fcc and bcc crystals [32]. This can be explained by the presence of microgels completely collapsed and microgels that still present a fuzzy external shell in the concentration range where the fcc–bcc coexistence is observed. This feature suggests a similarity between ultrasoft microgels and other soft objects; for instance, a shear-induced transition from stable fcc to bcc has been observed for star polymers [99].

So far, we have considered only microgels with peripheral charges. However, ionic microgels that change their size in response to changes in the suspension pH can be obtained by adding monomers with ionizable groups during the synthesis [5, 16, 100, 101]. Given the large-size difference between the counter-ions and the microgel, a Donnan potential that decays to zero in the vicinity of the particle can be used to describe the charge on the polymer network. Similar to the case of peripherally charged microgels, there are counter-ions in the particle periphery that, for entropic reasons, can escape the Donnan potential of the microgel. The fraction, Γ, of these ions can be estimated as [57]:

$$\Gamma = \frac{\left(\kappa^{-1} + \frac{d}{2}\right)^3 - \left(\frac{d}{2}\right)^3}{\left(\frac{d}{2}\right)^3} \approx 6\frac{\kappa^{-1}}{d} = \sqrt{3\frac{d/2}{l_B Q}} \tag{6.57}$$

The existence of a characteristic length scale $\kappa^{-1} = \sqrt{\frac{(d/2)^3}{2l_B Q}}$, with l_B the Bjerrum length, d the microgel diameter and Q its number of fixed charges, implies that the counter-ions attracted to the microgel with a strength $\lesssim k_B T$ can leave the microgel determining the value of the suspension osmotic pressure. These "free" counter-ions can be treated as an ideal gas and their contribution to the suspension osmotic pressure can be expressed in terms of the microgel volume fraction:

$$\pi_m(\phi) = k_B T \frac{\Gamma Q N}{V - N v_0} = k_B T \frac{\Gamma Q}{v_0} \frac{\phi}{1 - \phi} \tag{6.58}$$

where ΓQ is the number of free counter-ions per microgel contributing to π and $V - N v_0$ is the volume that is available to them. Here, we notice that it is the true volume fraction, ϕ, that enters in Eq. (6.58), and not the generalized volume fraction, ζ, given by the fully swollen size of the microgel. Eq. (6.58) is the basis of the method described in the introduction to map the generalized volume fraction, ζ, into ϕ [59].

6.5 Flow Properties

It is well known that the flow properties of colloidal suspensions strongly depend on colloid concentration. A textbook example is represented by the shear-thinning and shear-thickening observed in suspensions of hard-sphere colloids [102, 103]. For microgel suspensions, the viscosity, η, depends strongly on the microgel concentration and the applied shear rate, $\dot{\gamma}$. For very dilute microgel solutions, the Batchelor–Einstein equation [104] correctly describes how η depends on concentration. In addition, the fluid can be considered Newtonian, i.e. $\eta \neq \eta(\dot{\gamma})$. The Batchelor–Einstein equation correctly describes the behavior not only of regular microgels synthesized with different amounts of crosslinker [29, 30, 35–39, 41, 44, 57] but it is also valid for microgels with different internal structures and compressibility, such as ultralow-crosslinked microgels [31, 32, 69, 93] and hollow microgels [31, 33, 65]. We note that in the highly dilute limit, microgels do not change their size or shape, i.e. being soft has no effect, since hydrodynamically there is essentially no flow through the microgel, but only around the particle. Hence, in dilute concentration, microgels essentially behave as normal spheres.

Once the microgel concentration rises, microgel–microgel interactions are not negligible anymore and the viscosity of the solution depends on $\dot{\gamma}$. Similar to the case of hard spheres, there is a plateau for low $\dot{\gamma}$, corresponding to the zero-shear viscosity, η_0. With increasing $\dot{\gamma}$, the solution viscosity decreases and eventually reaches a second plateau, η_∞, for high enough values of $\dot{\gamma}$. In the literature, there are many different models that have been developed to capture this behavior [58, 105, 106]. One of the most commonly used is the Cross model [28, 93, 106–108]:

$$\eta = \eta_\infty + \frac{\eta_0 - \eta_\infty}{1 + (\frac{\dot{\gamma}}{\dot{\gamma}_c})^m} \tag{6.59}$$

where $m \in \mathbb{R}$ and $\dot{\gamma}_c$ is the critical value of the shear rate halfway between η_0 and η_∞.

Figure 6.11 shows the viscosity versus shear rate for (squares) hard, (triangles) soft, and (diamonds) ultrasoft microgels synthesized by adding 5, 2.5, and 0 mol% crosslinker. As can be seen, the Cross model (black lines) correctly captures the course of the data.

A question that still remains is whether particle deformability affects the flow behavior. In the past, Richtering and coworkers studied microgel suspensions under flow combining rheology and SANS. They found evidence for shear-induced de-mixing and the appearance of two separated phases [108]. In addition, the measurement of the particle arrangement under flow showed that on the one

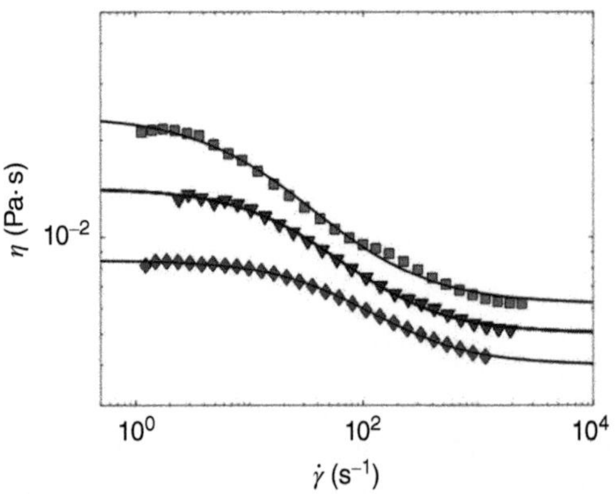

Figure 6.11 Viscosity, η, versus shear rate, $\dot{\gamma}$, for microgels with crosslink densities 5 mol% (squares) and 2.5 mol% (triangles), and for ultralow-crosslinked microgels (diamonds). All samples have $\zeta = 0.55$. Black lines are fits obtained with the Cross model. The fitting parameters are: $\eta_0 = 0.024$ Pa·s, $\eta_\infty = 0.006$ Pa·s, $\dot{\gamma}_c = 16.17$ s^{-1}, and $m = 0.84$ for the 5 mol% crosslinked microgels (squares); $\eta_0 = 0.014$ Pa·s, $\eta_\infty = 0.005$ Pa·s, $\dot{\gamma}_c = 36.28$ s^{-1} and $m = 0.94$ for the 2.5 mol% crosslinked microgels (triangles); and $\eta_0 = 0.008$ Pa·s, $\eta_\infty = 0.004$ Pa·s, $\dot{\gamma}_c = 79.04$ s^{-1} and $m = 0.96$ for the ultralow-crosslinked microgels (diamonds).

hand, these particles changed size, and on the other hand, that sliding hexagonal close-packed layers formed [108].

From the flow curves, one can obtain η_0 and how it changes with ζ. The increase in the viscosity of concentrated suspensions of soft and deformable objects is often described by a model originally thought of for describing molecular glass formers, the Vogel–Fulcher–Tammann (VFT) model [44, 109, 110]:

$$\eta = \eta_0 \exp\left(\frac{A\zeta}{\zeta_0 - \zeta}\right) \tag{6.60}$$

with ζ_0 the generalized volume fraction where $\eta \to \infty$, and A is a constant that controls the growth of η in its approach to ζ_0 [44, 109]. While the temperature, T, is the control variable in molecular glass formers, ζ is the control parameter for microgel suspensions. The suspension viscosity is expected to increase with ζ and eventually diverge at a critical value of ζ_0. We notice that this is the upper limit for ζ_g, the generalized volume fraction at the glass transition; ζ_0 sets the divergence of the structural relaxation time, while $\zeta_g < \zeta_0$ corresponds to the ζ where the configurational time scales matches the time scale associated with the rate of sample concentration (corresponding to the cooling rate when T is the control variable). Different rates of sample concentration thus affect the value of ζ_g; for slower concentration rates, ζ_g increases, progressively approaching ζ_0.

The $\eta_r = \eta/\eta_0$ versus ζ relation strongly depends on the microgel compressibility, estimated for instance by the swelling ratio, as shown in Figure 6.12. The behavior of hard microgels, synthesized with 10 and 5 mol% crosslinkers (circles and squares, respectively), appears to be similar to the one observed for hard incompressible colloids (dashed line), and the divergence of the viscosity is

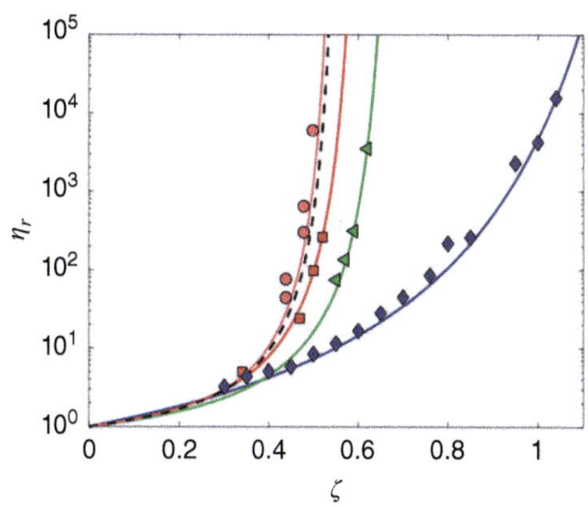

Figure 6.12 Relative viscosity, $\eta_r = \eta/\eta_0$, versus generalized volume fraction, ζ, for: ULC microgels (diamonds), 1 mol% crosslinked microgels (triangles), 5 mol% crosslinked microgels (squares), and 10 mol% crosslinked microgels (circles). Solid lines are fits to the data with the VFT model [93]. The fitting parameters as obtained using Eq. (6.60) are: $\zeta_0 = 1.45$ and $A = 3.77$ for the ULC microgels (diamonds); $\zeta_0 = 0.71$ and $A = 1.15$ for 1 mol% crosslinked microgels (triangles); $\zeta_0 = 0.63$ and $A = 1.19$ for the 5 mol% crosslinked microgels (squares); and $\zeta_0 = 0.57$ and $A = 0.98$ for the 10 mol% crosslinked microgels (circles). The dashed line is the viscosity for hard spheres according to the VFT model with $\zeta_0 = 0.58$.

observed in the proximity of $\zeta_0 \simeq \phi_{\mathrm{glass}} = 0.58$, the lowest concentration at which the glass transition happens for hard spheres [56]. For softer microgels, i.e. with a lower concentration of crosslinker agent used during the synthesis, the data still diverge but at a larger generalized volume fraction (triangles). Finally, for the softest pNIPAM-based microgels, one can obtain with precipitation polymerization, the ultralow-crosslinked microgels, the increase of the viscosity with ζ is much more gradual and the value of ζ_0 appears well above $\zeta = 1$. As mentioned above, these very soft microgels can deswell and deform and the real volume occupied by them in solution, ϕ, is much smaller than ζ.

At high concentration, microgels form a glass or are in a jammed state [56, 109, 111]. In a glass, the internal energy of the system is close to the thermal energy and thermal fluctuations are relevant. In contrast, other phenomena, such as particle deformation or interpenetration, dominate when the system is in a jammed state. As a consequence, the internal energy per degree of freedom in a jammed system is much larger than $k_BT/2$. The plateau of the shear modulus, G_p, that is the value of the shear modulus, G', corresponding to the minimum in the loss modulus, G'', measured in an oscillatory sweep when the system is in the linear viscoelastic regime, can be used to distinguish between a glass and a jammed state as proposed by Pellet and Cloitre. They rescaled G_p with k_BT/R_h^3, the entropic elasticity of a colloidal solution composed of particles of hydrodynamic radius R_h, to distinguish the glass and jammed states finding a cutoff value of $G_p^* = 100k_BT/R_h^3$ [58].

In Figure 6.13, $G_pR_h^3/(k_BT)$ is shown as a function of ζ/ζ_0 [93]. The horizontal dashed line at $100k_BT/R^3$ separates the glass and jammed states [58]. Data for hard and soft microgels in the glass state (circles, squares, and triangles) collapse together in a master curve. Thus, independently on the amount of crosslinker used

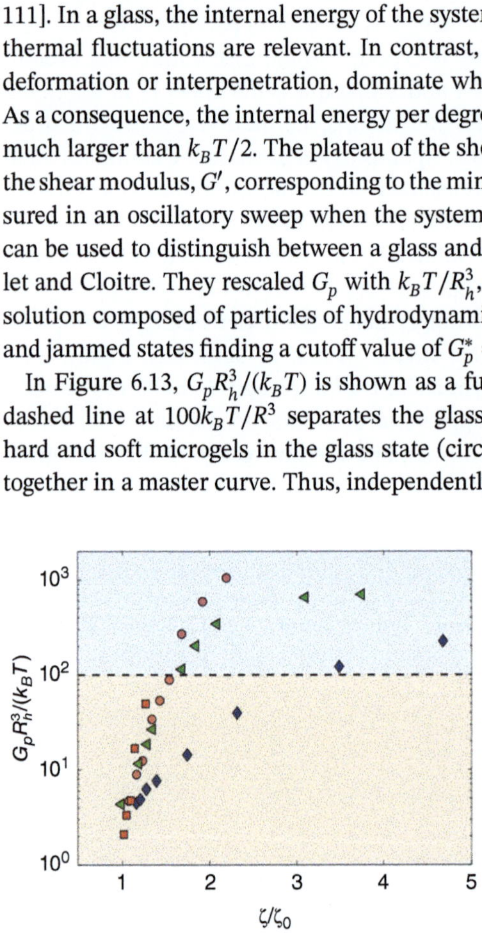

Figure 6.13 Rescaled plateau moduli, G_p, versus ζ/ζ_0 for ULC microgels (diamonds), 1 mol% crosslinked microgels (triangles), 5 mol% crosslinked microgels (squares), and 10 mol% crosslinked microgels (circles). G_p is rescaled with $R_h^3/(k_BT)$ and ζ is divided by the generalized volume fraction ζ_0 corresponding to the divergence of the relative viscosity according to the VFT model. The dashed line represents the transition between glass and jammed states according to Refs. [58, 93]. Source: Scotti et al. [93]/with permission of John Wiley & Sons.

in the synthesis, the internal energy of microgels with a similar architecture – a more crosslinked core surrounded by a fuzzy shell – has a similar dependence on concentration. In contrast, the behavior is different in the jammed phase. In these jammed states, the microgels are compressed, deformed, and/or interpenetrating and therefore, network compressibility strongly affects the elasticity of the suspension. Finally, the behavior of ultralow-crosslinked microgels (diamonds) differs at all ζ/ζ_0 from the behavior of the other microgels, as they have an approximately uniform and very low polymer density due to the absence of exogenous crosslinker during their synthesis [69]. Their increased softness and compressibility significantly lower $G_p R_h^3/(k_B T)$ compared to microgels synthesized with exogenous crosslinker [26]. This can be due to the fact that very soft microgels experience osmotic deswelling and/or faceting already a relatively low concentration due to the smaller bulk modulus compared to harder microgels.

6.6 Final Remarks

In this chapter, we have reviewed the basic principles of the swelling thermodynamics of microgels and discussed the basics of light and neutronscattering, focusing on 3D dynamic light scattering (3D-DLS) and SANS with contrast variation; both are used to study concentrated suspensions of colloids [26, 28–33, 41, 45, 58, 65, 112, 113]. These techniques both probe the structure and dynamics of microgel suspensions in crowded environments where these particles can deswell and/or facet. This offers the possibility to investigate the role of particle compressibility – or softness – on both their phase behavior and flow properties, as discussed in the last sections of the chapter.

One of the key aspects of microgel suspensions, and colloidal suspensions more generally, pertains to the origin of the suspension osmotic pressure [42, 59]. In many cases, this quantity results from the counter-ions associated with the microgels – both ionic or neutral – and determines whether and when isotropic deswelling takes place [29]. In dilute condition, the osmotic pressure is determined by counter-ions leaving the microgel to maximize their configurational entropy [42, 57–59]. In concentrated suspensions of "neutral" microgels, which only have peripheral charges, the overlap and percolation of counter-ion clouds lead to an abrupt increase in osmotic pressure [29, 38, 114]. Depending on the bulk modulus of the microgel, isotropic deswelling can happen even before direct contact between the particles occurs [38]. Alternatively, when $\pi < K_p$, microgels can come into direct contact and facet before deswelling [36, 37, 59]. In general, the key aspect determining the response of a microgel to crowding is the relation between suspension osmotic pressure and single microgel compressibility. The latter can be affected during the synthesis by controlling either the amount of crosslinker used during the synthesis or the particle internal architecture [31]. In our opinion, these aspects, in particular the role of counter-ions and internal architecture in the microgel deswelling, have almost not been considered explicitly in theoretical studies and computer simulations. We look with interest at the possibility of including these aspects in the more advanced simulation techniques that are able to reproduce the internal architecture of single microgels [61, 115] and simulate their behavior in concentrated solutions [116–118].

Another key aspect is the difference between generalized volume fraction and the true volume fraction occupied by the microgels in solution. The combination of osmotic pressure measurements [59] with SANS with contrast variation experiments [26] allows determining the volume occupied by the particles and hence ϕ, as well as the internal structure of the particles. We believe that this combined knowledge can be useful to achieve a unified description of the phase behavior and flow properties of microgel suspensions and also offer a better way to compare their properties to the ones of hard incompressible spheres so that the role of particle compressibility on the macroscopic properties of the suspension can be further understood. Furthermore, the use of SANS with contrast variation can also help in probing the individual microgel deformation and global microgel arrangement in solution under shear to understand whether the failure of the classical models used for hard spheres to describe shear-thinning and shear-thickening effects is related to microgel compressibility. These experiments will also shed light on the role of particle compressibility in the formation of glasses and the jammed state.

Another problem that we think can be solved using SANS with contrast variation is the determination of the microgel bulk modulus. Here, we can imagine experiments using deuterated large polymers in combination with a "few" microgels. These polymers will exert osmotic stress and progressively compress the microgels but at the same time, they can be contrast matched with the solvent, allowing for the direct measurement of the changes in microgel size and structure as a function of the exerted osmotic pressure. Similar methods were already used with light [119], but there the stressing polymer and microgel particles both contributed to the scattering intensity, making elucidating the microgel size difficult in certain experimental conditions. Additionally, the use of contrast variation will not only enable a clearer estimation of K_p but also distinguish the bulk modulus of the core and shell of microgels with internal morphology.

We believe that microgels will continue to be a widely used system to continue to address both relevant scientific questions where deformability and compressibility are both at play and applications where their responsiveness and softness offer unique potential to adapt and reorganize in response to environmental conditions.

References

1 Heyes, D. and Brańka, A. (2009). *Soft Matter* 5: 2681.

2 Lyon, L.A. and Fernandez-Nieves, A. (2012). *Annu. Rev. Phys. Chem.* 63: 25.

3 Brijitta, J. and Schurtenberger, P. (2019). *Curr. Opin. Colloid Interface Sci.* 40: 87.

4 Minami, S., Suzuki, D., and Urayama, K. (2019). *Curr. Opin. Colloid Interface Sci.* 43: 113.

5 Plamper, F.A. and Richtering, W. (2017). *Acc. Chem. Res.* 50: 131.

6 Karg, M., Pich, A., Hellweg, T. et al. (2019). *Langmuir* 35: 6231.
7 Oberdisse, J. and Hellweg, T. (2020). *Colloid Polym. Sci.* 298: 921–935.
8 Tanaka, T., Annaka, M., Ilmain, F. et al. (1992). Phase transitions of gels. T. K. Karalis. *NATO ASI series, H64, mechanics of swelling*. Berlin: Springer-Verlag, 683–703.
9 Chang, Y.-W., Dimitriyev, M.S., Souslov, A. et al. (2018). *Phys. Rev. E* 98: 020501.
10 Keidel, R., Ghavami, A., Lugo, D.M. et al. (2018). *Science advances* 4: eaao7086.
11 Wrede, O., Reimann, Y., Lülsdorf, S. et al. (2018). *Sci. Rep.* 8: 1.
12 Pelton, R. and Chibante, P. (1986). *Colloids Surf.* 20: 247.
13 Pelton, R. (2000). *Adv. Colloid Interface Sci.* 85: 1.
14 Stieger, M., Richtering, W., Pedersen, J.S., and Lindner, P. (2004). *J. Chem. Phys.* 120: 6197.
15 Stieger, M., Pedersen, J.S., Lindner, P., and Richtering, W. (2004). *Langmuir* 20: 7283.
16 Hoare, T. and Pelton, R. (2004). *Macromolecules* 37: 2544.
17 Mohanty, P.S. and Richtering, W. (2008). *J. Phys. Chem. B* 112: 14692.
18 Lietor-Santos, J.J., Sierra-Martin, B., Vavrin, R. et al. (2009). *Macromolecules* 42: 6225.
19 Lietor-Santos, J.J., Gasser, U., Vavrin, R. et al. (2010). *J. Chem. Phys.* 133: 034901.
20 Lietor-Santos, J.J., Sierra-Martin, B., Gasser, U., and Fernandez-Nieves, A. (2011). *Soft Matter* 7: 6370.
21 Flory, P.J. and Rehner, J. (1943). *J. Chem. Phys.* 11: 512.
22 Flory, P.J. and Rehner, J. (1943). *J. Chem. Phys.* 11: 521.
23 Flory, P.J. (1953). *Principles of Polymer Chemistry*. Ithaca: Cornell University Press.
24 Lopez, C.G. and Richtering, W. (2017). *Soft Matter* 13: 8271.
25 Romeo, G., Imperiali, L., Kim, J.-W. et al. (2012). *J Chem. Phys.* 136: 124905.
26 Scotti, A. (2021). *Soft Matter* 17: 5548.
27 Pich, A. and Richtering, W. (2010). Chemical Design of Responsive Microgels, p. 1.
28 Gasser, U., Hyatt, J.S., Lietor-Santos, J.-J. et al. (2014). *J. Chem. Phys.* 141: 034901.
29 Scotti, A., Gasser, U., Herman, E.S. et al. (2016). *Proc. Natl. Acad. Sci. U.S.A.* 113: 5576.
30 Mohanty, P.S., Nöjd, S., Gruijthuijsen, Kv. et al. (2017). *Sci. Rep.* 7: 1487.
31 Scotti, A., Denton, A.R., Brugnoni, M. et al. (2019). *Macromolecules* 52: 3995.
32 Scotti, A., Houston, J., Brugnoni, M. et al. (2020). *Phys. Rev. E* 102: 052602.
33 Scotti, A., Denton, A., Brugnoni, M. et al. (2021). *Phys. Rev. E* 103: 022612.
34 Iyer, A.S.J. and Lyon, L.A. (2009). *Angew. Chem.* 121: 4632.
35 De Aguiar, I.B., Van de Laar, T., Meireles, M. et al. (2017). *Sci. Rep.* 7: 1.
36 Conley, G.M., Aebischer, P., Nöjd, S. et al. (2017). *Sci. Adv.* 3. https://doi.org/10.1126/sciadv.1700969.
37 Conley, G.M., Zhang, C., Aebischer, P. et al. (2019). *Nat. Commun.* 10: 2436.

38 Scotti, A., Gasser, U., Herman, E. et al. (2017). *Phys. Rev. E* 96: 032609.
39 Senff, H. and Richtering, W. (1999). *J. Chem. Phys.* 111: 1705.
40 Paloli, D., Mohanty, P.S., Crassous, J.J. et al. (2013). *Soft Matter* 9: 3000.
41 Nöjd, S., Holmqvist, P., Boon, N. et al. (2018). *Soft Matter* 14: 4150.
42 Pelaez-Fernandez, M., Souslov, A., Lyon, L. et al. (2015). *Phys. Rev. Lett.* 114: 098303.
43 Sierra-Martin, B. and Fernandez-Nieves, A. (2012). *Soft Matter* 8: 4141.
44 Van Der Scheer, P., Van De Laar, T., Van Der Gucht, J. et al. (2017). *ACS Nano* 11: 6755.
45 Bergman, M.J., Gnan, N., Obiols-Rabasa, M. et al. (2018). *Nat. Commun.* 9: 1.
46 Otto, P., Bergmann, S., Sandmeyer, A. et al. (2020). *Nanoscale Adv.* 2: 323.
47 Wrede, O., Bergmann, S., Hannappel, Y. et al. (2020). *Soft Matter* 16: 8078.
48 Gelissen, A.P., Oppermann, A., Caumanns, T. et al. (2016). *Nano Lett.* 16: 7295.
49 Schulte, M.F., Bochenek, S., Brugnoni, M. et al. (2020). *Angew. Chem. Int. Ed.* https://doi.org/10.1002/anie.202011615.
50 Bochenek, S., Scotti, A., Ogieglo, W. et al. (2019). *Langmuir* 35: 16780.
51 Camerin, F., Gnan, N., Ruiz-Franco, J. et al. (2020). *Phys. Rev. X* 10: 031012.
52 Bochenek, S., Scotti, A., and Richtering, W. (2021). *Soft Matter* 17: 976.
53 Scheffold, F. (2020). *Nat. Commun.* 11: 1.
54 Cipelletti, L., Bissig, H., Trappe, V. et al. (2002). *J. Phys. Condens. Matter* 15: S257.
55 Scheffold, F. and Cerbino, R. (2007). *Curr. Opin. Colloid Interface Sci.* 12: 50.
56 Brambilla, G., El Masri, D., Pierno, M. et al. (2009). *Phys. Rev. Lett.* 102: 085703.
57 Borrega, R., Cloitre, M., Betremieux, I. et al. (1999). *EPL (Europhys. Lett.)* 47: 729.
58 Pellet, C. and Cloitre, M. (2016). *Soft Matter* 12: 3710.
59 Scotti, A., Pelaez-Fernandez, M., Gasser, U., and Fernandez-Nieves, A. (2021). *Phys. Rev. E* 103: 012609.
60 Lally, S., Bird, R., Freemont, T.J., and Saunders, B.R. (2009). *Colloid Polym. Sci.* 287: 335.
61 Rovigatti, L., Gnan, N., Ninarello, A., and Zaccarelli, E. (2019). *Macromolecules* 52: 4895.
62 Nikolov, S.V., Fernandez-Nieves, A., and Alexeev, A. (2020). *Proc. Natl. Acad. Sci. U.S.A.* 117: 27096.
63 Debord, S.B. and Lyon, L.A. (2003). *J. Phys. Chem. B* 107: 2927.
64 St. John, A.N., Breedveld, V., and Lyon, L.A. (2007). *J. Phys. Chem. B* 111: 7796.
65 Scotti, A., Brugnoni, M., Rudov, A. et al. (2018). *J. Chem. Phys.* 148: 174903.
66 Pusey, P.N. and Van Megen, W. (1986). *Nature* 320: 340.
67 Gasser, U. (2009). *J. Phys. Condens. Matter* 21: 203101.
68 Paulin, S., Ackerson, B.J., and Wolfe, M. (1996). *J. Colloid Interface Sci.* 178: 251.
69 Scotti, A., Bochenek, S., Brugnoni, M. et al. (2019). *Nat. Commun.* 10: 1.
70 Dusek, K. and Patterson, D. (1968). *J. Polym. Sci., Part B: Polym. Phys.* 6: 1209.
71 Tanaka, T. (1986). *Physica A* 140: 261.

72 Rubinstein, M. and Colby, R.H. (2003). *Polymer Physics*. Pennsylvania, USA: Oxford University Press.

73 Fernandez-Nieves, A., Wyss, H., Mattsson, J., and Weitz, D.A. (2011). *Microgel Suspensions: Fundamentals and Applications*. Wiley.

74 Nigro, V., Ruzicka, B., Ruta, B. et al. (2020). *Macromolecules* 53: 1596.

75 Nickel, A.C., Scotti, A., Houston, J.E. et al. (2019). *Nano Lett.* 19: 8161.

76 Dhont, J.K.G. (1996). *An Introduction to Dynamics of Colloids*. Elsevier.

77 Pusey, P.N. (1999). *Curr. Opin. Colloid Interface Sci.* 4: 177.

78 Philippe, A.-M., Truzzolillo, D., Galvan-Myoshi, J. et al. (2018). *Phys. Rev. E* 97: 040601.

79 Sakurai, J.J.S. and Napolitano, J.J. (2014). *Modern Quantum Mechanics*. Pearson.

80 Berne, B.J. and Pecora, R. (2000). *Dynamic Light Scattering: With Applications to Chemistry, Biology, and Physics*. Courier Corporation.

81 Zemb, T. and Lindner, P. (2002). *Neutron, X-Rays and Light. Scattering Methods Applied to Soft Condensed Matter*. Elsevier Science: North-Holl and Delta.

82 Varley, F.S. (1992). *Neutron News* 3: 29.

83 Furrer, A., Mesot, J.F., and Strässle, T. (2009). *Neutron Scattering in Condensed Matter Physics*, vol. 4. World Scientific Publishing Company.

84 Pedersen, J.S. (1997). *Adv. Colloid Interface Sci.* 70: 171.

85 Sjöblom, J., Lindberg, R., and Friberg, S.E. (1996). *Adv. Colloid Interface Sci.* 65: 125.

86 Fisher, M.E. (1964). *J. Math. Phys.* 5: 944.

87 Brugnoni, M., Scotti, A., Rudov, A.A. et al. (2018). *Macromolecules* 51: 2662.

88 Brugnoni, M., Nickel, A.C., Kröger, L.C. et al. (2019). *Polym. Chem.* 10: 2397.

89 Gasser, U., Lietor-Santos, J.-J., Scotti, A. et al. (2013). *Phys. Rev. E* 88: 054702.

90 Mohanty, P.S., Paloli, D., Crassous, J.J. et al. (2014). *J. Chem. Phys.* 140: 094901.

91 Hoover, W.G., Gray, S.G., and Johnson, K.W. (1971). *J. Chem. Phys.* 55: 1128.

92 Agrawal, R. and Kofke, D.A. (1995). *J. Phys. Chem. B* 85:

93 Scotti, A., Brugnoni, M., Lopez, C.G. et al. (2020). *Soft Matter* 16: 668.

94 Alder, B.J. and Wainwright, T.E. (1957). *J. Chem. Phys.* 27: 1208.

95 Hoover, W. and Ree, F. (1968). *J. Chem. Phys.* 49: 3609.

96 Frenkel, D. (1994). *J. Phys. Condens. Matter* 6: A71.

97 Pàmies, J.C., Cacciuto, A., and Frenkel, D. (2009). *J. Chem. Phys.* 131: 044514.

98 Frenkel, D. (2009). *Nature* 460: 465.

99 Ruiz-Franco, J., Marakis, J., Gnan, N. et al. (2018). *Phys. Rev. Lett.* 120: 078003.

100 Debord, J.D. and Lyon, L.A. (2003). *Langmuir* 19: 7662.

101 Pich, A., Tessier, A., Boyko, V. et al. (2006). *Macromolecules* 39: 7701.

102 Maranzano, B.J. and Wagner, N.J. (2001). *J. Rheol.* 45: 1205.

103 Gurnon, A.K. and Wagner, N.J. (2015). *J. Fluid Mech.* 769: 242.

104 Batchelor, G.K. (1977). *J. Fluid Mech.* 83: 97.

105 Gupta, S., Stellbrink, J., Zaccarelli, E. et al. (2015). *Phys. Rev. Lett.* 115: 128302.

106 Lopez, C.G., Colby, R.H., and Cabral, J.T. (2018). *Macromolecules* 51: 3165.

107 Sneff, H. and Richtering, W. (2000). *Colloid Polym. Sci.* 278: 830.

108 Stieger, M. and Richtering, W. (2003). *Macromolecules* 36: 8811.

109 Mattsson, J., Wyss, H.M., Fernandez-Nieves, A. et al. (2009). *Nature* 462: 83.

110 Shamana, H., Grossutti, M., Papp-Szabo, E. et al. (2018). *Soft Matter* 14: 6496.
111 Trappe, V., Prasad, V., Cipelletti, L. et al. (2001). *Nature* 441: 772.
112 Eckert, T. and Richtering, W. (2008). *J. Chem. Phys.* 129: 124902.
113 Joshi, R. and Tata, B. (2017). *Colloid Polym. Sci.* 295: 1671.
114 Gasser, U., Scotti, A., and Fernandez-Nieves, A. (2019). *Phys. Rev. E* 99: 042602.
115 Gnan, N., Rovigatti, L., Bergman, M., and Zaccarelli, E. (2017). *Macromolecules* 50: 8777.
116 Weyer, T.J. and Denton, A.R. (2018). *Soft Matter* 14: 4530.
117 Brito, M.E., Denton, A.R., and Nägele, G. (2019). *J. Chem. Phys.* 151: 224901.
118 Weight, B.M. and Denton, A.R. (2018). *J. Chem. Phys.* 148: 114904.
119 Sierra-Martin, B., Frederick, J.A., Laporte, Y. et al. (2011). *Colloid Polym. Sci.* 289: 721.

7

Structure and Properties of Smart Micro- and Nanogels Determined by (Neutron) Scattering Methods

Julian Oberdisse[1] *and Thomas Hellweg*[2]

[1] *Laboratoire Charles Coulomb (L2C), Université de Montpellier, CNRS, F-34095 Montpellier, France*
[2] *Universität Bielefeld, Department of Chemistry, Physical and Biophysical Chemistry, 33615, Bielefeld, Universitätsstr. 25, Germany*

7.1 Introduction

So-called smart microgels were first synthesized by Pelton and Chibante [1]. Microgels are colloidal particles exhibiting an internal gel structure. They have diameters in the range of 100 nm up to micrometers [2–9]. Below the 100 nm threshold, they are called nanogels [9]. Their most striking feature is the temperature-induced reversible collapse of the polymer network that is similar to macroscopic gels [10–13]. Besides a temperature-induced response, also particles with sensitivity to changes in pH, ionic strength, and other external fields can be made [14–19]. However, the internal structure of microgels seems to differ from macroscopic gels, which is due to differences in reaction kinetics of the crosslinker (in many cases *N,N′*-methylene-bis-acrylamide, BIS) and the monomer (very often *N*-isopropyl acrylamide, NIPAM).

Microgels are of steadily growing interest that is related to the plethora of potential applications ranging from drug and vaccine delivery [20–25], over smart catalytic systems [26–28] and photonic applications [29–31] to smart surfaces for vertebrate cell manipulation [32] and responsive membranes [33, 34]. For these purposes, different copolymer systems [35–38] and structures exhibiting separated compartments were designed [39–44]. Moreover, several of these systems are interesting model colloids allowing to study particles with tunable interaction potential [45, 46]. It is even possible to generate extremely soft self-crosslinked particles that are at the edge between star polymers and well-defined colloidal particles [45].

The overall size and shape of microgels are mainly studied by static and dynamic light scattering (DLS) (namely photon correlation spectroscopy [PCS]). Especially, PCS is very often used to follow the temperature or pH-dependent swelling

Smart Stimuli-Responsive Polymers, Films, and Gels, First Edition.
Edited by Liang Hu, Yongfeng Gao, and Michael J. Serpe.

and de-swelling of smart microgels. Such experiments are complemented by electron microscopy techniques and atomic force microscopy (AFM). However, these methods reveal only internal structure and dynamics of microgels. Such structures are in the range from some Ångströms to some nanometers. This length-scale domain is successfully studied by small-angle scattering techniques, such as small-angle X-ray scattering (SAXS) and small-angle neutron scattering (SANS) [47–49], neutron and X-ray reflectometry [50], and neutron spin-echo spectroscopy [51, 52]. Recently, also X-ray PCS has been employed to follow microgel dynamics [53].

In the last 5–6 years, a lot of progress was made in the use of the above-mentioned methods. Hence, it is thus time to review some of these advances. Besides giving a comprehensive update of recent work done in this area, the present chapter also tries to provide an introduction to the basics of scattering experiments on responsive microgels. Besides PCS, no inelastic experiments will be discussed. The contribution is structured as follows – in Section 7.2, scattering is introduced, covering both static and DLS, as well as small-angle scattering. In the next Section 7.3, recent progress in multi-compartment and multi-stimuli-responsive microgels is reviewed. Sections 7.4 and 7.5 then deal with time-resolved experiments on microgels, and a short overview of crowded microgel systems.

7.2 Scattering Techniques Applied to Microgels

Scattering of radiation is a prominent method for investigating the micro- and nanoscopic state of matter on the scale of the wavelength of the radiation in question, and such methods have been actively used for more than a century. Since World War II, the advent of powerful sources has triggered an enormous increase in such applications. Light, X-rays, and neutrons are the most common probes, and each is adapted for a different context. One generally differentiates static from dynamic scattering techniques, and both will be discussed in this chapter. Static scattering such as SANS or SAXS, - as well as static light scattering (SLS) - can be treated in a common theoretical formalism. Many excellent textbooks [54, 55] exist on such techniques, and the most relevant aspects for microgel studies will be described. In addition, a subsection about PCS (often just called DLS) has been included.

7.2.1 Static and Dynamic Light Scattering Applied to Microgels

Light scattering techniques are workhorse methods in the investigation of colloidal systems, such as responsive nano- and microgels. In nearly all studies of smart microgels, their hydrodynamic dimension and their swelling behavior are followed by photon correlation spectroscopy. However, different flavors of SLS are also used to gain information about size, molecular weight M_w, and shape of microgels. In this subsection, we will give a brief introduction to light scattering experiments and will summarize some recent works using these techniques.

7.2.1.1 Static Light Scattering (SLS)

In SLS, the intensity of the light scattered from a microgel sample is measured as a function of scattering angle θ. θ can be converted to the magnitude of the scattering vector q by

$$q = \frac{4\pi n_0}{\lambda} \sin\left(\frac{\theta}{2}\right) \tag{7.1}$$

Here, λ is the used wavelength and n_0 is the refractive index of the solvent. For neutron and X-ray scattering the same expression can be used assuming $n_0 \approx 1$. q has the unit of a reciprocal length and as a rule of thumb $1/q$ is approximately the length scale resolution of the scattering experiment. q can be used to bring different scattering experiments to a common abscissa scale. The most straightforward way to analyze SLS data is by means of using the Guinier approximation for the form factor of the scattering particles [56] that leads to the well known Zimm equation:

$$\frac{Kc}{R_\theta} = \frac{1}{M_w \exp\left(-\frac{1}{3}R_g^2 q^2\right)} + 2A_2 c \tag{7.2}$$

Here, R_θ is the Rayleigh ratio and R_g is the radius of gyration. K in Eq. (7.2) is given by

$$K = \frac{4\pi^2 \left(\frac{dn}{dc}\right)^2 n_0^2}{N_A \lambda^4} \tag{7.3}$$

N_A is Avogadro's number, c is the mass concentration of the scattering particles, $\frac{dn}{dc}$ is the refractive index increment, and A_2 is the second virial coefficient. Hence, SLS can provide the radius of gyration R_g, which is very helpful in expressing "typical" sizes of complicated objects, and it can be calculated for any known geometry of finite size. For instance, the R_g of a homogeneous sphere of radius R is $\sqrt{3/5}$ of R. Moreover, A_2 may be determined from such SLS measurements. This quantity is a measure of the interaction between the particles. Moreover, also the molecular weight of the nano- and microgels can be obtained from SLS using the Zimm equation [57, 58].

Besides the simple Guinier analysis in some cases also significant parts of the nano- and microgel form factors, $P(q)$, fall into the q range of SLS and can be analyzed by fitting the data with respective form factor models. This is nicely demonstrated in a work by Schneider et al. [59].

Figure 7.1 shows a comparison of the fit of SLS data using a homogeneous polydisperse sphere model with the famous fuzzy sphere form factor model for microgels that was introduced by Stieger et al. [60] already a while ago. The different form factors corresponding to the fits shown in Figure 7.1 is given by

$$P(q) = \left[3\frac{\sin(qR_{SLS}) - qR_{SLS}\cos(qR_{SLS})}{(qR_{SLS})^3}\right]^2 \tag{7.4}$$

and

$$P(q) = \left[3\frac{\sin(qR_{SLS}) - qR_{SLS}\cos(qR_{SLS})}{(qR_{SLS})^3} \times \exp\left(\frac{(q\sigma_{surf})^2}{2}\right)\right]^2 \tag{7.5}$$

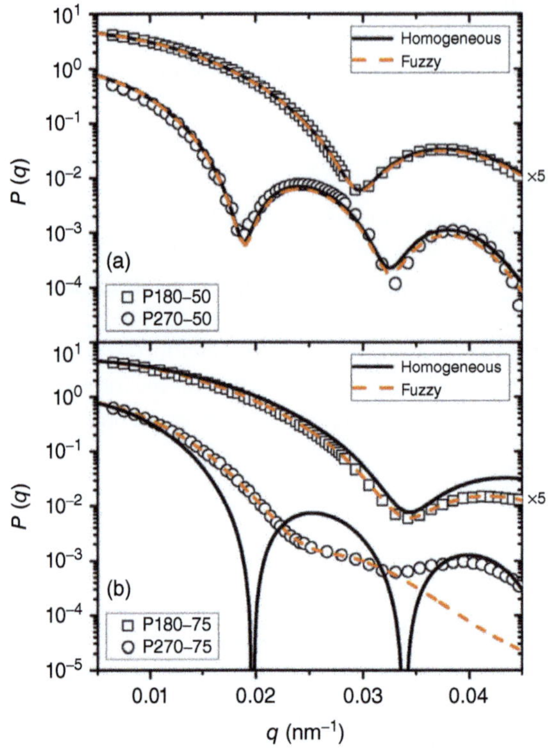

Figure 7.1 SLS data for four different microgels. The solid black lines correspond to a fit using a homogeneous sphere model. The dashed yellow lines are fits with the fuzzy sphere model. Source: Reproduced from Ref. [59] which was published under the license CC BY 3.0 by the Royal Society of Chemistry.

where Eq. (7.4) describes the form factor for a homogeneous monodisperse sphere, and in Eq. (7.5), the fuzzy sphere from Stieger et al. σ_{surf} represents the decay length of the fuzzy, hairy corona of the microgels. Note that we discuss in the Appendix the prefactor of the fuzzy sphere form factor, the deviation of which from one is important for mass determination using absolute units. In both cases, polydispersity can be taken into account by convolution of the relevant $P(q)$ with a Gaussian distribution of the radius R_{SLS} [59]. Looking at Figure 7.1, it is obvious that the fuzzy sphere model leads to a better description of the SLS data. Moreover, SLS can also be used to determine the long-range order and interaction between the colloidal gels. This is quantified in terms of the so-called structure factor $S(q)$ [59]. $S(q)$ will be discussed in more detail in the subsections dealing with small-angle scattering.

As already mentioned, using q as abscissa normalized absolute SLS intensities can be joined with absolute small-angle data. This was nicely shown in a recent work by Kozhunova et al. [56] on stimuli-responsive inter-penetrating network microgels. In Figure 7.2, an example of such a joined data set is shown.

In this figure, the series of points at low q was determined by means of SLS. The red curve represents a model that consists of the hard-sphere form factor in Eq. (7.4), which is supplemented by an Ornstein–Zernike type contribution given by

$$I_{OZ} = \frac{I_0}{1 + q^2\xi^2} \tag{7.6}$$

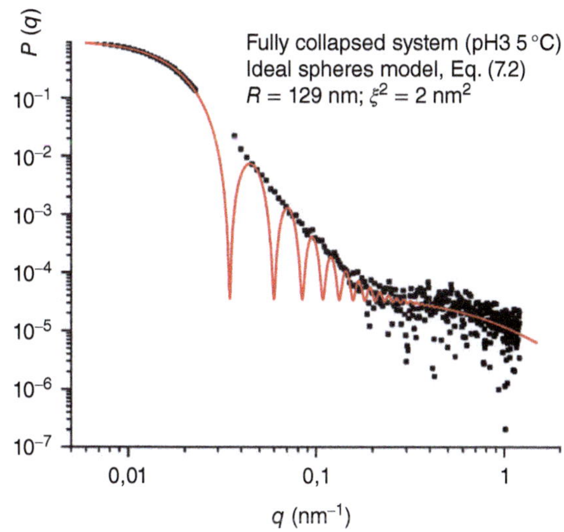

Figure 7.2 Joined SLS and SAXS data set for a collapsed microgel measured at 50 °C. Source: Reprinted from Publication [56] with permission from Elsevier.

In this equation, ξ is the local correlation length of the internal network of the microgel. In the example, in Figure 7.2, despite being in the collapsed state, the studied system still has some "porosity," and ξ is found to be ≈ 2 nm.

7.2.2 Dynamic Light Scattering (DLS/PCS)

In a DLS experiment, the time-dependent fluctuations of the scattering intensity $I(q)$ are analyzed using the formalism of intensity correlation functions generated by hard-wire correlators [54, 61]. This is why this method should be called PCS to distinguish from other quasi- or inelastic light scattering methods, such as depolarized dynamic light scattering (DDLS) that can also be done by looking at the laser line broadening in the frequency domain. The intensity fluctuations arise from the Brownian motion of the colloidal nano- or microgels in the sample solution. The standard PCS experiment yields the intensity time auto-correlation function $g_2(\tau)$ that can be converted into the time autocorrelation function of the electric field $g_1(\tau)$ using the Siegert relation

$$g_2(\tau) = 1 + |g_1(\tau)|^2 \tag{7.7}$$

Here, τ is the sample time. For ideal monodisperse samples, the function $g_1(\tau)$ is a single exponential decay of the form $g_1(\tau) = \exp(-\Gamma\tau)$ and the relaxation rate Γ can be used to compute the translational diffusion coefficient D according to $\Gamma = Dq^2$. This also indicates that a plot of Γ vs q^2 yields information about the type of motion that is effectively observed in a PCS experiment. If such a plot results in straight lines through the origin, this reveals that the experiment only "sees" center-of-mass diffusion of the colloidal gels. For most colloidal gels, this is fulfilled and a recent example is shown in Figure 7.3.

Figure 7.3 shows data for the already widely used and commercially available monomers NiPAM and *N*-isopropyl methacrylamide (NiPMAM), as

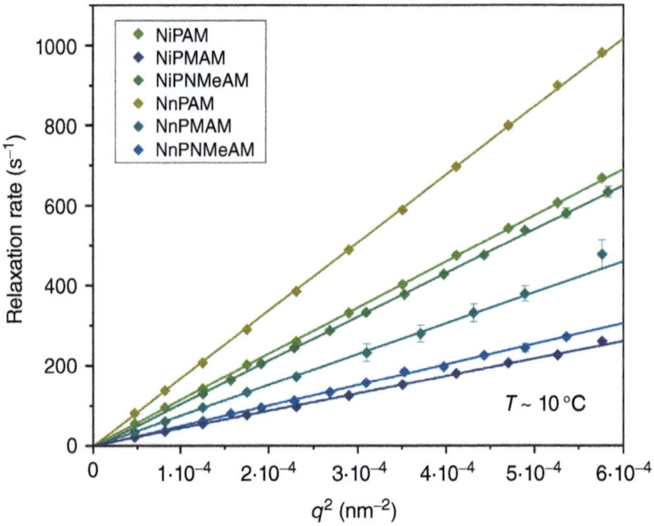

Figure 7.3 Averaged relaxation rates Γ for a series of homopolymer microgels made with alternative monomers [62] at $T = 10°C$: swollen state. Each correlation function was measured three times at 11 scattering angles. For all particles, the data follow the expected linear behavior indicating purely diffusional dynamics. Within the experimental precision, all fits go through the origin. Source: Republished from Hannappel et al. [62].

well as for the rarely used acrylamides, N,n-propylacrylamide (NnPAM) and *N*-*n*-propylmeth- acrylamide (NnPMAM), and the previously not used monomers *N*-isopropyl-*N*-methylacrylamide (NiPNMeAM), *N*-*n*-propyl-*N*-methylacrylamide (NnPNMeAM). Usually, the polydispersity of microgels is rather low and the decay of the correlation functions nearly follows a single exponential. However, in case of higher polydispersities, $g_1(\tau)$ has to be described by a weighted sum of exponentials

$$g_1(\tau) = \int_0^{\infty} G(\Gamma) \exp(-\Gamma\tau)\, d\tau \tag{7.8}$$

The relaxation rate distribution function $G(\Gamma)$ can be obtained by numerical Laplace inversion of $g_1(\tau)$ [63, 64] or by using the method of cumulants [65, 66]. In both cases, a mean value for Γ is obtained.

The slope of the straight lines in Figure 7.3 is the translational diffusion coefficient of each microgel. Knowing D, the hydrodynamic radius R_h is computed via the Stokes–Einstein equation:

$$D = \frac{kT}{6\pi\eta R_h} \tag{7.9}$$

As stated before, in most studies of microgels R_h is measured at different temperatures crossing the volume phase transition temperature (VPTT) of the particles. A recent example of the outcome can be found in a work by Freeman et al. [67].

Figure 7.4 shows the swelling curves for rather unusual polysaccharide-based microgels that also exhibit a volume phase transition like their polyacrylamide counterparts.

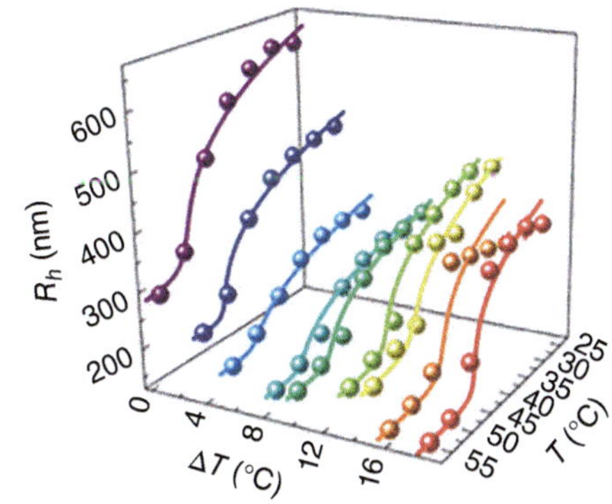

Figure 7.4 Swelling curves of polysaccharide microgels as measured by PCS. ΔT indicates the difference between the synthesis temperature and the LCST of the microgels. Obviously, the swelling behavior is impacted by ΔT. Source: Reproduced with permission from Freeman et al. [67]. American Chemical Society.

At several large-scale facilities, PCS can be used to follow microgel samples in situ during SANS experiments [68, 69]. Small-angle scattering experiments will be discussed in the following subsection.

7.2.3 Small-Angle Neutron and X-Ray Scattering Applied to Microgels

In this subsection, we will focus on the specific application of SANS (and SAXS) to unravel the structure of microgels suspended in solvents. The fundamental scattering process of neutrons (or X-ray photons interacting with electrons, a valid analogy throughout this text) is the creation of a spherical secondary wave by the presence of a nucleus hit by an incoming wave. The latter is created in the heart of a nuclear research reactor, or a spallation source, and transported and directed conveniently toward the sample by experimental installations, such as guides, collimations, and wavelength selectors, which we will take for granted in this introduction. In the end, a planar wave of wavelength λ hits the sample and is mostly transmitted without change in direction, sadly ending its short life on a beam-stop. Some of the neutrons, however, are scattered laterally under the form of the above-mentioned secondary waves, they thus leave the direct beam, and can be collected on a detector. These neutron waves coming from different nuclei in the sample interfere, and the detector thus shows an intensity pattern resulting from such interferences. The latter depends on the phase of each neutron, and thus on the distance the neutron has traveled, the intensity pattern encodes the distance distribution between nuclei in the sample.

The fundamental problem of the scientist analyzing such data is thus to work his/her way back from the intensity pattern to the distance distributions and thus to the microstructure of the samples. This is much more difficult than looking at an electron micrograph, but it contains valuable information. In particular, all the distances in the sample are averaged, and one thus does not only look at a very small sample portion as in a TEM picture. Scattering provides thus intrinsically an ensemble average, which is both an advantage (because it minimizes artifacts), and a disadvantage (because it only gives averages, e.g. over size distributions). And it is also considerably more complicated ...

To fully understand how microgel particles can scatter, one needs first to revisit the elementary scattering process. When writing down the equation of the spherical secondary wave, a prefactor describing the power of the interaction is needed. This prefactor has the unit of a length, and it is called the (coherent) scattering length *b* of the nucleus causing the process. It is not only specific to each element, but also each isotope. This implies that we can differentiate isotopes in the sample, which is one of the greatest features of neutron scattering. Indeed, replacing, e.g. a "normal" polymer (i.e., made of protons, like in polyethylene, $-CH_2-$), with its deuterated counterpart ($-CD_2-$) leaves the chemistry mostly unchanged, but totally modifies the scattering. By using deuterated and hydrogenated polymers (possibly within the same chain, e.g. to label and thus locate chain ends), it is thus possible to see which one is where. In SANS, one is usually interested in the overall, supra-molecular structure of objects. The individual scattering length *b* of each nucleus is then regrouped and expressed as a scattering length density ρ given by the sum of all the *b*'s in a volume, divided by that volume. A typical volume may be a chemical repeat unit. The interested reader should look up the coherent scattering lengths of H, D, and O, and calculate the scattering length densities of H_2O and D_2O (-0.5×10^{10} cm^{-2}, and 6.4×10^{10} cm^{-2}, respectively). This little exercise shows that changing the solvent is a way to completely change what is seen by the neutrons. The corresponding technique is called contrast variation, and an example is shown in Figure 7.5.

In Figure 7.5, the scattering length density of the solvent "water" is varied by mixing light and heavy water, and the polymer is seen to have varying visibility in each solvent, expressed by the square-root of the scattered intensity in some (small-angle)

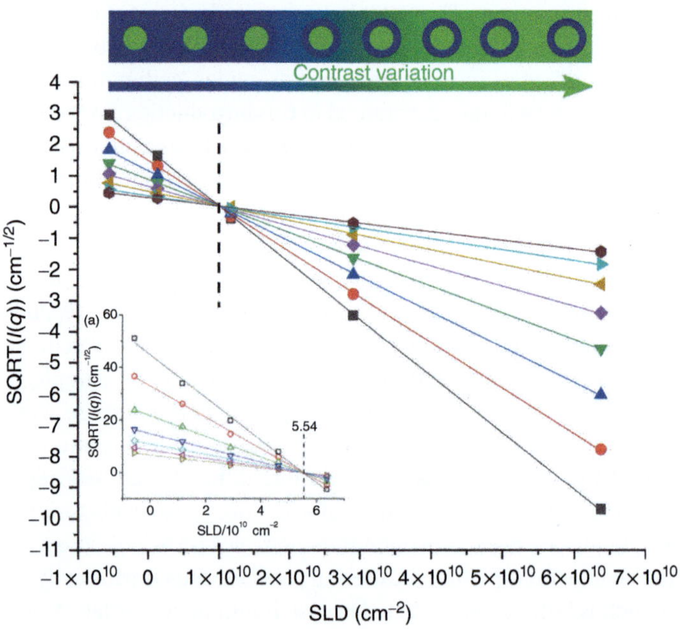

Figure 7.5 Contrast variation for several *q*-values. Inset: D-polymer. Sketch: visibility of D-core and H-shell in varying solvent mixtures. Source: Republished with permission of American Chemical Society from Cors et al. [70]. 2018 permission conveyed through Copyright Clearance Center, Inc.

direction. The most important feature in Figure 7.5 is that the scattered intensity vanishes at a given H_2O/D_2O ratio, which is called the match point. Here the scattering length density of the polymer is exactly equal to the one of the water mixture, i.e. we have actually measured it, and thus also its density (i.e., volumetric mass) within the solvent. If this polymer is now part of an assembly, it can be seen for mixtures far from the match point, and it will be invisible at the match point. It is good practice in neutron scattering experiments to start with a contrast variation of each ingredient and compare it to values expected on the basis of the chemical sum formula. As a result, the relevant quantity is the scattering contrast, i.e. difference in scattering length density between the object and the solvent (or matrix): $\Delta\rho = \rho - \rho_{\text{solv}}$. In Figure 7.5, we have also shown in the bottom inset the contrast variation for a D-polymer, and in the sketch an artist's view of the visibility of a particle made of H- and D-polymer, in various solvent mixtures. Toward the right, the H-shell is matched by the mostly H-solvent, and toward the left, the inverse is true. Contrast variation thus serves to highlight specific parts of the particle if isotopic substitution can be performed chemically.

If we now measure the entire scattered intensity collected by the detector (or sets of detectors to cover wider angles), we obtain a function of the magnitude of the scattering wave vector q. Examples of intensity are shown in Figure 7.6. Technically, $I(q)$ is a scattering cross section of the sample per unit volume, which is why its units are cm^{-1}. Once this measurement is done correctly – this implies a little extra work, such as exactly knowing the contrasts, concentrations, measuring transmissions, and incoherent backgrounds – the happy scientist can start thinking about the micro-structure of this sample.

In Figure 7.6, three intensities have been superimposed, corresponding to the same pNNPAM microgel at different temperatures – shifted vertically for clarity. At the highest T, the particles are collapsed and do not interact: form factor scattering is observed, corresponding to the Guinier plateau followed by oscillations

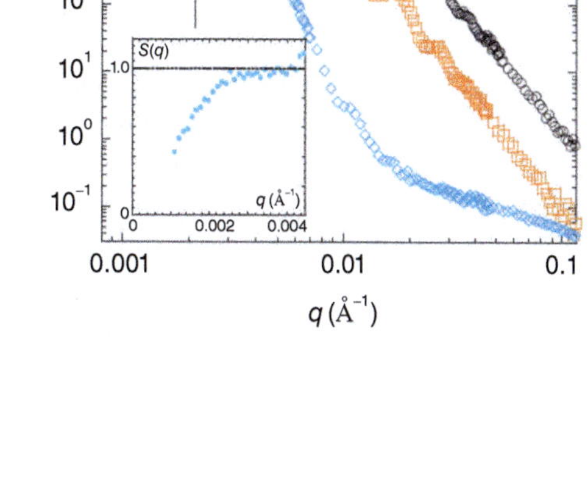

Figure 7.6 Structure of pNNPAM microgel in D_2O for three different temperatures as indicated in the legend. The SANS intensities at 32 and 55 °C have been vertically shifted by a factor of 10 (resp. 100). In the inset, an estimate of the structure factor is shown for the lowest temperature.

caused by interferences due to the well-defined particle geometry. As the particles swell with lower T, oscillations weaken, and steric interactions set in, leading to the structure factor effect at low q. Such interactions can be studied and described. They are interesting, e.g. if one wished to study squeezing of microgel particles [45].

Due to lack of space, we cannot focus on structure factors here and only introduce some basic notions. The fundamental concept is one of spatial correlations, which are the result of interparticle interactions. For hard spheres, e.g. the evolution of the low-q limit of $S(q)$ is well described by the Carnahan–Starling equation [71], and the entire function has been determined from integral equation theory, using the Percus–Yevick closure [72]. On the other hand, highly repulsive long-range interactions, like electrostatic ones, lead to ordering of the suspension. Suitable solutions to the integral equations exist in the literature, such as like Ref. [73]. The corresponding particle–particle correlation function then adopts a characteristic shape, usually peaks, which translate into peaks in the scattered intensity. Here a simple rule can often be used: $\frac{2\pi}{q_0}$, where q_0 is the 1st order peak position that gives the size of a cubic box containing one particle. Using volume conservation, it is thus possible to estimate the dry volume of the particle from a structure factor measurement. Repulsion also leads to a depression of the low-q intensity. Other interactions, such as attraction-inducing aggregation, typically lead to low-q increases, which can be interpreted as mass increases. It is thus possible to follow interactions by focusing on the low-q scattering, e.g. in terms of the second virial coefficient at high dilution. Alternatively, one may focus on the shape of the entire function, if models are available for comparison, such as for fractals [74]. Coming back to pure form factors, the usual method of measuring microgel particles only (i.e. without being perturbed by their interaction) is to lower the concentration and check that two successive measurements give the same $I(q)/\phi$. We have insisted on the unit of the intensity because knowing $I(q)$, one says "in absolute units" allows one to exploit the low-q limit $I_0 = I(q \to 0)$. The scattered intensity then reads at low angles:

$$I(q) = I_0 \exp\left(-\frac{q^2 Rg^2}{3}\right) \tag{7.10}$$

$$I_0 = \phi \Delta\rho^2 V_{dry} \tag{7.11}$$

Equation (7.10) is called the Guinier approximation, and it is valid (see Figure 7.6) at low angles in absence of interaction. It also corresponds to the low-q version of Eq. (7.4), or equivalently to Eq. (7.2) with $A_2 = 0$. The Guinier law gives the radius of gyration R_g, which describes the average spatial extent of the objects and has been introduced in Section 7.2. In Eq. (7.11), we have given the low-q limiting value, which is proportional to the volume fraction of objects, to their dry volume V_{dry}, and to the square of their contrast. This square is the reason for plotting the square root of the intensity in Figure 7.5.

The dry volume deserves some further comment. We have chosen to identify the object by its chemistry. Imagine a thin hollow shell made of dry polymer, with solvent in the core. The total object contains water and polymer, but V_{dry} only refers to the polymer part, ϕ only to the polymer volume fraction, and $\Delta\rho$ to the dry polymer contrast in water. The amount of polymer matter in the object can thus be compared to their spatial extent given by R_g. For a thin shell, R_g is given by the shell radius. Comparing V_{dry} to R_g thus gives a hint on how much matter is distributed where in space, and one can start imagining geometric models consistent with these two (independent) measurements. The simplest one could be a hydrated sphere: R_g can be converted into a sphere radius R, of (wet) volume $\frac{4\pi}{3}R^3$, and the latter compared to V_{dry}, the ratio giving the hydration. Below, we will discuss the principle, which has similar approach to internal microgel concentration profiles, where the two ingredients, local concentration at a given distance from the center, are resolved. As the last point, let us mention that particles are usually polydisperse, and polydisperse expressions involving moments of the size distributions are readily available for Eq. (7.11), see for instance Ref. [75].

Up to here, we have concentrated on the low-q scattering, which according to the q^{-1}-argument encodes the average large-scale structure of the microgel particles, i.e. global spatial extent and total (dry) mass. Of course, much more information is available: the form factor scattering in Figure 7.6 can be analyzed as a function of q. Going to higher q is equivalent to increasing the magnifying power of a microscope. In the intermediate q-domain, spherical particles usually display oscillations related to their size, and the damping of these oscillations indicates how homogeneous, monodisperse, and well-defined the particles are. At the highest q, a power law is usually found. In the case of microgels, it extends typically between a shallow q^{-1} power-law characteristic of linear chains, some intermediate fractal statistics of chains, e.g. $q^{-1.7}$ for chains in good solvent, or q^{-2} for ideal chains, up to q^{-4} (Porod decay) when all the chains collapse smoothly onto the particle and only a sharp interface remains. However, it can not be ruled out that especially the collapse of dangling ends at the surface of the microgels might lead to some surface roughness. This can be modeled by using modified Porod decays [76]. Measuring these power laws thus provides indications of the chain statistics, which in the case of microgels are strongly temperature dependent. Several expressions exist in the literature for chain scattering, the best-known one being the Debye equation for ideal chains [77]. The latter can be generalized to any power law if needed [78]. Depending on the total size of the microgel particles, the polymer network may also contribute characteristic scattering features. Indeed, due to fluctuations in crosslinking density, density fluctuations may be visible, as well as the network mesh size. The reader is referred to the relevant literature for details [79].

Once one knows the amount of matter to be distributed on average within microgel particles (via V_{dry}), their average spatial extent (R_g), and the local conformation of the chains making up the network, one is left with the question of the internal concentration profile of the microgels. There are essentially two approaches to this problem. The first one is called "model fitting", and it consists in starting from a hypothetical model structure (e.g. a homogeneous sphere of radius R). The form factor of such an object is analytically known, and one can thus adapt R until the theoretical intensity describes well the measured one. Due to the mysteries of scattering, one can then state that microgels may be homogeneous spheres of radius

R. In reality, one can never be completely sure, but if the dry volume, the R_g, the shape with its oscillations coincide, one may be pretty sure to have chosen the right model.

In the literature, several models have been proposed to describe microgel scattering – fuzzy spheres as discussed in 7.2 introduce a soft decay on the edge of the spheres [60, 80], and other models have been discussed by us [70], including an expression based on a Flory–Rehner approach by Boon and Schurtenberger [81]. The essence of all these models is that if they are chosen correctly, then the fits are good, and the structure is resolved. And it is the experimentalist's job to choose the right model.

We have recently applied a more general approach, where no geometry is a priori imposed, besides that the spherical symmetry of the objects, as well as their polydispersity, is known beforehand. Such a free-form analysis has been applied to block-copolymer micelles [82], and later to microgel particles [70, 83]. The method is based on a reverse Monte Carlo algorithm, which searches stochastically for the best fit of a multi-parameter model given by the concentrations in successive spherical shells. The concentration in each shell thus defines the average density profile of the particle, from the center to its surface. It is thus not much different from the above model-fitting, only that (many) more parameters need to be optimized. Such approaches are nonetheless known as "inversion" techniques, as they invert the intensity back into structure, whereas a model fit assumes a structure and compares to intensity. Inversion techniques are part of a larger class of techniques, like the (deterministic) indirect Fourier transform and its variants [84, 85], and (stochastic) reverse Monte Carlo calculations [86]. The difference here is that the concentration profiles are based on real monomers moving between the shells to optimize the fits and that the spherical symmetry is implemented from the beginning. The physics of the problem is thus taken into account from the construction of the method, as it is done in other fields where the basic building blocks, – such as nanoparticles, – are used [87, 88]. This means of course that for each problem, a new algorithm has to be set up. Finally, as with all such methods, they suffer from under-determination. In such so-called "ill-posed" problems, more than one solution exists. The standard way is to favor a solution by regularization, e.g. by demanding that the profile decays smoothly, and average over families of solutions obtained in the course of a simulation.

The geometry of the multi-shell model is defined by the number of monomers in an average particle consistent with Eq. (7.11), taking into account the crosslinkers and particle polydispersity in all calculations. These monomers are distributed over N_p shells, where N_p is typically 30, of width $\Delta R = 2$ nm for our particles, but this needs to be adapted to the particle size. Obviously, certain physical bounds need to be respected, like the minimum occupation of a shell (no monomers in a shell), but also the maximum filling (100% volume fraction). The idea of the algorithm is to start from some initial condition, which may be having all shells filled from the center of the particle until all monomers have been distributed. This initial condition should be varied to check the convergence of the procedure. The volume and number of monomers in each shell may be transformed into the scattering length density of each, and then the sum of each contribution may be calculated to obtain the total intensity:

$$I_{\text{shells}}(q,r) = \sum_{i=1}^{N_p} \frac{N}{V} \left[\frac{4\pi}{3} r(i)^3 \left(\Delta\rho(i) - \Delta\rho(i+1)\right) 3 \frac{\sin(qr) - qr\cos(qr)}{(qr)^3} \right]^2 \quad (7.12)$$

Here, N/V is the number of microgel particles per unit volume, r is the outer radius of each shell $r(i) = i\Delta R$, and $\Delta\rho(i)$ is the scattering length contrast of shell i. The contrast of the surrounding solvent shell $\Delta\rho(N_p + 1)$ is set to zero. The Monte Carlo part is then to modify the distribution of the monomers between the shells, e.g. by moves of some of them between randomly chosen shells (respecting the bounds), and check if the outcome of Eq. (7.12) agrees better with the experimental scattered intensity. Here, agreement is usually expressed in terms of χ^2, i.e. a normalized sum over squares of differences between the prediction I_{shells} and experiment, modulated by some error bar. Moves leading to decreasing χ^2 are always accepted, but to explore also some less satisfying monomer profiles in a so-called simulated annealing approach, slight increases of χ^2 may also be accepted. This latter process is driven by an effective temperature parameter χ_T, which in the beginning accepts large χ^2-increases, and progressively diminishes this amplitude. As the last point, the result needs to be regularized as outlined above, which amounts to simultaneous optimization of the agreement with the experimental intensity (low χ^2), and high smoothness.

In Figure 7.7a, the progression of the algorithm from the initial condition to the final outcome is shown. The black squares show the evolution on a logarithmic scale of χ^2 with the number of Monte Carlo steps. It is found to be satisfying after a few hundred thousand steps, and then all parameters are kept fixed and the solution is averaged. In the course of the annealing process, the effective temperature χ_T is seen to be pushed down until it is kept constant in the averaging phase. Simultaneously, the number of monomers N_{move} tentatively moved in each MC step is also decreased. While the simulation is running, the roughness of the profile stays under control (see original article). In the inset of Figure 7.7a, the experimental intensity to be fitted is shown, with the initial guess, and the final fit superimposed on the data.

In Figure 7.7b, the intensities for the same microgel particles are shown for various temperatures, together with the best RMC-fit. In Figure 7.7c, finally, for each T, the resulting profiles are shown for this p(NIPMAM) microgel, which afterward is used as the core in subsequently studied core–shell microgels (see Section 7.3). This is thus the real-space signature of the microgels as seen by small-angle scattering analyzed through the multi-shell model. The profiles give simultaneous information on the local density and spatial extent, the two being coupled by volume conservation. At the highest temperatures, dense and small cores are thus found, whereas swelling leads to much more diffuse, tenuous structures.

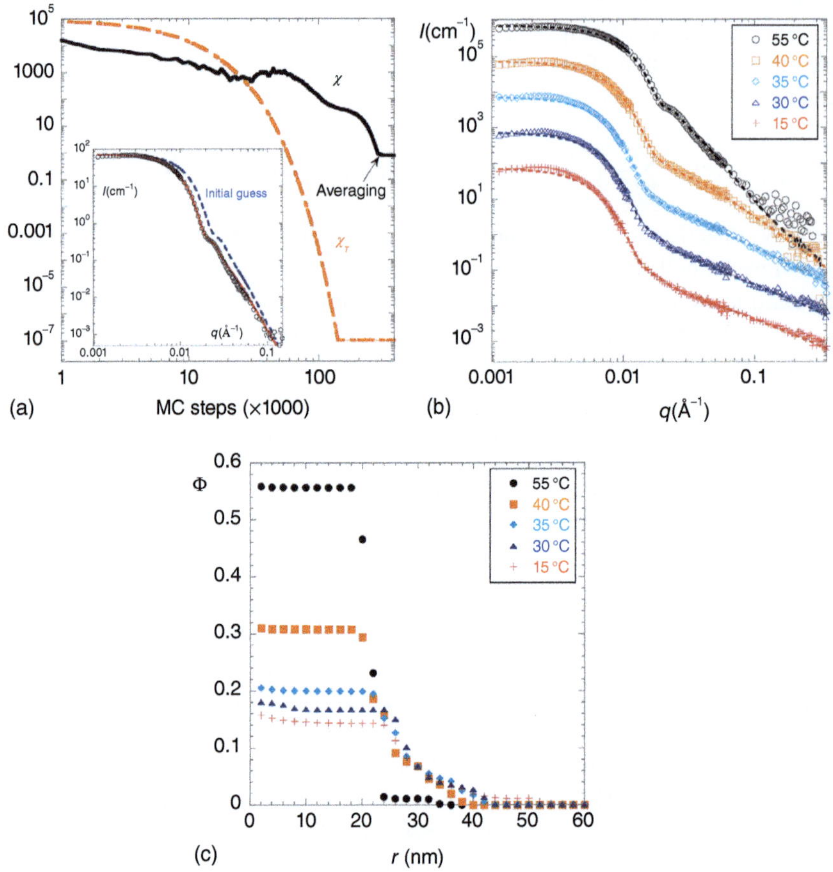

Figure 7.7 Structural analysis of a pNIPMAM core in D_2O. (a) The decrease of χ^2 as a function of the number of Monte Carlo steps, up to the moment when averaging is started. The process is driven by an effective temperature χ_T. The inset shows the improvement of the fit starting from the initial guess. (b) Intensities as a function of wave vector for a temperature series, shifted vertically by powers of 10 for clarity. The fit functions are superimposed. (c) The radial density profiles corresponding to the RMC-solutions. Source: Adapted with permission from [70]. Copyright 2018 American Chemical Society.

7.3 Multicompartment and Multi-Stimuli-Responsive Microgels

The thermoresponsiveness of microgel particles is only one of available smart properties. Sometimes the sensitivity to other stimuli, possibly within the same object, is of interest for fine-tuning morphologies, colloidal stability, and interaction with guest molecules. Introducing electrostatic charges within the microgel is an obvious choice, and copolymerization with ionic or ionizable monomers is a straightforward way to make the backbone carry charge, possibly with a pH-dependence as with for instance weak acids, which are neutral below their pK_a, and negatively charged above. The main principles of microgel response to electrostatics are that the pH can act as a trigger to switch on charges, the effect of which is possibly compensated by mobile ions screening electrostatic interactions on the scale of the Debye length. The ionic equilibrium between counterions and the charged backbone is governed by a compromise between electroneutrality and translational entropy of ions, and the resulting average spatial distribution of ions is described by the Poisson–Boltzmann theory – or Debye–Hückel in its linearized version. This results in swelling of the microgel by the osmotic pressure of the counter-ions. This swelling effect is thus both pH- and ionic strength-dependent.

Having different stimuli at hand, the next step in complexity is then to design multicompartment-multi-stimuli microgel particles, with a different thermosensitivity for each compartment. From the synthesis conditions of precipitation around nuclei, spherical symmetry emerges naturally, and multicompartments of different swelling properties are conveniently characterized by shell-like structures [89], the radial density profiles of which can be extracted from scattering profiles.

Density profile analysis of isotropic structures turns out to be a powerful method both for "simple" microgels (like pure crosslinked pNIPAM), where the crosslinking density varies with the radius due to the different reactivities of the crosslinker and the monomer, and purpose-synthesized hollow-shell, particle-shell, and other core–shell microgel particles.

Hyatt et al. have proposed a study of pNIPAM co-polymerized with relatively high amounts of acrylic acid (24 mol%) [90], complementing a previous study where the acid groups remained neutral at low pH [91]. At high pH and high temperature, pNIPAM deswell, while partially ionized acid groups tend to swell following the entropic pressure of the counterions. This competition leads to an electrostatically driven microphase separation with a dense core and a tenuous shell. This particular structure has been evidenced by light scattering, PCS providing the hydrodynamic radius R_h sensitive to the soft shell, and SLS the radius of gyration R_g, i.e. a measure of the mass distribution in space. Data are analyzed with a polydisperse fuzzy sphere form factor, which these authors refer to as core–shell due to the fuzzy surface; at higher q, a standard polymer network term is included, which we will not explicitly mention in all work discussed here. As a result, R_g/R_h is observed to become quite small (below $\frac{1}{5}$), indicating that most mass is concentrated in the center of the microgel and thus forms a dense core. Simultaneously, a decrease in mesh size accompanies the R_g/R_h decrease above the volume transition temperature. These findings are similar to the internal phase separation observed by Wu et al. [92], who synthesized microgel systems with ionic liquid comonomers. In their case, however, the alkyl side chains are distributed across the entire particle, and TEM micrographs and DSC show the formation of spherical crystalline mesophase zones exhibiting a radius in the 10 nm range, inside the particles. This might have been a promising system for a complete analysis of the total small-angle scattering, as done in the first system of the Richtering group some years ago [93]. The latter authors had added a local spherical term to the intensity model to account for what they call an imaginative picture "dirty snowballs."

An example of control of the colloidal stability via the addition of reactive macromolecular stabilizers has been provided by Etchenausia et al. [94]. This group has studied PVCL-based microgels and could show that an outer cationic stabilizing shell is formed, without losing temperature sensitivity. SANS fits with a sphere model at low q confirms the DLS results of swelling with T and surprisingly also with total concentration. It remains to be clarified as to which extent aggregation modifies the scattering signatures.

By adjusting both ionic stimuli – pH and ionic strength –, the morphology of microgels can be controlled as shown by Bergman et al. [95]. These authors have copolymerized acrylic acid and NIPAM, and they show that they can target swelling of the core and of the shell made of dangling ends by playing with both stimuli. Very nicely resolved SLS data including higher order form factor oscillations allow inverting the scattering profile into a density profile using Mie theory relevant for big objects that strongly scatter light. This procedure is also backed up by form-free modeling in an analogous way [83], as outlined in our own contributions below [70, 82, 96]. In parallel, PCS data gives access to the hydrodynamic radius, and thus to the extension of the shell into the solvent. The same group has previously set up a Flory–Rehner-based model accounting for the effect of crosslinking gradients on the microgel shape as seen by the density profile [81]. Here, they add an electrostatic osmotic pressure term to their model and show that it captures successfully the effect of the ionic stimuli. In particular, deionization induces shell swelling due to the reduced charge screening without affecting the core, whereas deprotonation by increasing the pH makes the core swell by a factor of two under the pressure of the counter-ions. Adding salt again at this high pH leaves the core swollen, but collapses the corona because the latter has a higher sensitivity to salt due to the lower density of backbone ions. This sequence of behavior is depicted in Figure 7.8, illustrating the effect of each stimulus individually.

In traditional pNIPAM microgels, the nonhomogeneous crosslinking density is a major property influencing the particle morphology. This is not necessarily always the case, as shown for another system by Schneider et al. [59], the analysis of which

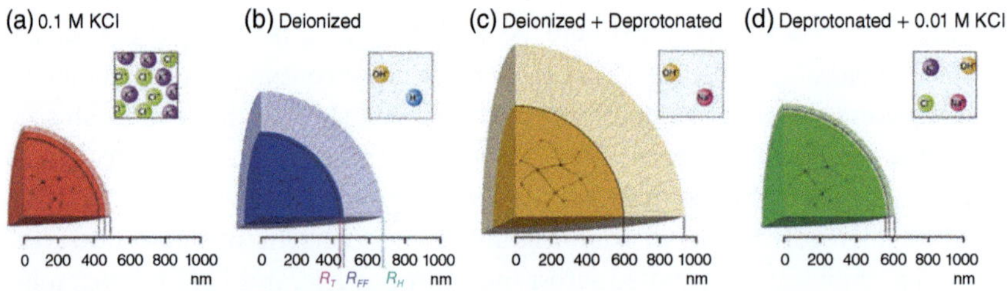

Figure 7.8 Radial measures of ionic microgels, indicating the swelling of the core and the corona in various ionic environments fixed by resins and salt. The figures (a–d) correspond to different conditions. The core radius R_{FF} as determined from SLS, as well as the hydrodynamic radius R_h, from PCS are shown, differences between them are used as a measure of the dangling-end contributions. Source: Republished with permission of Royal Society of Chemistry from Bergman et al. [95]. Royal Society of Chemistry (Great Britain), 06/01/2005 permission conveyed through Copyright Clearance Center, Inc.

has been used above to introduce form factors (Figure 7.1). This group has varied the crosslinking density of polystyrene particles by a factor of ten, screening the range from soft pNIPAM-like behavior to hard spheres. The similar reactivity of the crosslinker and the styrene makes these particles homogeneous, free of any crosslinking gradient. These authors are mainly interested in colloidal glass transitions and proceeded, therefore, in a mapping of these spheres of increasing softness onto the hard sphere phase diagram via the observation of $S(q)$. Only at the lowest crosslinking ratios, weakly fuzzy spheres have been observed by SLS, as shown by their rigorous study of form factor scattering in the limit of the lowest possible concentrations. Finally, fuzziness could also be tuned by reaction conditions, which were initially varied to ensure low polydispersity. The corresponding scattered intensities are shown in Figure 7.9, where the fits with homogeneous or fuzzy spheres are seen to be more and more differentiated as the monomer-crosslinker feed time is increased, indicating growing fuzziness.

Unlike the formation of fuzziness sometimes referred to as corona structure caused by the inevitable crosslinking gradient, purpose-designed core–shell microgels with true double thermoresponsiveness have been developed in the past few years. Brugnoni et al. have set up a system based on a hard silica core, which can be dissolved after shell synthesis to obtain hollow microgels [97]. They have subsequently polymerized two shells, the inner with pNIPAM, and the outer one with pNIPMAM that has a ca. 10 K higher volume transition temperature, for a typical swollen size of 250 nm characterized by DLS. The systems are thus silica-shell–shell, and hollow-shell–shell, while moreover the outer shell can be left out for comparison. Besides cryo-TEM, SANS is used to access the polymer density profiles, taking advantage of the solvent H_2O/D_2O to match the silica contribution if present. Core-multishell modeling with inter-penetrating interfaces and outer

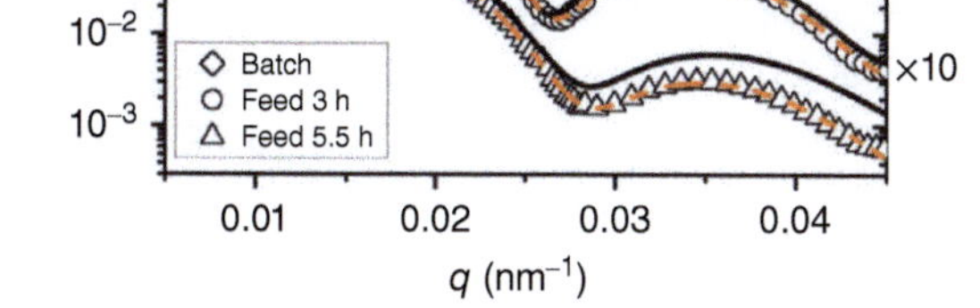

Figure 7.9 Comparison of the experimental form factors of three different samples of 1:75 crosslinked polystyrene microgel particles in the good solvent toluene. The diamonds represent the data for a sample made by standard batch synthesis. The other two data sets correspond to semi-batch synthesis with two different monomer feeding speeds. Slower feeding leads to increased "fuzziness" of the PS particles. For easier comparison $P(q)$ was multiplied by the factors indicated on the right side of the figure. Source: Reproduced from Ref. [59] with permission of the Royal Society of Chemistry.

fuzziness, which moreover is compared to computer simulations, provides density profiles. Their analysis provides the keys to understanding the impact of constraints within such doubly thermo-responsive microgels, which the authors refer to as "push–pull" effect. Indeed, attachment of the inner shell to the silica core limits swelling, and in this case, inter-penetration with the outer collapsed shell is observed. On the contrary, swelling is more pronounced for the hollow core, which is partially filled upon heating, avoiding shell inter-penetration. Also, PCS shows that the outer shell constrains the inner one between the two T_c, when the inner shell starts swelling, a phenomenon referred to as "corset effect" and that has attracted further attention as discussed below. In all cases, however, the interaction between the two shells has the effect of preserving the cavity, a case that may be compared to the inverted system studied by the same group before, with the outer shell swelling first upon heating [98]. The microgels with a collapsed inner and a swollen outer shell showed similar density profiles with and without silica core, with a slight increase in the cavity size. In the context of such studies, one may mention that most conclusions (such as inter-penetration) are drawn by comparison of the total density profiles. One should keep in mind that in absence of deuteration, SANS is not capable of differentiating between the two monomers. Some - usually plausible - assumptions are thus necessary for data interpretation.

In a very recent article, Brugnoni et al. [99] study hollow double-shell microgels designed to release possible guest molecules from the cavity by swelling of the shell, thereby providing permeability via the increase of the mesh size. Their system is based on pNIPAM co-polymerized with an ionic monomer in the inner shell, thereby lowering the transition temperature to below room temperature. They show by SLS via the radial density profiles obtained with a fuzzy core–shell model that the cavity remains preserved, while colloidal stability is maintained even in the shrunken state. Indeed, at room temperature where the pores toward the cavity are more closed to localize potential guest molecules, the steric protection of the outer and swollen pNIPAM shell remains effective.

Adding a new trigger is not always the only reason for introducing ionizable comonomers of weak acids. The same pathway has been explored by Wypysek et al. to circumvent the problem of too low T-responsiveness of highly crosslinked hollow shells, whereas too loose crosslinking introduced floppiness and did not preserve the hollow cavity [100]. As with the above-mentioned systems, the pH then controls if the system is responsive to temperature in a neutral state. In case of backbone ionization, here of co-polymerized itaconic acid at 10 mol%, at high pH sensitivity to ionic strength is introduced. The advantage of this molecule is to maintain colloidal stability of the silica during synthesis, as it can be copolymerized in the neutral state. The main effect of the backbone charge present at high pH is to add counter-ion osmotic pressure in the system, which acts as a radial force toward the outside, thus preserving the cavity. In this context, a Poisson–Boltzmann cell model has been set up to investigate the ionic profiles and resulting pressures. The corresponding scattering data has been analyzed by a fuzzy core–shell model. The first result of the introduction of the ionic stimulus is the suppression of thermo-sensitivity in the relevant temperature range. In the charged state, the system becomes sensitive to ionic strength. Screening by adding salt lowers the electrostatic osmotic pressure, leading to the surprising result of a less fuzzy outer surface, and a more fuzzy interior cavity wall. This is due to the expansion of the polymer toward the interior, whereas the outer surface becomes less stretched. Wypysek et al. thus evidence new ways of controlling multi-compartment microgel morphology, by either using T-sensitivity or charge screening.

Once the core–shell and potentially hollow architectures are well mastered, electrostatic charges can be employed to trigger uptake, storage, and release of polyelectrolyte as generic guest molecules. Naturally, other host–guest interactions may be chosen, such as hydrophobicity, but then the convenient pH trigger of electrostatics to control uptake, storage, and in particular release is usually lost. Gelissen et al. have studied the introduction of poly(styrene)-sulfonate of various masses into core–shell microgels designed with a cationic polymeric core to attract functional guest molecules [101]. Monomethylitaconate is copolymerized to introduce the positive charges. As cationic macromolecules exhibit cell toxicity due to interactions with negatively charged biomolecules, the authors have added an anionic APMH-shell, thus limiting toxicity. This has also the advantage of accumulating the guest molecules inside the particles, avoiding layer formation on the surface that could limit uptake. On the other hand, the total amount of incorporated guest molecules does not seem to be decreased by the presence of the shell, but is mostly controlled by the mass of the guest polyelectrolyte – smaller molecules are found to penetrate deeply in the core, while the longer ones remain into the periphery of the core. The latter provides a larger uptake, and due to the reduced interaction with the cationic polyelectrolytes, a more complete release. The corresponding internal microstructure has been shown by detailed SANS modeling based again on the fuzzy core–shell structure of all hydrogenated material – microgel and polyelectrolyte. The characteristic increase of the overall density profile above the value of the pure microgel indicates the presence of the guest molecules. If one excludes reorganization of the microgel morphology itself – this is one of the structural assumptions needed in absence of deuteration as mentioned above –, the effect of density added by chains is clearly seen in Figure 7.10, where the pure microgel volume fraction has been set to one in the center. Its increase shows thus the presence of the polyelectrolyte, which has different locations for short and long chains. As a result, uptake can be performed at low pH where the shell is neutral, and released at high pH. The latter seems to be more efficient for the longer chains, presumably because of their accumulation at the interface between the two shells. And finally, as often done by these groups, experiments are compared to computer simulations for structure and release kinetics, providing further insight into the mechanisms.

In a French–German collaboration, our own groups have recently revisited the structure of core–shell microgels where the core and the shell monomers, pNIPMAM and pNNPAM, respectively, have different volume transition temperatures, namely ca. 45 and 23 °C, respectively. This leads to a peculiar swelling behavior as measured by DLS, which reports the hydrodynamic radius as a function of temperature [102]. In spite of showing a jump in size at each transition temperature,

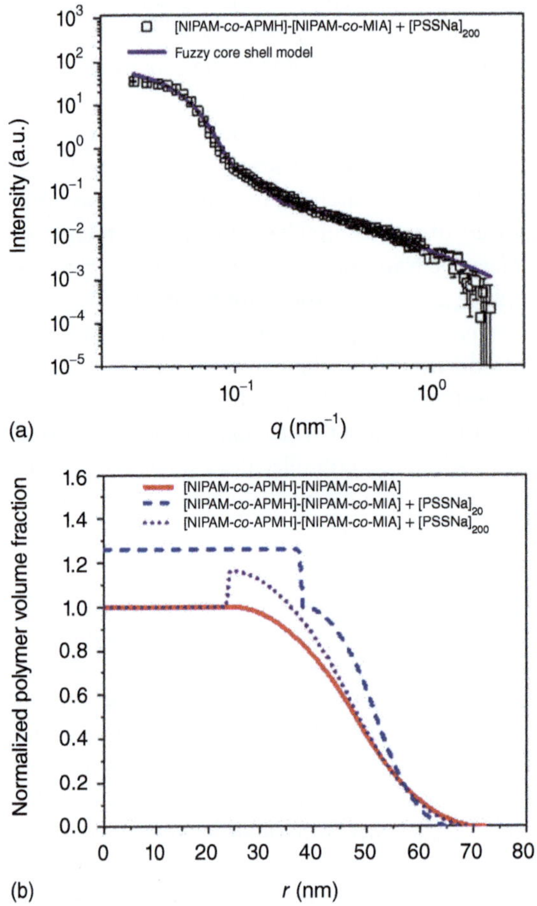

Figure 7.10 (a) Polyampholyte microgel with $[PSSNa]_{200}$ at pH 2 and I = 50 mM. SANS form factor and corresponding fit with a fuzzy core–shell model. (b) Radial profiles as a function of the microgel radius as obtained from fits in (a) for the polyampholyte microgel (solid line), polyampholyte microgel with $[PSSNa]_{20}$ (dashed line) and polyampholyte microgel with $[PSSNa]_{200}$ (dotted line). Note that the volume fraction has been normalized to one in the center of the bare microgel. Source: Reproduced with permission of the Royal Society of Chemistry from Ref. [101].

a linear swelling domain is found between the two transition temperatures. This was first explained by the above-mentioned "corset effect" [102], where the swelling of the core as the temperature is lowered from 55 °C is hindered by the still collapsed shell of pNNPAM. However, the complete picture is a bit different, as we could show by introducing deuteration of one of the monomers, pNNPAM. Deuteration gives access to the individual spatial distribution of each monomer, the quantitative determination of which is particularly well-performed using the form-free Monte Carlo simulations outlined in the "neutron section" above, and which will be applied here. In this context, it is worth mentioning that there has been some effort invested into characterizing the consequences of deuteration on the phases. Cors et al. have studied the volume transitions for different deuteration schemes by DLS [103]. Among the surprising results, it was shown that the transition temperature is not a monotonous function of the fraction of H-D-replacement. Brugnoni et al. arrived at a similar conclusion in their studies of ultralow crosslinked pNIPAM microgels, i.e. with synthesis without dedicated crosslinking molecules [104]. They found that the amplitude of the temperature shift depends on the specific group

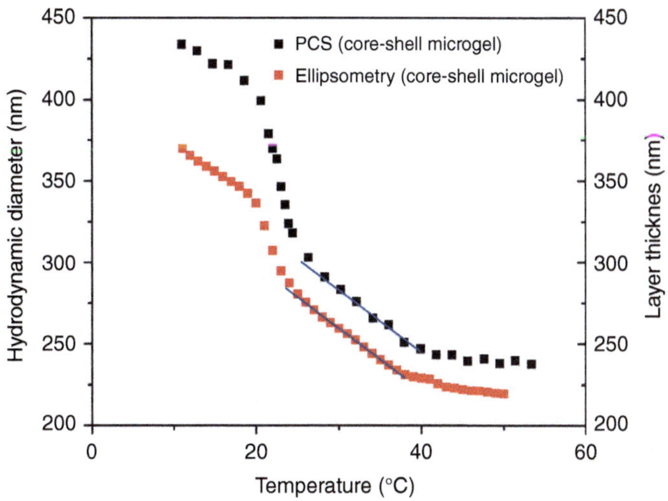

Figure 7.11 Comparison of the swelling of microgel particles in the bulk (in terms of R_h as measured by PCS, scale on the left, black symbols) to the same particles adsorbed on a surface (in terms of layer-thickness as measured by ellipsometry, scale on the right, red symbols). The linear swelling domain is indicated by the blueish lines. Both lines have the same slope. Source: Cors et al. [105]/with permission of American Chemical Society.

that is substituted, vinyl or isopropyl, only the latter one showing a strong impact of deuteration. Moreover, self-crosslinking was also found to be sensitive, and possibly hindered by deuteration, as found with a deuterated isopropyl group, which presumably is the group responsible for spontaneous crosslinking.

Coming back to the pNIPMAM-core/pNNPAM-shell system, we have first compared the qualitatively new linear swelling behavior in bulk (by PCS) to the one observed for microgel particles adsorbed onto a surface (by ellipsometry) and found that the swelling is robust, i.e. not affected by the surface [105]. In Figure 7.11, this is seen to be robust for a given core crosslinking ratio (BIS).

Using the form-free reverse Monte Carlo simulation, we have then solved the multi-shell model for the pNIPMAM cores, called "core-only" microgel particles [70]. The resulting volume fraction profile has already been shown in Figure 7.7. The decrease in temperature is seen to cause the swelling of the particle, with a decrease of its density in the core region, which extends to progressively higher radii. At 15 °C, the fully swollen state is characterized by an approximately constant central zone, which is surrounded by a more and more dilute corona. This core-corona structure is the natural result of the gradient in crosslinker, and not the core–shell structure purpose-synthesized successively with two different monomers, both crosslinked.

The microstructure of the core–shell microgels has also been studied by SANS, in separate experiments made with the same microgel particles, while adapting the solvent [106]. Using a H_2O/D_2O solvent rich in light water allows for matching the hydrogenated monomers (pNIPMAM), and the inverse is true for the pNNPAM shell. The scattered intensity thus reflects only the spatial distribution of the

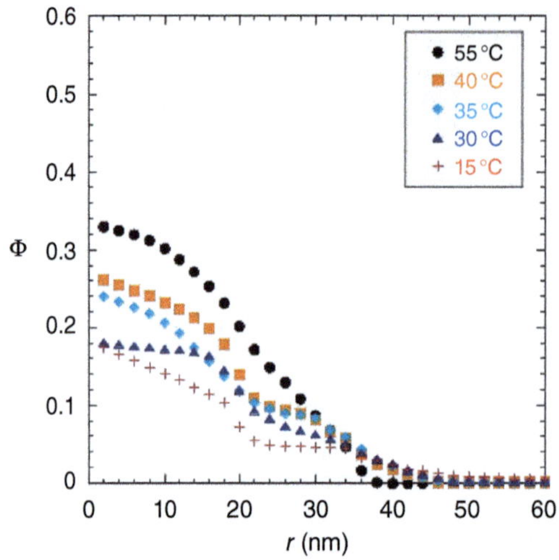

Figure 7.12 Density profile of the core monomers of the H-pNIPMAM-core D7-pNNPAM-shell system with the same core as in Figure 7.7, and contrast matched shell (CCC = 1.9 mol%). Source: Cors et al. [103]/Springer Nature/CC BY.

non-matched component. In Figure 7.12, the density profile of the core monomers within the core–shell microgel is plotted, for invisible but present shell monomers. While the set of curves follows mostly the same sequence as the "core-only" particles already shown in Figure 7.7, the surprising result is that the density of the core-monomers has not been increased by the surrounding deuterated pNNPAM shell, but actually decreased. This stands at odds with our own (naive) interpretation of the "corset effect", which states that the shell actually compresses the core, and should, therefore, increase its density.

The complementary SANS measurements of the pNIPMAM–pNNPAM microgels with a deuterated shell monomer in mostly hydrogenated solvent matching the pNIPMAM have provided molecular understanding of the above surprise: the now visible shell monomers do not form a homogeneous shell around the core, but rather penetrate into it down to the center of the microgel particle. In Figure 7.13a,

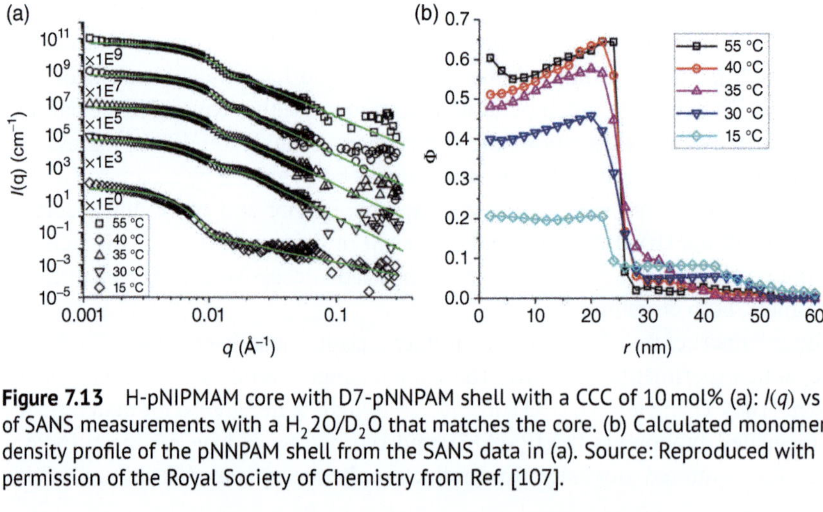

Figure 7.13 H-pNIPMAM core with D7-pNNPAM shell with a CCC of 10 mol% (a): $I(q)$ vs q of SANS measurements with a H_2O/D_2O that matches the core. (b) Calculated monomer density profile of the pNNPAM shell from the SANS data in (a). Source: Reproduced with permission of the Royal Society of Chemistry from Ref. [107].

the scattered intensities for different temperatures are found to be well described by the model. The corresponding shell monomer density profiles are reproduced in Figure 7.13. As a surprising result of monomer compatibility at high temperatures, the so-called shell monomer is actually more abundant in the core than the core monomers themselves. It looks like the second polymerization fills up the holes and meshes, before finishing its polymerization with a thin shell around the composite core. Obviously, this result must be system-dependent, because higher amounts of shell monomer would then induce the growth of a true shell. In the present case, however, the result is a monomer concentration gradient throughout the microgel, which presumably leads to different transition temperatures for each piece of material, depending on its distance from the center. This progressive evolution of the transition temperature is probably responsible for the striking linear swelling behavior shown in Figure 7.11.

7.4 Time-Resolved Small-Angle Scattering

According to Tanaka and Filmore [11], the swelling kinetics of gels scales with their largest dimension. Macroscopic gels are, therefore, rather slow in reaching their equilibrium state which limits their usefulness in sensor or actuator design. However, since the smallest dimension is decisive for the kinetics, microgels can have very short response times. A good way to measure the response time of colloidal gels is the use of classical relaxation methods, such as p- or T-jump and stopped-flow techniques, combined with fast scattering methods to follow the induced structural changes. The stopped-flow technique was successfully used by the Richtering group in the study of de-swelling of microgels triggered by mixing the aqueous microgel solution with alcohol. In this experiment, the co-nonsolvency effect is exploited to induce the volume transition of the particles in the mixing cuvette [108]. The experiments were done at the European Synchrotron Radiation Facility (ESRF) (Grenoble, France) using the ID02 SAXS beamline. The stopped-flow system was employed for ultrafast mixing of a microgel solution in pure methanol and a methanol–water mixture with $X_{MeOH} \approx 0.2$.

In Figure 7.14a, the SAXS data are plotted as a function of time. The statistics of the SAXS data are sufficiently good to allow a fit with a form factor model. In combination with simulations, the authors have shown that the collapse is a two-step process with two largely different time constants of $\tau_1 = 1.3$ ms and $\tau_2 = 2227$ ms. These results are in good agreement with recent stroboscopic p-jump SANS experiments by Wrede et al. [109]. The later experiments were performed on the D11 small-angle scattering machine at the Institute Laue Langevin that has a time resolution of 2 ms. The authors used a home-built pressure jump device allowing also 2 ms of dead-time. However, in contrast to the work by the Richtering group, here a poly(N-n-PAM) microgel was studied. Also in this case, two relaxation times were observed for the pressure-induced collapse. Swelling is characterized by only one relaxation time.

Also, the growth kinetics of microgels can be studied by time-resolved scattering techniques. The synthesis of nano and microgels is usually done using bad solvent

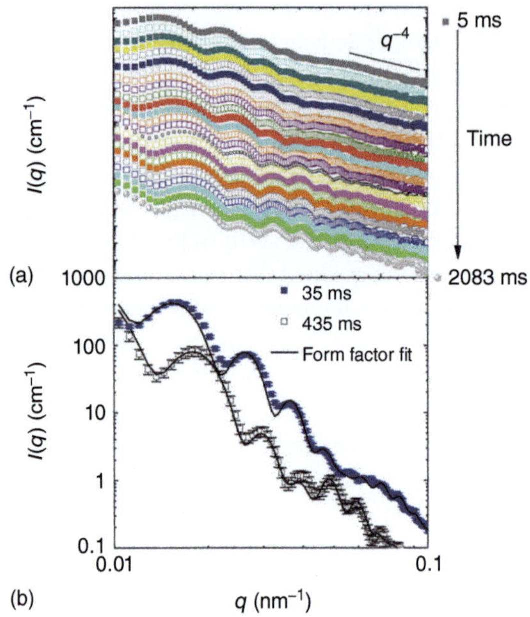

Figure 7.14 (a) SAXS patterns of PNIPAM microgels for the solvent composition change from pure MeOH to $X_{MeOH} \approx 0.2$ at different times. (b) Examples of form factor fit for two SAXS patterns obtained at 35 and 435 ms after mixing (for better visibility, the intensity of the SAXS pattern at 435 ms is shifted vertically by a factor of 0.3). Source: Reproduced with permission from Ref. [108].

conditions for the respective polymer which is employed. After initiation, the forming oligomers phase separates and forms a nucleus for the further precipitation of growing chains. However, many details of this so-called precipitation polymerization are still unknown. Here, time-resolved scattering experiments are of paramount importance to achieve a more detailed understanding of the process. An interesting example of such a study is the article by Virtanen et al. [110]. These authors investigated several initial conditions with respect to their influence on the formation of standard PNIPAM microgels.

Virtanen et al. have performed their experiments at the Heinz Maier-Leibnitz Zentrum (MLZ) (Munich, Germany) using the KWS-2 SANS machine. They used a stopped-flow device with the sample cell mounted in a heating block (see Figure 7.15a). Heat is used to initiate the reaction after rapid mixing. The use of such a device allows repeating the time-resolved recording of the scattering data until sufficiently good data statistics are obtained by summing up time binned data of several shots. This approach was introduced by Isabelle Grillo and coworkers [111] and has been used for the study of structural changes in a number of colloidal systems. The right-hand side of Figure 7.15 shows typical SANS scattering curves as obtained at different times in the course of the reaction. Already a qualitative look at the curves shows that the particles are indeed growing. An example of the obtained kinetic curves is shown in Figure 7.16.

The polymerization rate is determined by the concentration and transport of the unreacted monomer since the initiator concentration is nearly constant during the course of the reaction. Due to this, the mean particle volume is growing according to a pseudo-first-order kinetics. Virtanen et al. have confirmed this by calorimetric

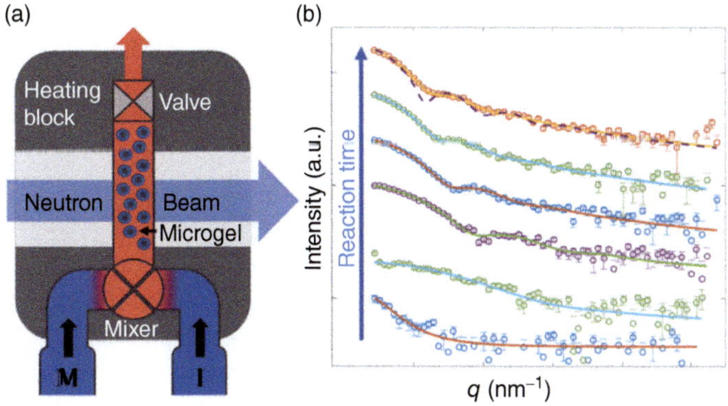

Figure 7.15 Scheme of the setup used by Virtanen et al. (a) and some examples for the measured form factors (b). Already by bare eyes, one can observe the increasing particle size. Source: Reproduced with permission from https://pubs.acs.org/doi/10.1021/acsomega.8b03461.

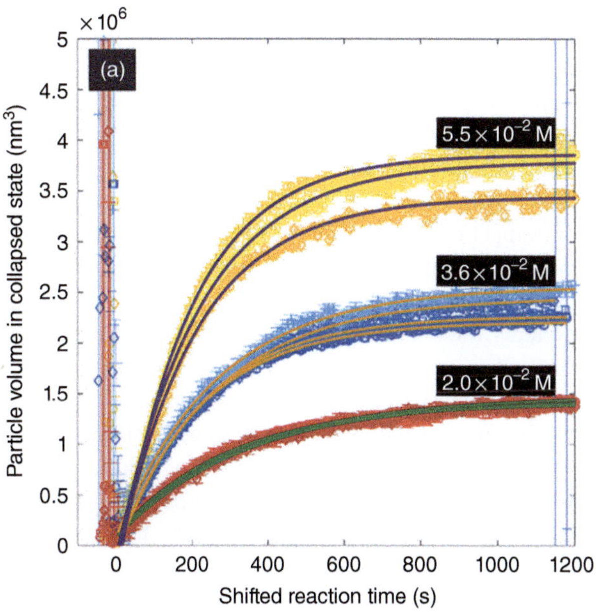

Figure 7.16 Series of kinetic data obtained for different initial PNIPAM concentrations. Source: Reproduced with permission from https://pubs.acs.org/doi/10.1021/acsomega.8b03461.

measurements of the polymerization process. Moreover, also the influence of the initiator and the crosslinker were studied in this work. The particle number density in the reaction volume is found to decrease due to peroxide initiator-induced self-crosslinking of NIPAM. The same phenomenon occurs upon increasing the amount of crosslinker. The most important outcome is related to the finding that a certain number of nuclei is created during the early stages of the reaction. Afterward the number density is found to stay constant. This explains the very low polydispersity usually achieved in precipitation polymerization.

7.5 Crowded Microgel Systems

Up to here, we have reported on microgel particles in dilute suspension, without interaction between neighboring particles possibly modifying their swelling behavior and morphology. The two main types of interaction are steric potentials that impede corona overlap, and electrostatics, which is caused by the ionic comonomers as already explored above. For low ionic strengths, electrostatic repulsion may have a range such that these interactions play a role. In this case, it is the ratio between Debye length and average inter-particle distance which determines if the regime is indeed dilute, or not. In any case, increasing the particle number density inevitably introduces interactions, which may thus be of electrostatic or steric origin, including soft-core repulsion. Several groups have focused on the morphology and swelling behavior of microgel particles embedded in a sea of other particles, or surrounded by other molecules, and some recent work is reviewed below.

As this chapter focuses mainly on small-angle scattering experiments, a key difference between SAXS and SANS should be emphasized here. In SAXS, it is complicated to modify particle contrast, although some possibilities using anomalous X-ray scattering exist [112] or in some cases, the electron density of the solvent can be changed by sugar addition [113, 114]. However, the sugar molecules drastically change the solvent properties and might also influence the particle interaction potential in an unwanted way.

In concentrated microgel systems, the interferences of waves scattered by different particles thus overlap with the "self-scattering", i.e. the form factor averaged over all particles [55]. The more concentrated the suspension, the stronger the perturbation of the form factor – which contains the morphological information – by the structure factor encoding the inter-particle correlations. On the contrary, SANS is very sensitive to deuteration, and synthesis of the same type of microgel particle, i.e. same size, monomer, and densities, in two versions, one deuterated, one hydrogenated, is a feasible task. Naturally, some minor discrepancies may remain due to polymerization issues, but also because hydrogenated and deuterated molecules do not have exactly identical physical properties. Nonetheless, it is often possible to either contrast match one of the two species via the solvent mixture, and thereby reduce the effective density of the visible species, or, even better, perform a zero-average contrast experiment [115, 116]. In the latter case, the structure factor cancels due to some "magic" compensation of interferences between H–D and H–H and D–D particles. The latter method needs to be carefully designed when preparing experiments but turns out to be extremely powerful for the analysis of concentrated suspensions.

The work by Nöjd et al. perfectly illustrates the usefulness of SAXS and SANS, respectively [117]. These authors have studied pNIPAM microgels copolymerized with acrylic acid. Before mixing fully deuterated and fully hydrogenated particles in water, dilute and partially charged microgels at low pH were compared by PCS, finding that they were close in size, of similar swelling properties, and only slightly shifted in transition temperature (see discussion of Refs. [106] and [99] in Section 7.3). In the following concentration series, both particles were mixed with H_2O/D_2O in the appropriate ratio to fulfill the zero-average contrast condition. In Figure 7.17a, the SANS intensities have been plotted, and they are seen to have no influence on structure factors, i.e. the matching worked perfectly. In Figure 7.17b, the same samples have been measured with SAXS. Here, the deuteration does not modify the electron-density distribution. Thus, the full intensity with form and structure is obtained. Very prominent structure factor peaks arise, which have been described by the authors by a variant of a well-known solution of integral equations called renormalized mean spherical approximation. The original form of the latter takes electrostatic interactions into account, and its modification by penetrating solvent typical for hydrogels. In both SAXS and SANS, the form factor can be described by a fuzzy sphere plus polymer network fluctuations. Intensity measurements in absolute values then provide a mass determination of the particles, using Eq. (7.11). At the lowest concentrations, the measurements are complemented by static and dynamic light scattering, both highlighting persistent interactions. To summarize, the authors report that these ionic microgels are quite swollen at close to infinite dilution due to electrostatic forces coming from partial protonation of acrylic acid. Via solution of a Poisson–Boltzmann cell model for the core, and Monte Carlo simulations of the dangling ends forming the corona, they highlight the electrostatic effect on the particle structure, termed "porcupine" in a zoological analogy. At intermediate concentrations, the microgels decrease in size and reach an almost constant core radius throughout the SANS regime. Only at the highest concentrations, when the effective volume fraction approaches one, a weak shrinking is observed.

In a follow-up, the same group has studied H- and D-microgels in presence of alternating electric fields, also under zero-average contrast [118]. The fuzzy sphere model successfully describes particle shapes for high and low fields, indicating that particles are not deformed by external electric forces. They are, however, found to align, as evidenced by laser scanning confocal microscopy. The SANS measurements thus allow to show that in spite of the decrease of interparticle distances in the strings, there is no particle compression or deformation, but apparently only particle interpenetration. The latter is probably the mechanism responsible for long-lived metastable crystalline states, which have to overcome chain entanglement between microgel particles to rearrange.

However, in recent work by Scotti [119], the conclusion is slightly different that is probably related to the fact that in the works by Nöjd and coworkers only linearly aligned particles were studied. Determination of the real volume fraction of microgels is not at all trivial and the chemical composition has a major influence on possible particle deformation. Scotti has addressed this problem also using contrast variation in SANS having partly deuterated microgels as a matrix for nearly identical hydrogenated particles. Using appropriate heavy and light water mixtures allows

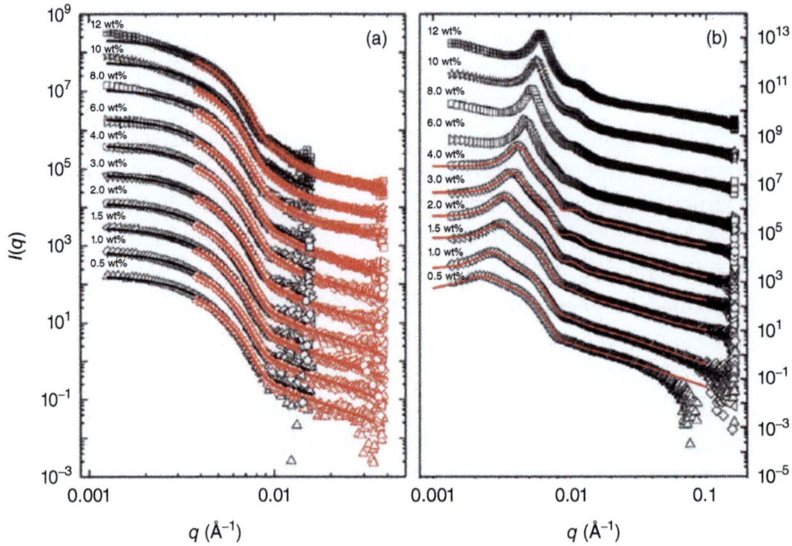

Figure 7.17 (a) SANS data obtained as a function of concentration as indicated in the legend. Solid lines are fits to the data using the fuzzy sphere model. (b) SAXS data obtained using the same samples as in (a). The solid lines are fits to the data based on input information from (a) taking the structure factor into account. All curves are offset along the ordinate for clarity. Source: Reproduced from Ref. [117] with permission of the Royal Society of Chemistry.

measuring an apparent form factor for dilute hydrogenated microgels that are nevertheless compressed by the matrix. For NIPAM microgels with 5 mol% crosslinker, he has observed that up to a volume fraction of 0.64 no significant deviation between the real and the so-called generalized volume fraction occurs. Beyond this value, the real volume fraction is smaller than the generalized volume fraction based on the radius of the diluted particles. Hence, at $\phi = 0.64$ for these microgels compression sets in. For ultralow crosslinked microgels compression starts even earlier at $\phi = 0.58$. The compression of microgels was found to lead to faceting [120] using super-resolution optical microscopy. In his work, Scotti uses this phenomenon to explain the observed increase in apparent polydispersity in the form factor fits.

As with monodisperse hard spheres, microgel particles may crystallize at high concentrations. As soon as polydispersity exceeds some 12% for hard spheres, crystallization is suppressed. But contrary to hard spheres, microgels may also adapt their shape and possibly size. Scotti et al. have followed this idea by comparing phase diagrams and scattering of microgel mixtures of controlled polydispersity, comparing in particular Gaussian polydispersity to bidispersity with a ratio in size of 1.3–1.4 [121]. They focus on a concentration study of the fully swollen state of pNIPAM accessible at room temperature. By introducing bidispersity, the size mismatch impedes in theory crystallization. Due to the presence of electrostatic charges, caused by the initiator molecules, the counter ion pressure is sensitive to the microgel concentration and responsible for shrinking at high concentrations [122]. By analyzing the apparent structure factors relying on polydisperse fuzzy sphere form factors measured at high dilution – it has been checked independently that this introduces only little error in peak shape and position –, this study finds that the bigger microgel particles shrink preferentially, thus reducing the polydispersity. The resulting S(q) resembles one of the small particles only, indicating that the bigger ones have deswollen, thus favoring crystallization up to a number fraction of ca. 30% of big particles. Concerning Gaussian polydispersity, the flexibility of the microgel allows accommodating up to almost 20% polydispersity and still crystallizing, presumably due to the shrinking of the large-size tail of the distribution.

The next step taken in concentration studies was to introduce a new degree of freedom, by taking advantage of the hollow microgels discussed in Section 7.3. The presence of the cavity naturally allows the microgel particles to adjust to external pressure exerted by the surrounding microgels, which have been chosen to be plain but of identical crosslinking and (almost identical) hydrodynamic radius. The group takes nicely advantage of the properties of neutron scattering, by contrast, matching via the solvent mixture of light and heavy water the plain microgel particle, and highlighting thus only the hollow ones [42]. As their concentration was deliberately kept low, the structure factor influence becomes negligible, and they measure directly the form factor, which has been described by the usual fuzzy hollow-shell model. It may be noted that the form-free model (see Section 7.2) has been applied as a cross-check, and provides virtually identical profiles. As a result, the cavity is seen to be preserved up to moderate volume fraction. At higher pressures, the polymer extends into the cavity, and simultaneously the outer fuzziness of the hollow microgels disappears.

In a subsequent step, Scotti et al. comparatively discuss the influence of microgel crosslinking and architecture on morphology under pressure at high concentrations [45]. They use the same contrast matching approach of the matrix of microgels exerting the pressure on some selected, nondeuterated microgels. They confirm the important role of the cavity, which in particular has a stronger contribution to deswelling than the decrease of the crosslink density by a factor of two.

As a final point in this review of recent work on concentrated microgels, we propose to highlight a completely different approach. The effect of increase in concentration of microgel particles is usually studied in a molecular solvent, as in the above studies. An interesting crossover has been proposed by Schneider et al., who studied the effect of added linear polymer chains replacing the solvent [123]. Their microgel particles are made of p(*N*-vinylcaprolactam), of swollen size in heavy water at about 200 nm. These particles are incorporated in PEG melts of different chain masses using a common solvent and subsequent evaporation. With dPEG, the microgel particles gain neutron visibility, and bumps in the SANS intensity can be described by a core-shell model with fuzziness. Surprisingly, the small polymer chains gain visibility also, indicating a molecular mixture, i.e. interpenetration into the microgels, for the smallest masses of below 3k. The larger masses do not show chain scattering if samples are fully equilibrated, and a strong low-q increase seems to indicate phase separation. At high temperatures, between 160 and 180 °C, the microgels are found to deswell, they thus seem to cross a critical temperature interpreted as LCST. This results in the expulsion of matrix chain molecules, with a corresponding decrease in the visibility of the latter. If this happens at high microgel concentrations, the particles appear to order in the majority phase in close contact, and the corresponding peak position shifts with the deswelling.

7.6 Conclusion and Outlook

Responsive micro- and nanogels are a very active research topic, and we have given a necessarily subjective view of recent progress in the field. Our focus was directed toward new particle morphologies, but also included evolution of particle structure with external stimuli, including solvent, pressure, and temperature jumps, as well as internal stimuli, such as the pressure exerted by neighboring particles. For such studies, the use of scattering techniques is crucial, due to their nondestructive and possibly time-resolved measurements of both particle architectures and interactions. In particular, neutron scattering is a very powerful method, as (partial) deuteration can be used to localize different monomers and particles in space via isotopic substitution, zero-average contrast, and in general contrast variation methods. Given the importance of these methods, we have also provided an overview of the analysis of static and dynamic scattering data. In particular, density profile approaches have provided deep insight into particle morphologies. With an admittedly scholarly presentation of both the underlying mechanisms and some "hands-on" tools on how to access the relevant information, we hope to have convinced the reader that scattering is a highly suitable tool for microgel studies.

It is noteworthy that only very few systems were characterized with respect to growth kinetics. Especially, for copolymer systems at least to our knowledge, no data are available, and we hope that new experiments with ever more powerful instruments will allow unraveling the details of such mostly unknown processes. We are looking forward to reading such results in the literature, and hope that our modest contribution here will be useful for scientists working in the field.

Appendix: Absolute Intensity for Fuzzy Sphere Form Factors

The expression in Eq. (7.5) is the product of the normalized hard-sphere form factor by a Gaussian. This product in q-space corresponds to a convolution in r-space by a Gaussian, and as the Gaussian is symmetric, the weighting by the volume is usually overlooked in the literature. This leads to absolute intensity predictions with slight errors, of the order of 10% depending on fuzziness. In other words, the integral over the profile describing the density function contains a term $4\pi r^2$, which is asymmetric around the nominal radius $r = R$, and increases the weight of the outer zones of higher r. It is straightforward to calculate that in the case of a 10% interfacial width ($\frac{\sigma}{R} = 10\%$), the correction factor is 1.03, and for 20%, it is ca. 1.12. The effect of fuzziness is to increase the particle size visible in the Guinier regime, and this should be accompanied by the above intensity correction, although it is usually not explicitly included in form factor equations, such as Eq. (7.5).

Moreover, the enveloping high-q power law is shifted to a slope steeper than the Porod law, thereby decreasing the higher-order form factor oscillations, as expected for fuzziness.

References

1 Pelton, R.H. and Chibante, P. (1986). Preparation of aqueous latices with *N*-isopropylacrylamide. *Colloids Surf.* 20 (3): 247–256. https://doi.org/10.1016/0166-6622(86)80274-8.

2 Saunders, B.R. and Vincent, B. (1999). Microgel particles as model colloids: theory, properties and applications. *Adv. Colloid Interface Sci.* 80 (1): 1–25. https://doi.org/10.1016/s0001-8686(98)00071-2.

3 Pelton, R. (2000). Temperature-sensitive aqueous microgels. *Adv. Colloid Interface Sci.* 85 (1): 1–33. https://doi.org/10.1016/s0001-8686(99)00023-8.

4 Gilányi, T., Varga, I., Meszáros, R. et al. (2000). Characterisation of monodisperse poly(*N*-isopropylacrylamide) microgel particles. *Phys. Chem. Chem. Phys.* 2 (9): 1973–1977. https://doi.org/10.1039/b000571l.

5 Nayak, S. and Lyon, L.A. (2005). Soft nanotechnology with soft nanoparticles. *Angew. Chem.* 44 (47): 7686–7708. https://doi.org/10.1002/anie.200501321.

6 Ballauff, M. and Lu, Y. (2007). "Smart" nanoparticles: preparation, characterization and applications. *Polymer* 48 (7): 1815–1823. https://doi.org/10.1016/j.polymer.2007.02.004.

7 Pich, A. and Richtering, W. (2010). Microgels by precipitation polymerization: synthesis, characterization, and functionalization. In: A. Pich and W. Richtering. *Chemical Design of Responsive Microgels*, 1–37. Berlin, Heidelberg: Springer-Verlag. https://doi.org/10.1007/978-3-642-16379-1.

8 Plamper, F.A. and Richtering, W. (2017). Functional microgels and microgel systems. *Acc. Chem. Res.* 50 (2): 131–140. https://doi.org/10.1021/acs.accounts.6b00544.

9 Karg, M., Pich, A., Hellweg, T. et al. (2019). Nanogels and microgels: from model colloids to applications, recent developments, and future trends. *Langmuir* 35 (19): 6231–6255. https://doi.org/10.1021/acs.langmuir.8b04304.

10 Flory, P.J. and Rehner, J. (1944). Effect of deformation on the swelling capacity of rubber. *J. Chem. Phys.* 12 (10): 412–414. https://doi.org/10.1063/1.1723884.

11 Tanaka, T. and Fillmore, D.J. (1979). Kinetics of swelling of gels. *J. Chem. Phys.* 70 (3): 1214–1218. https://doi.org/10.1063/1.437602.

12 Mallam, S., Horkay, F., Hecht, A.M., and Geissler, E. (1989). Scattering and swelling properties of inhomogeneous polyacrylamide gels. *Macromolecules* 22 (8): 3356–3361. https://doi.org/10.1021/ma00198a029.

13 Shibayama, M., Tanaka, T., and Han, C.C. (1992). Small angle neutron scattering study on poly(N-isopropyl acrylamide) gels near their volume-phase

transition temperature. *J. Chem. Phys.* 97 (9): 6829–6841. https://doi.org/10.1063/1.463636.

14 Snowden, M.J., Chowdhry, B.Z., Vincent, B., and Morris, G.E. (1996). Colloidal copolymer microgels of N-isopropylacrylamide and acrylic acid: pH, ionic strength and temperature effects. *J. Chem. Soc., Faraday Trans.* 92 (24): 5013. https://doi.org/10.1039/ft9969205013.

15 Saunders, B.R., Crowther, H.M., and Vincent, B. (1997). Poly[(methyl methacrylate)-co-(methacrylic acid)] microgel particles: swelling control using pH, cononsolvency, and osmotic deswelling. *Macromolecules* 30 (3): 482–487. https://doi.org/10.1021/ma961277f.

16 Kim, J.-H. and Ballauff, M. (1999). The volume transition in thermosensitive core-shell latex particles containing charged groups. *Colloid Polym. Sci.* 277 (12): 1210–1214. https://doi.org/10.1007/s003960050512.

17 Fernández-Nieves, A., Fernández-Barbero, A., Vincent, B., and de las Nieves, F.J. (2000). Charge controlled swelling of microgel particles. *Macromolecules* 33 (6): 2114–2118. https://doi.org/10.1021/ma991520l.

18 Kratz, K., Hellweg, T., and Eimer, W. (2000). Influence of charge density on the swelling of colloidal poly(*N*-isopropylacrylamide-co-acrylic acid) microgels. *Colloids Surf., A* 170 (2–3): 137–149. https://doi.org/10.1016/s0927-7757(00)00490-8.

19 Hoare, T. and Pelton, R. (2004). Highly pH and temperature responsive microgels functionalized with vinylacetic acid. *Macromolecules* 37 (7): 2544–2550. https://doi.org/10.1021/ma035658m.

20 Serpe, M.J., Yarmey, K.A., Nolan, C.M., and Lyon, L.A. (2005). Doxorubicin uptake and release from microgel thin films. *Biomacromolecules* 6 (1): 408–413. https://doi.org/10.1021/bm049455x.

21 Schmaljohann, D. (2006). Thermo- and pH-responsive polymers in drug delivery. *Adv. Drug Deliv. Rev.* 58 (15): 1655–1670. https://doi.org/10.1016/j.addr.2006.09.020.

22 Das, M., Sanson, N., Fava, D., and Kumacheva, E. (2007). Microgels loaded with gold nanorods: photothermally triggered volume transitions under physiological conditions. *Langmuir* 23 (1): 196–201. https://doi.org/10.1021/la061596s.

23 Blackburn, W.H., Dickerson, E.B., Smith, M.H. et al. (2009). Peptide-functionalized nanogels for targeted siRNA delivery. *Bioconjugate Chem.* 20 (5): 960–968. https://doi.org/10.1021/bc800547c.

24 Turcu, R., Craciunescu, I., Garamus, V.M. et al. (2015). Magnetic microgels for drug targeting applications: physical–chemical properties and cytotoxicity evaluation. *J. Magn. Magn. Mater.* 380: 307–314. https://doi.org/10.1016/j.jmmm.2014.08.041.

25 Pergushov, D.V., Sigolaeva, L.V., Balabushevich, N.G. et al. (2021). Loading of doxorubicin into surface-attached stimuli-responsive microgels and its subsequent release under different conditions. *Polymer* 213: 123227. https://doi.org/10.1016/j.polymer.2020.123227.

26 Mei, Y., Lu, Y., Polzer, F. et al. (2007). Catalytic activity of palladium nanoparticles encapsulated in spherical polyelectrolyte brushes and core-shell microgels. *Chem. Mater.* 19 (5): 1062–1069. https://doi.org/10.1021/cm062554s.

27 Welsch, N., Ballauff, M., and Lu, Y. (2010). Microgels as nanoreactors: applications in catalysis. In: A. Pich and W. Richtering. *Chemical Design of Responsive Microgels*, 129–163. Berlin, Heidelberg: Springer. https://doi.org/10.1007/978-3-642-16379-1.

28 Besold, D., Risse, S., Lu, Y. et al. (2021). Kinetics of the reduction of 4-nitrophenol by silver nanoparticles immobilized in thermoresponsive core–shell nanoreactors. *Ind. Eng. Chem. Res.* 60 (10): 3922–3935. https://doi.org/10.1021/acs.iecr.0c06158.

29 Debord, J.D. and Lyon, L.A. (2000). Thermoresponsive photonic crystals. *J. Phys. Chem. B* 104 (27): 6327–6331. https://doi.org/10.1021/jp001238c.

30 Shu, T., Shen, Q., Wan, Y. et al. (2018). Silver nanoparticle-loaded microgel-based etalons for H_2O_2 sensing. *RSC Adv.* 8 (28): 15567–15574. https://doi.org/10.1039/c8ra02215a.

31 Wei, M. and Serpe, M.J. (2018). Temperature-light dual-responsive Au@PNIPAm core-shell microgel-based optical devices. *Part. Part. Syst. Charact.* 36 (1): 1800326. https://doi.org/10.1002/ppsc.201800326.

32 Uhlig, K., Wegener, T., Hertle, Y. et al. (2018). Thermoresponsive microgel coatings as versatile functional compounds for novel cell manipulation tools. *Polymers* 10 (6): 656. https://doi.org/10.3390/polym10060656.

33 Bell, D.J., Ludwanowski, S., Lüken, A. et al. (2021). Hydrogel membranes made from crosslinked microgel multilayers with tunable density. *J. Membr. Sci.* 620: 118912. https://doi.org/10.1016/j.memsci.2020.118912.

34 Bookhold, J., Dirksen, M., Wiehemeier, L. et al. (2021). Smart membranes by electron beam cross-linking of copolymer microgels. *Soft Matter* 17 (8): 2205–2214. https://doi.org/10.1039/d0sm02041a.

35 Annegarn, M., Dirksen, M., and Hellweg, T. (2021). Importance of pH in synthesis of pH-responsive cationic nano- and microgels. *Polymers* 13 (5): 827. https://doi.org/10.3390/polym13050827.

36 Balaceanu, A., Mayorga, V., Lin, W. et al. (2012). Copolymer microgels by precipitation polymerisation of N-vinylcaprolactam and N-isopropylacrylamides in aqueous medium. *Colloid Polym. Sci.* 291 (1): 21–31. https://doi.org/10.1007/s00396-012-2659-1.

37 Lane, T., Holloway, J.L., Milani, A.H. et al. (2013). Double network hydrogels prepared from pH-responsive doubly crosslinked microgels. *Soft Matter* 9 (33): 7934. https://doi.org/10.1039/c3sm51356d.

38 Hertle, Y., Zeiser, M., Hasenöhrl, C. et al. (2010). Responsive p(NIPAM-co-NtBAM) microgels: flory–Rehner description of the swelling behaviour. *Colloid Polym. Sci.* 288 (10–11): 1047–1059. https://doi.org/10.1007/s00396-010-2232-8.

39 Königer, A., Plack, N., Köhler, W. et al. (2013). Thermophoresis of thermoresponsive polystyrene–poly(N-isopropylacrylamide) core–shell particles. *Soft Matter* 9 (5): 1418–1421. https://doi.org/10.1039/c2sm27417e.

40 Clarke, K.C., Dunham, S.N., and Lyon, L.A. (2015). Core/shell microgels decouple the pH and temperature responsivities of microgel films. *Chem. Mater.* 27 (4): 1391–1396. https://doi.org/10.1021/cm504649t.

41 Kureha, T., Nagase, Y., and Suzuki, D. (2018). High reusability of catalytically active gold nanoparticles immobilized in core–shell hydrogel microspheres. *ACS Omega* 3 (6): 6158–6165. https://doi.org/10.1021/acsomega.8b00819.

42 Scotti, A., Brugnoni, M., Rudov, A.A. et al. (2018). Hollow microgels squeezed in overcrowded environments. *J. Chem. Phys.* 148 (17): 174903. https://doi.org/10.1063/1.5026100.

43 Nickel, A.C., Scotti, A., Houston, J.E. et al. (2019). Anisotropic hollow microgels that can adapt their size, shape, and softness. *Nano Lett.* 19 (11): 8161–8170. https://doi.org/10.1021/acs.nanolett.9b03507.

44 Islam, M.R., Nguy, C., Pandit, S., and Lyon, L.A. (2020). Design and synthesis of core–shell microgels with one-step clickable crosslinked cores and ultralow crosslinked shells. *Macromol. Chem. Phys.* 221 (19): 2000156. https://doi.org/10.1002/macp.202000156.

45 Scotti, A., Denton, A.R., Brugnoni, M. et al. (2019). Deswelling of microgels in crowded suspensions depends on cross-link density and architecture. *Macromolecules* 52 (11): 3995–4007. https://doi.org/10.1021/acs.macromol.9b00729.

46 Scotti, A., Houston, J.E., Brugnoni, M. et al. (2020). Phase behavior of ultrasoft spheres show stable BCC lattices. *Phys. Rev. E* 102 (5). https://doi.org/10.1103/physreve.102.052602.

47 Seelenmeyer, S., Deike, I., Dingenouts, N. et al. (2000). Analysis of the volume transition in thermosensitive core-shell particles by synchrotron small-angle X-ray scattering. *J. Appl. Crystallogr.* 33 (3): 574–576. https://doi.org/10.1107/s0021889899013163.

48 Seelenmeyer, S., Deike, I., Rosenfeldt, S. et al. (2001). Small-angle X-ray and neutron scattering studies of the volume phase transition in thermosensitive core–shell colloids. *J. Chem. Phys.* 114 (23): 10471–10478. https://doi.org/10.1063/1.1374633.

49 Saunders, B.R. (2004). On the structure of poly(N-isopropylacrylamide) microgel particles. *Langmuir* 20 (10): 3925–3932. https://doi.org/10.1021/la036390v.

50 Widmann, T., Kreuzer, L.P., Hohn, N. et al. (2019). Hydration and solvent exchange induced swelling and deswelling of homogeneous poly(*N*-isopropylacrylamide) microgel thin films. *Langmuir* 35 (49): 16341–16352. https://doi.org/10.1021/acs.langmuir.9b03104.

51 Hellweg, T., Kratz, K., Pouget, S., and Eimer, W. (2002). Internal dynamics in colloidal PNIPAM microgel particles immobilised in mesoscopic crystals. *Colloids Surf., A* 202 (2–3): 223–232. https://doi.org/10.1016/s0927-7757(01)01077-9.

52 Scherzinger, C., Holderer, O., Richter, D., and Richtering, W. (2012). Polymer dynamics in responsive microgels: influence of cononsolvency and microgel architecture. *Phys. Chem. Chem. Phys.* 14 (8): 2762. https://doi.org/10.1039/c2cp23328b.

53 Frenzel, L., Lehmkühler, F., Koof, M. et al. (2020). The phase diagram of colloidal silica–PNIPAm core–shell nanogels. *Soft Matter* 16 (2): 466–475. https://doi.org/10.1039/c9sm01884k.

54 Berne, B.J. and Pecora, R. (2000). *Dynamic Light Scattering: With Applications to Chemistry, Biology, and Physics*. Dover Publications Inc. ISBN 978-0486411552. https://www.ebook.de/de/product/3344676/bruce:j_berne_robert_pecora_physics_dynamic_light_scattering_with_applications_to_chemistry_biology_and_physics.html.

55 Lindner, P. and Zemb, T (eds.) (2002). *Neutrons, X-rays and Light: Scattering Methods Applied to Soft Condensed Matter*. North-Holland. ISBN 978-0444511225.

56 Kozhunova, E.Y., Rudyak, V.Y., Li, X. et al. (2021). Microphase separation of stimuli-responsive interpenetrating network microgels investigated by scattering methods. *J. Colloid Interface Sci.* 597: 297–305. https://doi.org/10.1016/j.jcis.2021.03.178.

57 Franco, S., Buratti, E., Ruzicka, B. et al. (2021). Volume fraction determination of microgel composed of interpenetrating polymer networks of PNIPAM and polyacrylic acid. *J. Phys. Condens. Matter* 33 (17): 174004. https://doi.org/10.1088/1361-648x/abe1ec.

58 Micali, N., Bertoldo, M., Buratti, E. et al. (2018). Interpenetrating polymer network microgels in water: effect of composition on the structural properties and electrosteric interactions. *ChemPhysChem* 19 (21): 2894–2901. https://doi.org/10.1002/cphc.201800707.

59 Schneider, J., Wiemann, M., Rabe, A., and Bartsch, E. (2017). On tuning microgel character and softness of cross-linked polystyrene particles. *Soft Matter* 13 (2): 445–457. https://doi.org/10.1039/c6sm02007k.

60 Stieger, M., Richtering, W., Pedersen, J.S., and Lindner, P. (2004). Small-angle neutron scattering study of structural changes in temperature sensitive microgel colloids. *J. Chem. Phys.* 120 (13): 6197–6206. https://doi.org/10.1063/1.1665752.

61 Schärtl, W. (2007). *Light Scattering from Polymer Solutions and Nanoparticle Dispersions*. Berlin, Heidelberg: Springer-Verlag. https://doi.org/10.1007/978-3-540-71951-9.

62 Hannappel, Y., Wiehemeier, L., Dirksen, M. et al. (2021). Smart microgels from unconventional acrylamides. *Macromol. Chem. Phys.* 2100067. https://doi.org/10.1002/macp.202100067.

63 Provencher, S.W. (1982). CONTIN: A general purpose constrained regularization program for inverting noisy linear algebraic and integral equations. *Comput. Phys. Commun.* 27 (3): 229–242. https://doi.org/10.1016/0010-4655(82)90174-6.

64 Scotti, A., Liu, W., Hyatt, J.S. et al. (2015). The CONTIN algorithm and its application to determine the size distribution of microgel suspensions. *J. Chem. Phys.* 142 (23): 234905. https://doi.org/10.1063/1.4921686.

65 Koppel, D.E. (1972). Analysis of macromolecular polydispersity in intensity correlation spectroscopy: the method of cumulants. *J. Chem. Phys.* 57 (11): 4814–4820.

66 Frisken, B.J. (2001). Revisiting the method of cumulants for the analysis of dynamic light-scattering data. *Appl. Opt.* 40 (24): 4087–4091. https://doi.org/10.1364/AO.40.004087.

67 Freeman, K.G., Adamczyk, J., and Streletzky, K.A. (2020). Effect of synthesis temperature on size, structure, and volume phase transition of polysaccharide microgels. *Macromolecules* 53 (21): 9244–9253. https://doi.org/10.1021/acs.macromol.0c01605.

68 Nigro, V., Angelini, R., King, S. et al. (2021). Apparatus for simultaneous dynamic light scattering–small angle neutron scattering investigations of dynamics and structure in soft matter. *Rev. Sci. Instrum.* 92 (2): 023907. https://doi.org/10.1063/5.0035529.

69 Schmid, A.J., Wiehemeier, L., Jaksch, S. et al. (2021). Flexible sample environments for the investigation of soft matter at the European spallation source: Part I —the in situ SANS/DLS setup. *Appl. Sci.* 11 (9): 4089. https://doi.org/10.3390/app11094089.

70 Cors, M., Wiehemeier, L., Hertle, Y. et al. (2018). Determination of internal density profiles of smart acrylamide-based microgels by small-angle neutron scattering: a multishell reverse Monte Carlo approach. *Langmuir* 34 (50): 15403–15415. https://doi.org/10.1021/acs.langmuir.8b03217.

71 Hansen, J.-P. and McDonald, I.R. (2013). *Theory of Simple Liquids*. Elsevier. https://doi.org/10.1016/c2010-0-66723-x.

72 Botet, R., Kwok, S., and Cabane, B. (2020). Percus–Yevick structure factors made simple. *J. Appl. Crystallogr.* 53 (6): 1570–1582. https://doi.org/10.1107/s1600576720014041.

73 Hansen, J.-P. and Hayter, J.B. (1982). A rescaled MSA structure factor for dilute charged colloidal dispersions. *Mol. Phys.* 46 (3): 651–656. https://doi.org/10.1080/00268978200101471.

74 Teixeira, J. (1988). Small-angle scattering by fractal systems. *J. Appl. Crystallogr.* 21 (6): 781–785. https://doi.org/10.1107/s0021889888000263.

75 Oberdisse, J. and Demé, B. (2002). Structure of latex-silica nanocomposite films: a small-angle neutron scattering study. *Macromolecules* 35 (11): 4397–4405. https://doi.org/10.1021/ma0118419.

76 Kratz, K., Hellweg, T., and Eimer, W. (2001). Structural changes in PNIPAM microgel particles as seen by SANS, DLS, and EM techniques. *Polymer* 42 (15): 6631–6639. https://doi.org/10.1016/s0032-3861(01)00099-4.

77 Debye, P. (1947). Molecular-weight determination by light scattering. *J. Phys. Colloid Chem.* 51 (1): 18–32. https://doi.org/10.1021/j150451a002.

78 Hammouda, B. (2012). Small-angle scattering from branched polymers. *Macromol. Theory Simul.* 21 (6): 372–381. https://doi.org/10.1002/mats.201100111.

79 Kyrey, T., Witte, J., Feoktystov, A. et al. (2019). Inner structure and dynamics of microgels with low and medium crosslinker content prepared via surfactant-free precipitation polymerization and continuous monomer feeding approach. *Soft Matter* 15 (32): 6536–6546. https://doi.org/10.1039/c9sm01161g.

80 Berndt, I., Pedersen, J.S., Lindner, P., and Richtering, W. (2006). Influence of shell thickness and cross-link density on the structure of temperature-sensitive poly-*N*-isopropylacrylamide-poly-N-isopropylmethacrylamide core-shell microgels investigated by small-angle neutron scattering. *Langmuir* 22 (1): 459–468. https://doi.org/10.1021/la052463u.

81 Boon, N. and Schurtenberger, P. (2017). Swelling of micro-hydrogels with a crosslinker gradient. *Phys. Chem. Chem. Phys.* 19 (35): 23740–23746. https://doi.org/10.1039/c7cp02434g.

82 Lienafa, L., Oberdisse, J., Mora, S. et al. (2011). Rheology and SANS on PET-b-PLAc-b-p(DMAEMAq) triblock copolymers: impact of the PET and polyelectrolyte chain length. *Macromolecules* 44 (13): 5326–5335. https://doi.org/10.1021/ma200331b.

83 Virtanen, O.L.J., Mourran, A., Pinard, P.T., and Richtering, W. (2016). Persulfate initiated ultra-low cross-linked poly(*N*-isopropylacrylamide) microgels possess an unusual inverted cross-linking structure. *Soft Matter* 12 (17): 3919–3928. https://doi.org/10.1039/c6sm00140h.

84 Glatter, O. (1977). Data evaluation in small angle scattering: calculation of the radial electron density distribution by means of indirect fourier transformation. *Acta Phys. Austriaca* 47: 83–102.

85 Glatter, O. (1977). A new method for the evaluation of small-angle scattering data. *J. Appl. Crystallogr.* 10: 415–421.

86 McGreevy, R.L. and Pusztai, L. (1988). Reverse Monte Carlo simulation: a new technique for the determination of disordered structures. *Mol. Simul.* 1 (6): 359–367. https://doi.org/10.1080/08927028808080958.

87 Oberdisse, J., Hine, P., and Pyckhout-Hintzen, W. (2007). Structure of interacting aggregates of silicananoparticles in a polymer matrix: small-angle scattering and reverse monte carlo simulations. *Soft Matter* 3 (4): 476–485. https://doi.org/10.1039/b614957j.

88 Musino, D., Genix, A.-C., Chauveau, E. et al. (2020). Structural identification of percolation of nanoparticles. *Nanoscale* 12 (6): 3907–3915. https://doi.org/10.1039/c9nr09395h.

89 Hellweg, T. (2013). Responsive core-shell microgels: synthesis, characterization, and possible applications. *J. Polym. Sci., Part B: Polym. Phys.* 51 (14): 1073–1083. https://doi.org/10.1002/polb.23294.

90 Hyatt, J.S., Douglas, A.M., Stanley, C. et al. (2017). Charge segregation in weakly ionized microgels. *Phys. Rev. E* 95 (1). https://doi.org/10.1103/physreve.95.012608.

91 Hyatt, J.S., Do, C., Hu, X. et al. (2015). Segregation of mass at the periphery of *N*-isopropylacrylamide-co-acrylic-acid microgels at high temperatures. *Phys. Rev. E* 92 (3). https://doi.org/10.1103/physreve.92.030302.

92 Wu, Q., Lv, C., Zhang, Z. et al. (2018). Poly(*N*-isopropylacrylamide-co-1-vinyl-3-alkylimidazolium bromide) microgels with internal nanophase-separated structures. *Langmuir* 34 (31): 9203–9214. https://doi.org/10.1021/acs.langmuir.8b01575.

93 Keerl, M., Pedersen, J.S., and Richtering, W. (2009). Temperature sensitive copolymer microgels with nanophase separated structure. *J. Am. Chem. Soc.* 131 (8): 3093–3097. https://doi.org/10.1021/ja807367p.

94 Etchenausia, L., Deniau, E., Brûlet, A. et al. (2018). Cationic thermoresponsive poly(N-vinylcaprolactam) microgels synthesized by emulsion polymerization using a reactive cationic macro-RAFT agent. *Macromolecules* 51 (7): 2551–2563. https://doi.org/10.1021/acs.macromol.8b00155.

95 Bergman, M.J., Pedersen, J.S., Schurtenberger, P., and Boon, N. (2020). Controlling the morphology of microgels by ionic stimuli. *Soft Matter* 16 (11): 2786–2794. https://doi.org/10.1039/c9sm02170a.

96 Rufier, C., Collet, A., Viguier, M. et al. (2011). Influence of surfactants on hydrophobically end-capped poly(ethylene oxide) self-assembled aggregates studied by SANS. *Macromolecules* 44 (18): 7451–7459. https://doi.org/10.1021/ma201150g.

97 Brugnoni, M., Scotti, A., Rudov, A.A. et al. (2018). Swelling of a responsive network within different constraints in multi-thermosensitive microgels. *Macromolecules* 51 (7): 2662–2671. https://doi.org/10.1021/acs.macromol.7b02722.

98 Schmid, A.J., Dubbert, J., Rudov, A.A. et al. (2016). Multi-shell hollow nanogels with responsive shell permeability. *Sci. Rep.* 6 (1). https://doi.org/10.1038/srep22736.

99 Brugnoni, M., Fink, F., Scotti, A., and Richtering, W. (2020). Synthesis and structure of temperature-sensitive nanocapsules. *Colloid Polym. Sci.* 298 (9): 1179–1185. https://doi.org/10.1007/s00396-020-04686-5.

100 Wypysek, S.K., Scotti, A., Alziyadi, M.O. et al. (2019). Tailoring the cavity of hollow polyelectrolyte microgels. *Macromol. Rapid Commun.* 41 (1): 1900422. https://doi.org/10.1002/marc.201900422.

101 Gelissen, A.P.H., Scotti, A., Turnhoff, S.K. et al. (2018). An anionic shell shields a cationic core allowing for uptake and release of polyelectrolytes within core–shell responsive microgels. *Soft Matter* 14 (21): 4287–4299. https://doi.org/10.1039/c8sm00397a.

102 Zeiser, M., Freudensprung, I., and Hellweg, T. (2012). Linearly thermoresponsive core–shell microgels: towards a new class of nanoactuators. *Polymer* 53 (26): 6096–6101. https://doi.org/10.1016/j.polymer.2012.10.001.

103 Cors, M., Wrede, O., Wiehemeier, L. et al. (2019). Spatial distribution of core monomers in acrylamide-based core-shell microgels with linear swelling behaviour. *Sci. Rep.* 9 (1). https://doi.org/10.1038/s41598-019-50164-6.

104 Brugnoni, M., Nickel, A.C., Kröger, L.C. et al. (2019). Synthesis and structure of deuterated ultra-low cross-linked poly(*N*-isopropylacrylamide) microgels. *Polym. Chem.* 10 (19): 2397–2405. https://doi.org/10.1039/c8py01699b.

105 Cors, M., Wrede, O., Genix, A.-C. et al. (2017). Core–shell microgel-based surface coatings with linear thermoresponse. *Langmuir* 33 (27): 6804–6811. https://doi.org/10.1021/acs.langmuir.7b01199.

106 Cors, M., Wiehemeier, L., Oberdisse, J., and Hellweg, T. (2019). Deuteration-induced volume phase transition temperature shift of PNIPMAM microgels. *Polymers* 11 (4): 620. https://doi.org/10.3390/polym11040620.

107 Cors, M., Wiehemeier, L., Wrede, O. et al. (2020). Contrast variation SANS measurement of shell monomer density profiles of smart core–shell microgels. *Soft Matter* 16 (7): 1922–1930. https://doi.org/10.1039/c9sm02036e.

108 Keidel, R., Ghavami, A., Lugo, D.M. et al. (2018). Time-resolved structural evolution during the collapse of responsive hydrogels: the microgel-to-particle transition. *Sci. Adv.* 4 (4): eaao7086. https://doi.org/10.1126/sciadv.aao7086.

109 Wrede, O., Reimann, Y., Lülsdorf, S. et al. (2018). Volume phase transition kinetics of smart *N*-*n*-propylacrylamide microgels studied by time-resolved pressure jump small angle neutron scattering. *Sci. Rep.* 8 (1). https://doi.org/10.1038/s41598-018-31976-4.

110 Virtanen, O.L.J., Kather, M., Meyer-Kirschner, J. et al. (2019). Direct monitoring of microgel formation during precipitation polymerization of *N*-isopropylacrylamide using in situ SANS. *ACS Omega* 4 (2): 3690–3699. https://doi.org/10.1021/acsomega.8b03461.

111 Barth, A., Grillo, I., and Gradzielski, M. (2010). Dynamics of formation of vesicles studied by highly time-resolved stopped-flow experiments. *Tenside Surfactants Deterg.* 47 (5): 300–306. https://doi.org/10.3139/113.110081.

112 Ballauff, M. (2001). SAXS and SANS studies of polymer colloids. *Curr. Opin. Colloid Interface Sci.* 6 (2): 132–139. https://doi.org/10.1016/s1359-0294(01)00072-3.

113 Garcia-Diez, R., Gollwitzer, C., Krumrey, M., and Varga, Z. (2016). Size determination of a liposomal drug by small-angle X-ray scattering using continuous contrast variation. *Langmuir* 32 (3): 772–778. https://doi.org/10.1021/acs.langmuir.5b02261.

114 Dargel, C., Hannappel, Y., and Hellweg, T. (2020). Heating-induced DMPC/glycyrrhizin bicelle-to-vesicle transition: a X-ray contrast variation study. *Biophys. J.* 118 (10): 2411–2425. https://doi.org/10.1016/j.bpj.2020.03.022.

115 Cotton, J.P. (1996). Variations on contrast in SANS: determination of self and distinct correlation functions. *Adv. Colloid Interface Sci.* 69 (1–3): 1–29. https://doi.org/10.1016/s0001-8686(96)00306-5.

116 Banc, A., Genix, A.-C., Dupas, C. et al. (2015). Origin of small-angle scattering from contrast-matched nanoparticles: a study of chain and filler structure in polymer nanocomposites. *Macromolecules* 48 (18): 6596–6605. https://doi.org/10.1021/acs.macromol.5b01424.

117 Nöjd, S., Holmqvist, P., Boon, N. et al. (2018). Deswelling behaviour of ionic microgel particles from low to ultra-high densities. *Soft Matter* 14 (20): 4150–4159. https://doi.org/10.1039/c8sm00390d.

118 Nöjd, S., Hirst, C., Obiols-Rabasa, M. et al. (2019). Soft particles in an electric field – a zero average contrast study. *Soft Matter* 15 (31): 6369–6374. https://doi.org/10.1039/c9sm01208g.

119 Scotti, A. (2021). Characterization of the volume fraction of soft deformable microgels by means of small-angle neutron scattering with contrast variation. *Soft Matter* 17 (22): 5548–5559. https://doi.org/10.1039/d1sm00277e.

120 Conley, G.M., Aebischer, P., Nöjd, S. et al. (2017). Jamming and overpacking fuzzy microgels: deformation, interpenetration, and compression. *Sci. Adv.* 3 (10): e1700969. https://doi.org/10.1126/sciadv.1700969.

121 Scotti, A., Gasser, U., Herman, E.S. et al. (2017). Phase behavior of binary and polydisperse suspensions of compressible microgels controlled by selective particle deswelling. *Phys. Rev. E* 96 (3). https://doi.org/10.1103/physreve.96.032609.

122 Scotti, A., Gasser, U., Herman, E.S. et al. (2016). The role of ions in the self-healing behavior of soft particle suspensions. *Proc. Natl. Acad. Sci. U.S.A.* 113 (20): 5576–5581. https://doi.org/10.1073/pnas.1516011113.

123 Schneider, F., Balaceanu, A., Di, Z. et al. (2017). Internal structure and phase transition behavior of stimuli-responsive microgels in PEG melts. *Soft Matter* 13 (15): 2738–2748. https://doi.org/10.1039/c6sm02501c.

8

Stimuli-Responsive Fluorescent Polymeric Hydrogels

Wei Lu, Shuxin Wei, and Tao Chen

Chinese Academy of Sciences, Ningbo Institute of Materials Technology and Engineering, Key Laboratory of Marine Materials and Related Technologies, Zhejiang Key Laboratory of Marine Materials and Protective Technologies, S1219 Zhongguan West Road, Ningbo 315201, P. R. China

8.1 Introduction

Stimuli-responsive fluorescent polymeric hydrogels (FPHs) are 3D crosslinked hydrophilic polymer networks with tunable luminescent properties, which are in high contrast to those classic fluorescent materials used primarily in dry solid states or in solutions [1–3]. For example, FPHs primarily exist as the water-swollen quasi-solid state and thus hold the potential to integrate the promising features of both solid and solution, including tissue-like mechanical strength, intrinsic soft wet nature, biocompatibility, programmable self-healing/shape memory features, as well as responsive fluorescence and volume/shape changes. These appealing advantages have made FPHs a rising star of luminescent materials, which show great promise for many applications, such as optical sensing, bioimaging, smart display, information coding/encryption, and soft actuators/robotics [3].

Over the past decades, several strategies have been proposed to produce a large number of stimuli-responsive FPHs for various potential applications. These synthetic strategies can be generally divided into three types, that is, by physical incorporation or covalent immobilization of fluorogens into polymeric hydrogels, or through supramolecular polymerization of small-molecule monomeric fluorogens. The present chapter intends to summarize the recent impressive advances in the construction of stimuli-responsive FPHs, highlighting also some important demonstrated applications.

Smart Stimuli-Responsive Polymers, Films, and Gels, First Edition.
Edited by Liang Hu, Yongfeng Gao, and Michael J. Serpe.

8.2 Strategies for Preparing Fluorescent Polymeric Hydrogels (FPHs)

8.2.1 Physically Incorporating Fluorogens into Polymeric Hydrogels

Most of the well-known stimuli-responsive polymeric hydrogels (e.g. poly (*N*-isopropylacrylamide) [PNIPAM], poly(2-(dimethylamino)ethyl methacrylate) [PDMAEMA]) are intrinsically nonfluorescent because of the lack of emissive units [4]. Fluorogens are thus usually necessary to be incorporated to produce fluorescent hydrogels. To do this, the most straightforward strategy is to physically blend the fluorogens into the hydrogel matrix. For example, Kumar and coworkers reported a multicolor fluorescent protein hydrogel by directly adding the water-soluble blue (coumarin 460), green (fluorescein), and red (5(6)-carboxy-X-rhodamine) light-emitting fluorogens into the gelation solutions [5]. Remarkably, even if the fluorogens are not water soluble, they can still be employed as building blocks for FPHs by using this method. A recent example was demonstrated by Tang and coworkers [6]. The employed fluorophores were several hydrophobic aggregation-induced emissive fluorogens (AIEgens) (Figure 8.1). To make these AIEgens evenly distributed into the hydrogel matrix, aqueous suspensions of aggregation-induced emission (AIE) dots were first prepared by slowly dropping their organic solutions into deionized water. A series of different-colored fluorescent hydrogels were then obtained by simply mixing the acylhydrazine-terminated polyethylene oxide (PEO) and tetra aldehyde-terminated

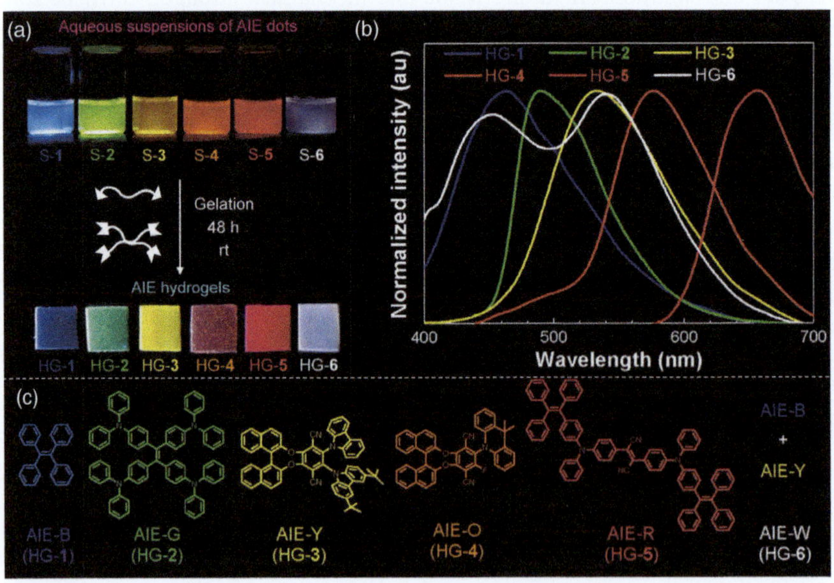

Figure 8.1 The preparation of different-colored FPHs by mixing the acylhydrazine-terminated PEO and tetra aldehyde-terminated PEO polymers (gelators) in the aqueous suspensions of different AIE dots. (a) The process to prepare these FPHs and (b) their corresponding fluorescence spectra. (c) Chemical structures of the embeded fluorogens. Source: Ji et al. [6]/John Wiley & Sons, Inc.

PEO polymers (gelators) in the aqueous suspensions of different AIE dots. By following a similar strategy, a large number of stimuli-responsive FPHs have been constructed over the past decades. These attempts have clearly demonstrated that this physical doping strategy could be generally applicable for various types of fluorogens, including organic fluorophores, fluorescent proteins [7], lanthanide complexes [8], and luminescent nanoparticles [9].

Despite these advantages, there are still some concerns about this strategy. One possible issue is the unexpected leakage of the embedded fluorogens, especially the small-molecule ones. To tackle this problem, several clever designs have been recently proposed. One is to functionalize the small-molecule fluorophores with suitable moieties that can strongly bind the polymer chain through multiple supramolecular interactions [10]. A typical example was shown in Figure 8.2a, the lanthanide coordinated fluorescent particles are engineered to bear high-density $-NH_2$ groups, which can thus be firmly anchored in the hydrogels by forming vast hydrogen bonds between the polyvinyl alcohol (PVA) polymer chains. Another important design is to produce internal microstructures (e.g. micelles) for fluorogens loading within the hydrogels [11]. As shown in Figure 8.2b, Li and coworkers

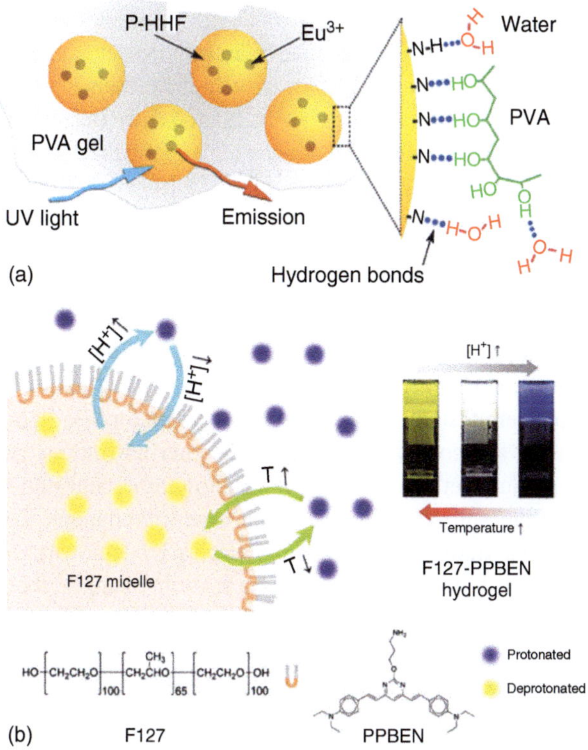

Figure 8.2 (a) Illustration of the FPH system prepared by blending the NH_2-functionalized fluorescent particles into the PVA hydrogel. Source: Yang et al. [10]/with permission of John Wiley & Sons. (b) Scheme of the FPH system prepared by firmly loading water-insoluble fluorophore into the hydrophobic cavities of F127 micelles in the hydrogel networks. Source: Wang et al. [11]/John Wiley & Sons, Inc.

presented a stimuli-responsive white-light-emitting hydrogel by the introduction of hydrophobic cavities of the F127 micelles to separate the aprotic regions of the water-insoluble fluorophores from the aqueous regions in the hydrogel networks.

8.2.2 Covalently Bonding Fluorogens into Polymeric Hydrogels

Another frequently used synthetic strategy is to covalently graft fluorogens into the hydrophilic crosslinked polymer networks, which is helpful to make the fluorophores evenly and firmly distributed in the hydrogel matrix. These fluorogens could be covalently linked onto the crosslinked polymer chains as pendant, terminal groups or incorporated into the main polymer chains, according to different synthetic procedures that are discussed below.

When fluorogen-containing vinyl monomers are subject to radically copolymerize with hydrophilic monomers and crosslinkers, these fluorogens have been anchored into the 3D crosslinked polymer network as pendant groups [12–14]. On the one hand, if the employed fluorogens are engineered to be sensitive to external stimuli (i.e. pH, ions), the intelligent FPHs with similar stimuli-responsive fluorescence responses are obtained. For instance, Chen, and coworkers presented a naphthalimide-based vinyl monomer DEAN (4-(*N*,*N*-dimethylaminoethylene) amino-*N*-allyl-1,8-naphthalimide) [12], which is nearly nonfluorescent under neutral conditions but emits intense yellowish light under acid conditions via the protonation of the dimethyl amino moieties to weaken its photoinduced electron transfer with the naphthalimide fluorogens. The subsequent copolymerization of DEAN into the chemically crosslinked poly(acrylamide) hydrogel produced an intelligent FPH with pH-responsive on–off fluorescence response. On the other hand, if the pendant fluorogens themselves are inert to external stimuli, the responsive FPH systems with tunable fluorescence features can still be produced, because the aggregation state of the grafted fluorogens may potentially be controlled by stimuli-responsive conformational change of the crosslinked polymer chains [13]. For example, García and coworkers have reported one vinyl-functionalized CdSe quantum dots (QDs) that emit static green luminescence under UV light irradiation [14]. But when they are copolymerized into the thermo-responsive poly(diethylene glycol methyl ether methacrylate), smart photoluminescent nano or microgel hybrids with dynamic and reversible emission intensity changes are produced. Further hydrodynamic diameter measurements clearly demonstrate that it is the temperature-triggered polymer chain conformation changes that induce the reversible aggregation/disaggregation of these grafted QDs. Besides temperature-responsive systems, other stimuli-responsive polymeric polymers have been demonstrated to be good candidates for the preparation of intelligent FPHs.

When the fluorogen-containing initiators are employed to initiate polymerization, these fluorogens will serve as the terminal ends of crosslinked polymer chains. One impressive system was recently developed by He and coworkers [15]. As illustrated in Figure 8.3, the intelligent hydrogel was prepared by utilizing a TPE-functionalized atom transfer radical polymerization (ATRP) agent to initiate the copolymerization of diacetone acrylamide (DAA) and *N*,*N*-dimethylacrylamide (DMA) by using

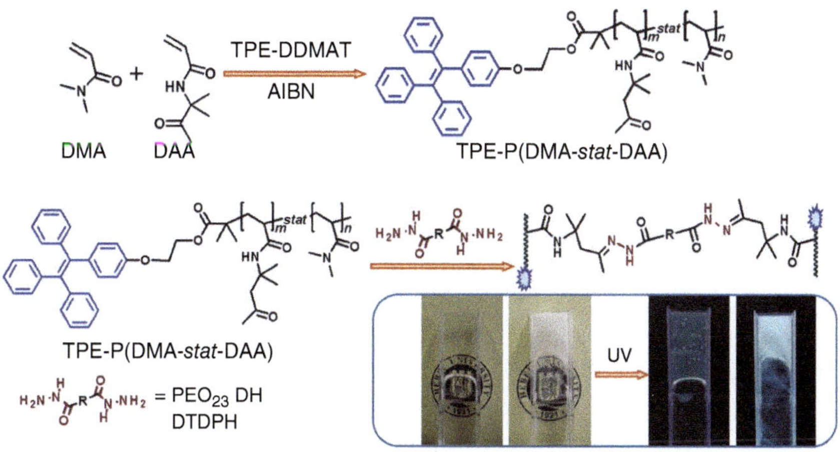

Figure 8.3 Preparation of the FPH system by using a TPE-functionalized ATRP agent to initiate the copolymerization of DAA and DMA, followed by the crosslinking reaction with diacylhydrazide. Source: Hou et al. [15]/Elsevier.

diacylhydrazide as the crosslinker. Since the poly(DMA-*co*-DAA) with balanced hydrophilic and hydrophobic segments was found to be thermo-responsive and had a lower critical solution temperature (LCST) near body temperature, the obtained hydrogel was endowed with appealing thermo-responsive fluorescence intensity change.

In addition to serving as pendant/terminal groups, the fluorogens have also been embedded into the main polymer chains of hydrogels. In this design, it is usually necessary to first modify the fluorogens with two or more polymerization sites (i.e. –OH, $–NH_2$), which are then subject to react with comonomers containing bi- or multifunctional –COOH or –NCO groups. Based on this synthetic procedure, Liow et al. presented a thermo-responsive FPH example consisting of hydrophilic poly(ethylene glycol) moieties and hydrophobic TPE and poly(propylene glycol) moieties [16]. The obtained multiblock amphiphilic macromolecules tend to self-assemble into fluorescent micelles in the hydrogel matrix, leading to the compact encapsulation of the hydrophobic toxic fluorophores by the hydrophilic crosslinked polymeric network, which is beneficial to improving their environmental friendliness and biocompatibility for many biorelated and soft robotic applications.

In this section, we have discussed some important advances in the preparation of fluorogen-grafted FPHs. Theoretically, all types of luminogens could be covalently linked into the main polymer chains or as pendant/terminal groups of the hydrophilic crosslinked polymer chains. But, in fact, most of the reported FPHs were prepared by labeling the fluorogens as pendant groups because of the relatively simple synthetic procedures. This is especially so for the fluorescent proteins and luminescent nanoparticles whose chemical structures cannot be easily modified on demand. More efforts are thus suggested to develop new and simple synthetic procedures or take advantage of well-organized supramolecular interactions for the purpose of constructing powerful FPHs.

8.2.3 Supramolecular Polymerizing/Crosslinking Monomeric Fluorogens

Stimuli-responsive FPHs have also been prepared by the supramolecular self-assembly of rationally designed small-molecule fluorogen-containing monomer units and other co-monomer molecules. These supramolecular polymerizing/crosslinking processes could be driven by various non-covalent interactions [17], including hydrogen bonds, host–guest interactions, electrostatic interactions, and metal coordination. Owing to the highly dynamic and reversible nature of these non-covalent interactions, the FPHs produced by supramolecular polymerization can be facilely tailored in terms of chemical structures, compositions, and responsiveness. Additionally, compared with the above-discussed chemically crosslinked systems, these hydrogel materials consisting of dynamic crosslinks are more favorable to be characterized by sensitive fluorochromic response, satisfying self-heal, as well as widely adjustable mechanical strength (e.g. gel-to-sol transition) [18].

Hydrogen bond, an electrostatic attraction between a hydrogen atom and a small electronegative atom (as of oxygen, nitrogen, or fluorine), is weak. It is usually necessary to specially modify the fluorogens with multiple donor and acceptor sites to form multiple hydrogen bonds that are strong enough to stabilize hierarchical supramolecular structures of FPHs. For example, Lin and coworkers have designed and prepared a host molecule, N^1, N^3, N^5-tri(pyridin-4-yl)benzene-1,3,5-tricarboxamide (DTA), which contains three primary amides and three pyridine moieties [19]. DTA was found to form vast hydrogen bonds with 3-fluorobenzaldehyde (3-FB) in water, leading to stable hydrogel with AIE-enhancement property (Figure 8.4a). Besides these relatively stronger NH—O, NH—F, and OH—N hydrogen bonds (3–7 kcal/mol), it was revealed that the weakest CH—O hydrogen bonds (0.3–0.8 kcal/mol) can still be capable of stabilizing the supramolecular structures of FPHs [20]. As shown in Figure 8.4b, the three coumarin-derived isomeric hydrogelators with different spatial structures can self-assemble into nanofibrous networks of blue fluorescent hydrogels through CH–O interactions. These recent advances highlight that the non-covalent hydrogen bonds, despite being weak, can still be synergistically utilized to create stable FPHs.

Compared with hydrogen bonds, electrostatic interactions between oppositely charged groups are much stronger and frequently used to construct FPHs. Since the electronic interaction strength is quite sensitive to competitive ions (especially charged biomolecules), these FPHs are usually endowed with bioresponsive fluorescence/mechanical properties. One typical example is illustrated in Figure 8.5, the hydrogel was spontaneously self-assembled from the tetraphenylethene derivative modified with two quaternary ammonium groups (tetraphenylethene derivative modified with two quaternary ammonium groups [TPE-Am]) and negatively charged poly(sodium *p*-styrenesulfonate) (PSS) through electronic interaction-driven supramolecular crosslinking [21]. The as-prepared hydrogels emit intense blue light of the aggregated TPE fluorogens. As expected, the addition of biomolecules, such as adenosine triphosphate (ATP), which interacted with TPE-Am more strongly than PSS, would break the crosslinked network,

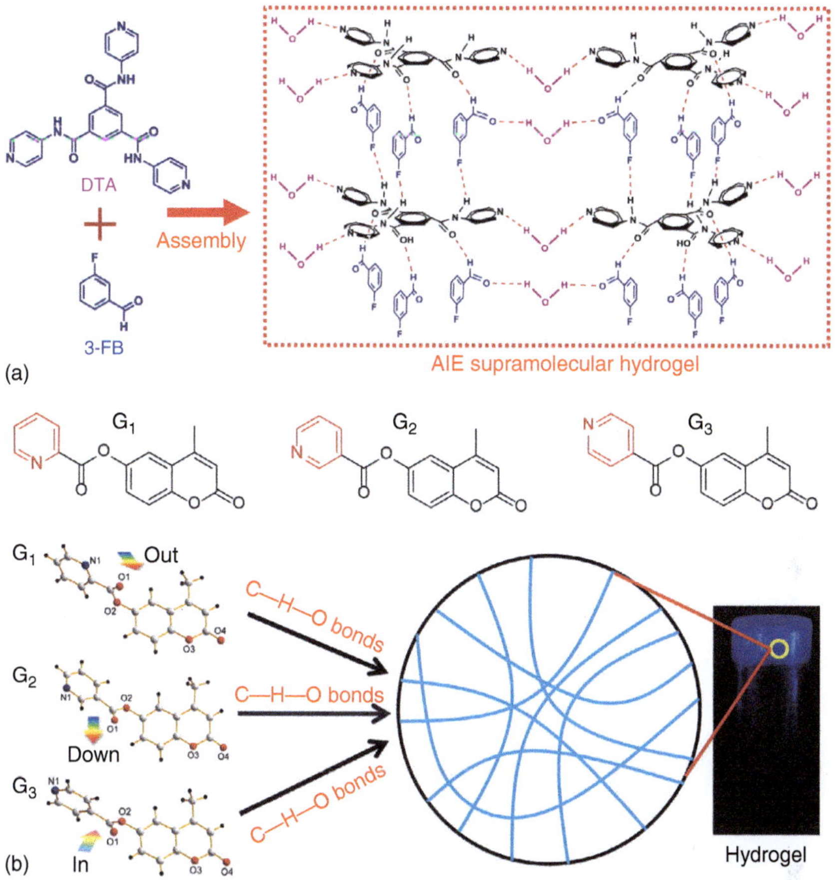

Figure 8.4 (a) Chemical structure of the FPH system prepared from the self-assembly of DTA, and 3-fluorobenzaldehyde (3-FB) in water. Source: Fan et al. [19]/with permission of Royal Society of Chemistry. (b) Chemical structure of three coumarin-derived isomeric hydrogelators and their self-assembled nanofibrous FPH through CH—O hydrogen bonds. Source: Ji et al. [20]/American Chemical Society.

causing simultaneous gel-to-sol transition and fluorescence intensity reduction. In addition to organic molecules, the charged inorganic components are also promising building blocks for FPHs. For instance, Li et al. presented a hierarchical organic–inorganic self-assembling strategy to produce robust luminescent hydrogels by starting from negatively charged exfoliated Laponite nanosheets and 3-methylimidazolium-modified fluorescent lanthanide complexes [22]. The hierarchical organic-inorganic supramolecular structures self-assembled via electronic interactions not only favor the improvement of mechanical strength but also endow the materials with room-temperature self-healing properties.

Host–guest interactions, involving two molecules that can form complexes, are also a common driving force for FPHs. Their strength largely depends on the shape

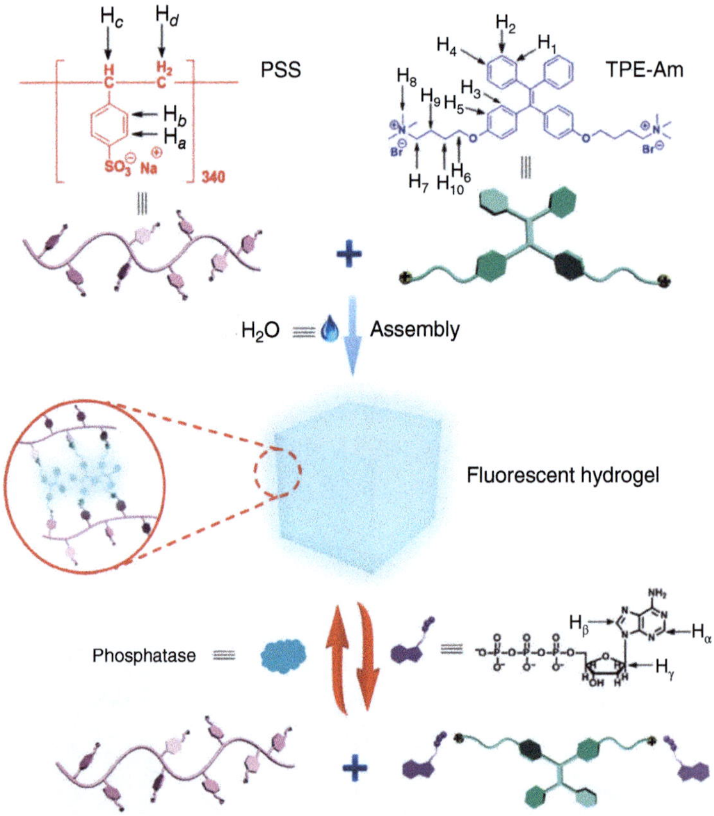

Figure 8.5 Scheme showing the preparation of FPH from the electrostatic interaction-based crosslinking between TPE-Am and PSS, as well as the reversible ATP-responsive sol–gel transition. Source: Wang et al. [21]/with permission of Royal Society of Chemistry.

complementarity between host and guest molecules. A well-known molecule pair is β-cyclodextrin (β-CD) and adamantane (Ad), which has matched shape complementarity and is favorable for the formation of stable complexes. Therefore, stable FPH systems could be produced in a straightforward way by the self-assembly of two monomeric molecules that are functionalized with β-CD and Ad moieties, respectively [23]. Other host–guest molecule pairs include α-CD and benzimidazole [24], Pillar[5]arene, and pyridine [25], which have also been used to construct FPHs.

Compared with the above-discussed supramolecular interactions, metal coordination ones usually have comparatively large binding strength, but also show highly dynamic and reversible nature. These characteristics made metal coordination interactions quite promising driving force to construct adaptive FPHs with tunable fluorescence and satisfying self-healing features. One typical example is shown in Figure 8.6, the supramolecular hydrogels were prepared by the lanthanide (Eu^{3+}/Tb^{3+})-directed self-assembly of 4,4′-(((pyridine-2,6-dicarbonyl) bis-(azanediyl))bis(methylene))-dibenzoic acid [26]. Their luminescence colors

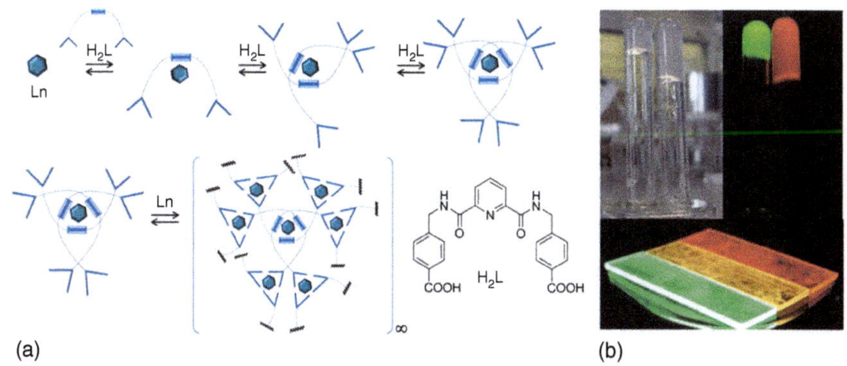

Figure 8.6 (a) Scheme showing the formation of FPH from the self-assembly of H_2L and lanthanide ions (Tb^{3+} and Eu^{3+}). (b) Photos of the Tb^{3+}- and Eu^{3+}-doped FPH gels taken under daylight and UV light illumination, as well as the Tb^{3+}-, Tb^{3+}/Eu^{3+}-, and Eu^{3+}-doped FPH gels on quartz plates. Source: Martinez-Calvo et al. [26].

could be continuously regulated from red to yellow or green by facilely varying the Eu^{3+}/Tb^{3+} stoichiometric ratios. Due to the dynamic nature of the lanthanide coordinated networks, intrinsic self-healing was also observed. If metal ions with adjustable valence states (e.g. Fe^{2+}/Fe^{3+}, Ru^{2+}/Ru^{3+}) are involved, the interesting redox responsive hydrogels will be produced. For example, Xu and coworkers took advantage of a Ru^{2+}-tris(bipyridine) complex to present the first supramolecular metallo-hydrogel that exhibits both fluorochromic response and gel–sol transition upon oxidation of the metal center [27]. These advances illustrate that metal coordination interactions, besides being able to impart rich fluorescent properties, may offer redox, mechanical, or self-healing properties to supramolecular hydrogels.

Generally speaking, supramolecular interactions are relatively weak, especially when compared with covalent bonds. Thus, there has been a recent research interest in synergistically combining two or more types of weak interactions for the construction of robust FPHs. One typical system was recently reported by Bryce and coworkers [28]. As shown in Figure 8.7, the designed monomer is an aggregation-induced emissive oligourethane gelator (OUG) containing both π-conjugated moieties and multiple hydrogen bond donors/acceptors. The OUG molecules tended to spontaneously form dynamic crosslinking networks that are stabilized through the combination of hydrogen bonding, π–π stacking, and van der Waals interactions. In parallel, Che and coworkers employed both the intra- and intermolecular Pt/Pt and π–π interactions as the driving force to produce a new class of luminescent hydrogels in aqueous dispersions based on rationally designed dinuclear cyclometalated platinum(II) complexes connected by oligo(oxyethylene) chains [29]. Li and coworkers presented a type of nanofibrillar fluorescent hydrogels, which are formed via one-step hydrothermal treatment of polyethyleneimine and cellulose nanocrystals through hydrogen bonding and hydrophobic interactions [30]. All of these impressive investigations have clearly demonstrated the vast possibility to construct multifunctional FPHs via a combination of various supramolecular interactions.

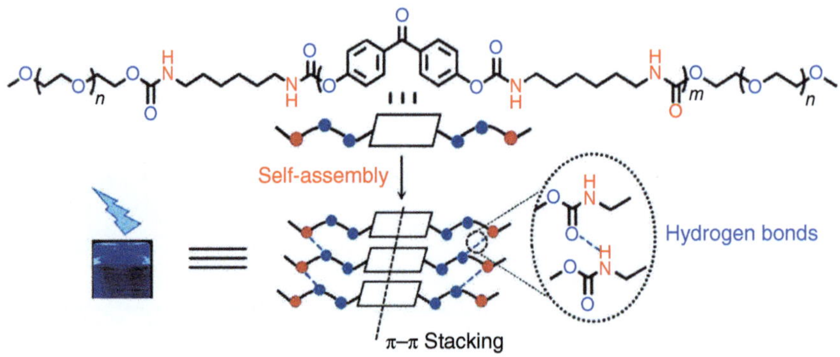

Figure 8.7 Scheme showing the preparation of the AIE-active FPH system from OUG via the combination of multiple supramolecular interactions. Source: Feng et al. [28]/with permission of Royal Society of Chemistry.

8.2.4 Comparison of Different Synthetic Strategies

As described above, three typical synthetic strategies have been proposed to prepare stimuli-responsive FPHs. Indeed, each synthetic procedure has its own advantages and disadvantages, which will be briefly summarized as follows.

- **Physically incorporating fluorogens into polymeric hydrogels**: straightforward and easy operation; generally applicable for myriad kinds of fluorescent molecules; possibly suffering from unexpected leakage of the embedded fluorogens.
- **Covalently bonding fluorogens into polymeric hydrogels**: enabling even and stable distribution of the fluorophores in the hydrogel matrix; suffering from the tedious synthetic procedures to prepare the fluorophore-containing monomers or initiators.
- **Supramolecular polymerizing/crosslinking monomeric fluorogens**: enabling sensitive fluorochromic response, satisfying self-heal, as well as widely adjustable mechanical strength; suffering from labile crosslinked supramolecular network that may be not stable in certain conditions.

8.3 Promising Applications

8.3.1 Optical Sensing and Bio-imaging

The classic fluorescent chemosensors can be divided into the solution or solid film ones, each of which has some specific advantages [31]. The FPHs have been recently considered as ideal materials to combine the merits of both solution and film sensors because the FPH film sensors can provide a liquid phase *in situ* sensing environment while allowing easy portability and good stability [32]. For example, Kumar and coworkers incorporated trichrometric fluorescent dyes into crosslinked protein matrix to produce a highly efficient white-emitting hydrogel for pH sensing (Figure 8.8) [5]. Owing to the inherent pH sensitivity of the protein hydrogel,

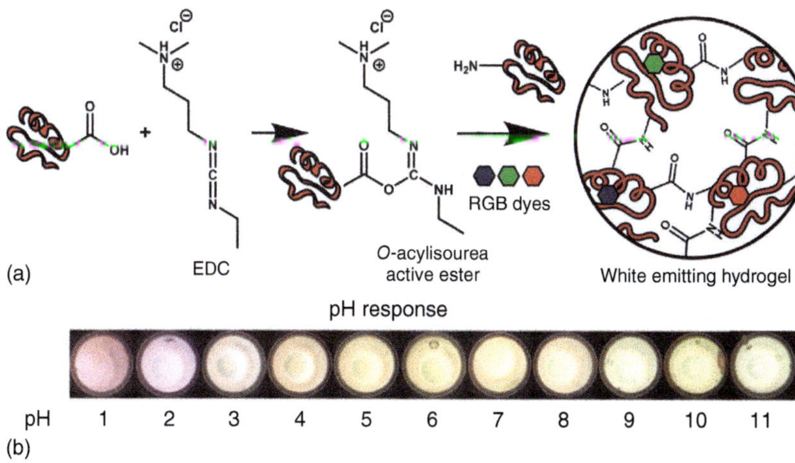

Figure 8.8 (a) Synthetic procedure of the white emitting hydrogel from the simultaneous physical incorporation of Red Green Blue (RGB) dyes. (b) Fluorescent images of the hydrogel in response to pH variation. Source: Benson et al. [5]. John Wiley & Sons, Inc.

pH-responsive swelling/deswelling processes were expected to change the dye–dye distances in the gel. As a result, ratiometric emission characteristics were observed, which brought desirable sensing advantages, such as high reproducibility and protection against baseline shifts. Additionally, this multichromophoric white-emitting hydrogel system also displayed a significantly broad linear range of detection (pH 1–11). In parallel, a large number of powerful FPH film sensors have been reported to allow sensitive and fast detection of many other important analytes (Fe^{2+} [33], Zn^{2+} [34], glucose [35], histamine [36]). These studies demonstrated the huge application potential of FPHs for optical sensing.

Besides these above-mentioned advantages in optical sensing, FPHs are also potentially endowed with good biocompatibility, intrinsic soft wet nature, and tissue-like mechanical properties [37]. These promising merits encouraged scientists to test the possibility of *in vitro* cell sensing and imaging. Many impressive advances have been made in the last two decades, which clearly demonstrated that FPHs could be ideal materials for *in vitro* cell sensing/imaging. On the basis of these results, great recent research attention has been further paid to the *in vivo* detecting and tracking studies. For example, Takeuchi and coworkers recently fabricated a hydrogel glucose sensor from boronic acid-functionalized fluorescent polymers (Figure 8.9), which are capable of displaying sensitive fluorescence response upon glucose binding [38]. To improve it's *in vivo* stability, superoxide dismutase and catalase (two antioxidant enzymes) were also introduced to protect the arylboronic acid from being degraded by hydrogen peroxide *in vitro*. Owing to this delicate design, this hydrogel sensor could be implanted in rats for steady and continuous glucose detection over 28 days. Besides glucose detection, FPH-based pH sensor was also demonstrated for *in vivo* pH imaging [39]. Lin and coworkers described a unique type of dual-emissive core/shell hydrogel nanoparticles, in which a red fluorescent rare-earth complex and a blue fluorescent tetraphenylethylene fluorogen

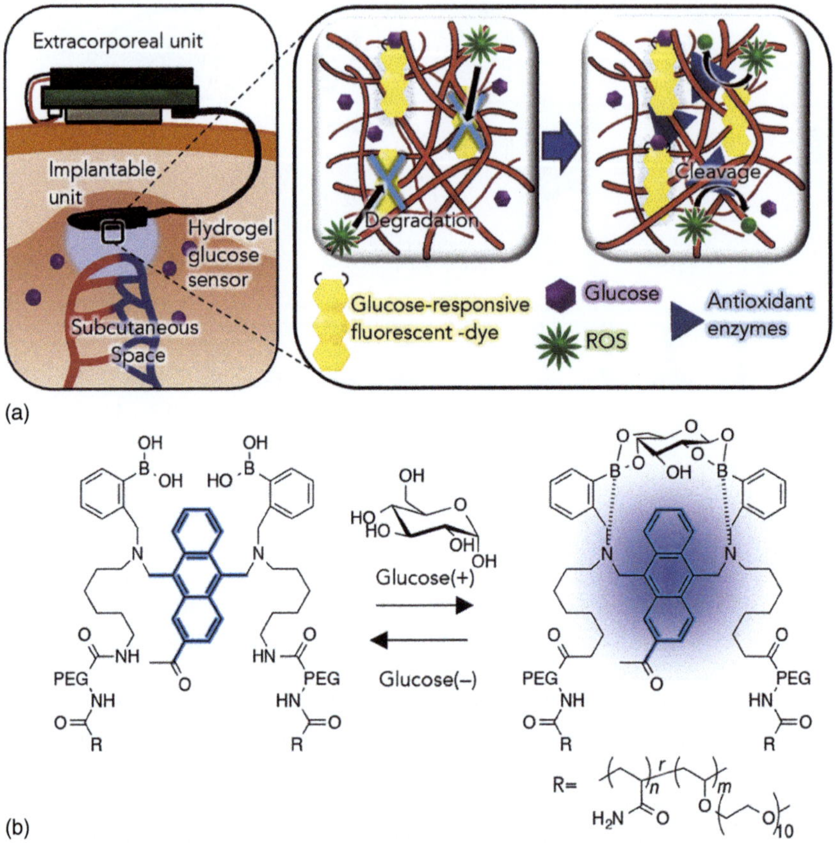

Figure 8.9 (a) Schematic illustration of the FPH-based glucose sensor and its *vivo* detecting process. (b) Chemical structure of the fluorescent hydrogel and its sensing mechanism. Source: Sawayama et al. [38]/with permission of Elsevier.

are introduced into the core and shell layers, respectively. The blue emission of the shell layer shows sensitive response to pH variation from 4 to 7.6, while the red emission in the core layer is pH-independent. Consequently, pH-responsive emission color change was observed. The developed hydrogel sensor was then used for *in vitro* and *in vivo* fluorescent imaging, the results of which exhibit different luminescent colors to clearly distinguish the cancer cells from normal ones. These impressive studies indicate that FPHs hold great potential to serve as ideal materials for *in vivo* detection.

FPHs have also been employed for *in vivo* monitoring of the implanted biomaterial degradation. A typical example was recently developed by Onofrillo and coworkers [40]. This fluorescently labeled sensitive hydrogel (FLASH) was constructed by covalently grafting the fluorescein isothiocyanate isomer (FITC) fluorophore to gelatin methacryloyl (GelMA). The implanted bioscaffolds were then prepared by casting different concentrations of FLASH GelMA, with and without human adipose-derived stem cells (hADSCs) undergoing chondrogenesis. Therefore, the fluorescence loss of the bioscaffolds could be directly correlated with their degradability extents, which was further demonstrated by the *in vivo* monitoring experiments. Meanwhile, Yu and coworkers presented another powerful FPH system [41], which could longitudinally and noninvasively track the biodegradation of synthetic polymer *in vivo* via the combination of the synergic tri-modal bioimaging, involving ultrasound, fluorescence, and magnetic resonance imaging. These technological platforms based on FPHs provide beneficial inspiration and powerful methodology to real-timely investigate *in vivo* state of implanted biomaterials.

8.3.2 Information Encoding and Encryption

Stimuli-responsive soft FPHs are also attracting increasing attention toward security protection uses for data encoding and encryption, because the encoded information cannot be revealed until some predesigned actions (e.g. under UV light illumination, exposure to certain stimuli) are taken [42]. The most convenient way is to encode and encrypt predesigned information into fluorescent patterns on 2D hydrogel surface. Figure 8.10 illustrates one typical example reported by Wu and coworkers [43]. This white fluorescent hydrogel was produced by the incorporation of one donor–acceptor fluorophore into the hydrophilic crosslinked poly(acrylamide) network. Fluorescent intensity/color of the grafted fluorophores was found to be well-tuned by varying their unimer/dimer ratio in response to the light and heat stimuli. Various 2D fluorescent patterns, which were invisible under daylight yet obvious under UV light, were then facilely produced on the hydrogel surfaces by photolithography for information protection.

To realize the desirable multistage information encryption, there has been a recent research interest to develop 3D encryption platforms. In this regard, the FPHs have been considered to serve as ideal materials, because the soft wet hydrogels have been demonstrated to be capable of being engineered into different 3D complex structures via the origami technique. For example, Chen and coworkers recently combined the dynamic boronate ester bonds and metal coordination interactions to present a multifunctional FPH system with Fe^{3+}-responsive fluorescence quenching [44], borax-triggered shape memory, and self-healing properties (Figure 8.11). The texts or patterns were then easily written or printed onto the 2D hydrogel blocks by using Fe^{3+} as the "ink." Subsequently, these 2D hydrogel blocks with pre-written information were programmed into various 3D hydrogel geometries by a functional synergy of shape memory and self-healing. In this way, the information was well-protected in 3D complex structures. The encrypted information can only be read after specially predesigned procedures (e.g. under UV light after treating the 3D structures in acid solutions for shape recovery), indicating much higher information security than those traditional 2D encryption platforms. Besides 3D structures, FPH-based 3D codes have also been developed as a means for better readout fidelity. For example, Sessler and coworkers prepared three different-colored FPH blocks by the radically copolymerizing three different fluorophore monomers into hydrophilic crosslinked polymer networks [45]. These colorful fluorescent hydrogel blocks could then be arranged into various 3D color code arrays via interfacial adhesion, each of which

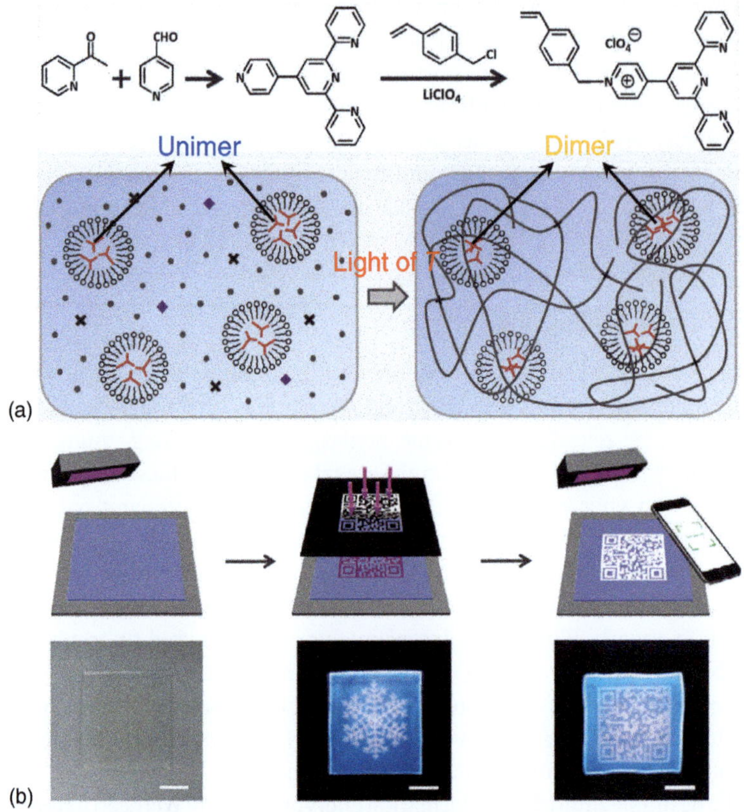

Figure 8.10 (a) Synthetic procedure of the fluorescent monomer and the schematic illustration showing the light- or temperature-responsive unimer-to-dimer transition. (b) Scheme showing the process to produce fluorescent patterns on the hydrogel surface by photolithography, as well as photos of different hydrogel patterns taken under daylight and UV light. Source: Zhu et al. [43]/American Chemical Society.

denotes different information that can be read by a mobile phone. Interestingly, the 3D color codes could be easily transformed into another one by both the physical and chemical approaches. Similar information encoding and transformation behaviors were demonstrated by using a set of FPHs with different fluorescent behaviors. These advanced studies illustrate a new approach to evolving the effective information encoding or encryption from 2D to 3D. Considering the soft flexible nature of FPHs, their further applications are expected to be expanded to smart, wearable materials.

8.3.3 Bioinspired Mechanosensing Systems and Soft Actuators/Robotics

Mechanosensing systems are ubiquitous in natural living organisms [46]. They are capable of sensing the environmental mechanical forces and then transducing mechanical forces into signals that control many important physiological functions (i.e. cell spreading and wound healing). Among them, the mechanofluorescent

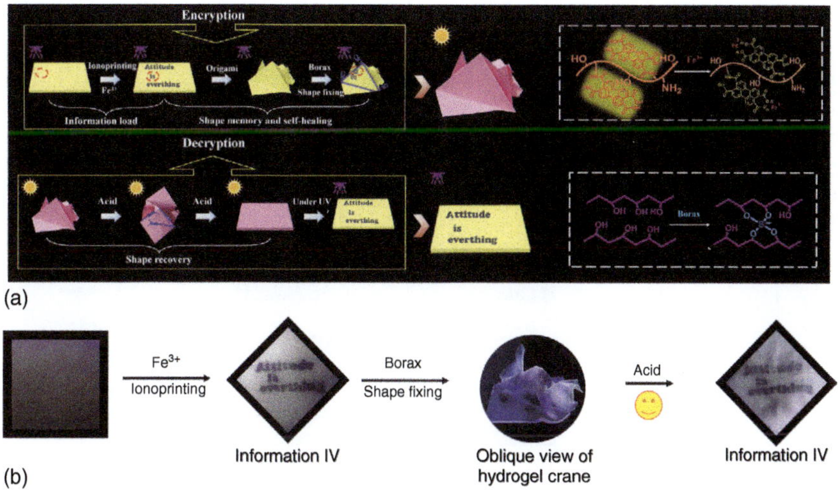

Figure 8.11 (a) Schematic illustration showing the preparation of 3D crane-shaped encryption platform by collective action of shape memory and self-healing. (b) The corresponding photos of the fluorescent hydrogel film, which show the process to encrypt information in the 3D crane-shaped hydrogel structure. Source: Zhang et al. [44]/John Wiley & Sons, Inc.

systems are of particular interest, because they can transduce the external forces into visible light signals, indicating their huge potential in strain visualization, dynamic display, and human–machine interactions [47]. Several impressive advances have recently been reported in this regard. One is reported by Professor Walther's group [48]. As shown in Figure 8.12a, this mechanofluorescent material was prepared by integrating a FRET (Forster resonance energy transfer)-based DNA tension probes into 3D crosslinked DNA hydrogels. Upon exposure to external forces, the distance between the fluorophore and quencher could be sensitively adjusted, resulting in varied FRET efficiencies and tunable emission colors. Owing to this delicate design, not only spatiotemporal resolution but also quantitative and modular force sensing were achieved in soft hydrogels. These advantages enabled the system to visualize both the strain fields in composites and freezing-induced strain patterns in hydrogels (Figure 8.12b). Another mechanofluorescent hydrogel example is reported by Miserez and coworkers, which is based on a double-layer design (Figure 8.12c) [49]. The two layers consist of two different-colored hydrogel films containing fluorescent carbon dots and lanthanide complexes, respectively. When being stretched by external forces, transmittance of the top layer was enhanced to regulate light emission from the bottom layer. Consequently, the overall hydrogel luminescence intensity/color was varied in response to mechanical stimuli (Figure 8.12d). This unique kind of strain-dependent mechanosensing hydrogels holds great application potential in dynamic display or material-based encryption devices. Generally speaking, the development of mechanofluorescent hydrogels is still in its infancy. The currently reported systems still suffer from low sensitivity and poor cycled performance. Many research efforts are highly desired in this field.

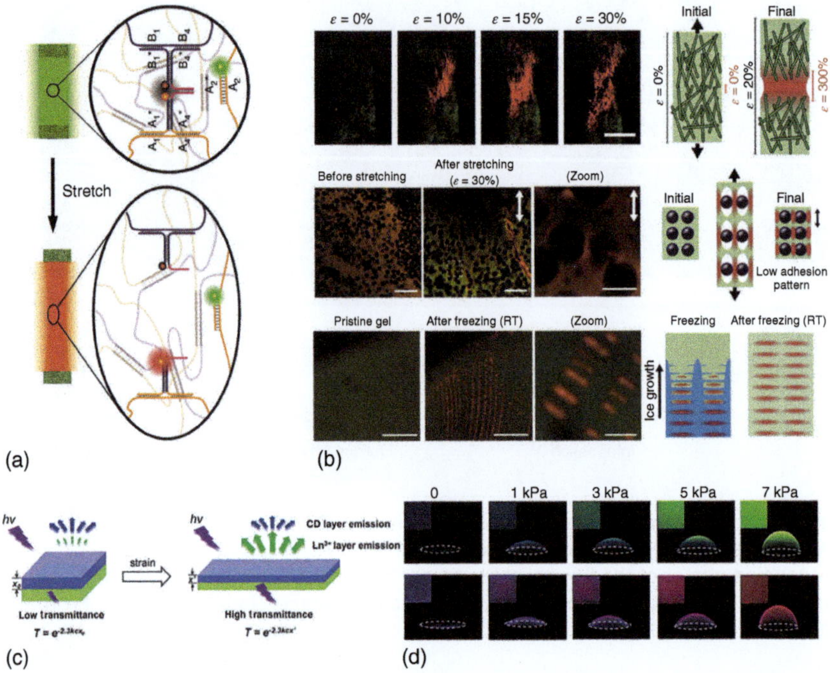

Figure 8.12 (a) Schematic illustration showing the mechanofluorescent response of the 3D crosslinked DNA hydrogel, and (b) its application for visualization of complex strain-field inhomogeneities. Source: Merindol et al. [48]/Springer Nature. (c) Schematic illustration showing the design principle of the double-layer mechanochromic hydrogel, and (d) photos of the mechanochromic hydrogel balls under different bulging pneumatic pressures. Source: Zhu et al. [49]/John Wiley & Sons, Inc.

Besides the wonderful sensing capacity, a rich variety of natural creatures can respond to environmental stimuli to show adaptive shape and/or skin color changes [50]. Representative examples include Octopus, Chameleon, and *Prunus armeniaca*. Synthetic actuators/robotics that is capable of mimicking these synergistic shape/color-changeable behaviors are highly expected. Considering the extreme similarity between polymeric hydrogels and bio-tissues in terms of soft wet nature and mechanical properties, there has been an intense research interest toward the fabrication of color-changing soft hydrogel actuators/robotics. One pioneering work was reported by Chen's group in 2018 [51]. As illustrated in Figure 8.13, the FPH-based soft actuator was prepared through macroscopic supramolecular assembly of two hydrogel layers, one thermo-responsive graphene oxide-poly(*N*-isopropylacrylamide) layer, and one pH-responsive perylene bisimide-functionalized fluorescent hydrogel layer. Owing to the mismatch between the swelling ability of these two layers upon temperature elevation, fluorescent hydrogel actuators with biomimetic complex shape deformable behaviors were obtained. Following this study, we further reported one more powerful kind of lanthanide coordinated FPH actuator that can display chameleon-like synergistic color/shape changeable behaviors in response to the subtle interplay of pH,

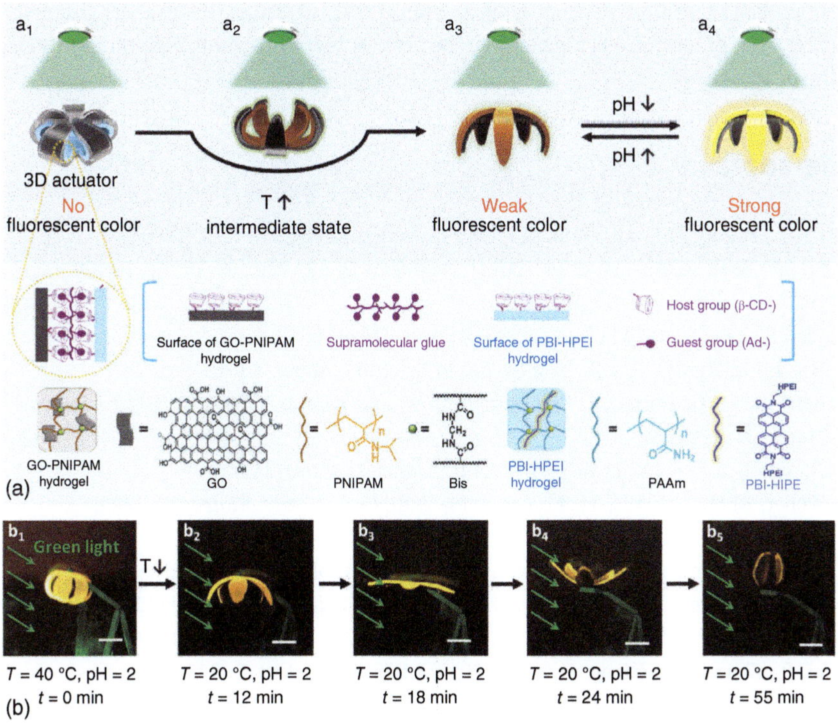

Figure 8.13 (a) Schematic illustration showing the preparation of the bilayer hydrogel actuator, and (b) the corresponding photos showing the thermo-responsive actuating process, accompanied by switchable fluorescence behaviors. Source: Ma et al. [51]/John Wiley & Sons, Inc.

temperature, and metal ion stimuli [52]. In parallel, Tang and coworkers utilized 3D/4D printing to report another impressive flower-shaped soft hydrogel robotics that displayed simultaneous fluorescence color, brightness, and shape-changing behavior in an acidic environment [53]. These proposed systems have demonstrated the possibility of color-changeable soft hydrogel robots, potentially offering wide applications, such as visual detection/display, biomimetic soft robotics, and camouflage.

8.4 Conclusions

In this chapter, we have summarized the recent progress in the intelligent FPHs with great emphasis on their synthetic strategies and demonstrated applications. As detailed above, impressive advances have been made in the young but flourishing field over the past decade, highlighting the significantly important research values of FPHs. Nevertheless, there are still some urgent and difficult challenges facing this field, including poor fluorescence efficiency, weak fatigue resistance, and lack of multifunctionality. Great and continuous research efforts are thus highly desired.

We hope that this chapter will provide an opportunity to attract the research interests of more researchers, so as to accelerate the development of stimuli-responsive FPHs in the near future.

References

1 Li, Y., Young, D.J., and Loh, X.J. (2019). *Mater. Chem. Front.* 3: 1489.
2 Lu, W., Wei, S., Shi, H. et al. (2021). *Aggregate* 2: e37 https://doi.org/10.1002/agt2.37.
3 Wei, S., Li, Z., Lu, W. et al. (2021). *Angew. Chem. Int. Ed.* 60: 8608.
4 Hu, L., Zhang, Q., Li, X., and Serpe, M.J. (2019). *Mater. Horiz.* 6: 1774.
5 Benson, K., Ghimire, A., Pattammattel, A., and Kumar, C.V. (2017). *Adv. Funct. Mater.* 27: 1702955.
6 Ji, X., Li, Z., Liu, X. et al. (2019). *Adv. Mater.* 31: e1902365.
7 Weber, M.D., Niklaus, L., Proschel, M. et al. (2015). *Adv. Mater.* 27: 5493.
8 Meng, K., Yao, C., Ma, Q. et al. (2019). *Adv. Sci.* 6: 1802112.
9 Li, C.Y., Zheng, S.Y., Du, C. et al. (2019). *ACS Appl. Polym. Mater.* 2: 1043.
10 Yang, C., Folens, K., Du Laing, G. et al. (2020). *Adv. Funct. Mater.* 30: 2003656.
11 Wang, J., Tang, F., Wang, Y. et al. (2020). *Adv. Opt. Mater.* 8: 1901571.
12 Li, P., Zhang, D., Zhang, Y. et al. (2019). *ACS Macro Lett.* 8: 937.
13 Liu, H., Wei, S., Qiu, H. et al. (2020). *Macromol. Rapid Commun.* 41: e2000123.
14 Liras, M., Quijada-Garrido, I., and García, O. (2017). *Polym. Chem.* 8: 5317.
15 Hou, F., Xi, B., Wang, X. et al. (2019). *Colloids Surf., B* 183: 110441.
16 Liow, S.S., Dou, Q., Kai, D. et al. (2017). *Small* 13: 1603404.
17 Lu, W., Le, X., Zhang, J. et al. (2017). *Chem. Soc. Rev.* 46: 1284.
18 Wang, H., Ji, X., Li, Z., and Huang, F. (2017). *Adv. Mater.* 29: 1606117.
19 Fan, Y.-Q., Huang, Q., Zhang, Y.-M. et al. (2019). *Polym. Chem.* 10: 6489.
20 Ji, W., Liu, G., Li, Z., and Feng, C. (2016). *ACS Appl. Mater. Interfaces* 8: 5188.
21 Wang, H., Ji, X., Li, Y. et al. (2018). *J. Mater. Chem. B* 6: 2728.
22 Li, Z., Hou, Z., Fan, H., and Li, H. (2017). *Adv. Funct. Mater.* 27: 1604379.
23 Li, B., Lin, C., Lu, C. et al. (2020). *Mater. Chem. Front.* 4: 869.
24 Yao, H., Kan, X.-T., Zhou, Q. et al. (2020). *ACS Sustainable Chem. Eng.* 8: 13048.
25 Zhao, Q., Gong, G.-F., Yang, H.-L. et al. (2020). *Polym. Chem.* 11: 5455.
26 Martinez-Calvo, M., Kotova, O., Mobius, M.E. et al. (2015). *J. Am. Chem. Soc.* 137: 1983.
27 Zhang, Y., Zhang, B., Kuang, Y. et al. (2013). *J. Am. Chem. Soc.* 135: 5008.
28 Feng, Y., Jiang, N., Zhu, D. et al. (2020). *J. Mater. Chem. C* 8: 11540.
29 Xiao, X.-S., Lu, W., and Che, C.-M. (2014). *Chem. Sci.* 5: 2482.
30 Sui, B., Zhang, Y., Huang, L. et al. (2020). *ACS Sustainable Chem. Eng.* 8: 18492.
31 Park, S.H., Kwon, N., Lee, J.H. et al. (2020). *Chem. Soc. Rev.* 49: 143.
32 Hu, X., Ma, Z., Li, J. et al. (2020). *Mater. Horiz.* 7: 3250.
33 Fan, K., Wang, X., Yu, S. et al. (2019). *Polym. Chem.* 10: 5037.
34 Nishiyabu, R., Ushikubo, S., Kamiya, Y., and Kubo, Y. (2014). *J. Mater. Chem. A* 2: 15846.

35 Bruen, D., Delaney, C., Diamond, D., and Florea, L. (2018). *ACS Appl. Mater. Interfaces* 10: 38431.
36 Xu, X.Y., Lian, X., Hao, J.N. et al. (2017). *Adv. Mater.* 29: 1702298.
37 Buenger, D., Topuz, F., and Groll, J. (2012). *Prog. Polym. Sci.* 37: 1678.
38 Sawayama, J., Okitsu, T., Nakamata, A. et al. (2020). *iScience* 23: 101243.
39 Zhao, Y., Shi, C., Yang, X. et al. (2016). *ACS Nano* 10: 5856.
40 Onofrillo, C., Duchi, S., Francis, S. et al. (2021). *Biomaterials* 264: 120383.
41 Chen, X., Zhang, J., Wu, K. et al. (2020). *Small Methods* 4: 2000310.
42 Wang, H., Ji, X., Page, Z.A., and Sessler, J.L. (2020). *Mater. Chem. Front.* 4: 1024.
43 Zhu, C.N., Bai, T., Wang, H. et al. (2018). *ACS Appl. Mater. Interfaces* 10: 39343.
44 Zhang, Y., Le, X., Jian, Y. et al. (2019). *Adv. Funct. Mater.* 29: 1905514.
45 Ji, X., Wu, R.T., Long, L. et al. (2018). *Adv. Mater.* 30: 1705480.
46 Wang, Y., Cui, H., Zhao, Q., and Du, X. (2019). *Matter* 1: 626.
47 Sagara, Y., Yamane, S., Mitani, M. et al. (2016). *Adv. Mater.* 28: 1073.
48 Merindol, R., Delechiave, G., Heinen, L. et al. (2019). *Nat. Commun.* 10: 528.
49 Zhu, Q., Vliet, K., Holten-Andersen, N., and Miserez, A. (2019). *Adv. Funct. Mater.* 29: 1808191.
50 Morin, S.A., Shepherd, R.F., Kwok, S.W. et al. (2012). *Science* 337: 828.
51 Ma, C., Lu, W., Yang, X. et al. (2018). *Adv. Funct. Mater.* 28: 1704568.
52 Wei, S., Lu, W., Le, X. et al. (2019). *Angew. Chem. Int. Ed.* 58: 16243.
53 Li, Z., Liu, P., Ji, X. et al. (2020). *Adv. Mater.* 32: e1906493.

9

The Fabrication and Applications of Bioinspired Hydrogel Actuators

Baoyi Wu, Jiawei Zhang, and Tao Chen

[1] Chinese Academy of Sciences, Ningbo Institute of Material Technology and Engineering, Key Laboratory of Marine Materials and Related Technologies, Zhejiang Key Laboratory of Marine Materials and Protective Technologies, 1219 Zhongguan West Road, Ningbo 315201, China
[2] University of Chinese Academy of Sciences, School of Chemical Sciences, 19A Yuquan Road, Beijing 100049, China

9.1 Introduction

Nature is an unlimited source of inspiration, which has guided the design and the fabrication of artificial intelligent materials [1–3]. Imitating the biological behaviors of nature, existing artificial intelligent materials are capable of deforming, moving, and even growing like a living organism [4–8]. As one of the most potential artificial intelligent materials, hydrogel actuators are able to change their shapes under external stimuli [9–11]. Due to the properties of soft and wet, hydrogel actuators could exhibit large-scale deformation [12–14]. Therefore, hydrogel actuators have been regarded as an ideal candidate for soft robots [15, 16], artificial muscles [17, 18], and biomimetic devices [19, 20].

Besides, hydrogel actuators also play an essential role in stimuli-responsive materials. Benefiting from the molecular designability, various stimuli-responsive materials, including polymers and additives, could be introduced into the hydrogel systems, which endows hydrogel actuators with the ability to respond to diverse external stimuli, such as thermal, light, pH, and ionic strength [21–23]. Based on stimuli-responsive swelling/shrinking property, hydrogel actuators could generate reversible deformation and convert internal chemical energy to mechanical energy [24, 25]. However, limited by the isotropic structure, traditional hydrogel actuators could only provide homogenous swelling or shrinking deformation [26]. To break the dilemma, many anisotropic structures, such as bilayer [27, 28], oriented [29, 30], gradient [31, 32], and pattern structures [33, 34], have been proposed, which endows the hydrogel actuators with various programmable deformations. Besides, during the fabrication process of anisotropic hydrogel actuators, many novel fabricating methods, including stepwise polymerization, three-dimensional (3D) printing, and molecular assembly, are reported in recent [35]. It is precise because of the aforementioned major achievements, hydrogel actuators are further applied as soft

Smart Stimuli-Responsive Polymers, Films, and Gels, First Edition.
Edited by Liang Hu, Yongfeng Gao, and Michael J. Serpe.

robots, artificial muscles, intelligent grippers, and play an indispensable role in the corresponding fields.

Although some reviews about hydrogel actuators have been published, the development of hydrogel actuators has already exceeded the expectations [36–38]. Especially, in the last five years, plenty of innovative fabrication strategies, including temporary anisotropic structure and novel 3D printing, have been proposed [39, 40], which further improve deformability of hydrogel actuators such as multidimensional and programmable deformation [41, 42]. Moreover, inspired by nature, lots of hydrogel actuators have been incorporated with other stimuli-responsive materials and derived new applications in the fields of walking robots, biomimetic devices, and information encryption [43–46]. All of these encourage us to review the rapid development of hydrogel actuators and remake the future perspectives. In this review, the classification and corresponding deformation mechanism of hydrogel actuators will be briefly introduced first. Then anisotropic structures of hydrogel actuators, especially some innovative fabrication strategies will be discussed. Finally, we will exhibit the newest applications of hydrogel actuators and propose our perspectives on the challenges and opportunities of hydrogel actuators in the future (Figure 9.1).

9.2 The Classification of Hydrogel Actuators

As an interdisciplinary subject in the field of robots and hydrogels, hydrogel actuators have set an excellent example so that artificial materials could also respond to external stimuli and change their shapes via self-deformation like living organisms. With the development of material science, various fabrication strategies have been proposed, which endows the hydrogel actuators with multidimensional deformation performances and multifunctions [50–53]. According to the actuating mechanisms, hydrogel actuators can be mainly divided into three types: addition of active ingredients, pneumatic/hydraulic actuators, and stimuli-responsive actuating derived from asymmetric swelling. In this section, we will exhibit and discuss the three types of hydrogel actuators.

9.2.1 Addition of Active Ingredient

In the ordinary fabrication process of intelligent materials, introducing the expected property via adding active ingredient during preparation, is an efficient and convenient strategy. Similarly, intelligent hydrogels could also be endowed with new functions, such as fluorescence, deformation, and motion in the same way [54, 55]. For example, Serra et al. synthesized poly(ethylene glycol) (PEG) capped iron oxide-based nanoparticles (IOPs) and blended them within a gelatin methacryloyl matrix [56]. With the assistance of a low-intensity magnetic field, the IOPs were aligned and an anisotropic hydrogel actuator was printed via direct ink writing (Figure 9.2a). Benefiting from the magnetic responsiveness of IOPs, the hybrid

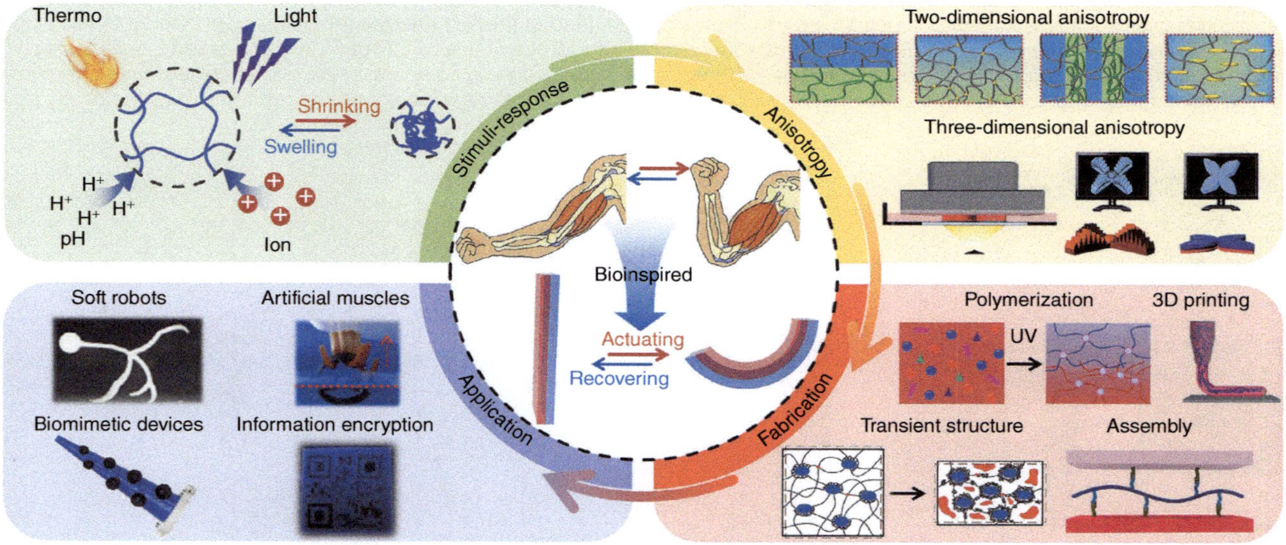

Figure 9.1 Overview of the recent progress of hydrogel actuators, including mechanism, anisotropic structures, fabrication strategies, and novel applications. Source: Reproduced with permission from Odent et al. [47]/Royal Society of Chemistry. Reproduced with permission from Han et al. [48]/American Chemical Society. Reproduced with permission from Le et al. [49]/Springer Nature.

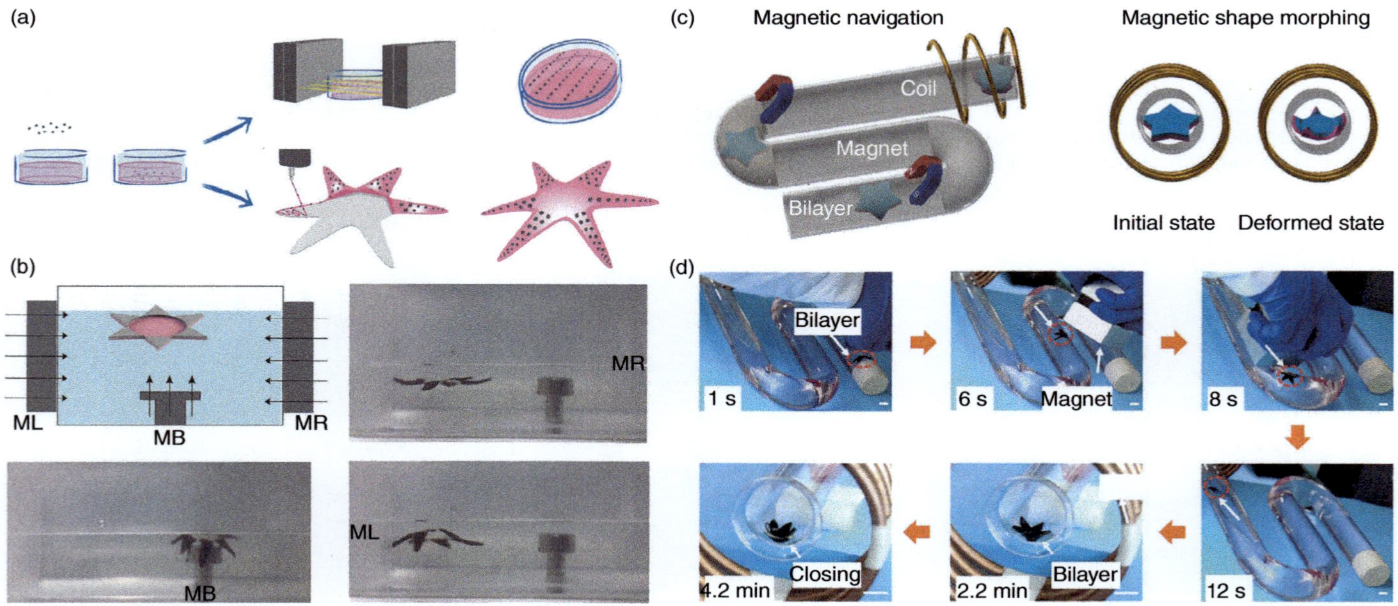

Figure 9.2 (a) Schematic illustration showing the fabrication process of nanocomposite hydrogel actuator. (b) Schematic illustration and images showing the remote control of nanocomposite hydrogel actuator by three electromagnets. Source: Reproduced with permission from Tognato et al. [56]/John Wiley & Sons, Inc. (c) Schematic illustration showing the deformation and motion of hydrogel actuator by magnetic navigation and magnetothermal effect. (d) Images showing the process of the hydrogel actuator passing through a narrow tube with the guidance of magnetic field. Source: Reproduced with permission from Tang et al. [57]/American Chemical Society.

hydrogel actuator was able to flappy move and catch object with the guidance of alternating magnetic field (Figure 9.2b).

Although the shape of hydrogel actuators could be changed by the attraction of magnetic field, the deformation was simple and involuntary. To enhance the deformation ability, further, Wang et al. introduced Fe_3O_4 nanoparticles into the network of poly(*N*-isopropylacrylamide) (PNIPAm) hydrogels and prepared a magnetic and thermal dual-responsive hydrogel (RH) actuator via *in situ* coprecipitation method (Figure 9.2c) [57]. As shown in Figure 9.2d, the hydrogel actuator could also be navigated by magnetic field and quickly pass through a closed rube. Besides, when an alternating magnetic field was applied, the hydrogel actuator was heated via magnet to thermal effect and self-deform from opening state to closing state. It is easy to imagine that a hydrogel actuator with adaptive shapes is going to expand its applications in the near future.

9.2.2 Pneumatic/Hydraulic Actuators

Moreover, addition of active ingredient is another efficient method to find out whether the pneumatic or hydraulic hydrogels with hollow structures are formed. The actuating mechanism can be simply understood as follows: when air or liquid is blown into the cavity of hollow hydrogel, the internal pressure is higher than the external pressure, which expands and actuates the hydrogel [58, 59]. Similarly, when the air or liquid is pumped out, the internal pressure decreases and the volume of hydrogel shrinks. Therefore, the key point to constructing pneumatic/hydraulic actuators is preparing a hollow hydrogel with connected cavity. Recently, a few strategies, such as 3D printing, sacrificial template, and microfluidic, have been applied to prepare hollow hydrogels [60–62]. In our previous work, many gelatin blocks containing Ca^{2+} were assembled and immersed in alginate solution [63]. Due to the osmotic pressure of Ca^{2+}, Ca^{2+} gradually diffused into alginate solution and form Ca^{2+}-alginate hydrogel on the surface of assembled gelatin via metal–ligand coordination. After melting and removing inner gelatin, the corresponding hollow Ca^{2+}-alginate hydrogel was obtained (Figure 9.3a). The hollow hydrogel could be utilized as a pneumatic actuator that was able to contract when air was pumped and stretched when air was blown (Figure 9.3b).

Besides, when the filling medium was replaced by liquid such as water, the pneumatic actuator could transfer to hydraulic actuator and obtain stronger actuating force. For instance, Zhao et al. designed and prepared a hydraulic actuator with asymmetric structure [64]. Compared with the isotropic pneumatic actuator mentioned above, the anisotropic hydraulic could bend rather than simply stretch when actuated by hydraulic. Besides, owing to the advantages of hydraulic actuating, the hydraulic actuator was able to complete a deformation process in one second, while a traditional hydrogel actuator would need two minutes to 20 hours (Figure 9.3c,d). Subsequently, the hydraulic actuator was further designed as a gripper and silently caught and lifted a moving goldfish due to the transparent property of hydrogel (Figure 9.3e).

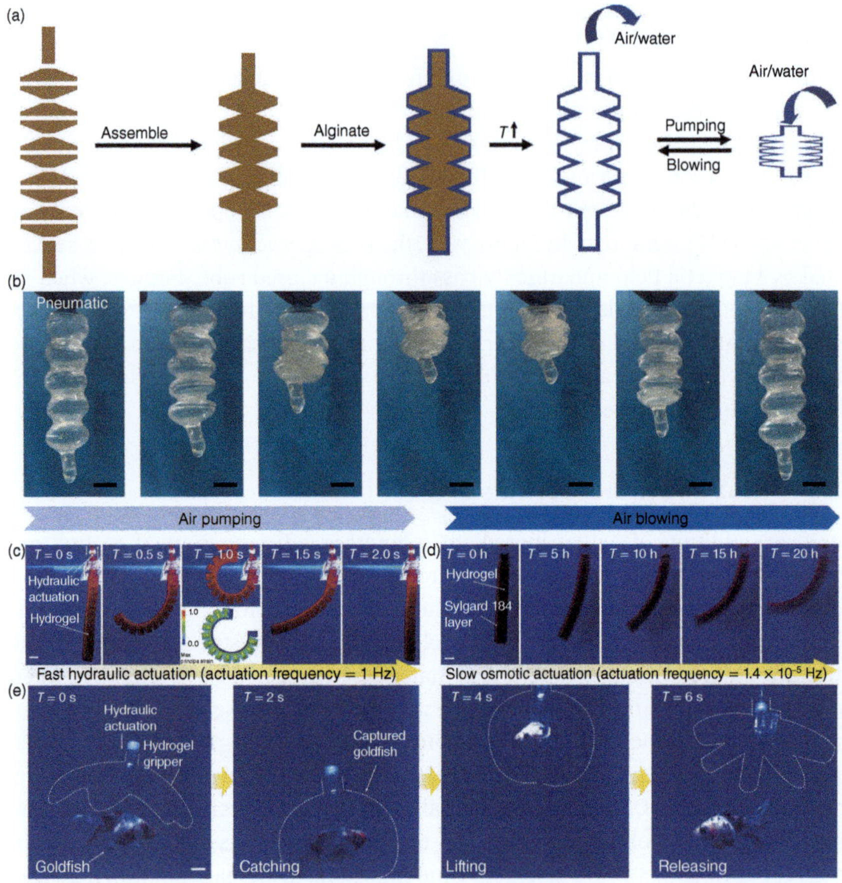

Figure 9.3 (a) Schematic illustration showing the fabrication process of hollow hydrogel actuator by hydrogel modular assembly and sacrifice template. (b) Images showing the pneumatic actuating process of a hollow hydrogel. Source: Reproduced with permission from Wu et al. [63]/American Chemical Society. (c) Images showing the fast hydraulic actuation. (d) Images showing the slow osmotic actuation. (e) Images showing the process of a hydrogel gripper catching goldfish by hydraulic actuation. Source: Reproduced with permission from Yuk et al. [64]/Springer Nature.

9.2.3 Stimuli-Responsive Hydrogel Actuator Derived from Asymmetric Swelling

Different from the above hydrogel actuators, which were actuated with the assistance of external field, stimulus-responsive hydrogel actuators could respond to the external stimuli and provide self-deformation properties, due to the transformation of their chemical networks [65–67]. According to the number of external stimuli, stimuli-responsive hydrogel actuators could be classified into single-stimulus-responsive hydrogel actuators that could only respond to one kind of external stimuli and multi stimuli-responsive hydrogel actuators that could respond to more than two kinds of external stimuli.

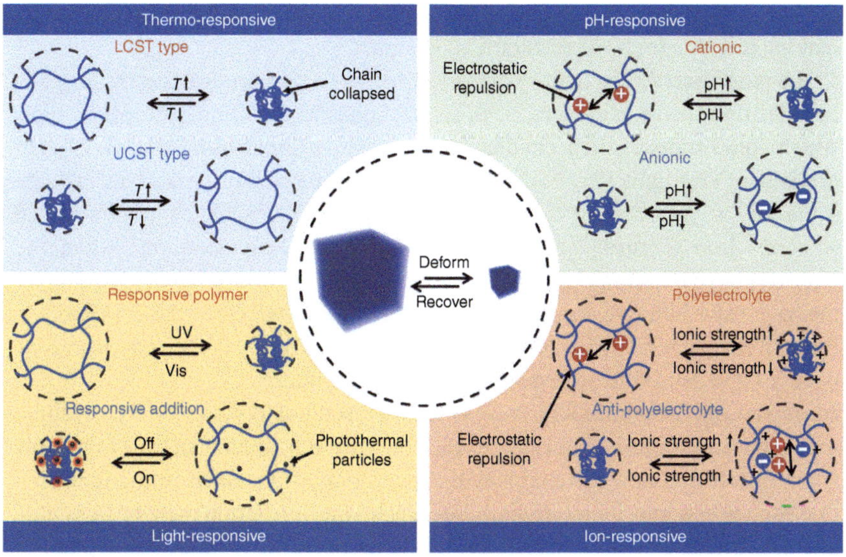

Figure 9.4 Schematic illustration showing the deformation mechanism of stimuli-responsive hydrogel.

9.2.3.1 Single-Stimulus-Responsive Hydrogel Actuators

Although there are many external stimuli that are capable of triggering the transformation of chemical chains inducing the deformation of hydrogel actuators, such as temperature, pH, light and ion strength (IS), the actuating mechanism is similar [68–70]. As shown in Figure 9.4, in the normal condition, the stimuli-responsive hydrogel actuators will swell in water due to the hydrophilic property of chemical chains. However, when the stimuli-responsive hydrogel actuators are trigged by external stimuli, the hydrophilic chemical chains would convert to hydrophobic, which induces collapse of chemical chains and further causes shrink of hydrogel. Herein, we will exhibit and discuss four kinds of major stimuli-responsive hydrogel actuators.

1. **Thermo-responsive:** Thermo-responsive hydrogel actuators including lower critical solution temperature (LCST) type and upper critical solution temperature (UCST) type [71]. The chemical networks of LCST-type hydrogel actuators usually have LCST polymers, such as PNIPAm, and the chemical networks of UCST-type hydrogel actuators usually have UCST polymers, such as poly(acrylamide-*co*-acrylic acid) (P(AAm-*co*-AAc)) [72]. When the external temperature is higher than the critical solution temperature, the chemical networks of LCST-type hydrogel actuator will collapse and the hydrogel will shrink, while the UCST-type hydrogel actuator will swell.
2. **pH-responsive:** The pH-responsive hydrogel actuators contain ionizable functional groups, such as carboxyl groups and amino groups [73, 74]. When the value of pH increases, the cationic polymer may become more difficult to ionize, which induces collapse of chemical networks. In contrast, when the value of pH

decreases, cationic polymer may ionize more cations and the chemical networks may expand due to the electrostatic repulsion.

3. **Light-responsive:** The light-responsive hydrogel actuators are usually designed by light-responsive polymers, such as azobenzene type and spiropyran type, which could transfer their chemical conformation under ultraviolet (UV) light or visible (Vis) light [75, 76]. Besides, the light-responsiveness could also be achieved by introducing light conversion materials, such as photothermal additives into thermo-responsive networks [77–79]. For example, when the light-responsive hydrogel actuator is exposed under near-infrared (NIR) light, the responsive additive, such as Fe_3O_4 nanoparticles, will convert light energy to thermal energy, which induces the volume-phase transition of thermo-responsive polymer and cause the deformation of hydrogel actuator.
4. **Ionic strength-responsive:** Polyelectrolytes, such as poly(methacryloxyethyltrimethyl ammonium chloride) usually (PMETAC), ionize in the low ionic strength solution and the chains will expand due to the electrostatic repulsion [31, 80]. When the ionic strength increases, the ionization will be inhibited and cause collapse of chains. Recently, some polyzwitterionic was applied to produce hydrogel actuator [81, 82], due to the antipolyelectrolyte effects, the polyzwitterionic hydrogel actuator will swell as the ionic strength increases and shrink as the ionic strength decreases.

9.2.3.2 Multi-stimuli-Responsive Hydrogel Actuators

With the development of stimuli-responsive hydrogel actuators, lots of stimuli-responsive polymers have been synthesized [51, 67, 73]. In the diversity of stimuli-responsive polymers, some stimuli-responsive polymers could respond to multi external stimuli, which motivates researchers to fabricate multi-stimuli-responsive hydrogel actuators [83–85]. For example, Zhao et al. prepared a PAAm–PAAc anisotropic hydrogel actuator with gradient structure (Figure 9.5a) [86]. The top layer of the hydrogel actuator contains interpenetrated PAAm and high concentration of PAAc, which would exhibit UCST property, while the bottom layer of hydrogel actuator contains PAAm and low concentration of PAAc, is non-thermo-responsive. The asymmetric thermo-responsive property induced the thermo-responsive deformation. As shown in Figure 9.5b, the hydrogel strip bent to a circle at a high temperature above the UCST and recovered to its original shape at a low temperature. It has been proven that urea is both a strong H-acceptor (–C=O) as well as H-donor ($-NH_2$). Therefore, when urea was added to the solution, the H-bonds between PAAm and PAAc were broken that significantly weakened the UCST effect of hydrogel actuator, and hydrogel actuator could recover to strip in higher concentration of urea solution at 5 °C (Figure 9.5c). Besides, hydrogel actuator was also responsive to strength and type of ions. As shown in Figure 9.5d, a hydrogel circle would unfurl when treated by Cl^- or SCN^- while further bend with the solution of SO_4^{2-}, and the trend would be enhanced with the increase of ionic strength.

Different from the system of PAAm–PAAc that could be responsive to multi external stimuli, multi-responsive hydrogel actuators could also be obtained by

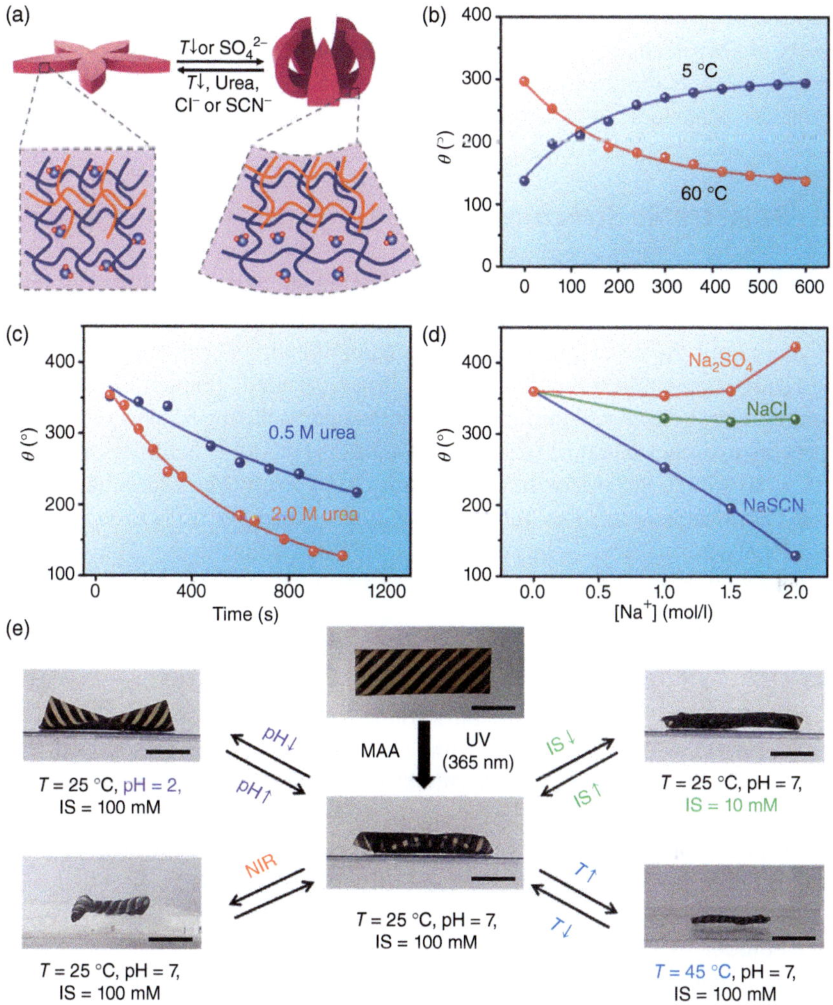

Figure 9.5 (a) Schematic illustration showing the deformation mechanism of bilayer hydrogel actuator with multi-responsive property. (b–d) Kinetic of actuating process trigged by temperature, urea, and ionic strength. Source: Reproduced with permission from Hua et al. [86]. Copyright 2019 American Chemical Society. (e) Images showing the multidimensional complex deformation of a multi-responsive hydrogel actuator. Source: Reproduced with permission from Ma et al. [87]/John Wiley & Sons, Inc.

incorporating two or more non-interfering stimuli-responsive polymers into one hydrogel system. In our previous work, a thermo-responsive PNIPAm hydrogel with pattern-reduced graphene oxide sheets (RGOs) was prepared first. Then hydrogel was immersed into the solution of the second pH-responsive monomer (methacrylic acid, MAA) and further polymerized via photopolymerization (Figure 9.5e) [87]. The hydrogel actuator could self-deform from a two-dimensional (2D) sheet to 3D complex shape in hot water or exposed to NIR due to the LCST property of PNIPAm and photothermal effect of the RGOs. Besides, the hydrogel actuator was able to respond to pH and ion strength due to the pH and ionic strength dual-responsive property of PMAA chain, and deform in lower pH or ionic strength. These works expand the scope for designing future hydrogel actuators with programmable deformation behaviors.

9.3 Anisotropic Structures

Limited by the isotropic structure, traditional hydrogel actuators could only provide simple homogenous swelling/shrinking property under external stimuli, which immensely limit the development and application of hydrogel actuators. Recently, some 1D/2D anisotropic structures such as bilayer, oriented, gradient, and pattern structures, have been introduced into hydrogel networks during fabrication, which endows the hydrogel actuators with more complex deformation behaviors from 1D to 2D or 2D to 3D. Besides, with the assistance of 3D printing technology, more complex anisotropic structures can be encoded into hydrogels, and provide higher dimensional deformations. In this section, we will discuss some types of anisotropic hydrogel actuators including their features and deformation behaviors.

9.3.1 1D/2D Anisotropic Structures

Although there are many types of anisotropic hydrogel actuators, their design and deformation mechanisms are similar. The reason why an anisotropic hydrogel actuator could provide asymmetric deformation is because of the different swelling/shrink properties in one hydrogel system, and the asymmetric swelling/shrink property can be caused by different types of chemical chains or different concentrations of polymers.

9.3.1.1 Bilayer

Bilayer hydrogel actuator usually contains two kinds of hydrogel layer with different swelling/shrinking property or stimuli-responsive behavior [88]. For example, a bilayer hydrogel actuator contains a thermo-responsive PNIPAm hydrogel layer, and a non-responsive PAAm hydrogel layer was prepared via stepwise polymerization. When hydrogel actuator was immersed in hot water at temperature above LCST, the PNIPAm hydrogel layer would immediately shrink due to the volume phase transformation of PNIPAm chains, while the PAAm hydrogel layer kept normal as in cold water. The asymmetric shrinking behavior induced the asymmetric deformation, and hydrogel actuator would deform from a strip to a circle. Similarly, when the deformed hydrogel actuator was transferred to cold water, the PNIPAm hydrogel layer would swell, while the PAAm hydrogel layer was kept in the same state, hydrogel actuator would recover to the original straight shape.

In past years, various bilayer hydrogel actuators have been fabricated, but most of them can only actuate in water [81, 86, 88]. Inspired by *Mimosa*, which would deform in the air via reversible water transfer between two compartments (Figure 9.6a) [89], we designed a bilayer hydrogel actuator with two complementary

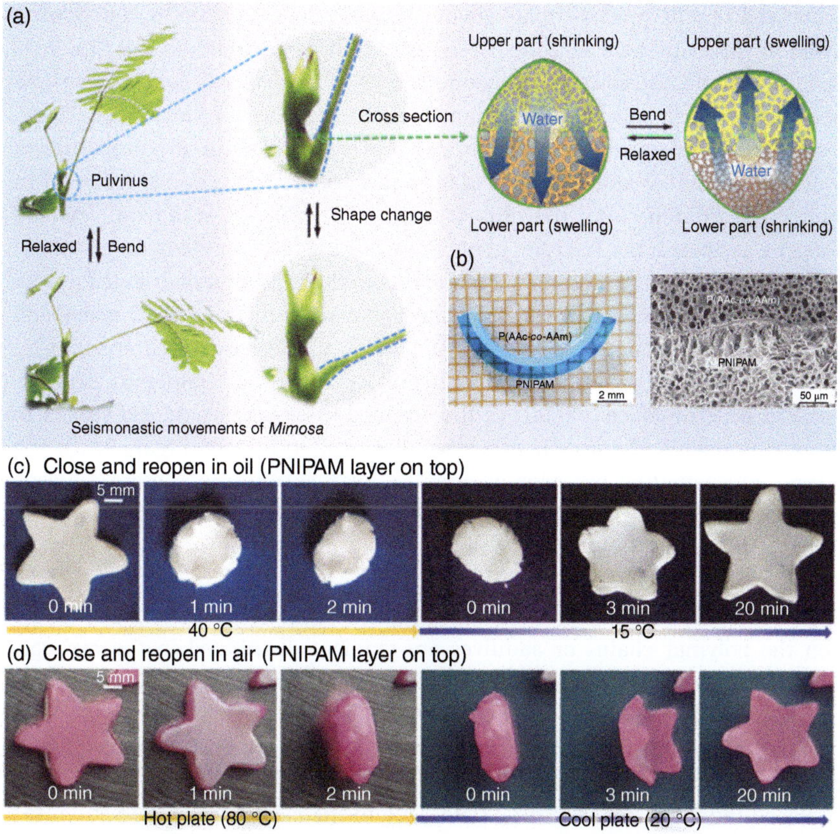

Figure 9.6 (a) Schematic illustration and images showing the bioinspired bilayer structure. (b) Images and SEM showing the bilayer hydrogel actuator. (c) The innovative actuating process of bilayer hydrogel actuator in oil and air. Source: Reproduced with permission from Zheng et al. [89]/Royal Society of Chemistry.

thermo-responsive hydrogel layers, a LCST layer contained PNIPAm and a UCST layer contained p(AAm-*co*-AAc) (Figure 9.6b). Different from the traditional bilayer hydrogel actuators, when the complementary hydrogel actuator was placed in a 40 °C oil, the LCST layer hydrogel would shrink and the extra water was absorbed by the UCST layer hydrogel via the internal water channels. On contrary, when the temperature of the oil decreased to 15 °C, the UCST layer would shrink and transferred the extra water to the swelling LCST layer (Figure 9.6c). Benefited by the innovative design, the hydrogel actuator could also generate reversible deformation in the air. This work provides a successful step for designing and fabricating of land deformable hydrogel actuator and expands the scope of their application scenario.

9.3.1.2 Oriented

Oriented hydrogel means that the hydrogel networks are oriented or there are oriented additives in the hydrogel [90]. According to the definition, oriented hydrogels are fabricated by orienting hydrogel via stretch or adding oriented additives and polymerized *in situ*. For example, Aida et al. placed the titanate(IV) nanosheets (TiNSs) in magnetic field and the TiNSs would aligned parallel to the flux lines [91]. Furthermore, the corresponding oriented PNIPAm hydrogel containing TiNSs was obtained via *in situ* polymerization (Figure 9.7a). Due to the aligned TiNSs, the oriented hydrogel exhibited totally different swelling/shrinking properties in the direction parallel and perpendicular to the oriented direction. As shown in Figure 9.7b, different from the ordinary PNIPAm hydrogel, when the oriented hydrogel was heated, the PNIPAm network would collapse but the hydrogel would stretch in parallel to the oriented direction due to the repulsion of oriented TiNSs. More specially, an oriented hydrogel strip would self-stretch in 50 °C water and recover to the original length at 15 °C (Figure 9.7c). Besides, when an oriented hydrogel block with scale of 10 mm × 10 mm × 1.2 mm was immersed in 45 °C water, the scale in *c*-face (parallel to the oriented direction) would self-stretch to 8 mm × 15 mm while the scale in *a*-face would shrink to 1.0 mm (Figure 9.7d). The extraordinary actuating behavior of oriented hydrogel motivates more fantastic applications, such as walking and wiggling, that are difficult to realize with ordinary hydrogel actuators.

9.3.1.3 Gradient

When the polymer chains or additives are gradient distributed in the hydrogel networks, the hydrogel actuators could also exhibit asymmetric swelling/shrinking property [92, 93]. In the past years, diverse gradient hydrogel actuators have been fabricated and exhibited programmable deformation. For example, Xu et al. injected the hydrogel precursor containing electric nanoparticles, Laponite to a mold, and placed the mold under electric field. Due to the electrostatic interaction, the Laponite tends to move under the direction of electric field, and many Laponite nanoparticles were aggregated on the positive side of the mode. After *in situ* polymerization, the Laponite nanoparticles were fixed in the hydrogel network and the oriented hydrogel was obtained (Figure 9.8a) [94]. During the fabricating process, the Laponite was applied as crosslinker, so in the cathode part with high concentration of Laponite, the crosslink density is high, while the crosslink density in the anode part is low. The gradient structure could be proven by the images of SEM (Figure 9.8b,c), and the average pore size increases from 0.35 μm on the anode side to 4.83 μm on the cathode side (Figure 9.8d). All of the evidence proved the successful fabrication of gradient hydrogel. Due to the gradient structure, the actuator could self-deform from a strip to a circle at 50 °C and quickly recover in 25 °C water. Benefiting from the quickly actuating and recovering property, the hydrogel actuator was able to be applied as an intelligent valve to effectively control the water flux based on temperature.

In addition to the mentioned method, gradient structure could also be obtained via asymmetric polymerization. In our previous work, a homogeneous PAAm hydrogel was prepared first [95]. Then, the PAAm hydrogel was immersed into a hydrogel precursor containing AAc as monomer. Due to the hindering effect of hydrogel on light transmission, the light intensity would gradient decrease along the thickness direction. Thus, when the treated hydrogel was exposed to a strong UV light, more

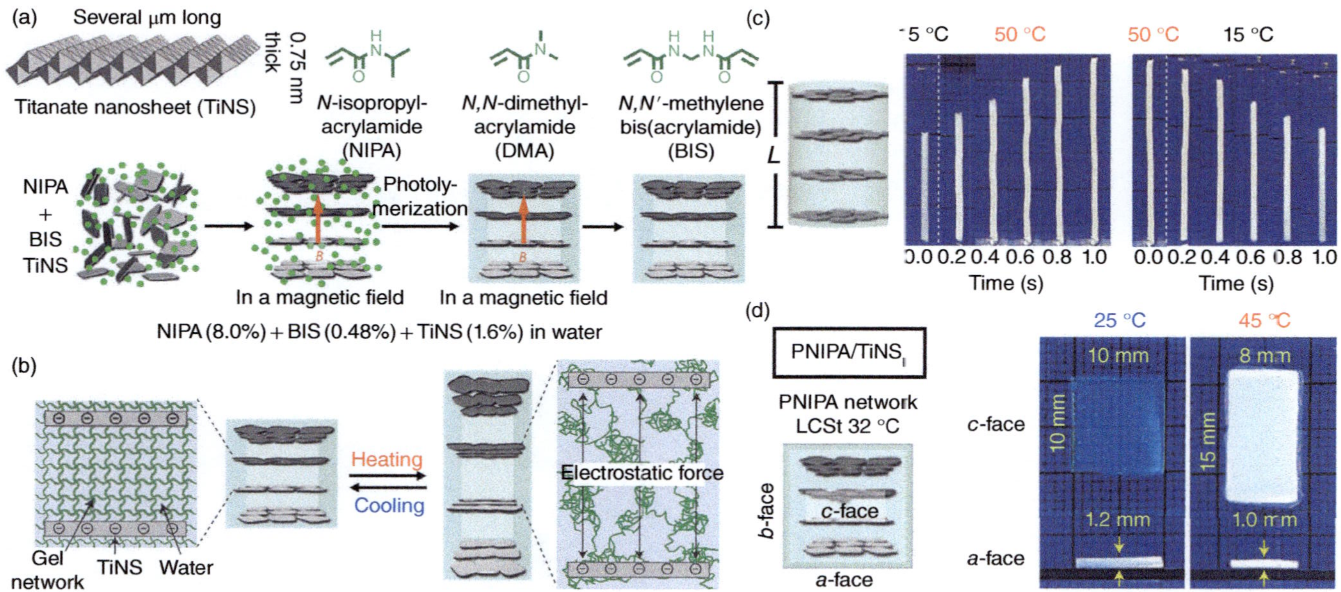

Figure 9.7 (a) Schematic illustration showing the fabrication process of oriented hydrogel. (b) Schematic illustration showing the actuating process of oriented hydrogel. (c and d) Images showing the anisotropic actuating process of oriented hydrogel actuator with one-dimensional and two-dimensional shapes, respectively. Source: Reproduced with permission from Kim et al. [91]/Springer Nature.

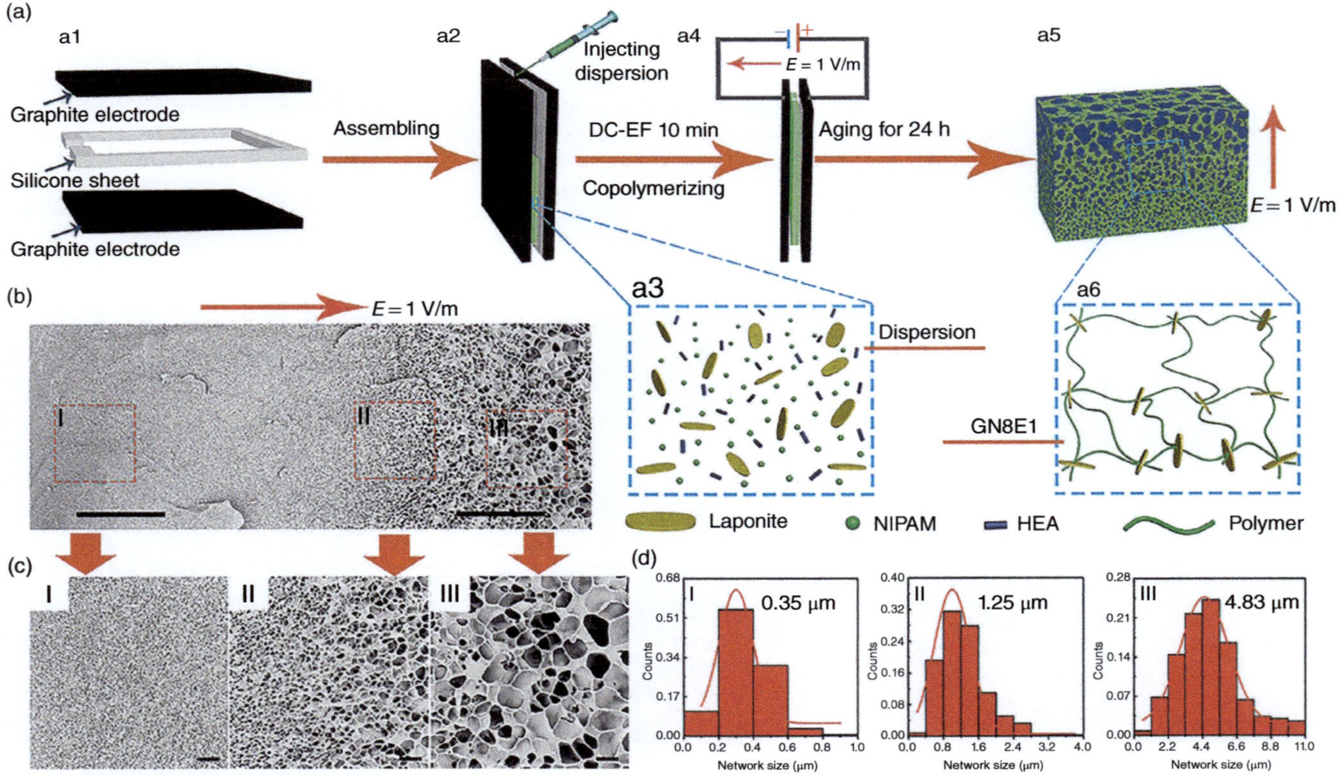

Figure 9.8 (a) Schematic illustration showing the fabrication process of gradient hydrogel. (b and c) SEM images showing microstructure of gradient hydrogel. (d) Network size statistics of gradient hydrogel in the gradient direction. Source: Reproduced with permission from Tan et al. [94]/American Chemical Society.

PAAc chains would be generated on the upper layer than the bottom layer, inducing the gradient distribution of PAAc chains along the thickness direction. The gradient hydrogel stripe would bend toward the top layer in acid solution or bend toward the bottom layer in alkaline solution. The design of gradient hydrogels, both in polymer chains and additives, enriches the fabrication of hydrogel actuators and makes a potential application in programmable deformation.

9.3.1.4 Patterned

As mentioned above, anisotropic structure induces asymmetric deformation during fabrication, and the anisotropic structure of hydrogel can also be encoded like a pattern and induce more complex and programmable deformation [96]. For example, Wu et al. fabricated a patterned hydrogel via photopolymerization [97]. The low swelling and patterned PAAm hydrogels were polymerized first under UV light, then a high swelling poly(acrylamide-*co*-2-acrylamido-2-methylpropanesulfonic acid) [p(AAm-*co*-AMPS)] hydrogels were polymerized and surrounded by the prepared PAAm hydrogel. Due to the patterned anisotropic structure, the patterned hydrogel would deform into alternating concave–convex configuration after asymmetric swelling. Yum et al. further proposed a continuous photopolymerization strategy and fabricated patterned thermo-responsive hydrogel in one step [98, 99]. As shown in Figure 9.9a, a hydrogel disk was exposed to a programmed UV light and encoded with corresponding pattern. Within the encoding process, the marginal part of disk was exposed quickly than the central part, which induced higher crosslink density and tougher modulus. After swelling, the hydrogel disk would deform to a 3D arched shape and the shape could further transform to saddled shape above critical temperature (T_c). Besides, more complex 3D deformation could be encoded after simulation calculation. As shown in Figure 9.9b, the target shape was proposed first, then the programmed photopolymerization strategy was encoded according to the concept of dynamic target metrics. Based on the dynamic theoretical model, the hydrogel sheet was able to self-deform to the same target 3D shape as the theoretical model.

After fabrication, the pattern structure can also be introduced into hydrogel. For instance, Wang et al. prepared poly(sodium acrylate) (PNaAAc) hydrogel first, then the patterned Fe^{3+} ink was printed on the surface of the hydrogel via ion inkjet printing technique [100]. Due to the coordination between Fe^{3+} and carboxyl, the pattern area of hydrogel became tough and hard to swell, while the unpattern areas kept normal. Therefore, after swelling the pattern hydrogel would self-deform to the target shape. Using pattern strategy, the deformation of hydrogel becomes more controllable and programmable, and these works motivate the development of hydrogel actuator to the next stage.

9.3.2 3D Anisotropic Structures

As mentioned above, endowing hydrogel actuators with 1D/2D anisotropic structures, such as bilayer, oriented, gradient, and pattern, could promote the deformation ability of hydrogel actuators. With the development of 3D printed

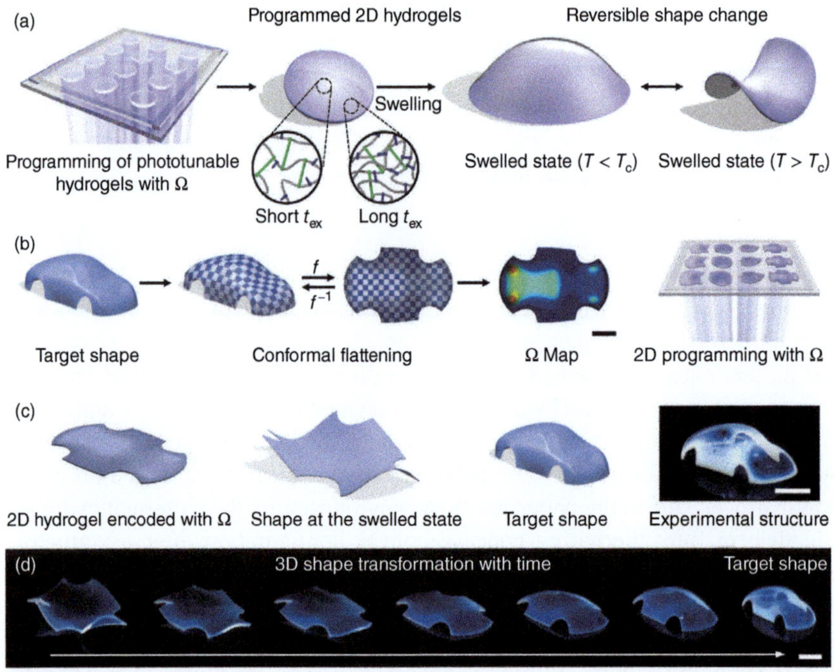

Figure 9.9 (a) Schematic illustration showing the fabrication process of patterned hydrogel via digital light 4D printing. Source: Reproduced with permission from Nojoomi et al. [98]. Copyright 2018 Amirali Nojoomi et al. (b) Schematic illustration showing the deformation mechanism of hydrogel with car pattern. (c and d) Schematic illustration and images showing the 3D complex deformation process of hydrogel with car pattern. Source: Reproduced with permission from Nojoomi et al. [99]/Springer Nature.

technology, more anisotropic structures could be introduced into hydrogel in 3D direction and exhibit more bioinspired deformations [101, 102]. For example, Ho et al. prepared a printable hydrogel ink containing NIPAm as monomer, alginate, carbon nanotube (CNT) as photothermal element, and 3-(trimethoxysilyl)propyl methacrylate (TMSPMA) as silane-coupling agent [103]. Then with the assistance of direct-ink-write (DIW) technology, the hydrogel ink was printed on the surface of PDMS. Benefited from the magnetic responsiveness of IOPs, the hybrid hydrogel actuator was able to flappy move and catch object with the guide of an alternating magnetic fields on the left (ML), right (MR), and bottom (MB) of the water pool (Figure 9.10a). Due to the anisotropic structure, the hydrogel–PDMS actuator could bend in hot water and the actuating ability could be further adjusted by changing the thickness of PDMS (Figure 9.10b). Subsequently, the 3D anisotropic structure could be encoded and an artificial tendril could be fabricated. Benefiting from the 3D anisotropic structure, the artificial tendril exhibited programmable phototropic motion like tendril in nature (Figure 9.10c). In detail, when a light shined from the west, only the west part of the artificial tendril irradiated. Therefore, the exposed part heated via photothermal effect of CNT, and the west part of hydrogel would shrink, which induced the artificial tendril to bend to the west. Similarly, when the

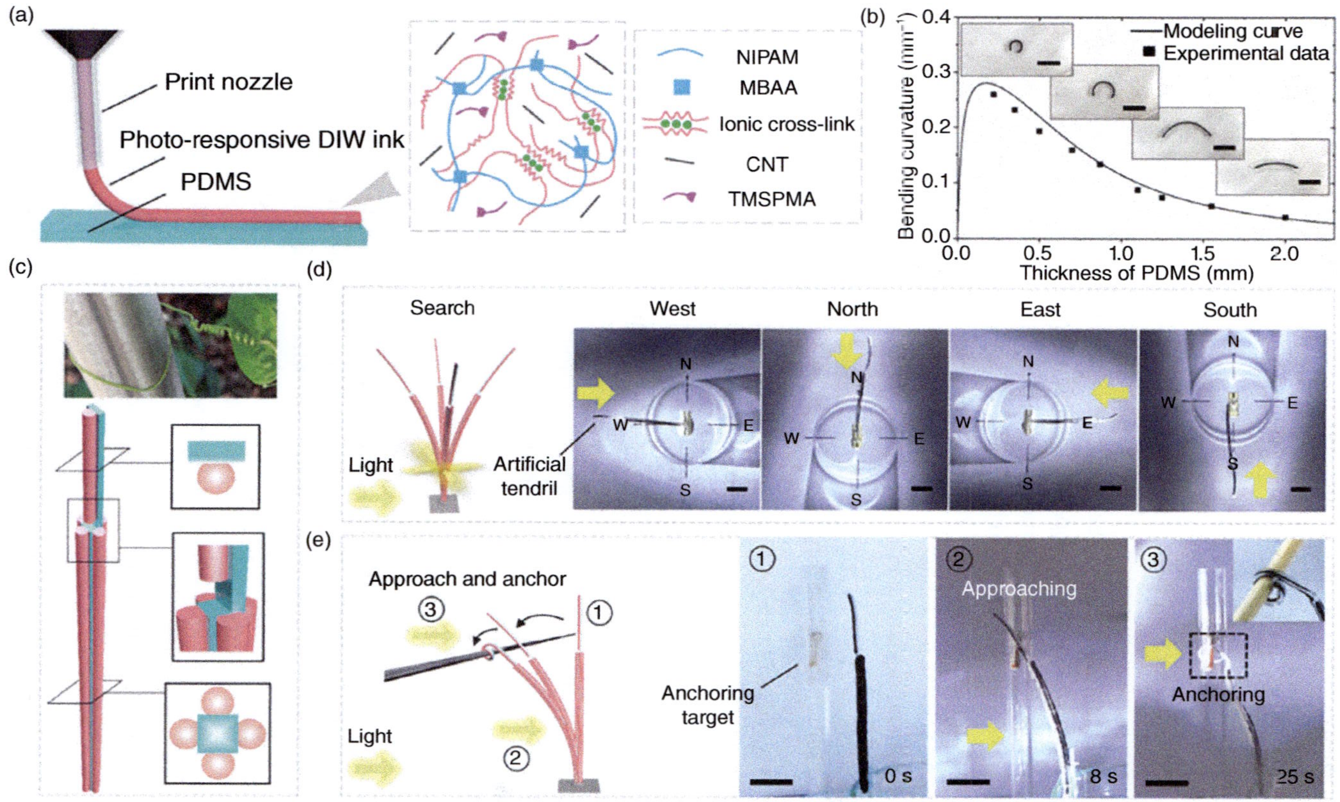

Figure 9.10 (a) Schematic illustration showing the fabrication process of composite hydrogel actuator by 3D printing technology. (b) The dependence of bending curvature with different thicknesses of PDMS. (c) Schematic illustration showing the 3D anisotropic structure of bioinspired artificial tendril. (d) Images showing the phototropism behavior of the bioinspired artificial tendril. (e) Images showing the search and catch behavior of the bioinspired artificial tendril. Source: Reproduced with permission from Cheng et al. [103]/American Chemical Society.

direction of light changed, the artificial tendril would bend to the corresponding directions (Figure 9.10d). Finally, the artificial tendril exhibited a more biological deformation behavior bent to the direction of light, approach, and anchored to the target object. Using the design of 3D anisotropic structure, hydrogel actuators will be able to imitate more 4D bionomic deformations in the near future.

9.4 Methods to Fabricate Anisotropic Structures

In traditional conditions, hydrogels were fabricated via the polymerization of precursor. Therefore, anisotropic hydrogels are fabricated in two major ways. (i) Introducing anisotropic additives or polymer chains, and polymerized *in situ*. (ii) Modularly assembling two or more prepared hydrogels. With the development of materials science, many innovative strategies have been proposed to prepare anisotropic hydrogels during or after the fabrication process. In this section, we will discuss some traditional and innovative methods.

9.4.1 Traditional Technology

In this part, we will exhibit some traditional technologies for the fabrication of hydrogel actuators and discuss some essential steps in detail.

9.4.1.1 Stepwise Polymerization

Taking the fabrication of bilayer hydrogels as an example. In ordinary, bilayer hydrogel contains two different types of hydrogel layers. Thus, stepwise polymerization is necessary [104]. As shown in Figure 9.11a, hydrogel precursor contains monomer 1, crosslinking agent and initiator should be prepared first. Then inject the hydrogel precursor into mold and heat the precursor to polymerization temperature (usually 60–80 °C, redox initiation system can trigger polymerization at room temperature, such as ammonium persulphate-*N*,*N*,*N*′,*N*′-tetramethylethylenediamine (APS-TEMED) system). In the middle of polymerization, open the mold and pour the second hydrogel precursor onto the surface of the first hydrogel layer. The second precursor will diffuse into the network of the first hydrogel layer and copolymerize with the unreacted monomer 1. After 24 hours due to the chemical covalent bond and physical entanglement force in the interface of two hydrogel layers, a bilayer hydrogel with stable interface bonding is obtained. Similarly, stepwise photopolymerization can also be used to fabricate bilayer hydrogel within 10 minutes [105, 106].

Recently, Zheng et al. proposed an innovative strategy to fabricate bilayer hydrogel in one pot [107]. As shown in Figure 9.11b, two kinds of hydrogel precursors are mixed to prepare a homogeneous solution. Especially, for continuous and efficient initiation of the polymerization, both the thermo- and photoinitiator are added to the homogeneous solution simultaneously. During thermo-polymerization and photopolymerization, most of the monomers 1 are polymerized first and separated from the solution. Then the retained monomers 2 are polymerized to form the bilayer

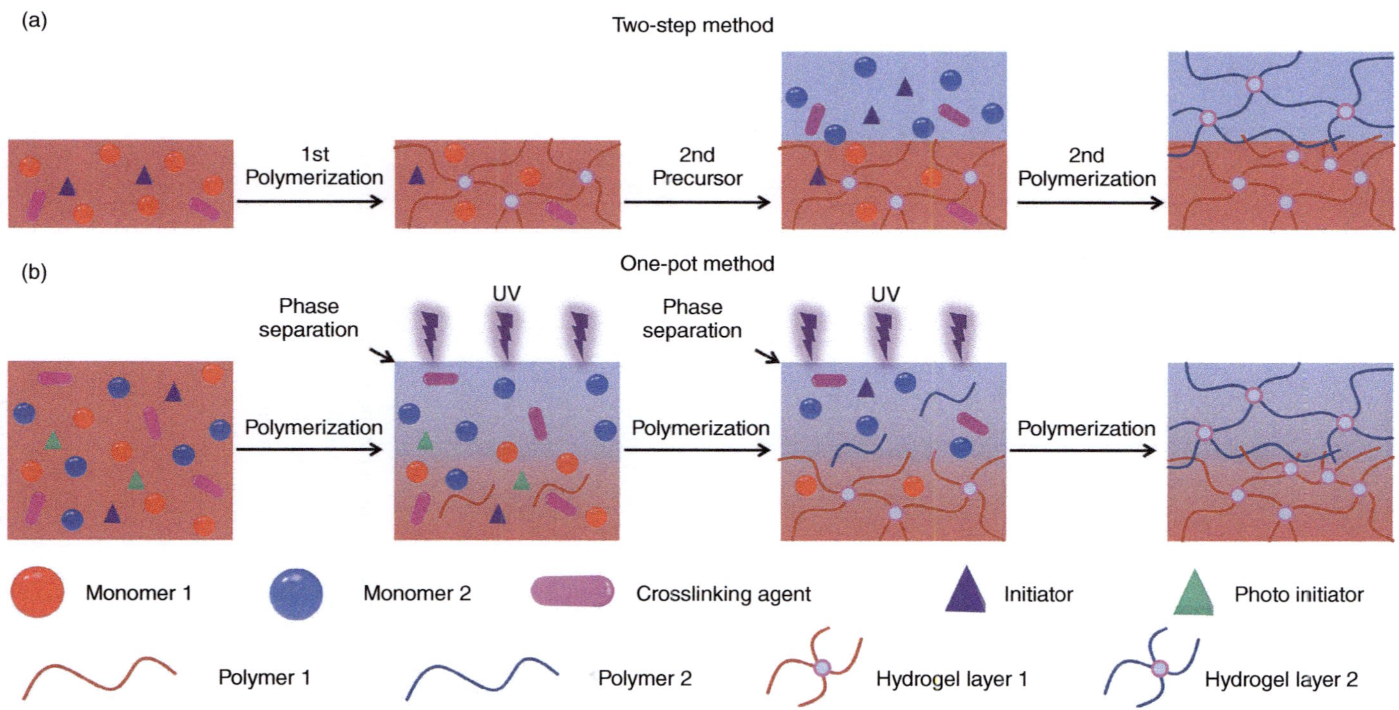

Figure 9.11 (a) Schematic illustration showing the fabrication mechanism of anisotropic hydrogels via two-step method. The bottom layer hydrogel was polymerized via thermal polymerization or photopolymerization first, and then the top layer hydrogel was polymerized on the top of the bottom layer hydrogel. (b) Schematic illustration showing the fabrication mechanism of hydrogel with anisotropic structure via one-pot method.

hydrogel. It is worth noting that only the special ratios of the two monomers could be used to prepare the bilayer hydrogel actuator in one pot. Compared to stepwise polymerization, it is more convenient and controllable to obtain a stable bilayer hydrogel by the one-pot method.

9.4.1.2 3D Printing

3D printing is an emerging technology for the additive manufacturing of hydrogels. Since the high precision and continued manufacturing, hydrogels with both 3D morphology and anisotropic structure can be printed [108, 109]. Therefore, 3D printing has been regarded as one of the most efficient technology for the programmable 4D shape deformation of hydrogel actuators. With the development of 3D printing technology, there are many types of 3D printing technology, such as DIW, stereolithography (SLA), and femtosecond laser, which have been successfully applied to fabricate hydrogel actuators with 3D shapes. For example, Lewis et al. introduced nanofibrillated cellulose (NFC) into hydrogel precursor and prepared viscoelastic ink [110]. When the ink was printed through the nozzle, the NFC would be stretched and oriented, which endowed the printed hydrogel silks with anisotropic swelling (Figure 9.12a). In addition, when the printed hydrogel silks were vertically arranged into bilayer structure, the hydrogel would deform from 2D sheet to 3D saddled shape. Besides, using the same way, an anisotropic flower with five petals was printed and generated 4D deformation with the petals gradually close when immersed in water (Figure 9.12b).

Combining with photopolymerization, 3D printing technology gradually evolved into SLA technology. Raquez et al. proposed a high-resolution SLA technology by Ember™ (Autodesk, Inc.) digital mask projection SLA printer [47]. As shown in Figure 9.12c, a photo-pattern UV light is projected through a transparent window first, then initiated by the photoinitiator, diphenyl(2,4,6-trimethylbenzoyl) phosphine oxide (TPO), to provide free radicals and initiate the local photopolymerization of hydrogel. After finishing the fabrication of bottom layer hydrogel, the pattern of UV light would change for the next photopolymerization cycle. Finally, a 3D shape hydrogel was obtained by layer-by-layer printing. During the fabrication process, the anisotropic swelling rates can be encoded by changing the dimensions, crosslink density, and chemical composition of every hydrogel layer (Figure 9.12d). On this basis, a hydrogel with 3D anisotropic structure could be obtained and exhibited multi-responsive and 4D deformation.

Recently, Zhang et el. further evolved the resolution of 3D printers and fabricated micro hydrogel actuators by femtosecond laser (Figure 9.12e) [111]. With the assistance of femtosecond laser direct writing, a microstructure flowered shape hydrogel with the size of 100 μm was printed, and the anisotropic structure was encoded into the petals (Figure 9.12f). Triggered by pH, the petals of micro hydrogel actuator would quickly close within 500 ms (Figure 9.12g). These works efficiently proved the advantages of 3D printers for the fabrication of hydrogel actuators, and 3D printing will surely promote the development of hydrogel actuators toward 4D bioinspired deformation.

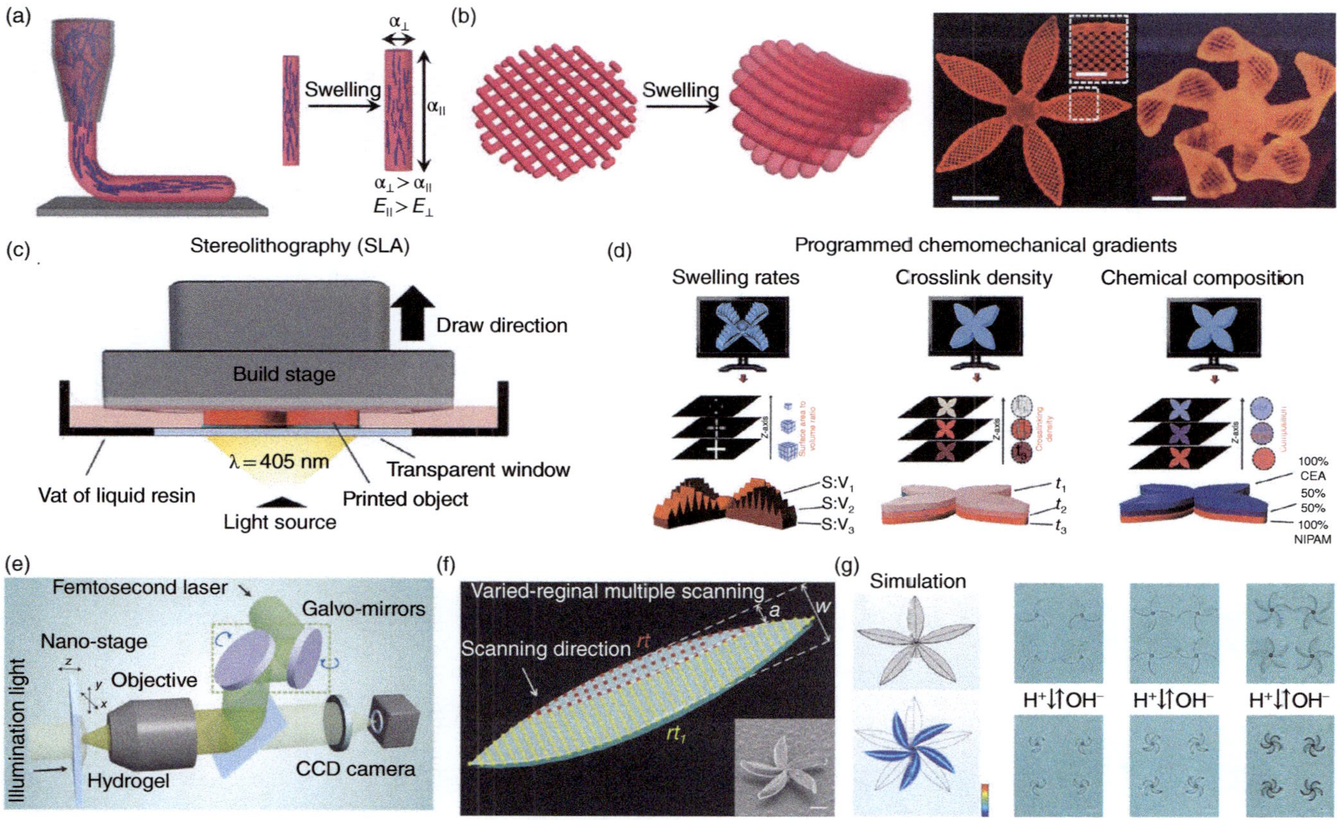

Figure 9.12 (a) Schematic illustration showing the printing mechanism of 3D printer via direct ink writing. (b) The complex deformation process of printed hydrogel actuator. Source: Reproduced with permission from Gladman et al. [110]/Springer Nature. (c) Schematic illustration showing the printing mechanism of 3D printer via stereolithography. (d) Schematic illustration showing the hydrogel fabrication process of printing programmed chemomechanical gradients. Source: Reproduced with permission from Odent et al. [47]/Royal Society of Chemistry. (e) Schematic illustration showing the printing mechanism of 3D printer via femtosecond laser. (f) Schematic illustration showing the anisotropic structure of printed hydrogel actuator. (g) Images showing the actuating process of micro hydrogel actuators. Source: Reproduced with permission from Hu et al. [111]/John Wiley & Sons.

9.4.1.3 Macromolecular Assembly

Similar to Lego, another innovative strategy to fabricate anisotropic hydrogel actuators is assembling two kinds of hydrogels with different swelling properties [112, 113]. Compared to other methods that introduce anisotropic structure during the fabrication process, macromolecular assembly is able to stably bond two hydrogel blocks after fabrication, which means the original statement of anisotropic hydrogel can be adjustable by assembling different hydrogel blocks with different swelling statements. As shown in Figure 9.13a, Chu et al. prepared homogeneous poly(*N*-isopropylacrylamide-*co*-acrylamide) (poly(NIPAM-*co*-AAm)) hydrogels crosslinked by clay nanosheets and named it as NC5[*x*] ("*x*" stands for the molar percentage of AAm) [114]. Then two NC5 hydrogel sheets were contacted and slowly dried to close the distance between neighboring polymer chains and clay nanosheets. After reswelling, plenty of new hydrogen-bonding interactions between polymer chains and clay nanosheets were rebuilt, which enhanced the bonding interaction of two hydrogel blocks and formed a stable bilayer hydrogel.

Similarly, Xie et al. prepared a responsive hydrogel (RH) that contains both carboxylic acid and b-CD groups, and non-responsive hydrogel (NGR) that only contains ferrocene groups (Figure 9.13b) [115]. When RH and NGR were closely contacted, a stable bilayer hydrogel was obtained via the host–guest interaction between b-CD and ferrocene in interface of two hydrogels. Due to the anisotropic structure, the bilayer hydrogel actuator was responsive to both pH and ionic strength and self-deformed from stripe to circle (Figure 9.13c).

In order to further enhance the bonding interaction, we recently introduced phenylboronic acid groups into thermo-responsive hydrogel (TRH) and ionic-responsive hydrogel (IRH) respectively [80]. Then using polyvinyl alcohol (PVA) as supra-molecular glue, the two hydrogels were assembled via dynamic boronic ester bond (Figure 9.13d). Benefiting from the different responsive properties of two hydrogel layers, the assembled hydrogel gripper exhibited thermo and ionic dual-responsive properties, and its gripping force could be adjusted by alternately triggering the thermo and ionic stimuli (Figure 9.13e). In addition, Zhou et al. assembled the hydrogels in multidimension from 1D to 3D by a similar strategy [116] and the hydrogels exhibited multidimensional deformations from 2D to 4D (Figure 9.13f). These works enrich the design and fabrication of hydrogel actuators and indicate the development of hydrogel actuators toward programmable deformation.

9.4.2 Innovative Technology

With the development of materials science, many efficient methods have been applied to prepare hydrogel actuators with fantastic abilities, such as fast response, multi-dimensional deformation, and biomimetic deformation. According to the discussion above, the anisotropic structures of traditional hydrogel actuators are permanent and deformation behaviors are unchangeable, which motivates us to temporarily encode the anisotropic structures of hydrogel actuators. Therefore, we proposed an innovative strategy to combine shape memory with actuating

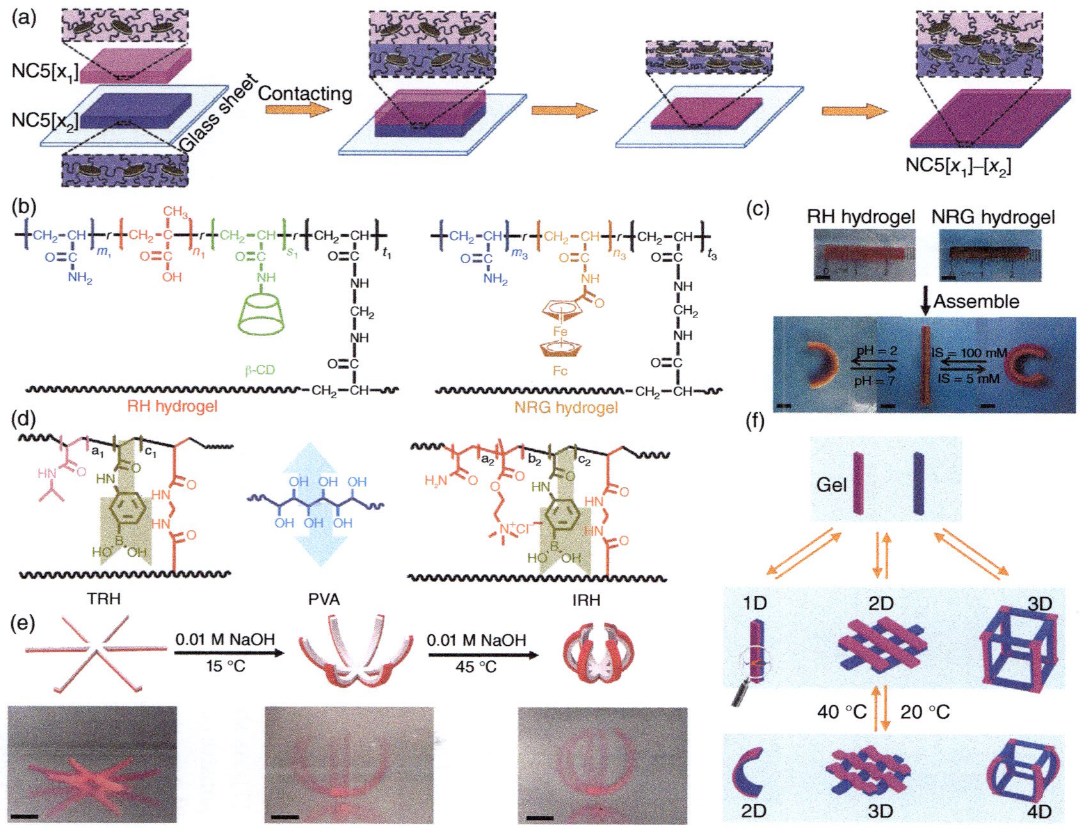

Figure 9.13 (a) Schematic illustration showing the modular assembly of two hydrogels via rearranged hydrogen bonding between hydrogel network and clay nanosheets. Source: Reproduced with permission from Yao et al. [114]. Copyright 2016 American Chemical Society. (b) Chemical structures of RH hydrogel and NRG hydrogel. (c) Images showing the modular assembly and multi responsive actuating process of RH hydrogel and NRG hydrogel. Source: Reproduced with permission from Ma et al. [115]/John Wiley & Sons, Inc. (d) Chemical structures of TRH hydrogel, IRH hydrogel, and supramolecular glue. (e) Schematic illustration and images showing the multi-responsive actuating process of a bilayer hydrogel actuator. (f) Schematic illustration showing the multidimensional modular assembly and multidimensional deformation. Source: Reproduced with permission from Sun et al. [116]. Copyright 2019 Royal Society of Chemistry.

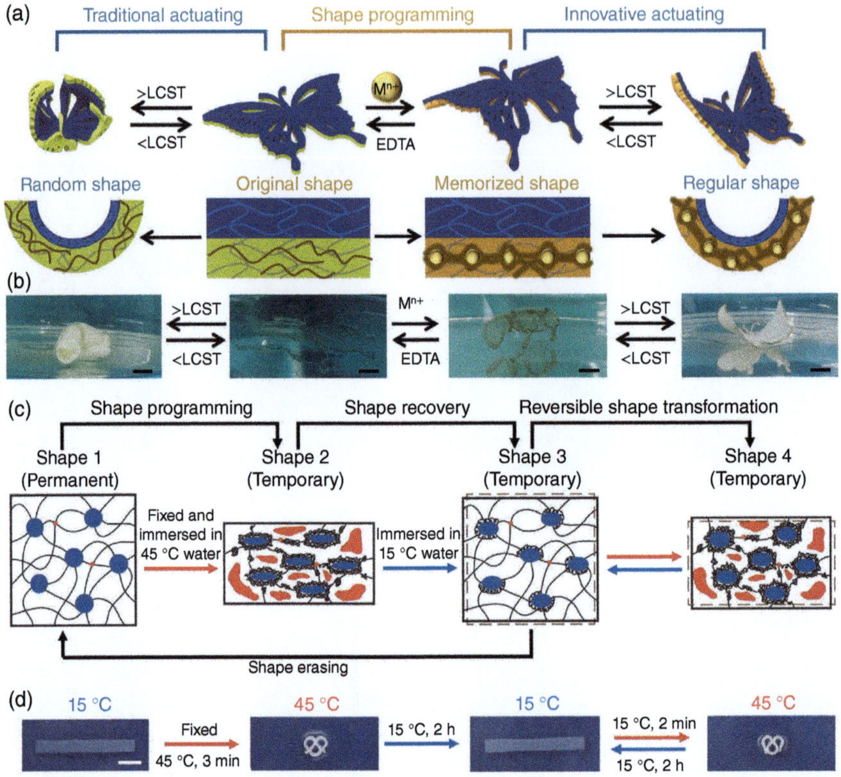

Figure 9.14 (a and b) Schematic illustration and images showing the traditional actuating process and programmable shape deformation by combining shape memory and actuating functions. Source: Reproduced with permission from Lu et al. [117]/John Wiley & Sons, Inc. (c and d) Schematic illustration and images showing the mechanism and actuating process of reversible shape transformation via transient structural anisotropy. Source: Reproduced with permission from Liu et al. [118]/John Wiley & Sons, Inc.

function [117]. As shown in Figure 9.14a,b, we fabricated a bilayer hydrogel with thermo-responsive actuating layer PNIPAm, and ionic-responsive shape memory layer Alg–PAAm, via stepwise polymerization. In traditional conditions, the 2D butterfly-shaped hydrogel would deform to random cluster because the anisotropic structure only exists in thickness direction, while the structure in the horizontal direction was isotropic. However, in the innovative actuating process, the body of the butterfly-shaped hydrogel was bent and the deformed shape was memorized via metal–ligand coordination between Alg chains and Ca^{2+}. Benefiting from the shape memory process, the structures of the hydrogel both in horizontal and vertical directions are anisotropic, which induced the directional deformation from bent shape to wings spread shape when immersed in hot water. Finally, using ethylene diamine tetraacetic acid (EDTA), the temporary shape of butterfly-shaped hydrogel could recover to the original shape and the temporary anisotropic structure could be erased with the shape recovery process.

Recently, Pan et al. proposed another efficient strategy to reprogram the deformation behavior of hydrogel actuators via temporary anisotropic structure [118]. First, the homogeneous hydrogel containing NIPAm and stearyl acrylate (SA) was prepared after polymerization in DMSO and solvent change in water (Figure 9.14c). Then the strip-shaped hydrogel was manually deformed to "Ω" shape and immersed in 45 °C water. Due to the thermo-responsive actuating property of PNIPAm and thermo-triggered crystal transition of poly(stearyl acrylate) (PSA), the temporary anisotropic structure was encoded into the hydrogel. After transforming the hydrogel back to 15 °C for two hours, the hydrogel would recover the original strip shape. But, when reheated the water to 45 °C, the hydrogel strip would self-deform to "Ω" shape, while the untreated hydrogel strip could only generate isotropic shrinking (Figure 9.14d). These works break the dilemma that the anisotropic structures of hydrogels are unchangeable, and will provide new ideas for the design of hydrogel actuators with programmable deformation.

9.5 Applications

Compared to the traditional mechanical devices, hydrogel actuators not only respond to various external stimuli but also provide programmable deformation. Therefore, many attempts are applied to explore the potential applications of hydrogel actuators. Based on the breakthrough in the design of anisotropic hydrogels, hydrogel actuators have gotten rid of traditional applications, such as grippers, and evolved into more fantastic applications, such as soft robots, artificial muscles, biomimetic devices, and information storage. In this section, we will discuss some design strategies and innovative applications of hydrogel actuators with programmable deformation.

9.5.1 Soft Robots

Hydrogel actuators have always been regarded ideal for soft robots, because they could transfer chemical energy to mechanical energy via shape deformation under external stimuli. In the early stage, the shape deformation is designed and utilized to catch or grip an object as gripper. To imitate more advanced biomimetic behaviors, such as walking and swimming, many structural hydrogels have been designed to transform the actuating behavior from *in situ* deformation to locomotion [88, 91]. For instance, Lee et al. prepared an electroactive hydrogel (EAH) that contains AAc by SLA [48]. According to Flory's theory and Donnan equilibrium, when the EAH was placed in the electric field, the carboxylic groups in the EAH network would be ionized and generated plenty of cations. Then the cations would be attracted to the cathode, which induced gradient osmotic pressure and caused EAH to bend toward the cathode (Figure 9.15a). With the assistance of micro-SLA, a human-shaped robot was obtained. When an electric field was applied, the soft robot would bend, which induced the forward movement of the legs and the backward movement of the center of gravity. In contrast, when the electric field was removed, the soft robot would

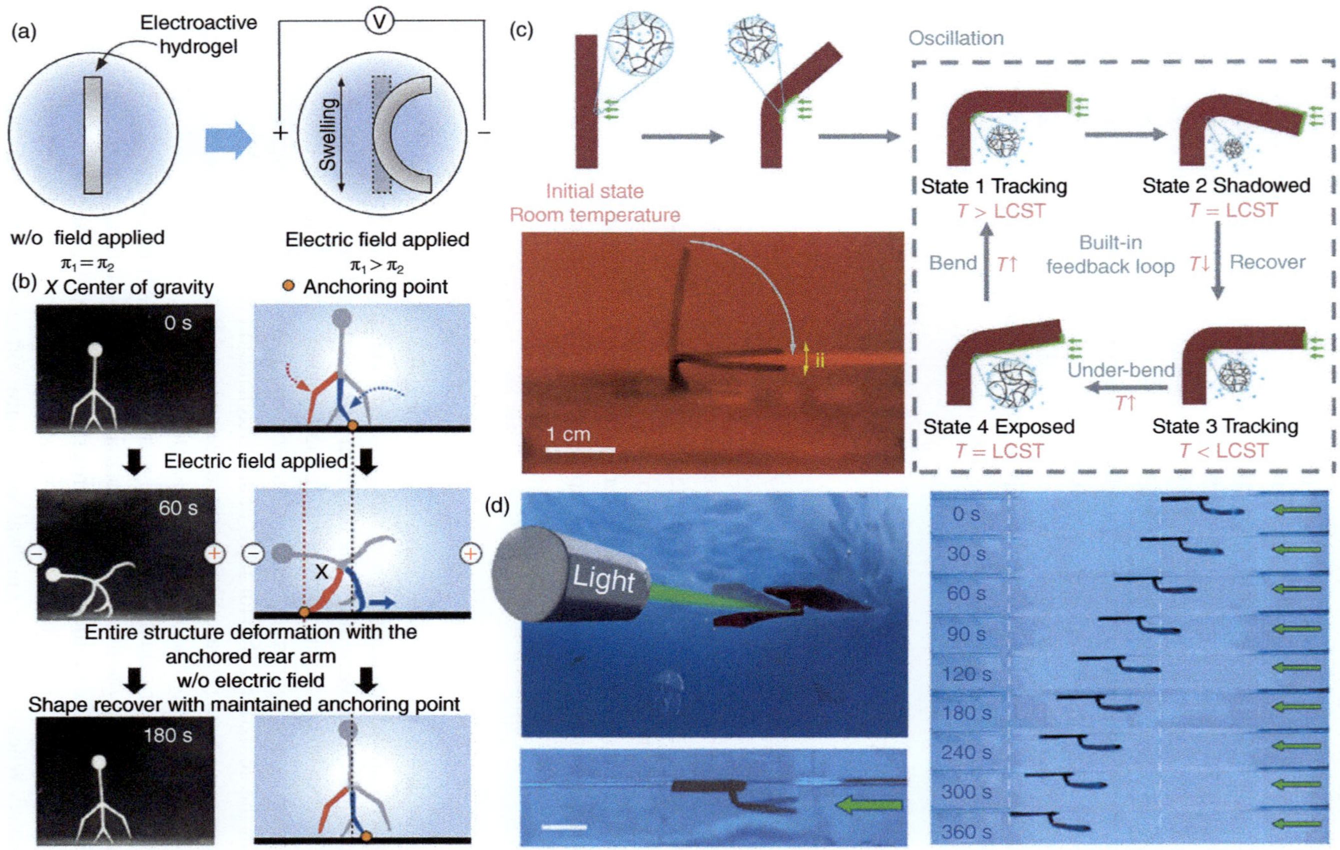

Figure 9.15 (a) Schematic illustration showing the deformation mechanism of electroactive hydrogel. (b) Schematic illustration and images showing the walking process of human-like hydrogel robot via electric field. Source: Reproduced with permission from Han et al. [48]/American Chemical Society. (c) Schematic illustration showing the oscillating mechanism of hydrogel actuator. (d) Images showing the swimming process of hydrogel oscillator. Source: Reproduced with permission from Zhao et al. [119]/American Association for the Advancement of Science - AAAS.

recover to the initial shape from bend shape, which induced the forward movement of the center of gravity and walking for a step (Figure 9.15b). Therefore, the soft robot could continuously walk under an alternating electric field.

Different from the structural design, recently, He et al. fabricated a hydrogel oscillator, which could swim in a continuous external stimulus [119]. As shown in Figure 9.15c, a PNIPAm hydrogel pillar contains gold nanoparticles (AuNPs) as photothermal additive. When a beam of UV light irradiated on the hydrogel pillar, the temperature of irradiated area would increase due to the photothermal effect of AuNPs, which induced the anisotropic deformation and the hydrogel pillar would over bend toward the direction of light source. Then the over bending hydrogel pillar would block the light source that would trigger the shape recovery process. During the shape recovery process, the hydrogel pillar would be exposed to the light source and began the above cycle. Therefore, based on the above mechanism, the hydrogel pillar could continuously oscillate under a stable light source. Furthermore, a hydrogel swimmer could be fabricated by connecting the above hydrogel oscillator to a hydrogel sheet, which was capable of swimming underwater via the oscillation of the hydrogel pillar (Figure 9.15d). These innovative works may inspire the fabrication of hydrogel actuators and the design of the actuating modes.

9.5.2 Artificial Muscles

Similar to muscles, hydrogel actuators are wet and soft, and can also export energy via deformation [120]. Therefore, utilizing hydrogel actuators as artificial muscles is an important application of hydrogel actuators [18]. Within the applications, hydrogel actuators could stretch, bend, or catch like muscles, but it is hard for hydrogel actuators to feedback on the actuating conditions. To combine the actuating and sensing in one hydrogel system, recently, we prepared an asymmetric CNTs-elastomer/hydrogel composite that consists of CNTs/Ecoflex as sensing layer and PNIPAm hydrogel as actuating layer (Figure 9.16a) [121]. In this system, when the composite actuator was exposed to NIR, it would be heated owing to the photothermal effect of CNTs. Due to the LCST property of PNIPAm hydrogel, the composite actuator would generate asymmetric deformation, which would stretch the sensing layer and induce the change of electric signal. According to the electric signal, computer could get the information on deformation conditions and decide the next instruction. As shown in Figure 9.16b, the composite actuator was applied as an artificial muscle by sticking the composite actuator on two PDMS tubes and keeping the value of electric signal at 0. Then the UV light is irradiated on the composite actuator. At first, the electric signal would decrease because of the photoelectric effect. Subsequently, with the actuating process, the electric signal would increase. After analyzing the electric signal, the UV light was turned off, and the electric signal would fall to 0 again with the shape recovery process.

Based on the design of structural hydrogels, such as tubular hydrogel, more types of artificial muscles have been fabricated. For instance, Zhou et al. used iron as template, and fabricated an asymmetric hydrogel tube containing thermo-responsive PNIPAm layer and non-responsive PAAc layer via surface catalytically initiated

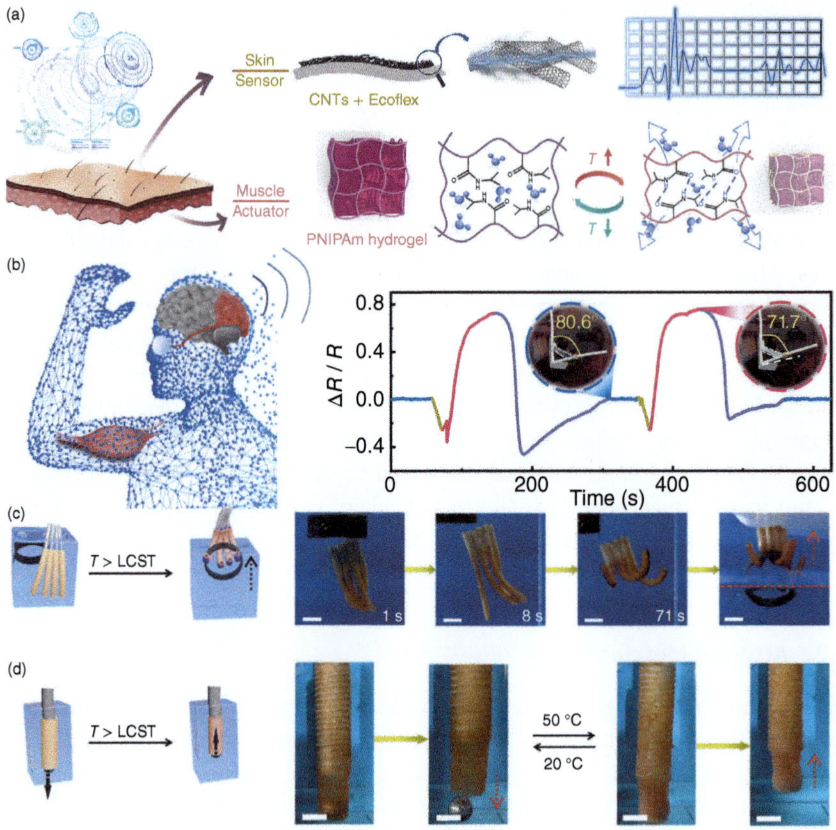

Figure 9.16 (a) Schematic illustration showing the structure of muscle-inspired hydrogel actuator with a hydrogel actuating layer and a sensing layer composed of CNTs and Ecoflex. (b) Schematic illustration showing the coupling of actuating function and self-sensing function. Source: Reproduced with permission from Li et al. [121]/Elsevier. (c) Schematic illustration and images showing lifting process of the four channels hydrogel. (d) Schematic illustration and images showing catching process of the single-channel hydrogel. Source: Reproduced with permission from Lin et al. [122]/American Chemical Society.

radical polymerization (SCIRP) [122, 123]. Benefiting from the hollow structure of the asymmetric hydrogel tube, hot water could flow through the hydrogel tube and trigger the thermo-responsive actuating process. Therefore, the asymmetric hydrogel tube could exhibit shape deformation in both water and air. Moreover, when four hydrogel tubes were arranged parallelly, the aligned hydrogel tubes were applied as artificial muscle and lifted an object from water to air via the asymmetric deformation (Figure 9.16c). Besides, the hydrogel tube could further imitate the function of muscle. As shown in Figure 9.16d, it is hard for a hydrogel gripper to catch a smooth ball because of the low friction force in ordinary conditions. However, thermo-responsive hydrogel tube could easily catch and lift the smooth ball because the diameter of the hydrogel tube would decrease, while the friction force increased due to the hydrophobicity of PNIPAm in the hot water.

9.5.3 Biomimetic Devices

By imitating the unmatched structure and biological behavior of living organisms, hydrogel actuators have been endowed with plenty of fantastic functions, such as tough, adhesion, and anti-freezing. Besides, some hydrogel actuators were further designed as devices for some application scenarios [20, 124]. For example, Liu et al. fabricated an organohydrogel containing hydrogel continuous phase of P(AAm-*co*-AAc) and microrganogel including polystearyl methacrylate (PSMA) via emulsion polymerization (Figure 9.17a) [125]. Inspired by the octopus sucker, which could stably catch an object with rough surface (Figure 9.17b), polypyrrole (PPy) was introduced into the network of organohydrogel, and the composite organohydrogel was stuck to a pneumatic gripper. When the novel pneumatic gripper was applied to catch an object with rough surface, the modulus of the composite organohydrogel would decrease due to the thermoelectric effect of PPy and thermo-responsive of PSMA, which improve the surface adaptability between the organohydrogel and the rough surface of the object. Thus, the object with rough surface could be firmly caught and lifted by the novel pneumatic gripper and turned the electronic switch off (Figure 9.17c).

During the application of hydrogel actuators, the poor environmental adaptability of hydrogel actuators has always hindered their development. For example, hydrogel actuators could not actuate in subzero environment because of the freeze of internal water. Recently, inspired by the anti-freezing behaviors found in nature, we fabricated a pH-responsive PAAm–PAAc bilayer hydrogel actuator with glycerol–water binary solvent [126]. Due to the strong hydrogen-bonding interactions between glycerol and water molecules, the bilayer hydrogel actuator could exist at subzero temperature without freezing as normal and exhibited asymmetric deformation in alkaline solution (Figure 9.17d). Subsequently, the hydrogel actuator with excellent anti-freezing properties was designed and applied as an intelligent valve. As shown in Figure 9.17e, the intelligent valve could work at subzero temperature, and the valve would open when the pH of the solution turned to alkaline and objects could pass through as a result. These works encourage the fabrication of biomimetic devices by designing multifunctional hydrogel actuators.

9.5.4 Information Storage Materials

With the development of technology and society, many new challenges and problems have broken in our life, which motivates the exploration of new applications of hydrogel actuators. For example, due to the growing problem of information leakage, how to find an efficient and reliable strategy to solve the problem is a tricky question. Benefiting from the unique deformation property and molecular designability, hydrogel actuators could also be applied to record and protect our information. Besides, when the deformation property cooperates with fluorescence behavior, the fluorescent hydrogel actuator could further exhibit multidimensional information encryption [127–129]. In our previous work, we copolymerized the fluorescence monomer, 1-pyrenylmethyl acrylate (PyMA), with AAc in DMSO solution and prepared p(PyMA-*co*-AAc) fluorescent hydrogel after solvent exchanging [49].

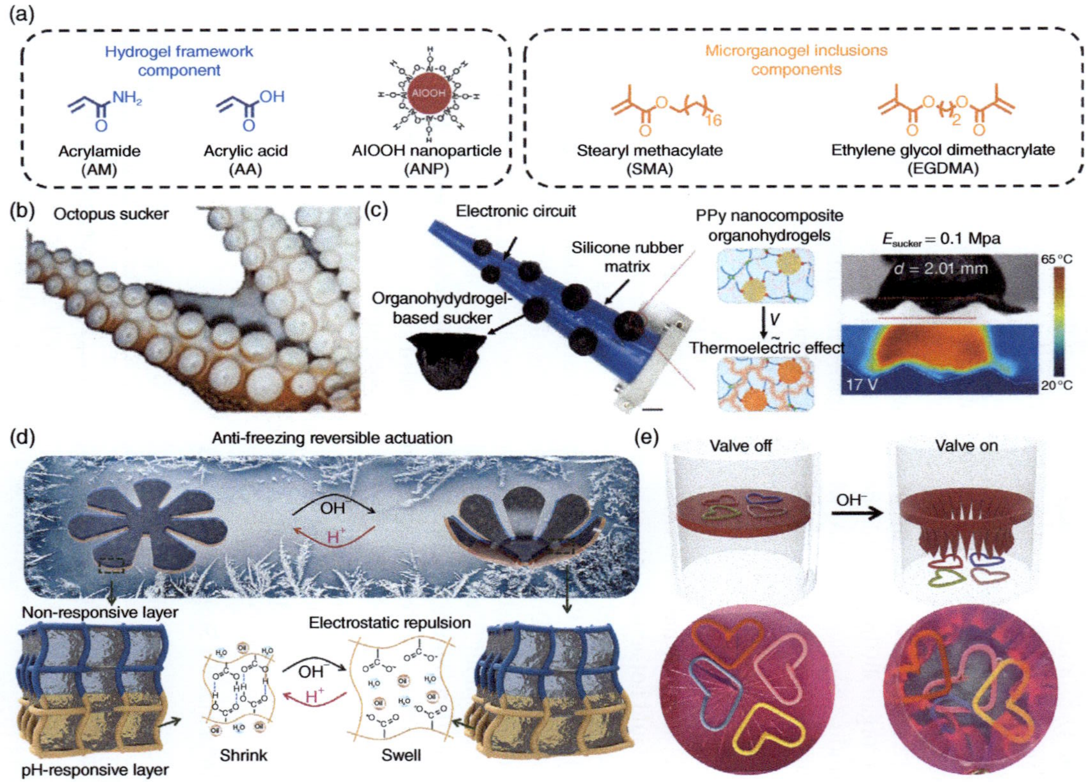

Figure 9.17 (a) Chemical structure of organohydrogel. (b) Image of the suckers of octopus. (c) Schematic illustration and images showing the catching process of biomimetic gripper. Source: Reproduced with permission from Zhuo et al. [125]/American Association for the Advancement of Science - AAAS. (d) Schematic illustration showing the deformation mechanism of anti-freezing hydrogel actuator. (e) Schematic illustration and images showing the working mechanism of artificial intelligence valve. Source: Reproduced with permission from Jian et al. [126]. Copyright 2019 Exclusive Licensee Science and Technology Review Publishing House.

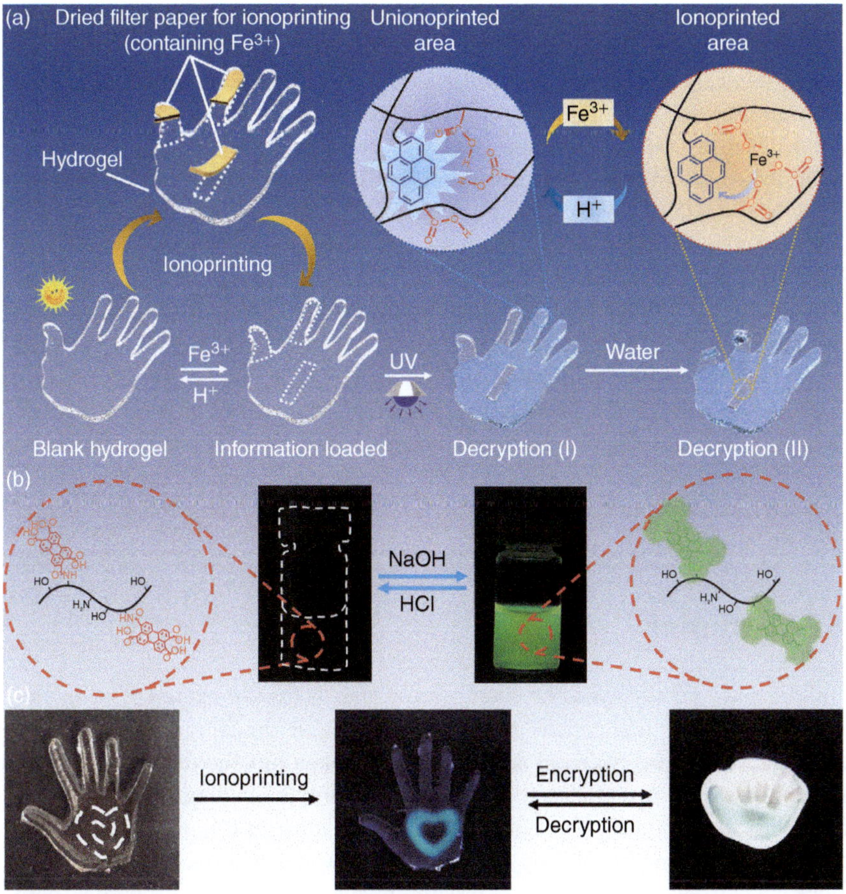

Figure 9.18 (a) Schematic illustration showing the ionoprinting controlled information encryption and deformation behavior. Source: Reproduced with permission from Le et al. [49]. Copyright 2018 Science China Press and Springer-Verlag GmbH Germany, part of Springer Nature. (b) Images showing the pH trigged fluorescence transition behavior. (c) Images showing the actuating and fluorescence induced multidimensional information encryption and decryption. Source: Reproduced with permission from Wu et al. [130]/John Wiley & Sons, Inc.

As shown in Figure 9.18a, Fe^{3+} was introduced into the hydrogel network by ionoprinting. On the one hand, the fluorescence of the hydrogel would be quenched because of the intramolecular charge transfer (ICT) between pyrene and Fe^{3+} and loaded the information (ionoprinting area). On the other hand, the Fe^{3+} would diffuse from the ionoprinting area to the hydrogel network and coordinated with the carboxyl groups of hydrogel network, which induced the gradient structure. Therefore, when the hand-shaped hydrogel was immersed in water, the fingers would bend and show the information of "OK," and the information of "I" can be readable under UV light. Finally, we can get the complete information: I am OK.

Besides, we further incorporated the actuating and fluorescence property [130]. At first, we synthesized a pH-responsive fluorescence molecule, perylene-tetracarboxylic-acid-modified gelatin (PTG), which could exhibit strong green fluorescence in alkaline solution while exhibiting non-fluorescence in acid solution (Figure 9.18b). Then the PTG was introduced into hydrogel network and fabricated the thermo-responsive PNIPAm–PAAm with gradient structure via photopolymerization. Finally, a heart-shaped filter paper with OH^- was ionoprinted on the central of the hand-shaped hydrogel and loaded the information. It is worth noting that the loaded information is first encrypted because it could only be readable under UV light. Besides, when the water was heated to the temperature above the LCST of PNIPAm, the fingers of the hand-shaped hydrogel actuator would bend and cover the loaded information. During this time, the loaded information is both unreadable in UV and natural light. The loaded information is second encrypted and could only be decrypted when the hydrogel actuator recovered to the initial state in cold water (Figure 9.18c). These works expand the application of hydrogel actuators and provide new designs for the combination of deformation and other functions.

9.6 Conclusion

In this review, we surveyed the literature related to the development of hydrogel actuators, from fabrications to applications. With the innovative design of anisotropic structures, hydrogel actuators have broken out the dilemma that the traditional hydrogel actuators could only exhibit isotropic swelling/shrinking properties. With the proposal of anisotropic structures, more and more breakthrough strategies have been successfully applied to endow the hydrogel actuators with various anisotropic structures and programmable deformation properties both during and after the fabrication process. Finally, inspired by nature, the novel hydrogel actuators have been applied as soft robots, artificial muscles, biomimetic devices, and information encryption materials. All of the discussions mentioned above have convincingly proven that hydrogel actuators are one of the most potential intelligent materials. Benefiting from the diverse molecular designable stage provided by hydrogel actuators, more emerging challenges and opportunities could be addressed and evolve the further development of hydrogel actuators in the near future research work.

First of all, to improve the deformability, including the velocity and degree of deformation, hydrogel actuators are usually designed to be thinner and softer, which causes the poor mechanical property of hydrogel actuators and hinders the development of corresponding applications. However, it is a tricky problem to trade off the modulus and deformability of hydrogel actuators, because the mechanical force exported by hydrogel actuator is too slight to actuate a tough hydrogel. Therefore, exploring and fabricating novel hydrogel actuators with powerful mechanical force exportability is necessary, and a tough hydrogel actuator with excellent deformability could be applied as a gripper to grasp heavier objects, which will greatly expand the applicative scope of hydrogel actuators and derived biomimetic devices.

Second, although there are diverse hydrogel actuators with programmable deformation, the deformation process is still uncontrollable. The existing hydrogel actuators have only two states, the original shape before actuating and the deformed shape after actuating. It means that it is hard to stop the deformation process and fix the temporary shape in real-time while the living organisms could. Therefore, incorporating the actuating property with shape memory property is necessary. Using shape memory to precisely control the deformation process will further improve the deformability and development of hydrogel actuators.

Third, the key point of the deformation mechanism of hydrogel actuators is the asymmetric swelling or shrinking, which means the hydrogel actuators could only actuate in water or water-rich environment. Although we have proposed the LCST–UCST bilayer hydrogel system, the deformability and cycles are still needed to be improved. Therefore, strategies that are more efficient need to be proposed to manage the water exchange within the actuating process.

At last but not least, incorporating the actuating property and sensing property, and constructing a complete closed-loop of actuating–sensing–actuating is a valuable challenge for the development of hydrogel actuators. In this closed-loop, the actuating condition of hydrogel actuators could be received by computer system via sensing, and the new instruction could also be fed back to hydrogel actuator and guide the next deformation. Therefore, the hydrogel actuators will become more controllable in the future.

Conflict of Interest

The authors declare no conflict of interest.

Acknowledgments

This work was supported by the National Natural Science Foundation of China (51873223, 52073295), Key Research Program of Frontier Science, Chinese Academy of Sciences (QYZDB-SSW-SLH036).

References

1 Fan, H.L. and Gong, J.P. (2020). *Macromolecules* 53: 2769.
2 Zhao, Z.G., Fang, R.C., Rong, Q.F., and Liu, M.J. (2017). *Adv. Mater.* 29: 1703045.
3 Chen, F., Xu, Z.Y., Wang, H.F. et al. (2020). *ACS Appl. Mater. Interfaces* 12: 55501.
4 Matsuda, T., Kawakami, R., Namba, R. et al. (2019). *Science* 363: 504.
5 Kwak, S.Y., Giraldo, J.P., Lew, T.T.S. et al. (2018). *Adv. Mater.* 30: 1804037.

6 Yu, K.H., Feng, Z.Z.R., Du, H.X. et al. (2021). *Proc. Natl. Acad. Sci. U.S.A.* 118: e2016524118.

7 Xue, L.L., Xiong, X.H., Krishnan, B.P. et al. (2020). *Nat. Commun.* 11: 963.

8 Ji, X.F., Li, Z., Hu, Y.B. et al. (2021). *CCS Chem.* 3: 1146.

9 Li, Z., Liu, P.C., Ji, X.F. et al. (2020). *Adv. Mater.* 32: 1906493.

10 Ko, H. and Javey, A. (2017). *Acc. Chem. Res.* 50: 691.

11 Wang, F.S., Chen, B.G., Wu, L. et al. (2020). *Cell Rep. Phys. Sci.* 1: 100011.

12 Yin, J.C., Zhang, D.L., Xu, Z.H. et al. (2020). *ACS Appl. Mater. Interfaces* 12: 49042.

13 Cui, H.Q., Zhao, Q.L., Zhang, L., and Du, X.M. (2020). *Adv. Intell. Syst.* 2: 2000138.

14 Peng, X. and Wang, H.L. (2018). *J. Polym. Sci., Part B: Polym. Phys.* 56: 1314.

15 Visentin, F., Babu, S.P.M., Meder, F., and Mazzolai, B. (2021). *Adv. Funct. Mater.* 31: 2101121.

16 Zhao, Q., Liang, Y.H., Ren, L. et al. (2018). *Nano Energy* 51: 621.

17 Matsubara, K., Tachibana, D., Matsuda, R. et al. (2020). *Adv. Intell. Syst.* 2: 2000008.

18 Park, N. and Kim, J. (2020). *Adv. Intell. Syst.* 2: 1900135.

19 Lo, C.Y., Zhao, Y., Kim, C. et al. (2021). *Mater. Today* 5: 8.

20 Zhao, Y., Lo, C.Y., Ruan, L. et al. (2021). *Sci. Robot.* 6: eabd5483.

21 Zhang, D.C., Zhang, J.W., Jian, Y.K. et al. (2020). *Adv. Intell. Syst.* 3: 2000208.

22 Kim, D., Kim, H., Lee, E. et al. (2016). *Chem. Mater.* 28: 8807.

23 Zhao, Q., Liang, Y.H., Ren, L. et al. (2018). *J. Mater. Chem. B* 6: 1260.

24 Dai, C.F., Zhang, X.N., Du, C. et al. (2020). *ACS Appl. Mater. Interfaces* 12: 53376.

25 Xu, Z.X. and Fu, J. (2020). *ACS Appl. Mater. Interfaces* 12: 26476.

26 Shi, Z., Peng, P., Strohecker, D., and Liao, Y. (2011). *J. Am. Chem. Soc.* 133: 14699.

27 He, X.M., Sun, Y., Wu, J.H. et al. (2019). *J. Mater. Chem. C* 7: 4970.

28 Zhou, S.Z., Wu, B., Zhou, Q. et al. (2020). *Macromol. Rapid Commun.* 41: 1900543.

29 Yang, Y., Tan, Y., Wang, X.L. et al. (2018). *ACS Appl. Mater. Interfaces* 10: 7688.

30 Sun, Z.F., Yamauchi, Y., Araoka, F. et al. (2018). *Angew. Chem. Int. Ed.* 57: 15772.

31 Cui, H.L., Pan, N., Fan, W.X. et al. (2019). *Adv. Funct. Mater.* 29: 1807692.

32 Luo, R.C., Wu, J., Dinh, N.D., and Chen, C.H. (2015). *Adv. Funct. Mater.* 25: 7272.

33 Wang, Z.J., Hong, W., Wu, Z., and Zheng, Q. (2017). *Angew. Chem. Int. Ed.* 56: 15974.

34 Hao, X.P., Li, C.Y., Zhang, C.W. et al. (2021). *Adv. Funct. Mater.* 31: 2105481.

35 Downs, F.G., Lunn, D.J., Booth, M.J. et al. (2020). *Nat. Chem.* 12: 363.

36 Le, X.X., Lu, W., Zhang, J.W., and Chen, T. (2019). *Adv. Sci.* 6: 1801584.

37 Ionov, L. (2014). *Mater. Today* 17: 494.

38 Shang, J.J., Le, X.X., Zhang, J.W. et al. (2019). *Polym. Chem.* 10: 1036.

39 Zhao, D.H., Liu, Y.D., Liu, B.H. et al. (2021). *ACS Appl. Mater. Interfaces* 13: 13714.
40 Chen, Z., Zhao, D.H., Liu, B.H. et al. (2019). *Adv. Funct. Mater.* 29: 1900971.
41 Hao, X.P., Xu, Z., Li, C.Y. et al. (2020). *Adv. Mater.* 32: 2000781.
42 Li, C.Y., Hao, X.P., Zheng, S.Y. et al. (2019). *Adv. Intell. Syst.* 1: 1900055.
43 Le, X.X., Shang, H., Yan, H.Z. et al. (2021). *Angew. Chem. Int. Ed.* 60: 3640.
44 Qiu, H.Y., Wei, S.X., Liu, H. et al. (2021). *Adv. Intell. Syst.* 3: 2000239.
45 Zhu, Q.L., Du, C., Dai, Y.H. et al. (2020). *Nat. Commun.* 11: 5166.
46 Wen, X.L., Sun, S.T., and Wu, P.Y. (2020). *Mater. Horiz.* 7: 2150.
47 Odent, J., Vanderstappen, S., Toncheva, A. et al. (2019). *J. Mater. Chem. A* 7: 15395.
48 Han, D., Farino, C., Yang, C. et al. (2018). *ACS Appl. Mater. Interfaces* 10: 17512.
49 Le, X.X., Lu, W., He, J. et al. (2018). *Sci. China Mater.* 62: 831.
50 Deng, H., Zhang, C., Sattari, K. et al. (2021). *ACS Appl. Mater. Interfaces* 13: 12719.
51 Song, M.M., Wang, Y.M., Wang, B. et al. (2018). *ACS Appl. Mater. Interfaces* 10: 15021.
52 Luo, X.X., Zhu, L.P., Wang, Y.C. et al. (2021). *Adv. Funct. Mater.* 31: 2104928.
53 Thérien-Aubin, H., Wu, Z.L., Nie, Z.H., and Kumacheva, E. (2013). *J. Am. Chem. Soc.* 135: 4834.
54 Wei, S.X., Li, Z., Lu, W. et al. (2020). *Angew. Chem. Int. Ed.* 59: 2.
55 Kim, Y., Yuk, H., Zhao, R. et al. (2018). *Nature* 558: 274.
56 Tognato, R., Armiento, A.R., Bonfrate, V. et al. (2019). *Adv. Funct. Mater.* 29: 1804647.
57 Tang, J.D., Yin, Q.F., Qiao, Y.C., and Wang, T.J. (2019). *ACS Appl. Mater. Interfaces* 11: 21194.
58 Siefert, E., Reyssat, E., Bico, J., and Roman, B. (2019). *Nat. Mater.* 18: 24.
59 Mirvakili, S.M., Sim, D., Hunter, I.W., and Langer, R. (2020). *Sci. Robot.* 5: eaaz4239.
60 Zhang, Q., Kuang, X., Weng, S.Y. et al. (2021). *Adv. Funct. Mater.* 31: 2010872.
61 Liang, S.M., Tu, Y.Q., Chen, Q. et al. (2019). *Mater. Horiz.* 6: 2135.
62 He, M., Zhao, Y.T., Duan, J.J. et al. (2014). *ACS Appl. Mater. Interfaces* 6: 1872.
63 Wu, B.Y., Jian, Y.K., Le, X.X. et al. (2019). *ACS Appl. Mater. Interfaces* 11: 48564.
64 Yuk, H., Lin, S.T., Ma, C. et al. (2017). *Nat. Commun.* 8: 14230.
65 Koetting, M.C., Peters, J.T., Steichen, S.D.S., and Peppas, N.A. (2015). *Mater. Sci. Eng., R* 93: 1.
66 Sun, Y.J., Chen, L.F., Jiang, Y. et al. (2019). *Mater. Horiz.* 6: 160.
67 Hubbard, A.M., Cui, W., Huang, Y. et al. (2019). *Matter* 1: 674.
68 Zheng, W.J., An, N., Yang, J.H. et al. (2015). *ACS Appl. Mater. Interfaces* 7: 1758.
69 Gao, G.R., Wang, L.F., Cong, Y. et al. (2018). *ACS Omega* 3: 17914.
70 Wang, E., Desai, M.S., and Lee, S.W. (2013). *Nano Lett.* 13: 2826.
71 Zhao, C.Z., Ma, Z.Y., and Zhu, X.X. (2019). *Prog. Polym. Sci.* 90: 269.
72 Zhao, C.Z., Dolmans, L., and Zhu, X.X. (2019). *Macromolecules* 52: 4441.

73 Dai, L., Ma, M.S., Xu, J.K. et al. (2020). *Chem. Mater.* 32: 4324.

74 Marcombe, R., Cai, S.Q., Hong, W. et al. (2010). *Soft Matter* 6: 784.

75 Yu, Q., Yang, X.J., Chen, Y. et al. (2018). *Angew. Chem. Int. Ed.* 57: 10192.

76 Li, C., Iscen, A., Palmer, L.C. et al. (2020). *J. Am. Chem. Soc.* 142: 8447.

77 Pan, D., Wu, D., Li, P.J. et al. (2021). *Adv. Funct. Mater.* 31: 2009386.

78 Geng, H.Y., Zhou, K., Zhou, J.J. et al. (2018). *Angew. Chem. Int. Ed.* 57: 15435.

79 Xue, P., Bisoyi, H.K., Chen, Y.H. et al. (2021). *Angew. Chem. Int. Ed.* 60: 3390.

80 Wu, B.Y., Xu, Y.W., Le, X.X. et al. (2019). *Acta Polym. Sin.* 50: 496.

81 Xiao, S.W., Yang, Y., Zhong, M.Q. et al. (2017). *ACS Appl. Mater. Interfaces* 9: 20843.

82 Li, X.H., Tang, C.J., Liu, D. et al. (2021). *Adv. Mater.* 33: 2102479.

83 Le, X.X., Lu, W., Xiao, H. et al. (2017). *ACS Appl. Mater. Interfaces* 9: 9038.

84 Gong, X.L., Xiao, Y.Y., Pan, M. et al. (2016). *ACS Appl. Mater. Interfaces* 8: 27432.

85 Yan, X.Z., Xu, D.H., Chi, X.D. et al. (2012). *Adv. Mater.* 24: 362.

86 Hua, L.Q., Xie, M.Q., Jian, Y.K. et al. (2019). *ACS Appl. Mater. Interfaces* 11: 43641.

87 Ma, C.X., Le, X.X., Tang, X.L. et al. (2016). *Adv. Funct. Mater.* 26: 8670.

88 Li, C., Xue, Y., Han, M. et al. (2021). *Matter* 4: 1.

89 Zheng, J., Xiao, P., Le, X.X. et al. (2018). *J. Mater. Chem. C* 6: 1320.

90 Zhu, Q.L., Dai, C.F., Wagner, D. et al. (2020). *Adv. Mater.* 32: 2005567.

91 Kim, Y.S., Liu, M.J., Ishida, Y. et al. (2015). *Nat. Mater.* 14: 1002.

92 Zhou, Y., Duque, C.M., Santangelo, C.D., and Hayward, R.C. (2019). *Adv. Funct. Mater.* 29: 1905273.

93 Zhu, C.N., Li, C.Y., Wang, H. et al. (2021). *Adv. Mater.* 33: 2008057.

94 Tan, Y., Wang, D., Xu, H.X. et al. (2018). *ACS Appl. Mater. Interfaces* 10: 40125.

95 Le, X.X., Zhang, Y.C., Lu, W. et al. (2018). *Macromol. Rapid Commun.* 39: 1800019.

96 Li, T.Z., Wang, J.H., Zhang, L.Y. et al. (2017). *J. Mater. Chem. B* 5: 5726.

97 Wang, Z.J., Zhu, C.N., Hong, W. et al. (2017). *Sci. Adv.* 3: e1700348.

98 Nojoomi, A., Arslan, H., Lee, K., and Yum, K. (2018). *Nat. Commun.* 9: 3705.

99 Nojoomi, A., Jeon, J., and Yum, K. (2021). *Nat. Commun.* 12: 603.

100 Peng, X., Liu, T.Q., Zhang, Q. et al. (2017). *Adv. Funct. Mater.* 27: 1701962.

101 Zheng, S.Y., Shen, Y.Y., Zhu, F.B. et al. (2018). *Adv. Funct. Mater.* 28: 1803366.

102 Arslan, H., Nojoomi, A., Jeon, J., and Yum, K. (2019). *Adv. Sci.* 6: 1800703.

103 Cheng, Y., Chan, K.H., Wang, X.Q. et al. (2019). *ACS Nano* 13: 13176.

104 Gao, G.R., Wang, Z.W., Xu, D. et al. (2018). *ACS Appl. Mater. Interfaces* 10: 41724.

105 Wang, L., Jian, Y.K., Le, X.X. et al. (2018). *Chem. Commun.* 54: 1229.

106 Li, J., Ma, Q.Y., Xu, Y. et al. (2020). *ACS Appl. Mater. Interfaces* 12: 55290.

107 Xiao, S.W., Zhang, M.Z., He, X.M. et al. (2018). *ACS Appl. Mater. Interfaces* 10: 21642.

108 Hong, S., Sycks, D., Chan, H.F. et al. (2015). *Adv. Mater.* 27: 4035.

109 Chen, S., Huang, T., Zuo, H. et al. (2018). *Adv. Funct. Mater.* 28: 1805108.

110 Gladman, A.S., Matsumoto, E.A., Nuzzo, R.G. et al. (2016). *Nat. Mater.* 15: 413.

111 Hu, Y.L., Wang, Z.Y., Jin, D.D. et al. (2019). *Adv. Funct. Mater.* 30: 1907377.

112 Yang, H.S., Li, Z.I., Sun, G.Q. et al. (2019). *Adv. Funct. Mater.* 29: 1901917.

113 Zhang, Q., Sun, Y.Z., He, C.Z. et al. (2020). *Adv. Sci.* 7: 2002025.

114 Yao, C., Liu, Z., Yang, C. et al. (2016). *ACS Appl. Mater. Interfaces* 8: 21721.

115 Ma, C.X., Li, T.F., Zhao, Q. et al. (2014). *Adv. Mater.* 26: 5665.

116 Sun, F., Yang, C.X., Xu, W.Y. et al. (2019). *J. Mater. Chem. B* 7: 1996.

117 Lu, H.H., Wu, B.Y., Yang, X.X. et al. (2020). *Small* 16: 2005461.

118 Liu, K.K., Zhang, Y., Cao, H.Q. et al. (2020). *Adv. Mater.* 32: 2001693.

119 Zhao, Y., Xuan, C., Qian, X.S. et al. (2019). *Sci. Robot.* 4: eaax7112.

120 Ma, Y.F., Hua, M.T., Wu, S.W. et al. (2020). *Sci. Adv.* 6: eabd2520.

121 Li, H.J., Liang, Y., Gao, G.R. et al. (2021). *Chem. Eng. J.* 415: 128988.

122 Lin, H., Ma, S., Yu, B. et al. (2019). *Chem. Mater.* 31: 4469.

123 Ma, S.H., Yan, C.Y., Cai, M.R. et al. (2018). *Adv. Mater.* 30: 1803371.

124 Zhao, Z.G., Li, C.X., Dong, Z.C. et al. (2019). *Adv. Funct. Mater.* 29: 1807858.

125 Zhuo, S.Y., Zhao, Z.G., Xie, Z.X. et al. (2020). *Sci. Adv.* 6: eaax1464.

126 Jian, Y.K., Wu, B.Y., Le, X.X. et al. (2019). *Research* 2019: 2384347.

127 Li, P., Zhang, D., Zhang, Y.C. et al. (2019). *ACS Macro Lett.* 8: 937.

128 Zhang, Y.C., Le, X.X., Jian, Y.K. et al. (2019). *Adv. Funct. Mater.* 29: 1905514.

129 Zhu, C.N., Bai, T.W., Wang, H. et al. (2021). *Adv. Mater.* 33: 2102023.

130 Wu, B.Y., Le, X.X., Jian, Y.K. et al. (2019). *Macromol. Rapid Commun.* 40: 1800648.

10

Hydrogels-Based Electronic Devices for Biosensing Applications

Quanduo Liang[1,2], Yuyuan Lu[3], and Qiang Zhang[1,2]

[1] *Changchun Institute of Applied Chemistry, Chinese Academy of Sciences, State Key Laboratory of Electroanalytical Chemistry, 5625 Renmin Street, Changchun, 130022, P. R. China*
[2] *University of Science and Technology of China, School of Applied Chemistry and Engineering, 96 Jinzhai Road, Hefei, 230026, P. R. China*
[3] *Changchun Institute of Applied Chemistry, Chinese Academic of Sciences, State Key Laboratory of Polymer Chemistry and Physics, 5625 Renmin Street, Changchun, 130022, P.R. China*

10.1 Introduction

Since the first synthetic hydrogel was produced 60 years ago, hydrogels have been used as one of the fundamental materials in prosthesis, tissue engineering, drug delivery, wound dressing, and contact lens. [1–5]. Hydrogels are comprised of water-rich three-dimension (3D) networks of hydrophilic polymers, which are crosslinked by covalent bonds, physical entanglement, electrostatic interactions, and molecular recognition [6]. In recent years, much effort has been devoted to exploring hydrogel electrodes as flexible wearable sensors and implantable devices [7–10]. Flexible wearable sensors show huge market prospects in the fields of medical diagnosis, health monitoring, soft robots, artificial intelligence, energy storage, and electronic skin [11–13]. Compared with traditional sensors with high Young's modulus, flexible wearable sensors show many unique advantages, such as high sensitivity, good flexibility, excellent stability, fast response, portability, and comfortable use. The flexible wearable sensors can fit on the human body, interact with an external computer or host, and convert human body or environmental stimuli into electrical signals, which can continuously, timely, and accurately monitor personal health parameters and environmental changes. As shown in Figure 10.1, flexible wearable sensors can be roughly divided into three types – physical sensors, chemical sensors, and electrophysiological sensors [22].

Traditional wearable sensors were prepared using conductive metal materials, which suffer from a mismatch with the human skin. That prevents the electric signal transfer from skins to sensors, which reduces the detection sensitivity. In addition, metal materials are not suitable for long-term implantation in the body due to inferior tissue compatibility [23]. To meet the requirements for flexible electronic devices, the use of flexible substrates to encapsulate conductive materials

Smart Stimuli-Responsive Polymers, Films, and Gels, First Edition.
Edited by Liang Hu, Yongfeng Gao, and Michael J. Serpe.

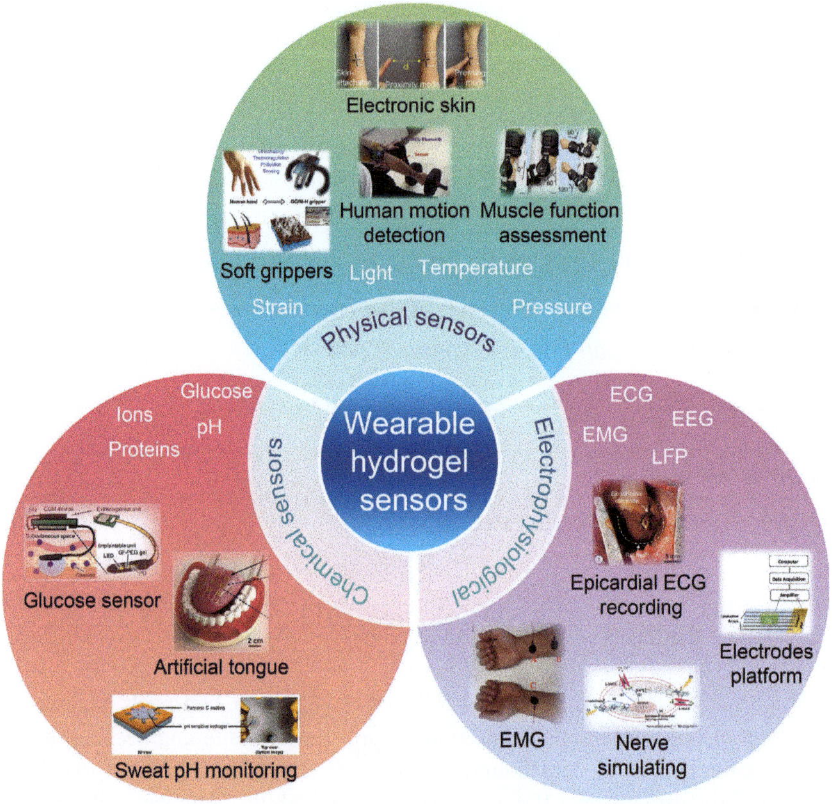

Figure 10.1 Types of wearable hydrogel-based sensors. The representative target analytes and applications are listed. Source: Reproduced from Jing et al. [12]/Royal Society of Chemistry. Reproduced from Cai et al. [13]/American Association for the Advancement of Science – AAAS/Commons Attribution NonCommercial License 4.0 (CC BY-NC). Reproduced from Wang et al. [14]/American Chemical Society. Reproduced from Hang et al. [8]/Elsevier. Reproduced from Sawayama and Takeuchi [15]/John Wiley & Sons, Inc. Reproduced from Yeom et al. [16]/American Association for the Advancement of Science – AAAS/Commons Attribution NonCommercial License 4.0 (CC BY-NC). Reproduced from Scarpa et al. [17]/Springer Nature/CC BY 4.0. Reproduced from Deng et al. [18]/Springer Nature. Reproduced from Li et al. [19]/American Chemical Society. Reproduced from Wu et al. [20]/American Chemical Society. Reproduced from Wang et al. [21]/John Wiley & Sons, Inc.

has once become a research hotspot. Flexible substrates need to have the characteristics of lightness, transparency, flexibility, insulation, and corrosion resistance. Polyurethane (PU), polymethylmethacrylate (PMMA), polydimethylsiloxane (PDMS), and other materials have been widely studied as candidate materials for flexible substrates [24–26]. However, these materials still have high modulus and low stretchability, which limits their use scenarios. To improve the performance, the precious metals commonly used in electrodes and nanofillers make costs rise sharply. In addition, the stability of the combination of the flexible substrates and the rigid conductive materials is also an urgent problem to be solved. Hydrogels have a 3D network structure similar to biological tissues and exhibit rich structural adjustability. Therefore, the hydrogels can easily obtain mechanical properties similar to tissues, skin, etc., which leads to better contact between the sensors and skins/tissues. The better contact will reduce interface resistance and amplify the targeted signals. For example, Jing et al. fabricated dopamine-talc-polyacrylamide (DTPAM) hydrogels as strain sensors [27]. The talc particles had a reinforcing effect, while the dopamine possessed a lubricating effect. The synergistic effect of dopamine and talc facilitates adjusting the mechanical properties of DTPAM hydrogels to obtain low modulus, high strength, and high stretchability similar to human skins. The DTPAM hydrogel showed a sensitivity with a gauge factor (GF) of 0.693 at 1000% strain. Xia et al. prepared a hybrid hydrogel composed of polyacrylamide (PAAm)/chitosan (CS)/multi-wall carbon nanotubes (MWCNTs), which contains hydrophobic associations, electrostatic interaction, and hydrogen bonds. The synergistic dynamic crosslinking formed by them could be easily broken and rebonded. That leads to a modulus of only 6–10 kPa and excellent stretchability. Also, MWCNTs as a conductive pathway lead to the fact that the conductivity decreases with strain increase, which is beneficial to improve the sensitivity of the sensor. The wearable sensor prepared from the hydrogel has a GF of 1.6 with a response time of only 180 ms [28].

When hydrogel-based electrodes were applied as implantable sensors, they also exhibited distinct advantages over metal probes. The modulus mismatch between tissue and the metal electrode will cause relative micromotions and shear motions during long-term applications, which will damage glial networks and the blood–brain barrier [29]. These motions will also reduce signal transfer across the interfaces between electrodes and tissues and affect the long-term application stability. In addition, the mechanical mismatch can exacerbate the foreign body reaction and cause the implanted devices to be encapsulated. For the central nervous system, this may lead to the loss of electrode functions used to treat diseases [30]. In the comparison of metal electrodes, the hydrogel-based electrodes have a similar stiffness to the living tissue. Thus, it is capable of conformally contacting the brain/tissue surfaces with sufficient adhesion. As such, the micromotion will be significantly suppressed and the long-term stability will be greatly enhanced. Nam et al. fabricated a conductive neural interface using a supramolecular βpeptide-based hydrogel. The hydrogel has a storage modulus and a loss modulus of ~1500 Pa, which matches the elastic modulus of neural tissues (100–1500 Pa). Furthermore, the viscoelastic nature of the designed hydrogel enabled seamless biointegration with surrounding neural tissue. Therefore, the neural interface not only exhibits excellent biocompatibility and biological stability but also achieves the amplification of neural signals with minimum signal dissipation [31].

This chapter mainly discusses the application of hydrogels in bioelectronics, including flexible wearable devices, *in vivo* electrochemical sensors, and tissue–machine interfaces. The preparation and sensing mechanism of these sensors will be introduced with recently reported examples. The goal of this chapter is to familiarize readers with the cutting-edge scientific theories and technologies as well as to encourage more scientists to push the research in this area forward.

10.2 Flexible Hydrogel-Based Sensors

The birth of flexible bioelectronics can be traced back to the invention of the pacemaker in 1957, which is an impressive application of electronic technology in biology [32]. Wearable bioelectronics has attracted people's attention since the beginning of the twenty-first century, thanks to the rapid development of computer technology, electronic communication technology, and materials science. On one hand, the development of computer technology can realize the miniaturization and portability of electronic devices. On the other hand, it enables human to recognize and process complicated physiological signals and obtain more valuable information. The development of electronic communication technology has undoubtedly a huge impact on bioelectronics. The birth of wireless transmission enables wearable devices to bid farewell to cumbersome wires and move toward a more portable direction. Real-time recording and transmission of signals obtained by bioelectronic devices to a computer network through a wireless network enable users to monitor the human body's movement or health status more intuitively and help them make corresponding countermeasures. The emergence of wireless charging technology will help implantable devices to obtain a longer service life, avoid repeated implantation operations, and help to alleviate users' pain and protect users' privacy. The emergence of new materials plays a vital role in the development of flexible electronic devices. Initially, the conductors widely used in flexible electronic devices were metal materials, such as copper, silver, gold, platinum, and aluminum. Metals generally have good electrical conductivity and ductility but have poor compatibility with organisms, poor adhesion to organic substrate materials, high prices, and poor weather resistance. In recent years, a large number of new materials used in flexible electronic devices have emerged, including liquid metals, polymer elastomers, and polymer nanocomposites, and each has its advantages [33]. Nevertheless, hydrogels are still the most powerful competitor for flexible electronic devices.

Most biological hydrogels are mechanically flexible but robust and durable and can adapt to transportation (e.g. convection and diffusion) and reactions of various life-endowing basic substances. These living substances have delicate functions, such as sensing, responding, self-healing, and self-enhancement [34]. The structural diversity and biomimetic characteristics of synthetic hydrogels have made them the best choice for biomedical materials since their inception. After the emergence of conductive hydrogels, researchers have conducted extensive research on hydrogel electronic devices, including energy-storage devices, human-motion detection, health monitoring, artificial intelligence, and soft robots [35]. Among them, flexible wearable sensors for artificial intelligence and health monitoring, with their high sensitivity, fast response, portability, and ease of use, have become a hot spot for scientific researchers. The huge market potential has stimulated both domestic and foreign researchers in the research and development of hydrogel flexible wearable electronic devices. The flexible wearable devices made of hydrogels overcome the shortcomings of traditional metal materials, which exhibited great flexibility, good biocompatibility, low price, small size, and simple manufacturing craft. The hydrogels-based flexible sensors can monitor the subtle changes in the external environment, such as temperature, humidity, pressure, and deformation, and convert them into recordable electronic signals, which is of great significance to meet the increasing demands of bioelectronics, electronic skin, health monitoring, and other modern electronic devices. This section mainly discusses flexible wearable electronic devices prepared using hydrogels.

10.2.1 Principles of Conductive Hydrogel Sensors

In general, there are four main types of conductive hydrogels – ionic conductive hydrogels, polyelectrolytes-based hydrogels, hybrid conductive hydrogels, and intrinsic conductive hydrogels. From the perspective of conduction mode, conductive hydrogels can be roughly divided into two types – electronic conductive and ionic conductive hydrogels [36]. Ionic conductive hydrogels are generally obtained by directly gelling monomers in conductive salt solutions. Ionic conductive hydrogels are easy to synthesize and are currently widely used in flexible wearable devices [37–39]. However, due to the unique ionic transport manner, various stimuli (e.g. strain, temperature, and humidity) will perturb the ion movement (speed, density, path, etc.) of the hydrogels, causing changes in resistance or capacitance on a macroscopic scale. Therefore, improving the stability of ionic conductive hydrogels is a challenge. Polyelectrolytes-based hydrogels are generally prepared by the polymerization of hydrophilic monomers containing ionic groups. The ionic groups are the key factor to obtain a high conductivity. The hybrid conductive hydrogels are generally comprised of polymeric networks and conductive fillers, such as graphene oxide (GO), carbon fiber, carbon nanotubes (CNTs), and metal particles [40–42]. These conductive fillers not only impart high conductivity to the hydrogels but also improve their mechanical properties. The intrinsic conductive hydrogels are comprised of 3D polymer networks with conductive polymer chains [38]. At present, a series of conductive polymers have been used to prepare intrinsic conductive hydrogels, such as polyaniline (PANI) [43, 44], polypyrrole (PPy) [45, 46], and polythiophene/poly(3,4-ethylenedioxythiophene) (PEDOT) [6, 47].

Hydrogels used in flexible wearable devices usually function as strain sensors. When a hydrogel-based sensor was placed on the joints, it can convert the mechanical deformations caused by human motions into electronic signals [43, 48]. According to the different response mechanisms to the external strain, hydrogel strain sensors can be divided into capacitive strain sensors, resistive strain sensors, and other strain sensors [49–53]. Capacitive strain sensors and resistive strain sensors have relatively simple test systems and can withstand a larger range of deformation, so they are the main types of hydrogel strain sensors [54]. Their working principles have been shown in Figure 10.2. For resistive strain sensors, to obtain a deformation linear response of sensors to resistance changes, the hydrogels were cut into a regular shape, such as a rectangle strip or a cylindrical shape. When a hydrogel strip was stretched or compressed uniaxially during the sensing process, its resistance R can be calculated by the formula:

$$R = \rho \frac{L}{A} \tag{10.1}$$

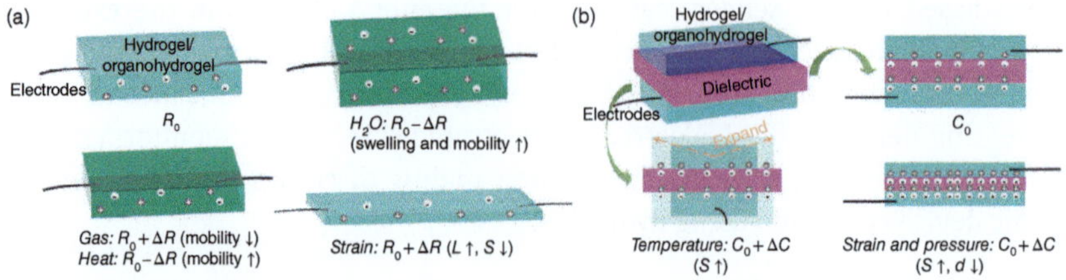

Figure 10.2 Schematics illustrating the working principles of hydrogel sensors based on (a) resistive- and (b) capacitance-type configurations. Source: Adapted from Wu et al. [36]. © 2021 American Chemical Society.

where ρ is the resistivity of the hydrogel, L is the length of the sensing region, and A is the crosssectional area of the deformation region. When the sensor is stretched, the length increases and the cross-sectional area decreases, resulting in an increase in resistance. It can be seen that the ρ is determined by the inherent conductivity and deformation of the hydrogels. The signals of resistive sensors are usually represented by the ratio of resistance changes (R_{CR}):

$$R_{CR} = \frac{R' - R_0}{R_0} \times 100\% \tag{10.2}$$

where R_0 is the initial resistance of the hydrogels and R' is the resistance during deformation.

To adapt to the activities of related biological tissues, flexible wearable devices usually use materials with low modulus and good elasticity. Therefore, in an ideal state, assuming that the hydrogel is an ideal elastomer, the total volume of the hydrogels does not change when deformation occurs. The inherent properties of hydrogels do not change with the occurrence of deformation, and the resistivity ρ can be considered unchanged. When the length of the hydrogels changes by ΔL, its deformation ε is:

$$\varepsilon = \frac{\Delta L}{L_0} \tag{10.3}$$

where L_0 is the initial length of the hydrogels. Assuming that the volume of the hydrogels is V, then

$$R_0 = \rho\frac{{L_0}^2}{V} \tag{10.4}$$

$$R = \rho\frac{(L_0 + \Delta L)^2}{V} \tag{10.5}$$

$$R_{CR} = \frac{\rho\frac{(L_0+\Delta L)^2}{V} - \rho\frac{{L_0}^2}{V}}{\rho\frac{{L_0}^2}{V}} = \varepsilon^2 + 2\varepsilon \tag{10.6}$$

It can be seen that when the hydrogel has rubber elasticity, the electrical resistance and the deformation value have a quadratic function relationship. For materials with zero Poisson's ratio:

$$R_{\mathrm{CR}} = \frac{\rho\frac{L_0+\Delta L}{A} - \rho\frac{L_0}{A}}{\rho\frac{L_0}{A}} = \varepsilon \tag{10.7}$$

In this case, the cross-sectional area of the hydrogels does not change with strain, so R_{CR} is proportional to ε. Therefore, the result of fitting the relationship between R_{CR} and ε in the experiment can usually be approximately fitted between a linear function and a quadratic function. The Poisson's ratio of most hydrogels is positive, but there are exceptions. For example, porous poly(vinyl alcohol) (PVA) hydrogels are a kind of negative Poisson's ratio materials with an auxetic effect [55]. The relationship between R_{CR} and ε will show different laws when fabricating sensors with this kind of hydrogels.

In general, flexible capacitive sensors have a sandwich structure composed of two conductive stretchable hydrogels and a dielectric middle layer, and the sensing mechanism is based on the regulation of the capacitance region by strain [56]. The response of the capacitive sensor is manifested by the changes in area and thickness of the dielectric layer, which makes the capacitance increase. For common parallel plate capacitors, the initial capacitance is defined as Eq. (10.8):

$$C_0 = \varepsilon_0 \varepsilon_\gamma \frac{l_0 w_0}{d_0} \tag{10.8}$$

where ε_0 and ε_γ represent the dielectric constant of the vacuum and the relative dielectric constant of the medium, respectively. l_0 is the initial length of parallel plate, w_0 is the initial width of parallel plate, and d_0 is the thickness of insulating layer. When a sensor is subjected to external tensile strain ε, the length of the capacitor increases to $l_0(1+\varepsilon)$, the corresponding width decreases to $w_0(1-\nu_e)$, and the thickness of dielectric layer decreases to $w_0(1-\nu_d)$. ν_e and ν_d are the Poisson's ratios of the stretchable electrode plate and the dielectric layer, respectively. Assuming that the stretchable electrode plate and the dielectric layer of the capacitive strain sensor have the same Poisson's ratio, the stretched capacitance is:

$$C = \varepsilon_0 \varepsilon_\gamma \frac{(1+\varepsilon) l_0 (1-\nu_e) w_0}{(1-\nu_d) d_0} = \varepsilon_0 \varepsilon_\gamma \frac{(1+\varepsilon) l_0 w_0}{d_0} = (1+\varepsilon) C_0 \tag{10.9}$$

It can be seen that the capacitance depends on the d and ε. Therefore, the capacitance of capacitive flexible sensors changes linearly with strain. It is worth noting that this linearity is only applicable to a small strain range. If the strain is too large, the linear relationship of Poisson's ratio between different axes will no longer be applicable.

The sensing performance of hydrogel strain sensors can be characterized by many indicators, including tensile strain range, limit of detection (LOD), sensitivity, linearity, stability, response, and recovery time. These parameters are critical to the application range of hydrogel-based strain sensors. The tensile strain range reflects the maximum strain that a sensor can withstand in normal working conditions. A good strain sensor not only withstands a large range of tensile force but also has stable performance even in a highly stretchable state. LOD reflects the minimum strain change that the sensor can detect, which is mainly determined by the sensitivity and the baseline stability of sensing signals. The calculation formula of the LOD is:

$$\mathrm{LOD} = \frac{3N}{S} \tag{10.10}$$

where N is baseline noise [57]. N is related to the precision of the sensor used to detect signals. To adapt to more application scenarios, researchers always pursue high sensitivity and low-detection limits. In practical use, the resistance of the hydrogels should be adjusted to match the measurement interval of instruments to obtain more accurate and stable signals. Sensitivity is usually characterized by gauge factor [58]. For resistive sensors:

$$\mathrm{GF} = \frac{R_c}{\varepsilon} \tag{10.11}$$

where R_c is the ratio of resistance changes and ε is the strain of the sensor. For capacitive sensors:

$$\mathrm{GF} = \frac{\Delta C/C_0}{\varepsilon} \tag{10.12}$$

where C_0 is the initial capacitance of sensors, ΔC is the change in capacitance value under external force, and ε is the strain of the sensor. GF is used to characterize the sensor's capability of sensing strain. Comparison of recently reported hydrogels-based strain sensors in terms of maximal strain and gauge factor has been shown in Figure 10.3. It can be seen that the maximum tensile strain of most current hydrogel sensors is less than 2000, and the GF is less than 12. Usually, hydrogels have low-detection strain and high GF or high detection strain and low GF. A few hydrogel sensors can obtain ultrastretchability or ultrasensitivity, but they cannot take into account the other performance of the sensor. It is a challenge to improve the stretchability and sensitivity of the hydrogel at the same time. As mentioned earlier, the relationship between R_{CR} and ε in practical applications is approximately linear. Therefore, researchers usually divide the sensor's stress or strain into an appropriate range so that the R_{CR} and the σ or ε are approximately linear. At this time, the inherent properties of the hydrogels are the main factors affecting the sensitivity of the sensor. At this time, the main factors affecting the sensitivity of the sensor are the inherent properties of the hydrogel, more specifically the Poisson's ratio and the change in conductivity during the deformation process. According to Hooke's law:

$$\sigma = E\varepsilon \tag{10.13}$$

where E is the elastic modulus. It can be seen that for the pressure sensor, the modulus also affects the sensitivity. Complex components and hierarchical structures have been designed to improve sensitivity of hydrogel-based strain sensors. Nanofillers, such as carbon materials and alloys, can significantly increase the resistivity of sensors in stretchable status. Some semiconductor materials also increase the resistance by several orders of magnitude under small deformations [59].

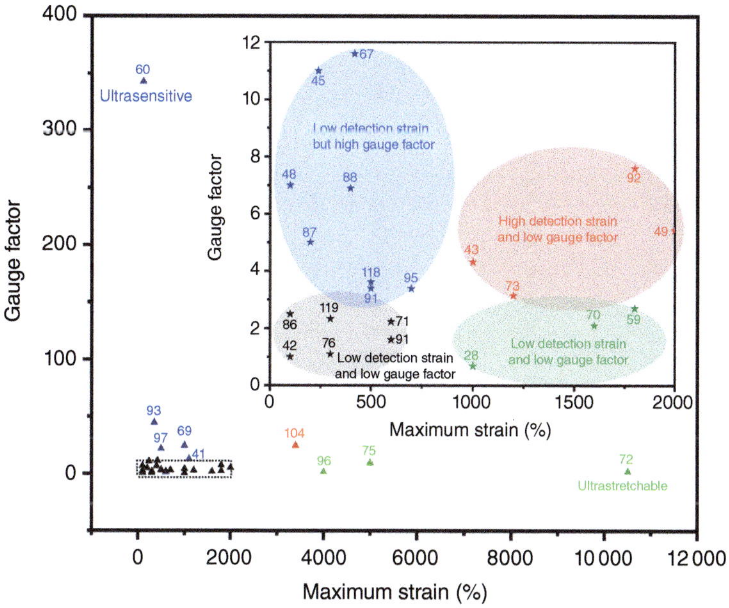

Figure 10.3 Comparison of reported hydrogels-based strain sensors in terms of maximum strain and GF.

10.2.2 Improved Mechanical Properties of Hydrogel-Based Sensors

To be used in flexible wearable devices, the performance of the hydrogels must be compatible with the use environments (e.g. skin, mussel, and cartilage). Therefore, hydrogels need to have the following characteristics – stretchable, tough, anti-fatigue, recoverable, nontoxic, and self-healable [60]. The mechanical properties can be improved by constructing multiple network structures or using inorganic nanofillers [61]. The structure and performance of some recently reported hydrogels-based strain sensors are listed in Table 10.1. A series of interactions can be used to construct multiple crosslinked networks, such as covalent bonds, hydrogen bonds, van der Waals forces, and polymer chain entanglement, which are commonly used to enhance the mechanical strength of hydrogels. Ni and coworkers prepared a poly(*N*-vinylpyrrolidone)/gallic acid (PVP/GA) composite hydrogel through hydrogen bonding between tertiary amide groups in PVP and hydroxyl groups in GA [75]. The carboxyl groups and hydroxyl groups in GA can be ionized to produce carboxylate anions, phenoxy anions, and protons, thereby giving the hydrogel high conductivity. The stress–strain performance of the PVP/GA composite hydrogels has been significantly improved due to the hydrogen bonding, showing good self-healing behavior after damage and excellent stretchability (1309%) (Figure 10.4a). Li and coworkers prepared a highly sensitive wearable stress sensor by crosslinking polyacrylic acid (PAA), chitosan, and GO via hydrogen bonding and coordination interaction (Figure 10.4b) [65]. The sensor exhibited excellent stretchability (≥1000%), repeatability, stability, and outstanding sensing performance in a wide temperature range (from −20 to 70 °C).

Table 10.1 Classification and summary of recently reported hydrogels-based strain and pressure sensors.

Hydrogel	Crosslinked network	Sensing type	Conductive type	Maximum tensile strain	Working range	Gauge factor	References
PAAm/PEGDA	—	Strain	Ion	900%	0–900%	1.94 (<100%), 6.44 (100–900%)	[8]
PAAm/carrageenan/CNT	DN	Strain	Electronic	120%	0–110%	343	[59]
PAA/gelatin/tannic acid (TA)/Al^{3+}	DN	Strain	Ion	1240%	0–100%	2.5 (0–100%)	[62]
PVA/sodium alginate (SA)/bacterial cellulose/CNTs	DN	Strain and pressure	Both	300%	0–200%	0.725 (<60%), 2.216 (60–145.2%), 5.01 (145.2–200%)	[63]
Chitosan/poly (hydroxyethyl acrylamide)	DN	Strain and pressure	Ion	700%	0–400%	6.9 (400%)	[64]
PAA/chitosan/GO/Gly	DN	Strain and pressure		1000%	0.25–1000%	1.138 (0.25–80%), 4.7 (80–1000%)	[65]
Zwitterionic poly(HEAA-*co*-SBAA)/PEDOT:PSS	IPN	Strain	Both	~5000%	—	2–10	[66]
PAAm/silk fibroin/GO/PEDOT:PSS	IPN	Strain and pressure	Ion	880%	2–600% 0.5–119.4 kPa	0.8 (<50%), 1.6 (50–400%), 0.6 (400–600%)	[67]
2-Methoxyethyl acrylate/acrylic acid/graphene	—	Strain and pressure	Electronic	1250%	0–500%	0.62 (<20%), 0.98 (20–100%), 1.83 (100–300%), 3.40 (300–500%)	[68]
PAA/cellulose nanosheet	—	Strain	Ion	1800%	0–1100%	4.37 (0–600%), 7.6 (600–1100%)	[69]
PAAm/PVA/MXene	DN	Strain	Electronic	—	0–350%	5.02 (0–200%), 44.85 (200–350%)	[70]
PAAm/oxidized CNTs	—	Strain	Electronic	1100%	0–700%	1.50 (0–250%), 3.39 (250–700%)	[71]
3,4-Dihydroxyphenyl/L-alanine acrylamide/polycaprolactone (PCL)	—	Strain	Ion	1140%	—	—	[72]
PAM/SA/TA	DN	Strain	Ion	~4000%	0.05–2100%	2	[73]
PAM/SA/NaCl	DN	Strain	Ion	3120%	0.3–1800%	2 (<2000%), 2.7 (200–1800%)	[58]
PVA/polyethyleneimine (PEI)	—	Strain	Ion	500%	0–450%	22 (450%)	[74]

Once the crosslinking structure was damaged by external forces, new crosslinking sites can be formed quickly. Therefore, hydrogels based on dynamic physical crosslinking often have good recoverability and stretchability [48]. However, only physical crosslinking structures, such as van der Waals forces, hydrogen bonds, and chain entanglement, cannot provide sufficient strength and stability. To achieve high strength and stretchability, dual-networks structures in hydrogels have been created using PVA, PAA, PAAm, etc. [41, 77–79], as the first brittle chemical networks to dissipate energy during the stretching process, and physical crosslinking sites, such as hydrogen bonding, perform as a malleable second network. Moreover, there are usually chemical interactions, entanglements, and other interactions between the first network and the second network [80–83]. Double network (DN) structure significantly improves the strength and stretchability of hydrogels. For example, a dual-networks glycol organogel reported by Niu and coworkers exhibited a stretchability of up to 21 000%, which is the highest value reported in the literature (Figure 10.4c) [76]. Constructing interpenetrating crosslinked networks (IPNs) is also a commonly used means to strengthen hydrogels [66, 84, 85]. IPN hydrogels are comprised of double or multiple network structures, which are prepared by the simultaneous or sequential crosslinking of two or more different polymer systems. IPN can be prepared using two kinds of polymers with different properties or different functions, so as to realize the function complementarity of these components. The interpenetrating structure and two-phase continuity of the IPN lead to a special synergy in performance or function, preventing phase separation, and greatly improving the mechanical strength of the hydrogels [86]. For example, Wang et al. prepared an inorganic/organic interpenetrating hydrogel by interpenetrating g-C_3N_4 and CNC in ionic crosslinked PVA networks [85]. The hydrogel exhibited excellent tensile strength (~2.55 MPa), high elongation (~1200%), and satisfactory tensile modulus (~3.65 MPa). Its mechanical properties far exceed those of single network hydrogels [85]. With deep research on the networks structure, a series of supramolecular interactions also emerge in hydrogels, including host–guest interactions and dynamic ionic interactions (Figure 10.4d) [28, 45, 87–93].

Inorganic nanofillers can endow hydrogels with excellent performance and unique functionality. Researchers have combined the advantages of nanofillers and hydrogels to prepare a variety of strain and pressure sensors based on conductive nanocomposite hydrogels [54, 94]. CNTs exhibited strong potential in electronic devices due to their superior electronic properties, such as nanosize, high thermal conductivity, and excellent charge-carrier mobility. Therefore, a lot of research has been conducted to combine these fascinating characteristics of CNTs with stretchable and biocompatible hydrogel sensors. Since CNTs are highly hydrophobic and insoluble in water, prepared CNTs are easy to aggregate and difficult to directly incorporate into hydrophilic polymer networks. In addition, the weak interface interaction between CNTs and polymer networks also limits the effective load transfer of the CNTs in hydrogels. To solve the problems, Yao and coworkers prepared conductive nanocomposite hydrogels with extremely high stretchability, high toughness, and fatigue resistance by integrating hydrophobic CNTs into hydrophobically associated PAAm hydrogels (Figure 10.4e) [42].

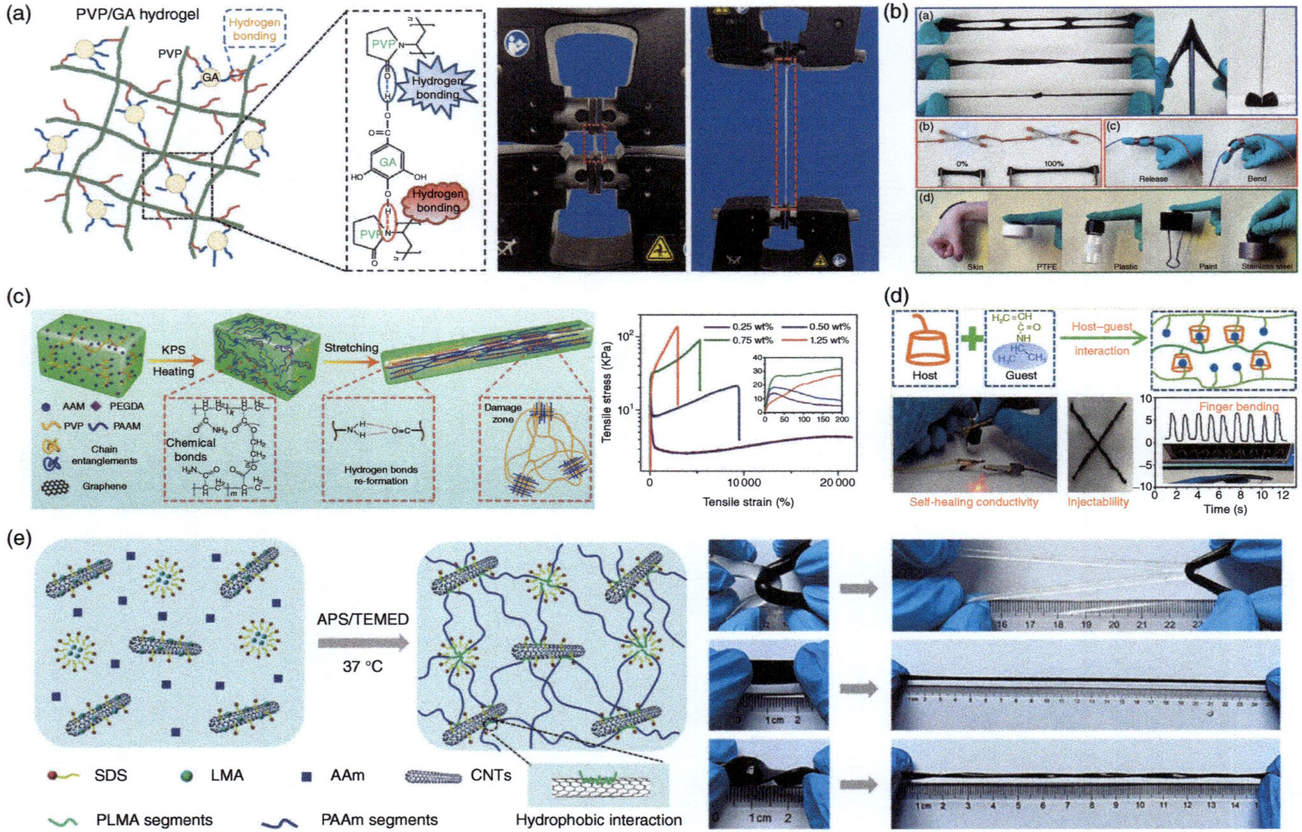

Figure 10.4 (a) Schematic diagram of hydrogen bonding between PVP and GA in PVP/GA composite hydrogel (left), and the excellent stretchability of the composite hydrogel (right) [75]. (b) PAA/chitosan/GO/Glycerol gel with extraordinary mechanical properties, adhesion, and conductivity [65]. (c) Schematic preparation process (left) and ultra-stretchability (right) of double-network PAAm/PVP/ethylene glycol (EG) gels [76]. (d) A self-repairing conductive hydrogel based on the host–guest interaction between β-cyclodextrin (β-CD) and *N*-isopropylacrylamide (NIPAM) reported by Guo and coworkers [45]. (e) Schematic illustration of the preparation process for the CNTs/HAPAAm hydrogel (left) and the mechanical performances of the hydrogel. AAM; acrylamide, KPS; potassium persulfate, PEGDA; polyethylene glycol diacrylate, APS; ammonium persulfate, TEMED; tetramethyl ethylenediamine, SDS; sodium dodecyl sulfate, LMA; lauryl methacrylate, PLMA; poly (lauryl methacrylate), PTFE; polytetrafluoroethylene [42]. Source: Adapted from Zhao et al. [75]. © 2020 Elsevier Ltd.; Xia et al. [65]/Royal Society of Chemistry; Zhang et al. [76]. © 2018 American Chemical Society; Deng et al. [45]/American Chemical Society; Reproduced from Qin et al. [42]/Elsevier.

In this study, amphiphilic sodium lauryl sulfate was used to form hydrophobic interactions between the hydrogel matrix and the CNTs surface ensuring uniform dispersion of CNTs in the hydrogel network, which thereby greatly improved the mechanical properties of hydrogels. Another effective method is to modify CNTs with hydrophilic functional groups, which can not only improve the dispersibility of CNTs in water but also enhance its interface interaction with the polymer network [71, 95]. Besides the CNTs, other carbon materials, such as GO, also have been used as nanofillers to increase the conductivity and mechanical properties of hydrogels. The GO exhibited excellent chemical stability, super-high thermal conductivity, and extremely high mechanical properties. More importantly, the surface and edges of GO nanosheets contain numerous hydroxyl, epoxy, and carboxyl groups, which allows the GO nanosheets to be easily and uniformly dispersed in water and other organic solvents, as well as form crosslinking networks with polymer chains in hydrogels. For example, a GO@dopamine/alginate/poly(acrylic acid-*co*-acrylamide) hydrogel has been prepared and exhibited the maximum adhesion strength and electrical conductivity of 36.9 kPa and 3.24 ± 0.12 S/m, respectively [96]. In addition to the aforementioned carbon materials, other nanomaterials also have been used as nanofillers in hydrogels, such as silver nanowires (AgNWs) [97], polymer nanofillers [98], and the emerging 2D $Ti_3C_2T_x$ (MXene) [70, 99–101]. It can be predicted that with the development of nanomaterials, more and more nanofillers with excellent properties will be incorporated into hydrogels.

10.2.3 Prolonged Longevity of Hydrogel Sensors

Flexible wearable devices need to follow the body movement for a long time, so they must have excellent biocompatibility, robust elasticity, and long-time lasting repeatability. One way to improve the biocompatibility of hydrogel sensors is to simulate the structure and performance of biological tissues [102, 103], and the other is to use biological sources or bio-based materials [104]. The introduction of some bio-derived materials can also bring unique properties to hydrogel-based sensors. For example, the catechol groups in tannic acid allow the hydrogels to directly adhere to human skin without residue and inflammation [105], and chitosan can provide the hydrogels frost resistance and fatigue resistance [64].

The improvement methods of mechanical strength and fatigue resistance have been discussed in Section 10.2.2. Although the improvement in the mechanical properties and fatigue resistance of hydrogels has significantly prolonged the longevity of hydrogel-based devices, the damage to the hydrogels caused by unavoidable external forces and other factors will seriously affect the signal stability of the device. However, considering the use of environment and the nature of biological tissues, it is necessary to allow the sensors to self-healable functions. It has been reported that the crosslinking by multiple hydrogen bonding, host–guest interaction, reversible chemical bonds, metal coordination complex, etc., leads to self-healing behaviors [95, 97, 106]. Zeng and coworkers developed a self-healable hydrogel using the multiple hydrogen bondings of 2-ureido-4[1*H*]-pyrimidinone (UPy) [43]. The UPy group acts as a crosslinking point to form an IPN structure

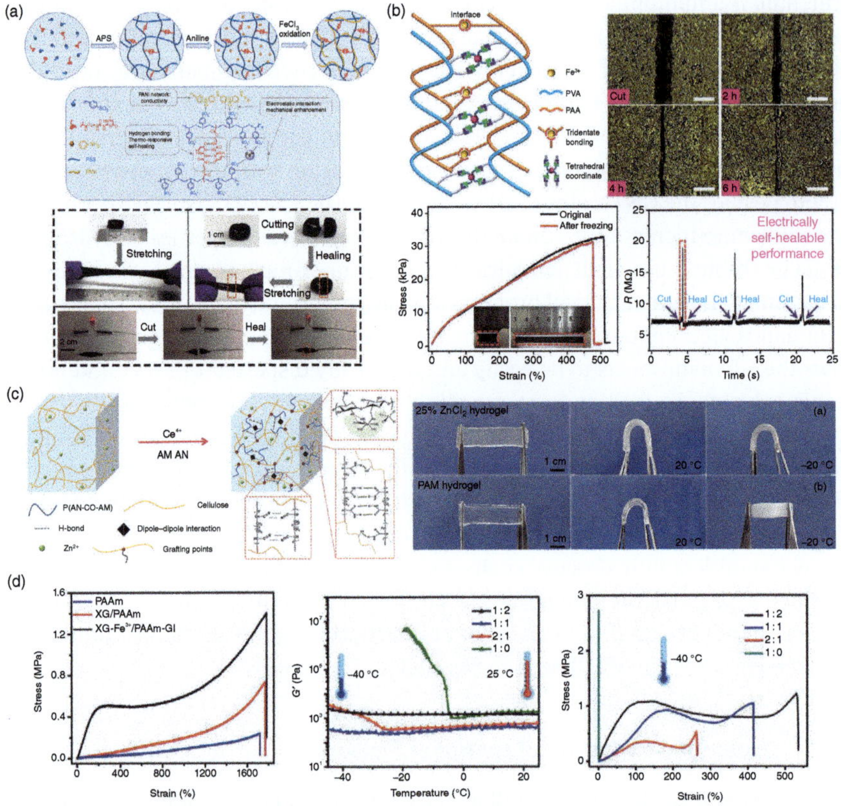

Figure 10.5 (a) Schematic illustration of synthesis process of the supramolecular conductive PANI/PSS-UPy hydrogels and the formation mechanism (top), and the excellent stretchability and self-healing properties of the hydrogel (bottom) [43]. (b) Schematic illustration of the self-healing mechanism and the self-healable performance of the hydrogel sensor based on metal-coordinated bonds and tetrahedral borate interactions [107]. (c) Schematic for the preparation (left) and anti-freezing performance (right) of the HPMC-*g*-AN/AM-$ZnCl_2$ hydrogel [108]. (d) The influence of XG-Fe^{3+} on the strength of hydrogel and the influence of water/glycerol ratio on the anti-freezing performance of XG-Fe^{3+}/PAAm-Glycerol organogel [109]. Source: Reproduced from Chen et al. [43]/American Chemical Society; Reproduced from Ge et al. [107]/Royal Society of Chemistry; Reproduced from Chen et al. [108]/American Chemical Society; Yu et al. [109]. © 2020 Elsevier B.V.

(Figure 10.5a). When an intact hybrid hydrogel was cut into two halves, the freshly generated surfaces were brought into contact to allow autonomous healing. The two segments adhered to each other instantly, and only after a few seconds, they merge into one piece with a well-bonded interface that could withstand stretching. The good self-recovery and self-healing of the hydrogel are attributed to the reversibility of non-covalent crosslinking and the continuous regeneration of multiple hydrogen bonds in the system. In another case, Si and coworkers used a hydrogel with a binary network structure constructed by the interaction of metal-coordinated bonds and tetrahedral borate to construct a piezoresistive sensor and achieved satisfactory stretchability, remarkable healed strain (497% after six hours), and high healing efficiency (90.4%) (Figure 10.5b) [107]. The self-healing of the hydrogel not only realizes the recovery of mechanical properties but also connects its conductive pathways [110]. However, since hydrogels are a water-containing crosslinked network structure, external stimuli may cause changes in the water content, which in turn affects the mechanical properties of the hydrogels. In addition, the crosslinked structure also restricts the movement of the hydrogel molecular chains, making it difficult for the molecular chains to entangle quickly in the molten state, such as linear polymers, which also limits the expansion of the hydrogel's self-healing mechanism. At present, the most common and effective mechanism for self-healing hydrogels is still hydrogen bonding and dynamic physical crosslinking.

10.2.4 Expanded Usage Scenarios of Hydrogel-Based Sensors

Traditional hydrogel-based sensors can only be used in a limited temperature range around room temperature. When the temperature drops below the freezing point (0 °C), the water molecules in hydrogels will inevitably freeze. This will cause the hydrogel to become rigid and lose its original stretchability. In addition, the conductivity also obviously decreases, especially for the ionically conductive hydrogel [65, 109]. These characteristics severely limit the applications of these materials as a sensor. There is a significant need to develop anti-freezing hydrogels that can be used in a wider temperature range [111]. The introduction of ionic compounds into hydrogels is a common method to improve the freezing resistance of hydrogels. The freezing tolerance derives from the colligative property of highly concentrated solutions, which suppress the generation of ice crystals in cells [62, 112, 113]. Jia and coworkers fabricated a cellulose hydrogel by grafting acrylonitrile and acrylamide copolymers onto cellulose chains in the presence of zinc chloride (Figure 10.5c) [108]. The dipole–dipole and multiple hydrogen-bonding interactions on the hydrogel networks allow the hydrogels ultrastretchability (1730%), high elasticity (90%), and good toughness (1074.7 kJ/m^3). The zinc chloride gave the hydrogels remarkable electric conductivity (1.54 S/m) and excellent anti-freezing performance (−33 °C). Besides giving the hydrogel conductivity and improving frost resistance, the ions can promote the entanglement and interaction between polymer segments, which is beneficial to improving the mechanical properties of hydrogels. In addition to the introduction of ionic compounds, the anti-freezing performance can also be achieved using binary-solvent system within hydrogel networks. For example, Yao and coworkers prepared xanthan gum-Fe^{3+}/PAAm-glycerol organogels by *in situ* polymerization and solvent exchange methods [109]. This organogels containing covalent crosslinking interactions, ionization coordination interaction, and hydrogen bonding, showed ultrastretchability (~1769%) and high strength (~1.5 MPa). In addition, the water–glycerol binary solvent leads to the excellent long-term anti-drying and anti-freezing performance of organogels. It can maintain excellent stretchability (>500%), good conductivity, and transparency even at −40 °C (Figure 10.5d).

10.2.5 Multifunctionalization and Expanding Application of Hydrogel Sensor

There has been a lot of work reported on how to improve the performance of hydrogels. Up to now, a series of hydrogels performances still need to be further explored for sensing applications, such as anti-swelling, self-adhesion, anti-fatigue, and self-healing. Conventional hydrogels inevitably "swell" under liquid or physiological conditions, thereby greatly affecting their mechanical properties. Especially for ion-conducting hydrogels, swelling will cause a decrease in ion concentration and signal drifts. Therefore, it is necessary to develop the anti-swelling properties of hydrogels [114]. Xu et al. prepared a cartilage-inspired anti-swelling conductive hydrogel. Due to the extremely dense crosslinking network structure, the hydrogel exhibits excellent anti-swelling capacity in different solutions (including H_2O, dimethyl sulfoxide [DMSO], physiological saline, seawater, and aqueous solutions in the pH range of 3–11) [115]. For some usage scenarios in the field of biomedicine, self-adhesion is an indispensable characteristic of hydrogel sensors. When self-adhesive hydrogels are applied for flexible wearable devices, especially electronic skins, the signal transfer from skin to sensors can be amplified, thanks to enough curvature contact [72, 116]. In addition, although some properties are contradictory to the intrinsic characteristics of hydrogels, a lot of work still needs to be done to prepare hydrogels with the characteristics of anti-fatigue, biocompatibility, self-healing, and conductivity to improve the durability and stability of hydrogel sensors [117]. The e-skins reported by Wu and coworkers exhibited the three functions of real skins: (i) UV triggered a color change with detectable resistance alterations; (ii) reactive oxygen species (ROS) were produced to defend against pathogens; and (iii) the piezoresistive hydrogels became highly stretchable to perceive all-scale human motions (Figure 10.6a) [118]. Zhang et al. reported chameleon-inspired electronic skins that can quantitatively feedback external stimuli through resistance signals and visually display the stimulating parts through color changes due to the cholesteric liquid crystal photonic structure contained in the hydrogels networks (Figure 10.6b) [119]. Multifunctionalization not only makes the sensor more intelligent but also makes it applicable to more scenarios. For example, a stimuli-responsive hydrogel-based sensor with multiple shape memory effects can be used to detect ion concentration in sweat through deformation (Figure 10.4c) [120]. A hydrogel-based sensor was prepared with temperature-sensitive polymers, such as poly(*N*-isopropylacrylamide), which has close to perfect stimulus discernibility for touch and temperature [38, 121]. When rare earth element-containing compounds were incorporated into hydrogels, they can impart fluorescence characteristics to the hydrogels [122]. In addition, the wireless transmission technology enables the hydrogel sensors to transmit signals remotely, realize real-time monitoring, and increase the portability and flexibility of the sensor [8, 123]. The emergence of new technologies will enable further development of implantable sensing devices.

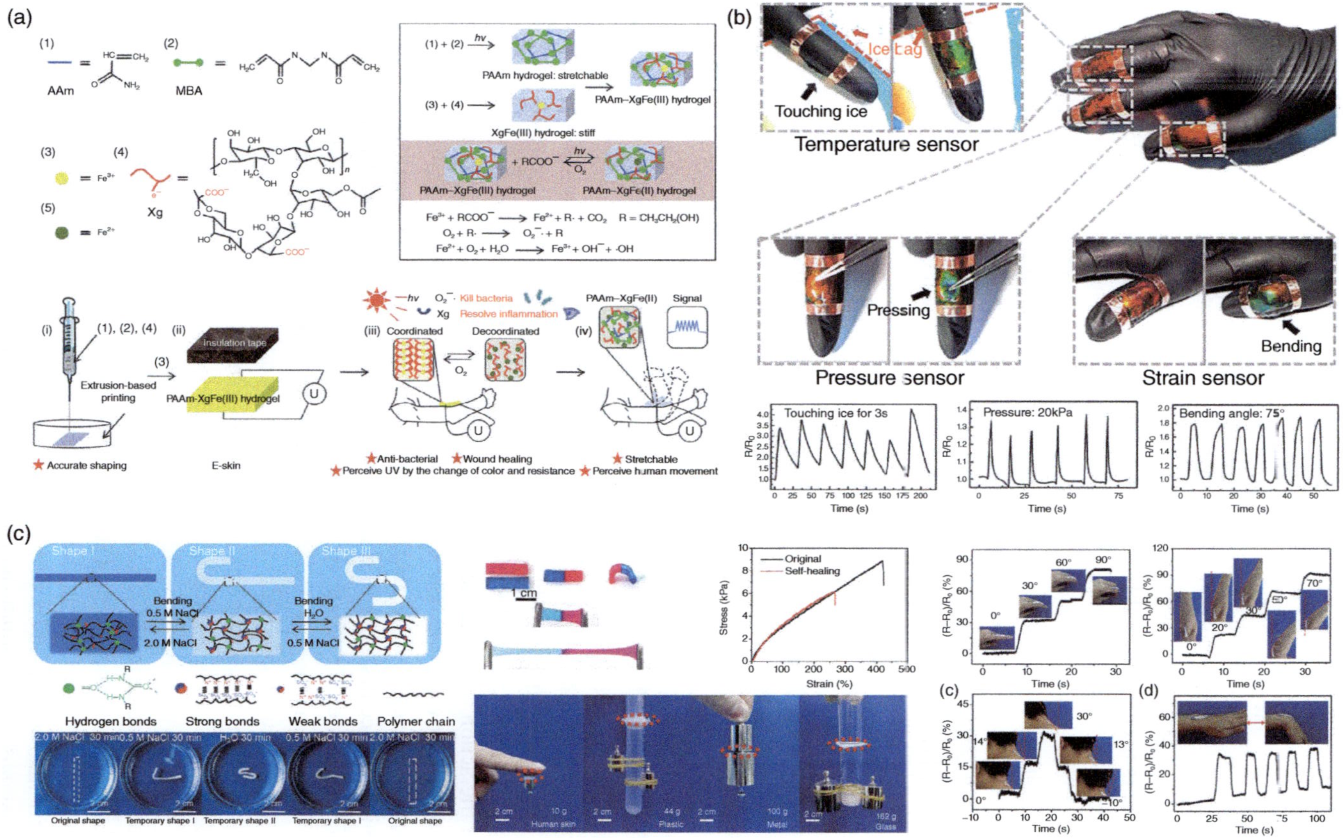

Figure 10.6 (a) The schematic diagram of the synthesis, fabrication, and application of the hydrogel e-skin, including the molecular formula of the chemical agents, the strategy for synthesis of the dual-network PAAm–XgFe(III) hydrogel, the approach achieving the reversible coordination of Xg/Fe(III) for the PAAm–XgFe(III) hydrogel by the redox of Fe(III)/Fe(II), and the chemical mechanism for the production of ROS during the de-coordination of Xg/Fe(III) [118]. (b)The application of the multi-functional e-skins attached to human fingers, including schematics and the optical images of the e-skins attached to human fingers (top), and the real-time signals of the variation of resistance of the e-skin under repeatedly touching of ice for three seconds (left), poking by tweezer with a pressure of 20 kPa (middle), and thumb bending with an angle of 75° (right) [119]. (c) A hydrogel flexible wearable electronic device with multiple shape memory effects, self-healing properties and human motion detection functions prepared by Wu et al. [120]. Source: Adapted from Huang et al. [118]. © 2021 The Royal Society of Chemistry; Zhang et al. [119]/Proceedings of the National Academy of Sciences PNAS; Wu et al. [120]/Royal Society of Chemistry.

10.3 Tissue–Machine Interfaces

Implantable sensors are of great significance to the development of biology and medicine. With the development of flexible and miniaturized electronic devices, implantable sensors are widely used in diagnosis and treatment. Implantable sensors can be placed in the body to identify and analyze electrophysiological signals, and deeply understand physiological phenomena. The core of monitoring bioelectric signals lies in the construction of tissue–machine interfaces. Since the implanted sensors need to be used in a very complex biological environment for a certain time, there are very strict requirements for the applied materials. Hydrogel is an excellent candidate for tissue–machine interfaces due to its excellent biomimetic and biocompatibility. Hydrogels have been applied to neural interfaces to achieve a range of goals, including improving mechanical mismatch between hard devices and soft tissues [124], biological functionalizations [125], and electrically active electrode coatings [126], which have exhibited broad development prospects.

10.3.1 Design and Mechanism of the Neural Interfaces

The most basic purpose of the neural interface is to realize the transformation of the field potential of a neuron into an electronic current in the detector. To this end, we first need to understand how the neural signals transmit. In the nervous system, the transmission of neural signals depends on ions and neurotransmitters [124]. Electrical signals are transmitted along axons and dendrites, while chemical signals are transmitted along synapses. When the action potential reaches a neuron through the dendrites, the signals are processed and sent to another neuron or effector through axons and synapses. The depolarized state of the cell membrane is transported along the unmyelinated axon, while ions exchange across the cell membrane occurs only at the Ranvier nodes to refresh the signals in the myelinated axon [127]. When the action potential reaches the chemical synapse, the voltage-gated Ca^{2+} channels are activated to allow Ca^{2+} to flow into the presynaptic terminal, which induces the release of neurotransmitters into the synapse. The released neurotransmitter molecules bind to the postsynaptic receptors, followed by the activation of ligand-gated Na^+ channels and the entry of Na^+ into the postsynaptic cell. Finally, the postsynaptic potential initiated by the increased Na^+ concentration is delivered to the neuron cell body. The signal transport in the nervous system induces the flux of ions through the neural cell membrane, which leads to changes in ions concentration in the extracellular fluid. Since ion is the charge carrier in extracellular fluid, its movement would modulate the electric field. The hydrogel sensor as a nerve interface needs to detect and record the electric field changes in the extracellular fluid, so the hydrogel is necessary to be electroactive and capable of charge transfer between ions and electrons. According to the difference in the charge transfer mechanism, electroactive materials can be divided into capacitive and faradaic materials. Ion conductive hydrogels and polyelectrolyte-based conductive hydrogels are capacitive materials, and their charge transfer mechanism is the redistribution of ions/electrons at the electrode/extracellular fluid interface.

The main factor affecting the electroactivity is the adsorption of ions to interfaces and the formation of electric double layers. Usually, capacitive materials have good electrochemical stability. Electron conductive hydrogels are similar to metal materials, which belong to Faradaic materials, and their charge transfer mechanism is attributed to electrochemical reactions between electrode surfaces and extracellular fluids. The main factor affecting the electroactivity of Faradaic materials is the efficiency of redox reactions. This type of material usually shows a high electroactivity unit per surface area. Suo and coworkers used artificial cerebrospinal fluid as a solvent to synthesize a poly(ethylene glycol) (PEG) hydrogel for nerve signal detection and explored the basic physical and chemical principles of the neuroelectronic interface (Figure 10.7a) [10]. In the described detector, the hydrogel contacts the cytomembrane of the neuron and a metallic electrode of the detector. Although the hydrogel is a soft solid, it can mimic the exchange of matters (e.g. ions, water, and possibly other molecules) between the extracellular fluid and neurons. This exchange is effective with or without extracellular fluid between the hydrogel and cytomembrane. In addition, there is no matter exchange through the electrical double layer between the metal electrode and the hydrogels [130]. The voltage associated with neural signals is too small to cause an electrochemical reaction occurring at the electrical double layer. The hydrogels acted as ionic conductors, and the metal is an electronic conductor. When the neuron fires an action potential, ion channels in the cytomembrane allow ions (mostly positive) to move from the hydrogels into the neuron. To regain electrical neutrality, in the hydrogels, some negative ions migrate toward the electrical double layer, and some positive ions migrate away from the electrical double layer. In the metal, electrons flow from the electrical double layer to the detector. A series of migration behaviors of ions and electrons enables the detector to record action potentials [102, 131–133]. To obtain long-term stable signal acquisition, the conductivity, swelling resistance, and biocompatibility of the hydrogel are all considerable challenges. In general, the bioelectric signals are very weak, for example, the hippocampus field potential is on the order of 100 μV [10]. With the development of electronic technology, more and more high-performance signal amplification systems have been developed, and electrical signal detectors have become more and more intelligent. The development of high-end electronic signal detection systems has promoted the progress of bioelectronics. However, to obtain more accurate and stable bioelectric signals, the materials used as the electrode must also have excellent conductivity and stability. Since the materials are used in the internal environment of organisms, it is impossible to increase the ion concentrations to improve the conductivity. In addition, due to the complex environment in the organism, there are also many challenges to the anti-swelling and acid and alkali resistance of the hydrogels. The commonly used method to improve the hydrogel tissue–machine interfaces is to introduce nanofillers or conductive polymers with excellent conductivity into the hydrogel [134]. Hirata and coworkers prepared an Au/AgNWs-based neural interface coated with PVA hydrogels (Figure 10.7b) [128]. Specifically, the neural interface is formed by plating a few nanometers of Au on AgNWs and establishing a transparent track. By coating the external microelectrodes with the PVA hydrogel, the neural interface can maintain a low impedance of 1.1–3.2 Ω cm^2 in the saline solution over five months. Liu and coworkers proposed a strategy to improve the neural interface by constructing a soft IPN on the electrode surface [135]. In this method, the soft alginic acid gel, and poly(3,4-ethylenedioxythiophene)/poly(styrenesulfonate) (PEDOT/PSS)-coated MWCNTs are firstly co-deposited on the neural electrode, and then PEDOT/PSS is electrochemically grown through the coating material. Compared with unmodified electrodes, soft IPN-modified electrodes have higher charge-storage capacity, lower electrochemical impedance, and lower noise floor. All these characteristics are greatly desired for the neural interface. In addition, another effective way to improve the efficiency of charge transfer is to amplify the signal through the design of the electrode shape and the construction of an array [136].

Unlike strain sensors or electrochemical sensors used outside the body, implantable sensors do not require high mechanical strength. On the contrary, implanting probes with high Young's modulus into living bodies may cause inflammation and damage biological tissues. Platinum (Pt), for example, is far stiffer than neutral tissues with an elastic modulus of 140–164 GPa. Silicon, metals, and metal oxides exhibit Young's moduli >50 GPa and rigidities between 10^{-3} and 10^{-2} Pa m^3 for films ≈100 μm in thickness, while conversely, excitable tissues targeted by sensors, such as the brain, spinal cord, and peripheral nerves, exhibit mechanical moduli on the order of 1–10 kPa. This disparity can lead to shear forces arising during movements and micromotions and consequently to a chronic inflammatory response occurring [129]. There have been many studies on applying biocompatible conductive hydrogels to neural tissue engineering and nerve regeneration due to the adjustability of hydrogels. To obtain the best biocompatibility, hydrogels for tissue interfaces need to have a modulus close to that of nearby tissues to mediate the mechanical mismatch. Lu and coworkers used MWCNTs, PEDOT/PSS, PAAm, and PVA to prepare hydrogels with IPNs, and used them to achieve reliable detection of physiological signals [137]. The hydrogel has extremely low Young's modulus, high stretchability, rapid self-healing, and injectability. During the experiment, the hydrogel can distinguish the breathing amplitude and frequency of mice during different activities. It can also produce pressure microsensors by printing injections to monitor the accurate blood pressure signals of the human radial artery and carotid artery. Pan et al. prepared a glycerol hydrogel by dispersing a mussel-inspired proanthocyanidins-coated cellulose nanofibrils nanocomposites in a solution of guar gum and glycerol [11]. This hydrogel shows excellent adhesion (7.9 KPa) and UV-blocking ability (82%). In addition, the borax solution used as a crosslinking agent also imparts high ionic conductivity to the hydrogel, and the strain sensor made of the hydrogel exhibits a low weight detection capability (200 mg) and a fast response speed (33 ms). The hydrogel can also be used to make portable and editable electrodes that can accurately detect electrophysiological signals of the human body (Figure 10.7d). The excellent biocompatibility of the hydrogels is also due to the extremely low modulus (less than 100 Pa). Another way to improve the biocompatibility of implantable sensors is through the introduction of chemical ingredients with an anti-inflammatory function into interface materials. For example, Lee and

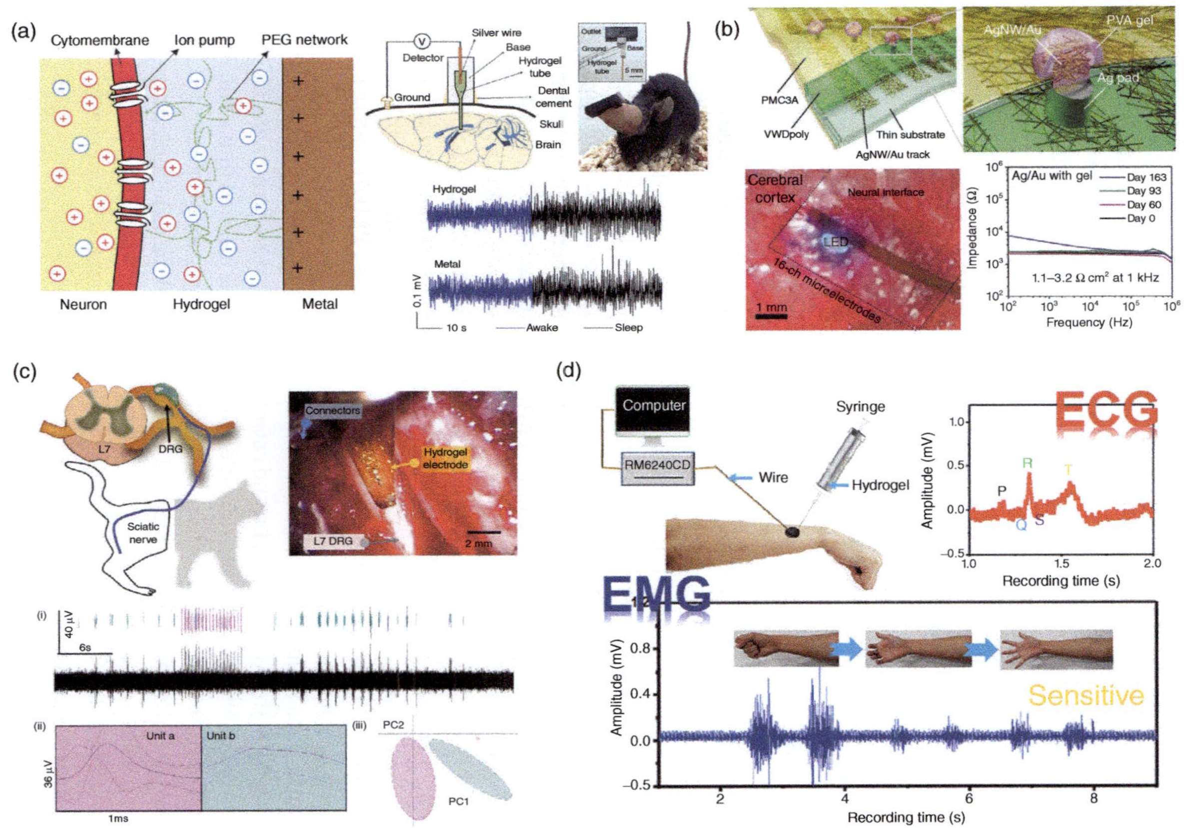

Figure 10.7 (a) The working principle of the hydrogel as an electronic interface proposed by Wang and coworkers and the application of the hydrogel prepared by them in the detection hippocampal field potential [10]. (b) Schematic of and photograph of a stereoscopic multifunctional neural interface Au/AgNW and PVA gel with low impedence [128]. (c) Schematic drawing (left) and optical image (right) of the dorsal root ganglia that the hydrogel microelectrode array (MEA) was placed on, and the obtained *in-vivo* neural recording signals (bottom) [129]. (d) Application of electrodes made of injectable hydrogel prepared by Pan et al. in human electrocardiograph (ECG) and electromyography (EMG) detection [11]. Source: Reproduced from Sheng et al. [10]/Elsevier; Araki et al. [128]/John Wiley & Sons, Inc.; Huang et al. [129]/John Wiley & Sons, Inc.; Pan et al. [11]/American Chemical Society.

coworkers developed a neural electrode coated with a functionalized PEG hydrogel [138]. The PEG hydrogel contains poly(lactic-*co*-glycol acid) (PLGA) microspheres (MS) loaded with cyclosporine A (CsA). The electrode substrate was chemically treated to make it contain micropores to improve the adhesion of the hydrogel and the CsA loading. Subsequently, PEDOT/PSS was formed on the hydrogel layer to improve the quality of nerve signals. The tissue deposition on the surface of the electrode fibers coated with hydrogels was significantly reduced in animal tests. The flexibility and low modulus of the PVA hydrogel can effectively alleviate the mismatch between the mechanical properties of the tissue and electrode. The drug's slow-release function can minimize tissue damage.

10.3.2 Multifunctional Applications of Biointerfaces

Implantable devices often need to destroy the tissues during insertion into the human body, which will increase safety risks. Therefore, the biological interface materials used in the body should be used as long as possible to reduce the number of implantation and removal. In addition, the integration of multiple functions can effectively reduce the volume of *in vivo* sensors and simplify operations, which is of great significance for improving sensing efficiency. Since the direction of charge flow in the conductive hydrogel is arbitrary, the neural interface used to detect potential can usually also be used for electrical stimulation. Function electrical stimulation, that is, applying a specific power and waveform current or voltage to the neural tissue of the organism, which then triggers various neurophysiological reactions. So far, in addition to pain relief, electrical stimulation has also been developed for the recovery of motor function and diagnosis of neurological diseases. For instance, Khodagholy and coworkers used aloe vera hydrogel loaded with PEDOT/PSS to prepare a composite electrode [139]. Plant-based aloe vera hydrogel provides composite materials with excellent biocompatibility and stability, while conductive PEDOT/PSS brings high conductivity and charge capacity (Figure 10.8a). The optimized composite hydrogel was compared with the crosslinked PEDOT/PSS and bare Au through experiments, and it was found that the composite hydrogel has high volumetric capacitance, in contrast to the formation of an electrical double layer capacitance on the Au interface. Lu and coworkers designed and synthesized a conductive, stretchable, self-adhesive, and self-healing hydrogel inspired by natural mussels (Figure 10.8b) [140]. The hydrogel was synthesized by a simple three-step method. Firstly, dopamine was polymerized to form a polydopamine molecular chain, then GO was added and reduced by polydopamine, and finally, acrylamide was added to form a double network hydrogel. The reduced graphene oxide (rGO) wrapped by polydopamine (PDA) is well dispersed in the network, resulting in high conductivity. In addition, the unreduced GO and PDA chains interact with PAAm through hydrogen bonds, π–π stacking, electrostatic interactions, and other non-covalent bonds, giving the hydrogel self-healing and self-adhesive properties. Moreover, the hydrogel also adhered to bone marrow stem cells (BMSCs) for electrical stimulation experiments. Various voltages were applied to the BMSC for 30 minutes per day to examine the electrical response of BMSC on the conductive

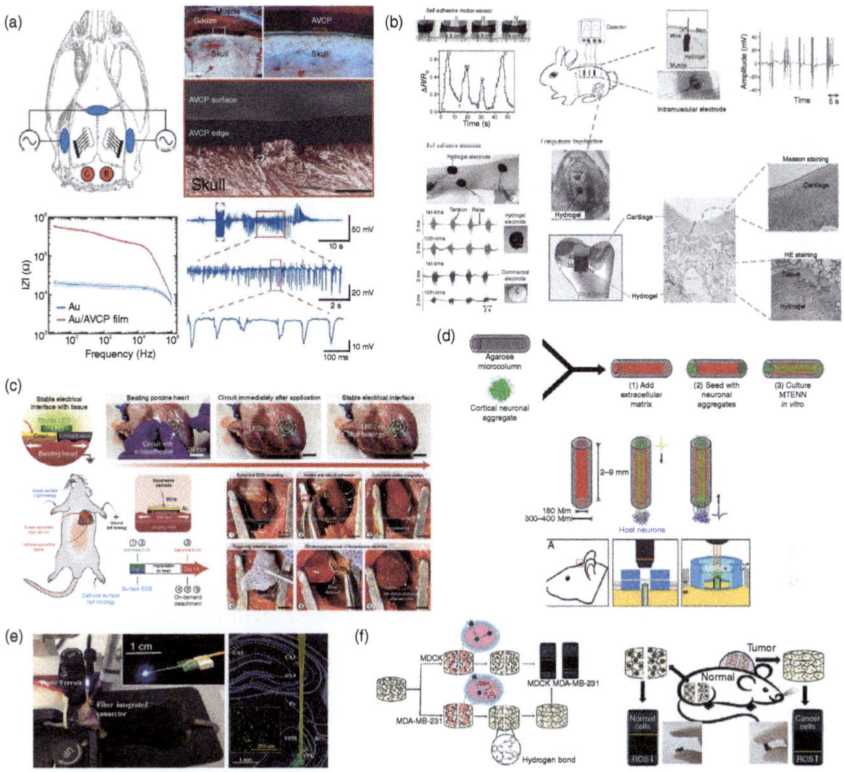

Figure 10.8 (a) AVCP-based TES electrodes enable focal seizure stimulation *in vivo* [139]. (b) Application of self-adhesive hydrogel as motion-sensor, electrophysiological signal sensor, and implantable electrical stimulation electrodes [140]. (c) Display of the biocompatibility, applicability, mechanical and electrical stability, and recording and stimulation functionalities of the electronic bioadhesive interface based *ex vivo* porcine and *in vivo* rat models [18]. (d) Construction method of micro-tissue engineering neural network [141]. (e) *In-vivo* optogenetics experiment of gene-embedded nanostructural biotic–abiotic optoelectrode arrays reported by Chen and coworkers [142]. (f) Self-healing mechanism of the wireless hydrogel biosensor and its response to cancer cells reported by Ryu and Park [143]. Source: Reproduced from Spyropoulos et al. [139]/John Wiley & Sons, Inc.; Han et al. [140]/John Wiley & Sons, Inc.; Deng et al. [18]/Springer Nature; Adewole et al. [141]. © 2021 The Authors, some rights reserved; exclusive licensee American Association for the Advancement of Science; Huang et al. [142]/American Chemical Society; Won et al. [143]/American Chemical Society.

hydrogels. The GO content in the hydrogels will not affect the cell behaviors without electric stimulation (ES), while the cell growth rate will be affected when ES is on. The ES voltage of 300–600 mV is beneficial to cell growth, and the growth rate increases with the GO content increasing. However, a higher ES voltage (900 mV) decreased the cell growth rate. Therefore, this kind of hydrogels can be used to precisely adjust cell fate by tuning the ES voltage.

Surgical operations with sutures can stably integrate bioelectronic devices with tissues. This method is particularly effective for non-thin-film devices. However, sutures can cause tissue damage and stress concentration at the piercing point, thereby forming a scar [144]. Recently, bioadhesives have been explored to integrate bioelectronic devices with tissues and organs. However, the poor adhesive properties and limited electrical properties of existing bioadhesives make them unable to serve as a perfect bridge between bioelectronic devices and wet dynamic tissues. Conductive hydrogel with a self-adhesive function is a suitable candidate to solve this problem [18]. By introducing GO into PVA hydrogel, and then forming IPNs with PAA grafted with *N*-hydroxysuccinimide ester, Guo and coworkers prepared an electrical bioadhesive interface. The described interface material can provide rapid (adhesion formation within five seconds), robust (interfacial toughness >400 J/m^2), and on-demand detachable integration of bioelectronic devices with diverse wet dynamic tissues. The electrical conductivity (>2.6 S/m) of the interface allows bidirectional bioelectronic communications. Moreover, the interface shows good biocompatibility and current stability in both *ex vivo* porcine and *in vivo* rat experiments, which can be used as implantable electrodes to obtain epicardial electrocardiograms stably for a long time in mice (Figure 10.8c).

Optogenetics is a new technology proposed by Deisseroth et al. in 2006 that can combine genetics technology with a light-controlled operation [145]. This emerging technology has been used in the research of Alzheimer's disease, epilepsy, Parkinson's disease, and other diseases. By adjusting the structure of conductive hydrogels, excellent light transmission performance can be obtained. Therefore, the implanted hydrogel electrode can be used as an optical fiber to stimulate biological cells while detecting bio-electrophysiological signals [10, 146]. Cullen and coworkers constructed a "living electrode" composed of microtissue-engineered neural networks (μTENNs), where μTENNs are discrete population(s) of neurons connected by long bundles of axons protected within a microscopic hydrogel cylinder (Figure 10.8d) [141]. The biofabrication and functional validation *in vitro*, as well as the targeted delivery, survival, and continued function posttransplant after transplantation in rat cortex of the electrode have also been demonstrated. The results indicate that the μTENNs can quickly and consistently form similar functional architecture and can be maintained over weeks *in vitro*, which is important for the biofabrication and scale-up of experimentally useful constructs. Therefore, μTENNs may serve as an ideal system for studying neuronal growth, maturation, and network dynamics due to their "abstraction" of mammalian brain connectome into individual units. Chen and coworkers developed a new gene-embedded optoelectrode array for neural implantation to spatiotemporal electroporation for gene delivery/transfection, photobiomodulation, and synchronous electrical monitoring of neural signals in the brain via one-time implantation (Figure 10.8e) [142]. Among them, amphiphilic chitosan modified by rGO and 3,4-ethylenedioxythiophene is the core part of the photoelectrode. Nonviral gene vectors formulated by neurotensin and polyethylenimine-coupled plasmid DNA are encapsulated in the hydrogel. The microelectrodes exhibit biomimetic mechanical and structural properties suitable for brain tissues and provide a gene-loaded platform that can electrically control the delivery of nonviral gene vectors on demand. The optogenetic plasmids from the microelectrodes on the flexible neural optoelectrode probe can be introduced into the brain cells by applying a short electrical pulse, thereby effectively expressing genes and regulating neural activity *in vivo* under optical stimulation via one-time implantation. This optoelectrode array provides precise neurotherapy without the need for multiple implants and extensive surgeries while allowing more biocompatibility and controllable optogenetics.

Hydrogel sensors as a tissue–machine interface can also find applications in the field of cancer detection. Lee and coworkers reported stimuli-responsive electrochemical wireless hydrogel biosensors [143]. The hydrogel used ureidopyriminone-conjugated gelatin (Gel-UPy) as a matrix and incorporate diselenide-containing carbon dots (dsCDs) for cancer detection. Glutathione (GSH) or ROS can cause the cleavage of diselenide bonds and the formation of hydrogen bonds in the hydrogel, thereby affecting the self-healing ability, conductivity, and adhesion of the hydrogel. The correlation between self-healing rate and cancer cells and the influence of diselenide bond cleavage on electrochemical signals enable the hydrogel to detect the presence of cancer cells and display them through wireless transmission.

10.4 The Prospects of Hydrogel Bioelectronic Devices

While significant progress has been made in the research on hydrogels-based electronic devices, there are still many remaining problems in practical applications. The development direction in the future of hydrogel electronic devices will be toward the following aspects: (i) Improving biocompatibility and avoiding allergic reactions. Implantable sensors should also have antibacterial and anti-inflammatory functions; (ii) Improving the biological and chemical stability, ensuring that the device maintains long-term and stable signal acquisition capabilities in the body, and avoid repeated implantation operations; (iii) Making hydrogels and the surrounding tissues have similar mechanical properties so that users can hardly feel the existence of the device, and the user's daily life is not affected at all. In view of sensing performance, the sensitivity and signal stability of hydrogel-based sensors are also foremost properties for wide practical applications. The exploration of novel structures and new sensory materials to enhance sensitivity and signal stability is still undergoing.

The development of anti-swelling, non-volatility, and low-impedance hydrogels can effectively improve the sensing performance of hydrogel sensors. The currently reported hydrogels still have a large space for improvement in this regard. Finally, with the further development of microelectronics technology, the development direction of hydrogel electronic devices is bound to be multifunctional and miniaturized. Miniaturization can be realized by improving the binding ability of microcircuits and hydrogel-based materials and reducing the volume of hydrogel electrodes, both of which propose high requirements for the preparation process of hydrogel-based materials. However, it is expected that in the future, electronic devices used for biological sensing will greatly improve people's life quality. More diverse signal acquisition, recognition, and analysis capabilities will push the development of modern medicine to a new stage.

Acknowledgments

The authors gratefully acknowledge the funding from the Jilin Province Science and Technology Development Plan (20200801008GH, 2021SYHZ0038, and 20210509036RQ), the National Key Research and Development Program of China (2018YFD1100503, 2020YFA0713601), and the National Natural Science Foundation of China (22073092).

References

1 Wichterle, O. and Lím, D. (1960). Hydrophilic gels for biological use. *Nature* 185: 117–118.
2 Hoffman, A.S. (2012). Hydrogels for biomedical applications. *Adv. Drug Delivery Rev.* 64: 18–23.
3 Cascone, S. and Lamberti, G. (2020). Hydrogel-based commercial products for biomedical applications: a review. *Int. J. Pharm.* 573: 118803.
4 Caló, E. and Khutoryanskiy, V.V. (2015). Biomedical applications of hydrogels: a review of patents and commercial products. *Eur. Polym. J.* 65: 252–267.
5 Keum, D.H., Kim, S.-K., Koo, J. et al. (2020). Wireless smart contact lens for diabetic diagnosis and therapy. *Sci. Adv.* 6 (17): eaba3252.
6 Walker, B.W., Lara, R.P., Mogadam, E. et al. (2019). Rational design of microfabricated electroconductive hydrogels for biomedical applications. *Prog. Polym. Sci.* 92: 135–157.
7 Heo, M., Lee, S.J., Lee, D. et al. (2017). Preparation of mechanically enhanced hydrogel scaffolds by incorporating interfacial polymer nanorods for nerve electrode application. *Fibers Polym.* 18 (11): 2248–2254.
8 Hang, C.-Z., Zhao, X.-F., Xi, S.-Y. et al. (2020). Highly stretchable and self-healing strain sensors for motion detection in wireless human–machine interface. *Nano Energy* 76: 105064.
9 Liu, T., Zhu, C., Wu, W. et al. (2020). Facilely prepared layer-by-layer graphene membrane-based pressure sensor with high sensitivity and stability for smart wearable devices. *J. Mater. Sci. Technol.* 45: 241–247.
10 Sheng, H., Wang, X., Kong, N. et al. (2019). Neural interfaces by hydrogels. *Extreme Mech. Lett.* 30: 100510.
11 Pan, X., Wang, Q., He, P. et al. (2019). Mussel-inspired nanocomposite hydrogel-based electrodes with reusable and injectable properties for human electrophysiological signals detection. *ACS Sustainable Chem. Eng.* 7 (8): 7918–7925.
12 Jing, L., Hsiao, L.-Y., Li, S. et al. (2021). 2D-Material-integrated hydrogels as multifunctional protective skins for soft robots. *Mater. Horiz.* 8 (7): 2065–2078.
13 Cai, Y., Shen, J., Yang, C. et al. (2020). Mixed-dimensional MXene-hydrogel heterostructures for electronic skin sensors with ultrabroad working range. *Sci. Adv.* 6 (48): eabb5367.

14 Wang, C., Qu, X., Zheng, Q. et al. (2021). Stretchable, self-healing, and skin-mounted active sensor for multipoint muscle function assessment. *ACS Nano* 15 (6): 10130–10140.

15 Sawayama, J. and Takeuchi, S. (2020). Long-term continuous glucose monitoring using a fluorescence-based biocompatible hydrogel glucose sensor. *Adv. Healthc. Mater.* 10 (3): 2001286.

16 Yeom, J., Choe, A., Lim, S. et al. (2020). Soft and ion-conducting hydrogel artificial tongue for astringency perception. *Sci. Adv.* 6 (23): eaba5785.

17 Scarpa, E., Mastronardi, V.M., Guido, F. et al. (2020). Wearable piezoelectric mass sensor based on pH sensitive hydrogels for sweat pH monitoring. *Sci. Rep.* 10 (1): 10854.

18 Deng, J., Yuk, H., Wu, J. et al. (2021). Electrical bioadhesive interface for bioelectronics. *Nat. Mater.* 20: 229–236.

19 Li, X., He, L., Li, Y. et al. (2021). Healable, degradable, and conductive MXene nanocomposite hydrogel for multifunctional epidermal sensors. *ACS Nano* 15 (4): 7765–7773.

20 Wu, C., Liu, A., Chen, S. et al. (2019). Cell-laden electroconductive hydrogel simulating nerve matrix to deliver electrical cues and promote neurogenesis. *ACS Appl. Mater. Interfaces* 11 (25): 22152–22163.

21 Wang, C., Rubakhin, S.S., Enright, M.J. et al. (2021). 3D Particle free printing of biocompatible conductive hydrogel platforms for neuron growth and electrophysiological recording. *Adv. Funct. Mater.* 31 (14): 2010246.

22 Ling, Y., An, T., Yap, L.W. et al. (2020). Disruptive, soft, wearable sensors. *Adv. Mater.* 32 (18): e1904664.

23 Prasad, A., Xue, Q.-S., Sankar, V. et al. (2012). Comprehensive characterization and failure modes of tungsten microwire arrays in chronic neural implants. *J. Neural Eng.* 9 (5): 056015.

24 Huang, H.-J., Ning, X., Zhou, M.-B. et al. (2021). A three-dimensional printable liquid metal-like Ag nanoparticle ink for making a super-stretchable and highly cyclic durable strain sensor. *ACS Appl. Mater. Interfaces* 13 (15): 18021–18032.

25 Zhu, L., Zhou, X., Liu, Y., and Fu, Q. (2019). Highly sensitive, ultrastretchable strain sensors prepared by pumping hybrid fillers of carbon nanotubes/cellulose nanocrystal into electrospun polyurethane membranes. *ACS Appl. Mater. Interfaces* 11 (13): 12968–12977.

26 Feng, B., Jiang, X., Zou, G. et al. (2021). Nacre-inspired, liquid metal-based ultrasensitive electronic skin by spatially regulated cracking strategy. *Adv. Funct. Mater.* 31 (29): 2102359.

27 Jing, X., Mi, H.-Y., Lin, Y.-J. et al. (2018). Highly stretchable and biocompatible strain sensors based on mussel-inspired super-adhesive self-healing hydrogels for human motion monitoring. *ACS Appl. Mater. Interfaces* 10 (24): 20897–20909.

28 Xia, S., Song, S., Jia, F., and Gao, G. (2019). A flexible, adhesive and self-healable hydrogel-based wearable strain sensor for human motion and physiological signal monitoring. *J. Mater. Chem. B* 7 (30): 4638–4648.

29 Lacour, S.P., Courtine, G., and Guck, J. (2016). Materials and technologies for soft implantable neuroprostheses. *Nat. Rev. Mater.* 1 (10): 372.

30 Moshayedi, P., Ng, G., Kwok, J.C.F. et al. (2014). The relationship between glial cell mechanosensitivity and foreign body reactions in the central nervous system. *Biomaterials* 35 (13): 3919–3925.

31 Nam, J., Lim, H.-K., Kim, N.H. et al. (2020). Supramolecular peptide hydrogel-based soft neural interface augments brain signals through a three-dimensional electrical network. *ACS Nano* 14 (1): 664–675.

32 Aquilina, O. (2006). A brief history of cardiac pacing. *Images Paediatr. Cardiol.* 8 (2): 17–81.

33 Byun, S.-H., Sim, J.Y., Zhou, Z. et al. (2019). Mechanically transformative electronics, sensors, and implantable devices. *Sci. Adv.* 5 (11): eaay0418.

34 Lin, S., Yuk, H., Zhang, T. et al. (2016). Stretchable hydrogel electronics and devices. *Adv. Mater.* 28 (22): 4497–4505.

35 Kim, H., Ahn, S.-k., Mackie, D.M. et al. (2020). Shape morphing smart 3D actuator materials for micro soft robot. *Mater. Today* 41 (26): 243–269.

36 Wu, Z., Yang, X., and Wu, J. (2021). Conductive hydrogel- and organohydrogel-based stretchable sensors. *ACS Appl. Mater. Interfaces* 13 (2): 2128–2144.

37 Di, X., Ma, Q., Xu, Y. et al. (2021). High-performance ionic conductive poly(vinyl alcohol) hydrogels for flexible strain sensors based on a universal soaking strategy. *Mater. Chem. Front.* 5 (1): 315–323.

38 Chen, Z., Liu, J., Chen, Y. et al. (2021). Multiple-stimuli-responsive and cellulose conductive ionic hydrogel for smart wearable devices and thermal actuators. *ACS Appl. Mater. Interfaces* 13 (1): 1353–1366.

39 Wang, Y., Tebyetekerwa, M., Liu, Y. et al. (2020). Extremely stretchable and healable ionic conductive hydrogels fabricated by surface competitive coordination for human-motion detection. *Chem. Eng. J.* 2019: 127637.

40 Wang, Y., Gao, G., and Ren, X. (2021). Graphene assisted ion-conductive hydrogel with super sensitivity for strain sensor. *Polymer* 215 (19): 123340.

41 Wang, X., Wang, X., Pi, M., and Ran, R. (2022). High-strength, highly conductive and woven organic hydrogel fibers for flexible electronics. *Chem. Eng. J.* 428: 131172.

42 Qin, Z., Sun, X., Yu, Q. et al. (2020). Carbon nanotubes/hydrophobically associated hydrogels as ultrastretchable, highly sensitive, stable strain, and pressure sensors. *ACS Appl. Mater. Interfaces* 12 (4): 4944–4953.

43 Chen, J., Peng, Q., Thundat, T., and Zeng, H. (2019). Stretchable, injectable, and self-healing conductive hydrogel enabled by multiple hydrogen bonding toward wearable electronics. *Chem. Mater.* 31 (12): 4553–4563.

44 Wang, Z., Chen, J., Cong, Y. et al. (2018). Ultrastretchable strain sensors and arrays with high sensitivity and linearity based on super tough conductive hydrogels. *Chem. Mater.* 30 (21): 8062–8069.

45 Deng, Z., Guo, Y., Zhao, X. et al. (2018). Multifunctional stimuli-responsive hydrogels with self-healing, high conductivity, and rapid recovery through host–guest interactions. *Chem. Mater.* 30 (5): 1729–1742.

46 Darabi, M.A., Khosrozadeh, A., Mbeleck, R. et al. (2017). Skin-inspired multi-functional autonomic-intrinsic conductive self-healing hydrogels with pressure sensitivity, stretchability, and 3D printability. *Adv. Mater.* 29 (31): 1700533.

47 Wang, Q., Pan, X., Lin, C. et al. (2019). Biocompatible, self-wrinkled, antifreezing and stretchable hydrogel-based wearable sensor with PEDOT:sulfonated lignin as conductive materials. *Chem. Eng. J.* 370: 1039–1047.

48 Xia, S., Zhang, Q., Song, S. et al. (2019). Bioinspired dynamic cross-linking hydrogel sensors with skin-like strain and pressure sensing behaviors. *Chem. Mater.* 31 (22): 9522–9531.

49 Zhang, H., Yan, Q., Horvat, J., and Lewis, R.A. (2021). Piezoresistive and electrical properties of a catecholic amino acid–polyacrylamide single-walled carbon nanotube hydrogel hybrid network. *ACS Appl. Polym. Mater.* 3 (2): 671–678.

50 Wei, J., Xie, J., Zhang, P. et al. (2021). Bioinspired 3D printable, self-healable, and stretchable hydrogels with multiple conductivities for skin-like wearable strain sensors. *ACS Appl. Mater. Interfaces* 13 (2): 2952–2960.

51 Han, X., Lv, Z., Ran, F. et al. (2021). Green and stable piezoresistive pressure sensor based on lignin-silver hybrid nanoparticles/polyvinyl alcohol hydrogel. *Int. J. Biol. Macromol.* 176: 78–86.

52 Barty-King, C.H., Chan, C.L.C., Parker, R.M. et al. (2021). Mechanochromic, structurally colored, and edible hydrogels prepared from hydroxypropyl cellulose and gelatin. *Adv. Mater.* 33: e2102112.

53 Xu, Y., Zeng, S., Xian, W. et al. (2021). Transparency change mechanochromism based on a robust PDMS-hydrogel bilayer structure. *Macromol. Rapid Commun.* 42 (1): e2000446.

54 Sun, X., Yao, F., and Li, J. (2020). Nanocomposite hydrogel-based strain and pressure sensors: a review. *J. Mater. Chem. A* 8 (36): 18605–18623.

55 Ma, Y., Zheng, Y., Meng, H. et al. (2013). Heterogeneous PVA hydrogels with micro-cells of both positive and negative Poisson's ratios. *J. Mech. Behav. Biomed. Mater.* 23: 22–31.

56 Pierre Claver, U. and Zhao, G. (2021). Recent progress in flexible pressure sensors based electronic skin. *Adv. Eng. Mater.* 23 (5): 2001187.

57 Long, G.L. and Winefordner, J.D. (1983). Limit of detection a closer look at the IUPAC definition. *Anal. Chem.* 55 (7): 712A–724A.

58 Zhang, X., Sheng, N., Wang, L. et al. (2019). Supramolecular nanofibrillar hydrogels as highly stretchable, elastic and sensitive ionic sensors. *Mater. Horiz.* 6 (2): 326–333.

59 Han, S., Liu, C., Lin, X. et al. (2020). Dual conductive network hydrogel for a highly conductive, self-healing, anti-freezing, and non-drying strain sensor. *ACS Appl. Polym. Mater.* 2 (2): 996–1005.

60 Zhang, C., Wu, B., Zhou, Y. et al. (2020). Mussel-inspired hydrogels: from design principles to promising applications. *Chem. Soc. Rev.* 49 (11): 3605–3637.

61 Wang, Z., Cong, Y., and Fu, J. (2020). Stretchable and tough conductive hydrogels for flexible pressure and strain sensors. *J. Phys. Chem. B* 8 (16): 3437–3459.

62 He, Z. and Yuan, W. (2021). Adhesive, stretchable, and transparent organohydrogels for antifreezing, antidrying, and sensitive ionic skins. *ACS Appl. Mater. Interfaces* 13 (1): 1474–1485.

63 Huang, J., Zhao, M., Cai, Y. et al. (2020). A dual-mode wearable sensor based on bacterial cellulose reinforced hydrogels for highly sensitive strain/pressure sensing. *Adv. Electron. Mater.* 6 (1): 1900934.

64 Yang, Y., Yang, Y., Cao, Y. et al. (2021). Anti-freezing, resilient and tough hydrogels for sensitive and large-range strain and pressure sensors. *Chem. Eng. J.* 403: 126431.

65 Xia, S., Song, S., Li, Y., and Gao, G. (2019). Highly sensitive and wearable gel-based sensors with a dynamic physically cross-linked structure for strain-stimulus detection over a wide temperature range. *J. Mater. Chem. C* 7 (36): 11303–11314.

66 Zhang, D., Tang, Y., Zhang, Y. et al. (2020). Highly stretchable, self-adhesive, biocompatible, conductive hydrogels as fully polymeric strain sensors. *J. Mater. Chem. A* 8 (39): 20474–20485.

67 He, F., You, X., Gong, H. et al. (2020). Stretchable, biocompatible, and multifunctional silk fibroin-based hydrogels toward wearable strain/pressure sensors and triboelectric nanogenerators. *ACS Appl. Mater. Interfaces* 12 (5): 6442–6450.

68 Liu, X., Zhang, Q., and Gao, G. (2020). Solvent-resistant and nonswellable hydrogel conductor toward mechanical perception in diverse liquid media. *ACS Nano* 14 (10): 13709–13717.

69 Lu, F., Wang, Y., Wang, C. et al. (2020). Two-dimensional nanocellulose-enhanced high-strength, self-adhesive, and strain-sensitive poly(acrylic acid) hydrogels fabricated by a radical-induced strategy for a skin sensor. *ACS Sustainable Chem. Eng.* 8 (8): 3427–3436.

70 Liao, H., Guo, X., Wan, P., and Yu, G. (2019). Conductive MXene nanocomposite organohydrogel for flexible, healable, low-temperature tolerant strain sensors. *Adv. Funct. Mater.* 29 (39): 1904507.

71 Sun, X., Qin, Z., Ye, L. et al. (2020). Carbon nanotubes reinforced hydrogel as flexible strain sensor with high stretchability and mechanically toughness. *Chem. Eng. J.* 382: 122832.

72 Zhang, X., Chen, J., He, J. et al. (2021). Mussel-inspired adhesive and conductive hydrogel with tunable mechanical properties for wearable strain sensors. *J. Colloid Interface Sci.* 585: 420–432.

73 Qiao, H., Qi, P., Zhang, X. et al. (2019). Multiple weak H-bonds lead to highly sensitive, stretchable, self-adhesive, and self-healing ionic sensors. *ACS Appl. Mater. Interfaces* 11 (8): 7755–7763.

74 Wang, C., Hu, K., Zhao, C. et al. (2020). Customization of conductive elastomer based on PVA/PEI for stretchable sensors. *Small* 16: e1904758.

75 Zhao, M., Tang, Z., Zhang, X. et al. (2020). A self-healing, stretchable, and conductive poly(*N*-vinylpyrrolidone)/gallic acid composite hydrogel formed via hydrogen bonding for wearable electronic sensors. *Compos. Sci. Technol.* 198: 108294.

76 Zhang, H., Niu, W., and Zhang, S. (2018). Extremely stretchable, stable, and durable strain sensors based on double-network organogels. *ACS Appl. Mater. Interfaces* 10 (38): 32640–32648.

77 Fei, R., Means, A.K., Abraham, A.A. et al. (2016). Self-cleaning, thermoresponsive P (NIPAAm-*co*-AMPS) double network membranes for implanted glucose biosensors. *Macromol. Mater. Eng.* 301 (8): 935–943.

78 Lei, H., Zhao, J., Ma, X. et al. (2021). Antibacterial dual network hydrogels for sensing and human health monitoring. *Adv. Healthc. Mater.* 10: e2101089.

79 Wang, T., Zhang, Y., Liu, Q. et al. (2018). A self-healable, highly stretchable, and solution processable conductive polymer composite for ultrasensitive strain and pressure sensing. *Adv. Funct. Mater.* 28 (7): 1705551.

80 Cui, C., Shao, C., Meng, L., and Yang, J. (2019). High-strength, self-adhesive, and strain-sensitive chitosan/poly(acrylic acid) double-network nanocomposite hydrogels fabricated by salt-soaking strategy for flexible sensors. *ACS Appl. Mater. Interfaces* 11 (42): 39228–39237.

81 Chen, G., Huang, J., Gu, J. et al. (2020). Highly tough supramolecular double network hydrogel electrolytes for an artificial flexible and low-temperature tolerant sensor. *J. Mater. Chem. A* 8 (14): 6776–6784.

82 Hou, W., Sheng, N., Zhang, X. et al. (2019). Design of injectable agar/NaCl/polyacrylamide ionic hydrogels for high performance strain sensors. *Carbohydr. Polym.* 211: 322–328.

83 Wei, P., Chen, T., Chen, G. et al. (2020). Conductive self-healing nanocomposite hydrogel skin sensors with antifreezing and thermoresponsive properties. *ACS Appl. Mater. Interfaces* 12 (2): 3068–3079.

84 Zhu, T., Cheng, Y., Cao, C. et al. (2020). A semi-interpenetrating network ionic hydrogel for strain sensing with high sensitivity, large strain range, and stable cycle performance. *Chem. Eng. J.* 385: 123912.

85 Wang, H., Li, J., Yu, X. et al. (2021). Cellulose nanocrystalline hydrogel based on a choline chloride deep eutectic solvent as wearable strain sensor for human motion. *Carbohydr. Polym.* 255: 117443.

86 Wang, A., Wang, Y., Zhang, B. et al. (2021). Hydrogen-bonded network enables semi-interpenetrating ionic conductive hydrogels with high stretchability and excellent fatigue resistance for capacitive/resistive bimodal sensors. *Chem. Eng. J.* 411 (16): 128506.

87 Xia, S., Song, S., and Gao, G. (2018). Robust and flexible strain sensors based on dual physically cross-linked double network hydrogels for monitoring human-motion. *Chem. Eng. J.* 354: 817–824.

88 Kim, D., Ahn, S.K., and Yoon, J. (2019). Highly stretchable strain sensors comprising double network hydrogels fabricated by microfluidic devices. *Adv. Mater. Technol.* 4 (7): 1800739.

89 Chen, F., Tang, Z., Lu, S. et al. (2019). Fabrication and mechanical behaviors of novel supramolecular/polymer hybrid double network hydrogels. *Polymer* 168: 159–167.

90 Wu, J., Han, S., Yang, T. et al. (2018). Highly stretchable and transparent thermistor based on self-healing double network hydrogel. *ACS Appl. Mater. Interfaces* 10 (22): 19097–19105.

91 Cai, Y., Qin, J., Li, W. et al. (2019). A stretchable, conformable, and biocompatible graphene strain sensor based on a structured hydrogel for clinical application. *J. Mater. Chem. A* 7 (47): 27099–27109.

92 Wang, T., Ren, X., Bai, Y. et al. (2021). Adhesive and tough hydrogels promoted by quaternary chitosan for strain sensor. *Carbohydr. Polym.* 254: 117298.

93 Pan, X., Wang, Q., Ning, D. et al. (2018). Ultraflexible self-healing guar gum-glycerol hydrogel with injectable, antifreeze, and strain-sensitive properties. *ACS Biomater. Sci. Eng.* 4 (9): 3397–3404.

94 Zhang, D., Ren, B., Zhang, Y. et al. (2020). From design to applications of stimuli-responsive hydrogel strain sensors. *J. Mater. Chem. B* 8: 3171–3191.

95 Deng, Z., Hu, T., Lei, Q. et al. (2019). Stimuli-responsive conductive nanocomposite hydrogels with high stretchability, self-healing, adhesiveness, and 3D printability for human motion sensing. *ACS Appl. Mater. Interfaces* 11 (7): 6796–6808.

96 Jin, X., Jiang, H., Zhang, Z. et al. (2021). Ultrastretchable, self-adhesive, strain-sensitive and self-healing GO@DA/alginate/P(AAc-*co*-AAm) multifunctional hydrogels via mussel-inspired chemistry. *Carbohydr. Polym.* 254: 117316.

97 Jing, X., Wang, X., Mi, H., and Turng, L.-S. (2019). Stretchable gelatin/silver nanowires composite hydrogels for detecting human motion. *Mater. Lett.* 237: 53–56.

98 Lin, F., Wang, Z., Shen, Y. et al. (2019). Natural skin-inspired versatile cellulose biomimetic hydrogels. *J. Mater. Chem. A* 7 (46): 26442–26455.

99 Wang, Q., Pan, X., Lin, C. et al. (2020). Modified Ti_3C_2TX (MXene) nanosheet-catalyzed self-assembled, anti-aggregated, ultra-stretchable, conductive hydrogels for wearable bioelectronics. *Chem. Eng. J.* 401 (6): 126129.

100 Wu, X., Liao, H., Ma, D. et al. (2020). A wearable, self-adhesive, long-lastingly moist and healable epidermal sensor assembled from conductive MXene nanocomposites. *J. Mater. Chem. C* 8 (5): 1788–1795.

101 Zhang, Y.-Z., Lee, K.H., Anjum, D.H. et al. (2018). MXenes stretch hydrogel sensor performance to new limits. *Sci. Adv.* 4 (6): eaat0098.

102 Pan, X., Wang, Q., He, P. et al. (2020). A bionic tactile plastic hydrogel-based electronic skin constructed by a nerve-like nanonetwork combining stretchable, compliant, and self-healing properties. *Chem. Eng. J.* 379: 122271.

103 Chang, Q., Darabi, M.A., Liu, Y. et al. (2019). Hydrogels from natural egg white with extraordinary stretchability, direct-writing 3D printability and self-healing for fabrication of electronic sensors and actuators. *J. Mater. Chem. A* 7 (42): 24626–24640.

104 Li, F., Liu, T., Gu, W. et al. (2021). Bioinspired super-tough and multifunctional soy protein-based material via a facile approach. *Chem. Eng. J.* 405: 126700.

105 Yin, J., Pan, S., Wu, L. et al. (2020). A self-adhesive wearable strain sensor based on a highly stretchable, tough, self-healing and ultra-sensitive ionic hydrogel. *J. Mater. Chem. C* 8 (48): 17349–17364.

106 Ge, G., Lu, Y., Qu, X. et al. Muscle-inspired self-healing hydrogels for strain and temperature sensor. *ACS Nano* 14: 218–228.

107 Ge, G., Yuan, W., Zhao, W. et al. (2019). Highly stretchable and autonomously healable epidermal sensor based on multi-functional hydrogel frameworks. *J. Mater. Chem. A* 7 (11): 5949–5956.

108 Chen, D., Zhao, X., Wei, X. et al. (2020). Ultrastretchable, tough, antifreezing, and conductive cellulose hydrogel for wearable strain sensor. *ACS Appl. Mater. Interfaces* 12 (47): 53247–53256.

109 Yu, Q., Qin, Z., Ji, F. et al. (2021). Low-temperature tolerant strain sensors based on triple crosslinked organohydrogels with ultrastretchability. *Chem. Eng. J.* 404: 126559.

110 Murphy, E.B. and Wudl, F. (2010). The world of smart healable materials. *Prog. Polym. Sci.* 35 (1, 2). 223–251.

111 Chen, F., Zhou, D., Wang, J. et al. (2018). Rational fabrication of anti-freezing, non-drying tough organohydrogels by one-pot solvent displacement. *Angew. Chem. Int. Ed.* 57 (22): 6568–6571.

112 Morelle, X.P., Illeperuma, W.R., Tian, K. et al. (2018). Highly stretchable and tough hydrogels below water freezing temperature. *Adv. Mater.* 30 (35): e1801541.

113 Pan, S., Xia, M., Li, H. et al. (2020). Transparent, high-strength, stretchable, sensitive and anti-freezing poly(vinyl alcohol) ionic hydrogel strain sensors for human motion monitoring. *J. Mater. Chem. C* 8 (8): 2827–2837.

114 Kamata, H., Akagi, Y., Kayasuga-Kariya, Y. et al. (2014). "Nonswellable" hydrogel without mechanical hysteresis. *Science* 343 (6173): 873–875.

115 Xu, J., Jin, R., Ren, X., and Gao, G. (2019). Cartilage-inspired hydrogel strain sensors with ultrahigh toughness, good self-recovery and stable anti-swelling properties. *J. Mater. Chem. A* 7 (44): 25441–25448.

116 Wang, Z., Chen, J., Wang, L. et al. (2019). Flexible and wearable strain sensors based on tough and self-adhesive ion conducting hydrogels. *J. Mater. Chem. B* 7 (1): 24–29.

117 Liao, M., Wan, P., Wen, J. et al. (2017). Wearable, healable, and adhesive epidermal sensors assembled from mussel-inspired conductive hybrid hydrogel framework. *Adv. Funct. Mater.* 27 (48): 1703852.

118 Huang, J., Liu, Y., Chi, X. et al. (2021). Programming electronic skin with diverse skin-like properties. *J. Mater. Chem. A* 9 (2): 963–973.

119 Zhang, Z., Chen, Z., Wang, Y., and Zhao, Y. (2020). Bioinspired conductive cellulose liquid-crystal hydrogels as multifunctional electrical skins. *Proc. Natl. Acad. Sci. U.S.A.* 117 (31): 18310–18316.

120 Wu, S., Shao, Z., Xie, H. et al. (2021). Salt-mediated triple shape-memory ionic conductive polyampholyte hydrogel for wearable flexible electronics. *J. Mater. Chem. A* 9 (2): 1048–1061.

121 Kweon, O.Y., Samanta, S.K., Won, Y. et al. (2019). Stretchable and self-healable conductive hydrogels for wearable multimodal touch sensors with thermoresponsive behavior. *ACS Appl. Mater. Interfaces* 11 (29): 26134–26143.

122 Ren, Y. and Feng, J. (2020). Skin-inspired multifunctional luminescent hydrogel containing layered rare-earth hydroxide with 3D printability for human motion sensing. *ACS Appl. Mater. Interfaces* 12 (6): 6797–6805.

123 Han, L., Huang, H., Fu, X. et al. (2020). A flexible, high-voltage and safe zwitterionic natural polymer hydrogel electrolyte for high-energy-density zinc-ion hybrid supercapacitor. *Chem. Eng. J.* 392: 123733.

124 Chen, N., Tian, L., Patil, A.C. et al. (2017). Neural interfaces engineered via micro- and nanostructured coatings. *Nano Today* 14: 59–83.

125 Nguyen, L.H., Gao, M., Lin, J. et al. (2017). Three-dimensional aligned nanofibers-hydrogel scaffold for controlled non-viral drug/gene delivery to direct axon regeneration in spinal cord injury treatment. *Sci. Rep.* 7: 42212.

126 Goding, J., Gilmour, A., Martens, P. et al. (2017). Interpenetrating conducting hydrogel materials for neural interfacing electrodes. *Adv. Healthc. Mater.* 6 (9): 1601177.

127 Hodgkin, A.L. and Huxley, A.F. (1939). Action potentials recorded from inside a nerve fibre. *Nature* 144: 710–711.

128 Araki, T., Yoshida, F., Uemura, T. et al. (2019). Long-term implantable, flexible, and transparent neural interface based on Ag/Au core–shell nanowires. *Adv. Healthc. Mater.* 8 (10): e1900130.

129 Huang, W.-C., Ong, X.C., Kwon, I.S. et al. (2018). Ultracompliant hydrogel-based neural interfaces fabricated by aqueous-phase microtransfer printing. *Adv. Funct. Mater.* 28 (29): 1801059.

130 Keplinger, C., Sun, J.-Y., Foo, C.C. et al. (2013). Stretchable, transparent, ionic conductors. *Science* 341 (6149): 984–987.

131 Castagnola, E., Ansaldo, A., Maggiolini, E. et al. (2013). Biologically compatible neural interface to safely couple nanocoated electrodes to the surface of the brain. *ACS Nano* 7 (5): 3887–3895.

132 Sheng, X., Qin, Z., Xu, H. et al. (2020). Soft ionic-hydrogel electrodes for electroencephalography signal recording. *Sci. China Technol. Sci.* 61: 371.

133 Oribe, S., Yoshida, S., Kusama, S. et al. (2019). Hydrogel-based organic subdural electrode with high conformability to brain surface. *Sci. Rep.* 9 (1): 13379.

134 Torris, A. and Badiger, M.V. (2019). Polysaccharide hydrogel incorporated carbon nanofiber microelectrode for designing neural interfaces. *J. Bionic Eng.* 16 (4): 696–710.

135 Wang, K., Tian, L., Wang, T. et al. (2019). Electrodeposition of alginate with PEDOT/PSS coated MWCNTs to make an interpenetrating conducting hydrogel for neural interface. *Compos. Interfaces* 26 (1): 27–40.

136 Staples, N.A., Goding, J.A., Gilmour, A.D. et al. (2017). Conductive hydrogel electrodes for delivery of long-term high frequency pulses. *Front. Neurosci.* 11: 748.

137 Ye, F., Li, M., Ke, D. et al. (2019). Ultrafast self-healing and injectable conductive hydrogel for strain and pressure sensors. *Adv. Mater. Technol.* 4 (9): 1900346.

138 Heo, D.N., Song, S.-J., Kim, H.-J. et al. (2016). Multifunctional hydrogel coatings on the surface of neural cuff electrode for improving electrode-nerve tissue interfaces. *Acta Biomater.* 39: 25–33.

139 Spyropoulos, G.D., Savarin, J., Gomez, E.F. et al. (2020). Transcranial electrical stimulation and recording of brain activity using freestanding plant-based conducting polymer hydrogel composites. *Adv. Mater. Technol.* 5 (3): 1900652.

140 Han, L., Lu, X., Wang, M. et al. (2017). A mussel-inspired conductive, self-adhesive, and self-healable tough hydrogel as cell stimulators and implantable bioelectronics. *Small* 13 (2): 1601916.

141 Adewole, D.O., Struzyna, L.A., Burrell, J.C. et al. (2021). Development of optically controlled "living electrodes" with long-projecting axon tracts for a synaptic brain–machine interface. *Sci. Adv.* 7: eaay5347.

142 Huang, W., Chi, H., Lee, Y. et al. (2019). Gene-embedded nanostructural biotic–abiotic optoelectrode arrays applied for synchronous brain optogenetics and neural signal recording. *ACS Appl. Mater. Interfaces* 11 (12): 11270–11282.

143 Won, H.J., Ryplida, B., Kim, S.G. et al. (2020). Diselenide-bridged carbon-dot-mediated self-healing, conductive, and adhesive wireless hydrogel sensors for label-free breast cancer detection. *ACS Nano* 14 (7): 8409–8420.

144 Roche, E.T., Horvath, M.A., Wamala, I. et al. (2017). Soft robotic sleeve supports heart function. *Sci. Transl. Med.* 9: eaaf3925.

145 Deisseroth, K., Feng, G., Majewska, A.K. et al. (2006). Next-generation optical technologies for illuminating genetically targeted brain circuits. *J. Neurosci.* 26 (41): 10380–10386.

146 Tsai, M.-J., Cheng, H., Ho, H.-H. et al. (2020). Hydrogel-based zinc ion sensor on optical fiber with high resolution and application to neural cells. *Biosens. Bioelectron.* 162: 112230.

Index

a

ab initio steered molecular dynamics (AISMD) 10
acylhydrazine-terminated PEO 282
adamantane (Ad) 39, 41, 288
adamantanecarboxylic acid 39
adenosine triphosphate (ATP) 286
aggregation-induced emission (AIE) 89, 282
aldehyde-poly(oligoethylene glycol methacrylate) (POEGMA) 166
α-particles 136–137
aminophenyl fluorescein (APF) 143
aminopyridine 69
aminothiophene 69
amphiphilic chitosan 362
amphiphilic sodium lauryl sulfate 351
anils 69, 70
anisotropic hydrogel actuators 301, 302, 308, 310, 322
anisotropic structures
  innovative technology 322–325
  macromolecular assembly 322
  stepwise polymerization 318–320
  3D printing 320–321
anthracene derived Diels–Alder adducts 20
9-anthracene ester-terminated polydimethylsiloxane 101, 102
anthracene-maleimide adduct-embedded polymer chain 20
anticounterfeiting/authentication technologies 56, 61
anti-head-to-head dimer 75, 78
artificial muscles 301, 302, 325, 327–328, 332
asymmetric CNTs-elastomer/hydrogel composite 327
Au/AgNWs-based neural interface coated with PVA hydrogels 357
autocorrelation experiments 214, 217, 245
averaged single-scattered intensity 216
average ionization energy (W_e) 136
azobenzene 54, 63, 68, 95, 96, 98, 101, 103–109, 112, 114, 115, 139
azobenzene-containing polymers 101
4,4-azodibenzoic acid 109

b

Batchelor–Einstein equation 231
Bell's model 911
benzoin acid (BA) 141
benzopyran 16, 17, 63, 64
benzoxazole aliphatic ester 29
benzyl-based PPGs 71, 72
β-cyclodextrin (β-CD) 105, 288
Bethe–Bloch formula 136, 137
bilayer hydrogel actuator 297, 309–311, 320, 322, 329
bioadhesives 362
bio-inspired mechanosensing systems and soft actuators/robotics 294
biological hydrogels 342
biomimetic devices 301, 325, 329, 332
bis(naphthopyran) 18, 19

Smart Stimuli-Responsive Polymers, Films, and Gels, First Edition.
Edited by Liang Hu, Yongfeng Gao, and Michael J. Serpe.

bis(adamantyl)-1,2-dioxetane
mechanoluminophore
functionalized polymers 28
bond rupture probability densities 13
borate derivatives 61
4-bromobenzaldehyde 62
2-bromo-5-hydroxybenzaldehyde 62, 85
α-bromonaphthalene 86
4-bromo-1,8-naphthalic anhydride 62, 86
bulk polymers 12, 22, 30, 35, 38
butyl methacrylate (BA) 88, 205

c

capacitive strain sensors 343, 345
carbazole 61, 111
carbon-based nanomaterials 181
carbon fiber 343
carbon nanosheets 181
carbon nanotube (CNT) 169, 316, 341, 343
carbonyl-based PPGs 71, 72
5(6)-carboxytetramrthylrhodamine (TAMRA) 143
cell culture 111, 162–167, 176
cellulose nanocrystals (CNCs) 161, 289
chemically cross-linked poly(acrylamide) hydrogel 284
chemical potential of solvent molecule 209
chemical sensors 339
chemical vapor deposition (CVD) 158
chromene 63
cinnamate derivatives 54, 71
cis-trans isomerization 54, 63, 101
CNTs/Ecoflex 327
coefficient of thermal expansion (CTE) 158
coherent scattering 138, 219, 224, 225, 248
colloidal core-shell particles 222
colloidal suspensions 205, 220, 231, 235
color change and emission 117
color-changing mechanoresponsive fibers 14
composite organohydrogel 329
compressibility of microgels 204, 228
swelling thermodynamics 207–210
Compton effect 138–139
Compton scattering 138, 139
conductive hydrogel 342, 343, 354, 356, 357, 358, 360, 362
conjugated polymer sensors 89
constant thermal fluctuations 11
constrained geometry simulating external force (CoGEF) 15
contrast variation 205, 211, 222–227, 235, 236
controlled/living radical polymerization (CRP) 70
conventional emulsion polymerization and SEP 184–186
core–shell microgels 182, 184, 189, 190, 195, 253, 255, 257, 259, 261, 262
coumarin 20, 21, 54, 57–59, 71, 72, 73, 75, 77, 78, 84, 93, 95, 96
coumarin-3-carboxylic acid (CCA) 141, 143
coumarin-containing polymers 84, 117
coumarin-derived isomeric hydrogelators 286, 287
coumarin-functionalized telechelic associative model polymer 101
coumarin-4-ylmethyl derivatives 54, 71
COVID-19 pandemic 175
cross correlation and 3D-DLS experiments 214–216
cryo-electron tomography (Cryo-ET) techniques 191
CT dyes 57
cyanines 58, 89
cyanostilbene 89
cyclobutane mechanophore-fused lactone polymer (poly(CBL)) 22
cyclobutenes 23
cyclopropane 23
cyclosporine A (CsA) 360

d

dendrimer based sensors 89
deswollen state 203, 204, 209
diacetone acrylamide (DAA) 284
diarylbibenzofuranone (DABBF) 25
diarylethenes 54, 63, 68, 71, 106
dibenzothiophene 61
di(1-hydroxylundecyl) diselenide 147
difluorenylsuccinonitrile (DFSN) 26
difluoroboron dibenzoylmethane 62
N,N-dimethylacrylamide (DMA) 284
N,N'-dimethylaminoethyl methacrylate 95
dinitrobenzylpyridine (DNBP) 63, 68, 70
diphenyl(2,4,6-trimethylbenzoyl) phosphine oxide (TPO) 320
direct-ink-write (DIW) 316, 320
diselenide-containing carbon dots (dsCD) 363
diselenide-containing polymer 147, 148
dissipative particle dynamics simulations 206
9,10-distrylanthracene (DSA) 89
1,2-disubsituted benzocyclobutenes (BCB) 8
diverse gradient hydrogel actuators 312
Donnan equilibrium 325
Donnan potential 231
dopamine/alginate/poly(acrylic acid-co-acrylamide) hydrogel 351
dopamine-talc-polyacrylamide (DTPAM) hydrogels 341
dosimeters 136, 139–141, 143, 149, 150
DOX-loaded nanocarriers 147
d3-pNIPAM-based microgels 227
drug-delivery 33, 34, 57, 71, 111
dual mechanophore-5,5-dichlorotricyclo (7.2.0.0)undecane (DCTCU) 22
dual-networks glycol organogel 349
dye dosimeters 139–141
dynamic light scattering (DLS) 195, 204, 205, 211–216, 213, 241, 242–245

e

E-4031 164
electrical bio-adhesive interface 362
electric field auto-correlation function 213
electroactive hydrogel (EAH) 325, 326
electroactive surface area (ESA) 173
electrochemical sensors 173, 175, 341, 358
electronic communication technology 342
electronic conductive hydrogel 343
electronic energy transfer (EET) 89
electron-ion pairs 136
electrons 17, 24, 25, 56–58, 70, 89, 110, 137, 138, 139, 147
emulsions stabilized by nanocomposite microgels 194
energy of the electrons 9
epoxy-functionalized fluorescent polymer nanoparticles 84
ethyl acrylate (EA) 205
ethylene diamine tetraacetic acid (EDTA) 324
3,4-ethylenedioxythiophene 343, 358, 362
europium-doped barium fluorochloride 140
excitation 20, 29, 57, 58, 60, 86, 91, 103, 136, 137
excited-state intramolecular proton transfer (ESIPT) 70, 89
exponential attenuation law 138

f

femtosecond laser 320, 321
Fermi pseudo-potential 217
ferrocene mechanopolymers 30
film dosimeters 139
films 83, 86, 107, 110, 112, 157, 158, 160, 161–167, 169
flexible hydrogels-based sensors
  expanded usage scenarios 353
  mechanical properties 347–351

flexible hydrogels-based sensors (*contd.*)
  multi-functionalization and expanding application 354–355
  principles of conductive hydrogel sensors 343–347
  prolonged longevity of 351–353
flexible wearable sensors 339
flexible wrinkled carbon nanotube (CNT)-based sensor 169
Flory–Rehner equation of state 210
Flory–Rehner theory 203
Flory solvency parameter 208
Flory's theory 325
flow properties of colloidal suspensions 231
fluorene derivatives 61
fluorescein 57, 58, 60, 88, 89, 93, 143, 282, 292
fluorescence resonance energy transfer (FRET) 89
fluorescent chemosensors 89, 290
fluorescent compounds 57, 58, 59, 141
fluorescent ink pattern 84
fluorescently labelled sensitive hydrogel (FLASH) 292
fluorescent merocyanine (MC) 63
fluorescent polymer dosimeters 141–143
fluorescent polymeric hydrogels 281–298
fluorescent polymers 56–58
fluorogen-containing vinyl monomers 284
fluorogen-grafted FPHs 285
force dependency of reaction rate 11
form factor 169, 220–223, 227, 243, 244, 251, 256, 257, 261
four-armed poly(ethylene glycol) 107
free radical polymerization 25, 70
fulgide 54, 63
fumaric acid (FAc) 188
furylfugides 63

g

γ-ray-responsive diselenated amphiphilic block copolymer 147
gelatine methacryloyl (GelMA) 292
gelochromism phenomenon 117
gene-embedded optoelectrode array 362
glucose sensors 174, 176, 291, 292
glutathione (GSH) 363
glycerol hydrogel 358
glycidyl methacrylate (GMA) 84, 88, 186
gold nanoparticles (AuNPs) 39, 143, 146, 175, 327
graft copolymers 95, 106
graphene oxide (GO) 343

h

hemithioindigo 63, 66
Hertzian potential 228
hexaarylbiimidazole (HABI) 27, 68, 79
hexamethylene diisocyanate (HDI) 109
hollow microgels 207, 224, 232, 257, 269
homogeneous PAAm hydrogel 312
homogeneous poly(*N*-isopropylacrylamide-*co*-acrylamide) (poly(NIPAM-co-AAm)) hydrogels 322
hybrid conductive hydrogels 343
hydrazide-POEGMA hydrogels 166
hydrogel 339
  actuators
    addition of active ingredient 302–305
    anisotropic structures 310–318
    artificial muscles 327–328
    biomimetic devices 329
    information storage materials 329–332
    methods to fabricate anisotropic structures 318–325
    overview of 303
    pneumatic/hydraulic actuators 305–306
    soft robots 325–327
    stimuli-responsive 306–310

stimuli-responsive
swelling/shrinking property 301
3D anisotropic structures 315–318
bioelectronic devices 363
hydrogel-based dosimeters 140
hydrogen bond 29, 30, 39, 61, 84, 98, 147, 283, 286, 289, 329, 341, 347
hydrogen transfer 54, 63, 69
hydrophilic/hydrophobic block copolymers 95
hydrophilic poly(ethylene glycol) moieties 285
hydrophobic aggregation-induced emissive fluorogens (AIEgens) 282
hydrophobically associated PAAm hydrogels 349
hydrophobic monomers 182, 184, 186, 192, 195
hydrophobic TPE 285

i

imidazole containing brush copolymers (ICPs) 34
implanted bioscaffolds 292
imprinted polymer-based sensors 89
indolinofulgides 63
information storage materials 329–332
inherent smart polymers 53
inorganic nanofillers 347, 349
in-situ radical solution polymerization 181
intensity auto-correlation function 213
intermolecular cyclization 74
interpenetrating cross-linked networks (IPN) 349
intramolecular charge transfer (ICT) 57, 89, 331
intrinsic conductive hydrogels 343
ionic conductive hydrogels 343
ionic dissociation 54, 63
ionic effects 209–210
ionic microgels 205, 209, 231, 256, 267
ionic strength-responsive hydrogel actuator 308
ionizing radiation 135
dosimetry
fluorescent polymer dosimeters 141–143
polymer-based dosimeters 139
polymer/dye dosimeters 139–141
polymer/metal nanomaterials dosimeters 143–146
ionizing-radiation-responsive hydrogels 147
ionizing radiation-responsive polymer systems for therapy 146–149
ion pairs 136
irreversible photolabile protecting groups 73
isophorone diisocyanate (IPDI) 147
isoprenaline 164
isotropic deswelling 206, 207, 229, 235

j

Jablonski diagram 57, 58

k

keto–enol phototautomerism 70

l

ladderenes 15, 23–24
lamina 2–4, 160, 161
lanthanide coordinated fluorescent particles 283
Laponite nanoparticles 312
leucomalachite green (LMG) 139
light-actuating shape-changing polymer systems 106
light-induced healable polymers 103
light-responsive cross-linkable compounds 74
light-responsive hydrogel actuators 308
light-responsive polymers 53, 55, 116, 308
light-sensitive molecules 71, 72
living electrode 362
luminescent carbon dots (CDs) 56

m

MAc comonomers 184
macrogels 203
macromolecular assembly 322
magnetic and thermal dual-responsive hydrogel actuator 305
magnetic nanoparticles 181
manually stretched DABBF-linked polymer 27
mechanochemistry 1
  in biological systems 2–7
  mechanistic view of 7–15
  polymer covalent 15–29
  polymer noncovalent mechanochemistry 29–41
mechanochromic bis(naphthopyran) (BNP) 18
mechanochromism 17, 19, 25, 30, 32
mechanofluorescent hydrogels 295
mechanophore oxanorbornadiene 20
mechanophores 2, 13, 15, 17, 18, 20, 22–30, 41
mechano-responses of metal-ligand bonds 30
mechanotransduction 1–4, 6, 7, 41
membranes 2, 4, 7, 54, 112, 114–116, 167
mercaptoethylamine 205
merocyanine (MC) 16, 63, 101, 140
metal–ligand charge transfer (MLCT) 89
metal-ligand mechanopolymers 32, 33
metal particles 343
7-methoxycoumarinyl-4-methyl derivatives 72
methylene blue (MB) 95, 139
methyl methacrylate (MMA) 38, 86, 205
methyl orange (MO) 139
microgel elasticity 209
microgel hydrodynamic radii 229
microgels 181–196, 203–236
microgel-stabilized emulsions (MS-Es) 193
microrganogel 329
micro tissue engineered neural networks (μTENNs) 362
molecular dynamics 206
multi-layered nanocomposite microgels 189–190
multi stimuli-responsive hydrogel actuators 306, 308–310
mussel-inspired proanthocyanidins-coated cellulose nanofibrils nanocomposites 358

n

nanoclay 141
nanocomposite/composite hydrogels 181
nanocomposite microgels
  applications of 193–195
  background on 182–184
  characterization of 190–192
  conventional emulsion polymerization and SEP 184–186
  internal structure design 188–189
  multi-layered 189–190
  prepared by SEP in the presence of microgels 186–187
  seeded emulsion polymerization (SEP) 181
nanofibrillar fluorescent hydrogels 289
nanofibrillated cellulose (NFC) 320
nanoparticles 20, 39, 56, 84, 86, 96, 136, 139, 143, 145, 146, 149, 175, 181
naphthalene imide 62
naphthopyran (NP) 17–19, 18, 63
2-naphthoquinone-3-methides groups 54
negatively charged exfoliated Laponite nanosheets 287
negatively charged poly(sodium *p*-styrenesulfonate) (PSS) 286
neutral microgels 235
2-nitrobenzyl (NB) series 71
nitro blue tetrazolium (NBT) 139
nitro blue tetrazolium (NBT)-gelation system 140
nitrogen-containing heteroaromatic derivatives 61
non-fluorescent spiropyran (SP) 63

non-responsive hydrogel (NGR) 322
non-spherical nanocomposite microgels 186
nonviral gene vectors 362
non-zero wavevectors 213

o
oligourethane gelator (OUG) 289
ONB derivatives 71, 72
1D/2D anisotropic structures
  bilayer hydrogel actuator 310
  gradient 312–315
  oriented hydrogel 311
  patterned hydrogels 315
o-nitrobenzyl acrylate 95
optogenetics 361–363
organic dyes 57, 139
organic fluorescent compounds 58, 59
organic fluorescent dyes 56, 57
organic phosphorescent compounds 55, 62
organo-boron complexes 89
organohydrogel 329, 330
oriented hydrogel 311–313
Ornstein–Zernicke equation 222
over-crowed alkene 63
oxazine 18, 54, 63
oxazine (OX)-derived mechanophores 17
oxazolidine 63, 84, 85

p
PAA-alginate hydrogel 37
PAAm-*co*-PAAc 329
PAAm-PAAc anisotropic hydrogel actuator 308
pair production 138, 139
para-hydroxyphenylacetic acid 72
patterned hydrogels 315, 316
P(BCH-Naph) 24
PEG capped iron oxide-based nanoparticles (IOPs) 302
PEG-PUSeSe-PEG 147
peptide target sequence Cys-Lys-Lys(Ac)-d-Ala-d-Ala (DADA) 39
persulfurated derivatives 61
perylene-tetracarboxylic-acid-modified gelatin (PTG) 332
P(PyMA-*co*-AAc) fluorescent hydrogel 329
phenacyl esters 54, 71, 72
phosphorescence phenomena 60
phosphorescent polymers 55, 58–62
photoadaptable polymers 78–83
photochromic polymer inks 86
photochromic polymers 62–71
photochromic sensors 91
photochromism 62, 63, 68–70, 83, 84
photocleavable polymers 71–74
photoconductive polymers 54, 71, 109–111
photocromic polymeric chemosensors 89
photocross-linking reaction 54, 74, 75, 103, 106
photocycloaddition reactions 75
photodegradable polymers 71, 72
photodimerizable polymers 74–78
photodynamic covalent polymers 79, 83
photoelectric effect 138
photoinduced cyclization 74
photoinduced electron transfer (PET) 70, 89, 284
photoisomerizable arylazopyrazole (*trans*-AAP) 105
photolabile *ortho*-nitrobenzyl ester (ONB) 54
photolabile protecting groups (PPGs) 71, 73
photolithography 72, 93–95
photoluminescent cellulosic papers 84
photoluminescent polymers 55–62
photomelting azobenzene monomer 107
photons 55–57, 137–139, 143
photoresponsive O/W emulsions 95
photoresponsive polymer composites 53

photoresponsive polymers
- color change and emission 117
- drug-delivery 111–112
- films 112–115
- membranes 112–115
- photoadaptable polymers 78–83
- photochromic polymers 62–71
- photocleavable polymers 71–74
- photoconductive polymers 109–111
- photodimerizable polymers 74–78
- photolithography 93–95
- photoluminescent polymers 55–62
- photorheological polymers 98–103
- polymer sensors 89–93
- self-healing polymers 103–106
- shape-changing polymers 106–109
- smart polymeric inks 83–88
- sol-gel transition 117
- surface active agents 95–98
- textiles 112–115
- viscosity variation 116–117
- water contact angle variation 116

photoreversible cross-linked epoxy polymers 106
photorheological polymers 98–103
photoswitchable polyacrylic nanofibers 86
pH-responsive fluorescence molecule 332
pH-responsive hydrogel actuators 307
pH-responsive PAAm-PAAc bilayer hydrogel actuator 329
pH-responsive polymers 53
physical sensors 339
physical vapor deposition (PVD) 157
Piezo proteins 7
π-extended anthracene-maleimide adducts 20
[2π+2π] or [4π +4π] cycloaddition mechanism 74
[4π +4π] or [2π +2π] cycloaddition mechanisms 54
Planck's constant 217
plant-based aloe vera hydrogel 360
pneumatic/hydraulic actuators 305
pNIPAM-based microgels 234
POEGMA-CNC fibers 166
polyacid derivatives 61
poly(acrylamide-*co*-2-acrylamido-2-methylpropanesulfonic acid) [P(AAm-*co*-AMPS)] hydrogels 315
poly(acrylamide-*co*-methacrylic acid (MAc)) microgels 182
poly(acrylamide-*co*-*N*-(3-amino-propyl) methyl acrylamide) (PAAm-*co*-APMA) nanogels 143
polyacrylamide (PAAm)/chitosan (CS)/multi-wall carbon nanotubes (MWCNTs) 341
poly(benzoLDE) 24
polycyclic aromatics 71
poly(2-(dimethylamino)ethyl methacrylate) (PDMAEMA) 282
polydimethylsiloxane (PDMS) 30, 101, 340
polydopamine (PDA) adhesive layer 175
polyelectrolytes-based hydrogels 343
poly(3,4-ethylenedioxythiophene)/ poly(styrenesulfonate) (PEDOT/PSS)-coated MWCNTs 358
poly(ethylene glycol) (PEG) 101, 147
- hydrogel 101, 302, 357

poly(ethylene oxide)-*b*-PSPA (PEO-*b*-PSPA) 112
poly(3-hexylthiophene) (P3HT) 111
poly(lactic acid) 61
poly(lactic-*co*-glycol acid) (PLGA) microspheres (MS) 360
polyladderene (PLDE) 23, 24
polymer-based dosimeters 136, 139
polymer-based phosphorescent materials 61
polymer covalent mechanochemistry
- ladderenes 23–24
- mechanophores 27–29
- pyran-based mechanochromophores 16–19

retro-cycloadditions 20–22
stable radical systems 24–27
polymer dots (PDs) 53, 56
polymer/dye dosimeters 139–141
polymer gel-based dosimeters 140
polymeric fluorescence chemosensors 89
polymeric hydrogels 105, 140, 281–298
polymer mechanochemistry 2, 15, 27, 30, 37
polymer/metal nanomaterials dosimeters 143–146
polymer nanofillers 351
polymer noncovalent mechanochemistry
functional polymers 37–41
mechano-responses of metal-ligand bonds 30–37
polymers containing pendant ligand groups 89
polymer sensors 71, 89–93
polymer/solvent mixing 208
poly(methacryloxyethyltrimethyl ammonium chloride) (PMETAC) 308
polymethyl acrylate (PMA) matrix 20
poly(methyl methacrylate) (PMMA) 26, 61, 143, 340
microspheres 186
poly(methyl methacrylate-*co*-*n*-butyl acrylate) 38
poly(NIPAm-*co*-FAc) microgels 188, 191
poly(NIPAm-*co*-MAc) microgels 188, 189
poly(*N*-isopropyl acrylamide) (pNIPAm) 20, 41, 91, 204, 307
hydrogels 20, 41, 305
microgels 182, 186
microgel suspensions 217
poly(*N*-isopropylacrylamide) (PNIPAAm) 20, 41, 91
poly(*N*-isopropylacrylamide-*co*-acrylic acid) (PNIPAm-*co*-AAc) 230
poly(*N*-vinyl carbazole) 109
poly(*N*-vinylpyrrolidone)/gallic acid (PVP/GA) composite hydrogel 347
poly(oligo(ethylene glycol) methyl ether methacrylate) (POEGMEMA) polymer chains 39
poly(propylene glycol) (PPG) 101, 285
polypyrrole (PPy) 329, 343
poly(sodium acrylate) (PNaAAc) hydrogel 315
polystearyl methacrylate (PSMA) 329
polystyrene (PS) 27, 87, 186
films 160
polystyrene (PS)-based composite microfibers 140, 141
polystyrene-poly(NIPAm-*co*-FAc) nanocomposite microgels 188, 193
polythiophene/poly(3,4-ethylenedioxythiophene) (PEDOT) 343
polyurethane (PU) 25, 27, 28, 101, 102, 139, 340
poly(vinyl acetate) 109
poly(vinyl alcohol) (PVA) 283, 322
hydrogels 61, 109, 345
poly(vinyl carbazole) 110
polyvinyl chloride (PVC) films 143
poly(vinyl fluoride) 109
poly(vinyl pyrene) 110
polyzwitterionic hydrogel actuator 308
precipitation polymerization 182–184, 188, 189, 190, 204
PRESAGE® dosimeters 139
protein unfolding by force 4–6
pure organic room-temperature phosphorescence luminophores 61
PVP/GA composite hydrogels 347
pyran-based mechanochromophores 16–19
pyrenylmethyl 54, 71, 72, 329
1-pyrenylmethyl acrylate (PyMA) 329

pyrenylmethyl esters 72
pyromellitic acid (PMA) 141
pyrrolidine-appended naphthopyrans 18

q
quantum dots (QDs) 56, 284
quinoline-malononitrile (QM) 89

r
radiation dose 135, 139, 140, 143, 149
radiation-matter interaction
  α-particles 136
  electrons 137
  photons 137–139
radical-based mechanochromophores 25, 26
raspberry-shaped composite microgels 186, 187
ratiometric fluorescence dosimeter 143
reaction pathway 7, 8
reduced graphene oxide (rGO) 309, 360
resistive strain sensors 343
resonant dyes 57
responsive hydrogel (RH) 39, 147, 305, 322
retro-cycloadditions 15, 19–22
reversible chromophore-containing polymeric surfactants 95
reversible photodimerization reactions 75
reversible photoresponsive chromophores 112
rhodamine 28, 57, 58, 84, 89
ring formation/cleavage 54
ring-opening metathesis polymerization (ROMP) 23, 91, 107

s
salicylideneaniline 63, 67, 69
SANS setup 218
SARS-CoV-2 spike protein 175
scattering angle 211, 212, 214, 215, 217
scattering length density (SLD) 219, 223, 224, 227, 248, 249, 253
scattering theory 218–220
security ink 55, 71, 84
seeded emulsion polymerization (SEP) 181–196
seeded precipitation polymerization 182, 184, 189
self-healing mechanism 38, 229, 352, 353, 361
self-healing polymers 37, 54, 71, 103–106
semiconductor dosimeters 139
sensing by ion channel opening 6–7
Se-polymer 147, 149
shape-changing polymers 106–109
silole 89
silver nanowires (AgNWs) 351
single-molecule spectroscopy 12
single scattered intensity 216
single stimulus-responsive hydrogel actuators 307–308
small-angle neutron scattering (SANS) 205, 216–228, 235
small-molecule fluorogen-containing monomer units 286
smart photoluminescent nano- or microgel hybrids 284
smart photoresist polymers 93
smart polymeric inks 54, 83–88
smart polymers 53, 54, 58, 83, 89
soft actuators/robotics 281, 294–297
soft alginic acid gel 358
soft robots 285, 297, 325–327
sol-gel transition 117
solid-like nanocomposite microspheres 186
solvent-based polymeric inks 83
spherical polystyrene-core/poly(NIPAm-*co*-FAc)-hydrogel-shell core-shell microgels 195
spirooxazine 54, 63, 88
spiropyran (SP) 16, 54, 63, 68, 71, 86, 91, 139, 140
spiropyran-containing polymer 91, 92, 101, 117
spirothiopyran (STP) 17, 63

stable radical systems 15, 24–27
static light scattering (SLS) 213, 242–245
stereocilia 6, 7
stereolithography (SLA) 320, 321
stilbene 54, 63, 66, 68
stimuli-responsive block copolymer-grafted CNC 97
stimuli-responsive fluorescent polymeric hydrogels (FPHs)
  advantages 281
  bio-inspired mechanosensing systems and soft actuators/robotics 294–297
  covalently bonding fluorogens 284–285
  information encoding and encryption 293–294
  optical sensing and bio-imaging 290–293
  physically incorporating fluorogens 282–284
  supramolecular polymerizing/crosslinking monomeric fluorogens 286–290
  synthetic strategies 290
  tunable luminescent properties 281
stimuli-responsive hydrogel actuators 306–310
stimuli-responsive photoresist polymers 93
stimuli-responsive polymers 53, 55, 117, 135, 136
strain-dependent mechanosensing hydrogels 295
strain-induced elastic buckling instability for mechanical measurements (SIEBIMM) 160
stressful environment during heart development 3–4
stress-mitigation by tissue 6
structured thin films 157–158
structure factor 220–222, 224, 228, 229, 244
styrene 85, 86, 88, 93, 98, 184, 186–189
styrene nanoparticles 86, 186–188, 190, 191, 195
sulfur-containing heteroaromatic derivatives 61
sulfur-nitrogen-containing heteroaromatic derivatives 61
super-resolved fluorescence microscopy (SRFM) 205
supramolecular β-peptide-based hydrogel 341
supramolecular conductive PANI/PSS-UPy hydrogels 352
supramolecular polymerizing/crosslinking monomeric fluorogens 286–290
supramolecular structures 56, 286, 287
surface active agents 54, 71, 95–98
suspension phase behavior 207, 228–231
swelling ratio 204, 224
syn head-to-head dimer 75, 78
synthetic hydrogel 339, 342
synthetic polymer mechanochemsitry 30

t

temperature-responsive polymers 53, 72
terephthalic acid (TPA) 141
tetraaldehyde-terminated PEO polymers 283
tetraarylsuccinonitrile (TASN) 25
tetraphenylethene (TPE) 89, 286
tetraphenylhydrazine 70
tetraphenylpyr-azine (TPP) 89
textiles 112–115
thermally expanded poly(dimethylsiloxane) (PDMS) substrate 158
thermal wrinkling 159–176
thermoluminescence dosimeters 139
thermo-responsive hydrogel actuators 307
thermoresponsive microgels 204
thermo-responsive poly(diethylene glycol methyl ether methacrylate) 284

thermoresponsive polymer 41
thiazine derivatives 70
thin films
cell culture 162–167
current challenges and future perspectives 175–176
structured 157–159
thermal wrinkling 159–176
for wearable electronics 167–173
wrinkled electrochemical sensors 173–175
thiomerocyanine 17
3D anisotropic structures 315–318
3D cross-linked polymer network 284
3D dynamic light scattering (3D-DLS) 235
3D nanocomposite structures 190
3D printing 163, 301, 302, 310, 320–321
3D shape hydrogel 320
three-layered nanocomposite microgels 189
tissue-machine interfaces
design and mechanism of the neural interfaces 356–360
multi-functional applications 360–363
titanate(IV) nanosheets (TiNSs) 312
traditional wearable sensors 339
transient body-centered cubic (bcc) crystals 230
trimesic acid (TMA) 141
3-(trimethoxysilyl)propyl methacrylate (TMSPMA) 316
1,3,3-trimethylindolino-6′-nitrobenzopyrylospiran
6-nitro BIPS 140
triphenylimidazole 63
triphenylimidazolyl dimer (TPID) 70
triphenylimidazolyl (TPI) radicals 27
triphenylimidazolyl radicals (TPIR) 70
tubular hydrogel 327
“turn-on” thiol-responsive fluorescence probe 89
twisted intramolecular charge transfer (TICT) 89
two-dimensional (2D) cell culture 162
two-layered core-shell microgels 189

u
ultra-low crosslinked microgels 230, 235
ultrasonication-induced benzoxazole (Bz)- appended aliphatic esters 28
upconverting nanoparticles (UCNPs) 56
ureidopyriminone-conjugated gelatin (Gel-UPy) 363

v
vancomycin (Van) 39
vinyl-functionalized CdSe quantum dots (QDs) 284
viologen 63, 68, 70
viscosity variation 98, 101, 116–117
Vogel–Fulcher–Tammann (VFT) model 233
von Willebrand Factor (vWF) 4
protein 5

w
water-based colloidal polymers 83
water contact angle variation 116
water-soluble biopolymers 161
wearable bioelectronics 342
wearable device 168, 170, 341–344, 347, 354
wearable electronics, wrinkled conductive thin films for 167–173
wearable hydrogels-based sensors 340
white fluorescent hydrogel 293
Woodward–Hoffmann (WH) rules 8, 74
wrinkled conductive thin films for wearable electronics 167–173
wrinkled electrochemical sensors 173–175

X

xanthan gum-Fe^{3+}/PAAm-glycerol organogels 353
X-ray-responsive diselenide polymer-based nanocarriers 147

Z

zero average contrast (ZAC) method 223, 266, 267, 270
zero-dimensional carbon-based fluorescent nanomaterials 56
zwitterionic indolium species 18